Biotechnology and Bioengineering Symposium **No. 8**

Biotechnology in Energy Production and Conservation

Proceedings of the First Symposium on
Biotechnology in Energy Production and Conservation
Held at Gatlinburg, Tennessee, May 10–12, 1978
Sponsored by the
Department of Energy and the Oak Ridge National Laboratory

Editor
Charles D. Scott
Oak Ridge National Laboratory

an Interscience® Publication
published by John Wiley & Sons

BIOTECHNOLOGY AND BIOENGINEERING

SYMPOSIUM NO. 8

E. L. GADEN, JR., *Editor*

Charles D. Scott has been appointed Editor for this Symposium by the Editorial Board of *Biotechnology and Bioengineering*.

This book constitutes a part of the annual subscription to *Biotechnology and Bioengineering*, Vol. XX, and as such is supplied without additional charge to subscribers. Single copies can be purchased from the Subscription Department, John Wiley & Sons, Inc.

Subscription price, *Biotechnology and Bioengineering*, Vol. XXI, 1979: $140.00 per volume. Postage and handling outside U.S.A.: $16.00.

Printed in the United States of America.

Contents

Introduction

In the past, biotechnology has been successfully applied to a variety of process problems and is still extensively used in areas such as pharmaceutical manufacture and wastewater treatment; however, it currently does not have a significant impact on major commercial operations. In fact, until recently, the application of modern process development techniques to biotechnology was primarily carried out in only a few academic institutions and certain progressive companies. Now there are numerous indications that the interest in utilizing biotechnology is gaining momentum, both nationally and internationally. Process development engineers are joining biological scientists, environmental engineers, and others in investigating innovative and productive ways of using biological systems. This is particularly true in the field of energy production and conservation, especially if the total effort, including environmental control and fuel production, is considered.

This First Symposium on Biotechnology in Energy Production and Conservation was sponsored by the Department of Energy and the Oak Ridge National Laboratory to serve as a forum for this emerging area of technology. In the context of the symposium, biotechnology was defined as any process or process step in which a biological agent or concept was used, or any technology that is directly supportive of the biological process. Applied science in direct support of biotechnology was also considered as a proper subject. In organizing this symposium (the first of a series), the goal was to bring together the various disciplines and institutions involved in developing viable new biotechnology for conversion of biomass to fuels and energy-intensive chemical feedstocks; for direct conversion to electrical power; for alternatives to energy-intensive resource recovery, including fuel values; and for environmental control technology associated with energy production. Various segments of the technical community were represented, including those involved in academic research and development, governmental agencies and laboratories, and the industrial sector.

The symposium was organized and directed by:

Chairman

Charles D. Scott, Oak Ridge National Laboratory

Organizing Committee

C. W. Hancher, Oak Ridge National Laboratory
W. W. Pitt, Jr., Oak Ridge National Laboratory
R. Rabson, Department of Energy
G. E. Stapleton, Department of Energy

Session Chairmen and Co-Chairmen

Session I. Advanced Biotechnology Concepts

S. W. Drew, Virginia Polytechnic Institute and State University

S. E. Shumate II, Oak Ridge National Laboratory

Session II. Bioconversion for Chemicals and Intermediates

E. K. Pye, University of Pennsylvania School of Medicine

G. W. Strandberg, Oak Ridge National Laboratory

Session III. Recovery of Gaseous Fuels and Other Resources

D. K. Walter, Department of Energy

R. K. Genung, Oak Ridge National Laboratory

Session IV. Environmental Control Technology

J. H. Koon, Associated Water and Air Resources Engineers, Inc.

J. A. Klein, Oak Ridge National Laboratory

Session V. Biomimetic and *in vitro* Processes

J. J. Katz, Argonne National Laboratory

R. M. Pearlstein, Oak Ridge National Laboratory

We are very pleased to have the proceedings of the symposium published as a supplement to *Biotechnology and Bioengineering*. A few of the original papers are not included in the proceedings due to the unavailability of manuscripts or to editorial considerations. Comments or discussion relative to the format of succeeding symposia in this series and to future participation are encouraged.

CHARLES D. SCOTT

Characteristics of Tapered Fluidized Reactors: Two-Phase Systems

H. W. HSU*

Department of Chemical, Metallurgical, and Polymer Engineering, The University of Tennessee, Knoxville, Tennessee 37916, and Chemical Technology Division, Oak Ridge National Laboratory, Oak Ridge, Tennessee 37830

INTRODUCTION

Reactor systems based on tapered fluidized beds are being developed in the Biotechnology and Environmental Program at the Oak Ridge National Laboratory [1–3]. The systems are being used for aqueous bioprocesses in which adhering microorganisms or immobilized active biological fractions are employed. The use of a fluidized bed prevents biomass buildup, accommodates particulates in the feed stream, is compatible with gas sparging, and allows easy removal or addition of the active materials. The tapered reactor tends to stabilize the fluidized bed, thus allowing a much wider range of operating conditions than for a cylindrical reactor. The tapered fluidized bed differs from the conventional fluidized bed by a configuration that resembles an inverted, truncated cone rather than a constant cross-sectional-area column. Thus, the reactor taper reduces the superficial velocity by increasing the cross-sectional area of the reactor with increasing column height, therefore increasing the range of fluidization velocities and reducing instabilities such as plugging.

The objective of this investigation was to estimate the volume fraction (or void fraction) and the particle size distributions in a two-phase fluidized bed consisting of solid-phase particles fluidized by an upward flow of liquid. The liquid forms the continuous phase, while solids are the discontinuous phase. Quantitative evaluations of the effect of reactor taper on volume fraction and particle size distributions will permit the rational selection of operating conditions for the use of fluidized reactors. The analysis may be applied to the hydrogenation of liquid petroleum fractions, the hydrogenation of unsaturated fats, and coal conversion processes as well as to biological reactors, because the analysis is in general terms.

* H. W. Hsu is with Oak Ridge National Laboratory as a part-time adjunct research participant.

Biotechnology and Bioengineering Symp. No. 8, 1–11 (1978)

FLUIDIZATION MODEL

We consider a reactor loaded with a known weight, W, of various sizes of spherical particles of some substance, which is the solid phase fluidized by the upward flow of liquid. The liquid forms the continuous phase, while the solids are a discontinuous phase distributed in the continuous phase. Further, we assume that a steady state has been achieved and that the liquid is in one-dimensional plug-flow. The equations describing the above fluidized process are the equations of continuity, eq. (1), and momentum, eq. (2), for each phase; for the fluid phase with no particles present they are

$$\frac{d}{dz}(\rho_f Q) = 0 \tag{1}$$

$$\rho_f u_z \frac{du_z}{dz} = -\frac{dP}{dz} + \rho_f g \tag{2}$$

where Q is the volumetric flow rate and ρ_f is the fluid local density. If particles are now introduced into the model, then ρ_f is related to a fluid material density ρ_f (constant for incompressible fluids) and the volume fraction of fluid, ϵ_f, in such a way that $\rho_f = \bar{\rho}_f \epsilon_f$. The quantity u_z is the superficial velocity, which is also related to ϵ_f and to the local fluid velocity, u_f, in such a way that $u_z = u_f \epsilon_f$. In a cylindrical reactor, the cross-sectional area of the reactor is a constant. Thus, u_z is constant for a given constant volumetric flow rate, that is, equal to the superficial velocity at the inlet, u_{z0}. The relationships between u_f and u_z are such that

$$Q/\pi R_0^2 = u_{z0} = u_f \epsilon_f \tag{3}$$

For a tapered reactor, the radius is no longer constant but is dependent on the height of the reactor, z, which is demonstrated by Figure 1. Thus, the radial coordinate becomes

$$r = R_0 (1 + \zeta) \tag{4}$$

where R_0 is the radius at the inlet of the reactor bottom and the quantity ζ is defined as

$$\zeta = z/z_0 = (z/R_0) \tan \theta \tag{5}$$

in which z_0 represents the axial distance from the bottom of the tapered reactor to the hypothetical apex of the inverted cone.

The pressure drop in a fluidized bed is obtained by a force balance around an increment at a given height of the reactor. The pressure gradient obtained from the force balance on the particles in a cylindrical bed is given by Kunii and Levenspiel [4] as

$$-\frac{dP}{dz} = (1 - \epsilon_f)(\bar{\rho}_s - \bar{\rho}_f) g \tag{6}$$

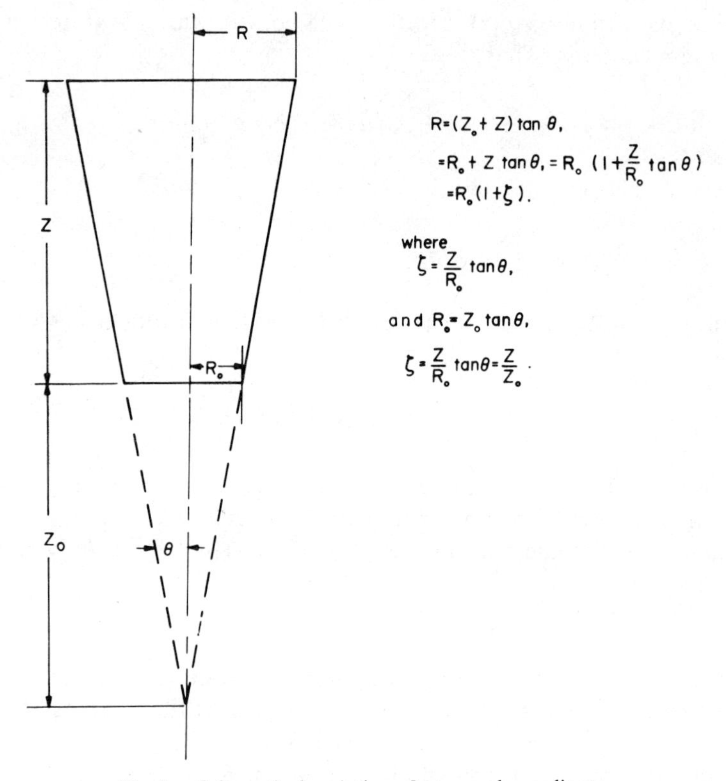

Fig. 1. Schematic description of a tapered coordinate.

where $\bar{\rho}_s$, a constant, is the solid particle material density and the term $(1 - \epsilon_f) = \epsilon_s$, the volume fraction of solid. For a tapered bed, a force balance around a height z gives

$$-\frac{dP}{d\zeta} = \frac{1}{3}(\bar{\rho}_s - \bar{\rho}_f)(1 - \epsilon_f)\left(1 + \frac{2}{(1+\zeta)^3}\right) g \tag{7}$$

Equation (7) reduces to eq. (6), if $\zeta \to 0$. In deriving eqs. (6) and (7), an assumption was made that the quantity ϵ_f does not change significantly with a small change in the height.

VOLUME FRACTION DISTRIBUTION OF FLUID

Characteristics of the fluid phase in both types of reactors, eqs. (3)–(7), were substituted into equations of continuity and momentum to obtain an equation describing each type of the reactor.

Cylindrical Reactors

The differential equation describing the volume fraction distribution in a cylindrical reactor was obtained by substituting eq. (3) into eq. (1), and the resulting expression of $u_f(z)$, together with eq. (6), was substituted into the momentum equation. Thus,

$$\bar{\rho}_f u_{z0}^2 \frac{d\epsilon_f}{dz} = (1 - \epsilon_f)(\bar{\rho}_s - \bar{\rho}_f)\cdot g + \bar{\rho}_f \epsilon_f g \tag{8}$$

The relation between W and ϵ_s is such that

$$W = \int_0^z \epsilon_s \bar{\rho}_s \pi R_0^2 dz \qquad (1 - \epsilon_f) = \epsilon_s = \frac{W}{\bar{\rho}_s \pi R^2 z} \tag{9}$$

Substitution of eq. (9) into eq. (8) and rewriting in dimensionless form yields

$$\frac{d\epsilon_f}{d\zeta} + \frac{1}{\text{Fr}} \epsilon_f = -\left(\frac{4}{\pi} F_\epsilon\right) \frac{1}{\zeta} \tag{10}$$

which is the differential equation describing the volume fraction of fluid in a fully fluidized cylindrical reactor. Equation (10) is a first-order ordinary differential equation, and the boundary condition (BC) for a fully fluidized steady state is

$$\text{BC} \qquad \zeta = 0, \qquad \epsilon_f = 1 \tag{11}$$

If a bed is not fully expanded, an appropriate scale factor may be incorporated into eq. (11). (In this case the boundary condition becomes $\zeta = 0$, $\epsilon_f = \epsilon_0$. The quantity ϵ_0 has to be determined experimentally.)

In eqs. (10) and (11), the dimensionless variables are defined as follows:

$$\zeta = z/z_0 = \text{reduced height } (z_0 = \text{an arbitrary characteristic length}) \tag{12a}$$

$$\text{Fr} = u_{z0}^2/gz_0 = \text{Fraude number} \tag{12b}$$

$$F_\epsilon = \text{Ar}/\text{Re}^2 = \text{fluidization number} \tag{12c}$$

$$\text{Ar} = W(\bar{\rho}_s - \bar{\rho}_f)\bar{\rho}_f g/\bar{\mu}_f^2 \bar{\rho}_s = \text{Archimedes number} \tag{12d}$$

$$\text{Re} = D_0 \bar{\rho}_f u_{z0}/\bar{\mu}_f = \text{Reynolds number} \tag{12e}$$

The integration of eq. (10) with the boundary condition, eq. (11), gives the volume fraction of fluid as a function of reduced height in terms of parameters Fr and F_ϵ as follows:

$$(\epsilon_f)_c = \exp\left(-\frac{\zeta}{\text{Fr}}\right)\left(1 - \frac{4}{\pi}F_\epsilon \sum_{j=1}\left\{\frac{1}{j}\left[1 + \frac{1}{j!}\left(\frac{\zeta}{\text{Fr}}\right)^j\right] + \frac{(-1)^{j+1}}{j}(\zeta - 1)^j\right\}\right) \tag{13}$$

In obtaining eq. (13), a power series expansion for the natural logarithm function

$$\ln \zeta = (\zeta - 1) - \tfrac{1}{2}(\zeta - 1)^2 + \tfrac{1}{3}(\zeta - 1)^3 - \cdots \tag{14}$$

$$= \sum_{j=1} \frac{(-1)^{j+1}}{j}(\zeta - 1)^j; \qquad 0 < \zeta < 2$$

was used. Then, setting $\zeta = 0$, the limiting value of the boundary condition was obtained.

Tapered Reactors

With a given constant Q, the superficial velocity u_z can be expressed in terms of u_f and ϵ_f as a function of the reactor height by use of eqs. (3) and (4), which gives

$$\frac{Q}{\pi R_0^2 (1+\zeta)^2} = \frac{u_{z0}}{(1+\zeta)^2} = u_f \epsilon_f \tag{15}$$

If eq. (15) is substituted into eqs. (1) and (2) and if eq. (7) is substituted into eq. (2), then the differential equation describing the volume fraction distribution in a tapered reactor can be given in a dimensionless form as follows:

$$\frac{d\epsilon_f}{d\zeta} + \left(\frac{(1+\zeta)^4}{\mathrm{Fr}} - \frac{2}{(1+\zeta)}\right)\epsilon_f = -\frac{4}{3\pi} F_\epsilon \left(\frac{(1+\zeta)^2}{\zeta} + \frac{2}{(1+\zeta)\zeta}\right) \tag{16}$$

where the reduced distance ζ is defined as in eq. (5) instead of eq. (12a). Equation (16) is also a first-order ordinary differential equation, and the general solution can be obtained as

$$(\epsilon_f)_T = \exp\left(-\int\left[\frac{(1+\zeta)^4}{\mathrm{Fr}} - \frac{2}{(1+\zeta)}\right]d\zeta\right)\left\{-\frac{4}{3\pi} F_\epsilon \int\left[\frac{(1+\zeta)^2}{\zeta} + \frac{2}{(1+\zeta)\zeta}\right]\right.$$
$$\left. \cdot \exp\left(\int\left[\frac{(1+\zeta)^4}{\mathrm{Fr}} - \frac{2}{(1+\zeta)}\right] d\zeta\right) d\zeta + C\right\} \tag{17}$$

Using the relation of the power series expansion for an exponential function,

$$\exp\left(\frac{(1+\zeta)^5}{5\mathrm{Fr}}\right) = \sum_{j=0} \frac{1}{j!}\left[\frac{(1+\zeta)^5}{5\mathrm{Fr}}\right]^j \tag{17a}$$

the bracket in eq. (17) was integrated together with the boundary condition given in eq. (11) to give

$$(\epsilon_f)_T = (1+\zeta)^2 \left[\exp\left(\frac{1-(1+\zeta)^5}{5\mathrm{Fr}}\right)\right.$$
$$-\frac{4}{3\pi} F_\epsilon \left\{\frac{3+2\zeta}{(1+\zeta)^2} - 3 + 3\sum_{j=1}\left[\frac{1}{j} + \frac{(-1)^{j+1}}{j}(\zeta-1)^j\right]\right\}$$
$$\times \exp\left(-\frac{(1+\zeta)^5}{5\mathrm{Fr}}\right) + \left[1 - \exp\left(-\frac{(1+\zeta)^5}{5\mathrm{Fr}}\right)\right]$$
$$\times \left\{3\sum_{j=1}\left[\frac{\zeta}{(1+\zeta)^{5j}} + \frac{(-1)^{j+1}(\zeta-1)^j}{j(1+\zeta)^{5j}} + \frac{1}{5j} + \sum_{n=1}^{5j-n=2}\frac{1}{(5j-n)(1+\zeta)^n}\right]\right.$$

$$- \frac{2}{(1+\zeta)^3} \sum_{j=1} \left[\frac{(1+\zeta)^2}{5j-1} + \frac{1+\zeta}{5j-2} + \frac{1}{(5j-3)}\right]\Bigg\}$$

$$- \left[\exp\left(\frac{1-(1+\zeta)^5}{5\mathrm{Fr}}\right) - \exp\left(\frac{-(1+\zeta)^5}{5\mathrm{Fr}}\right)\right]$$

$$\times \Bigg\{3 \sum_{j=1} \left[\frac{1}{5j} - \frac{1}{j} + \sum_{n=1}^{5j-n=2} \frac{1}{5j-n}\right]$$

$$- 2 \sum_{j=1} \left[\frac{1}{3j-1} + \frac{1}{5j-2} + \frac{1}{j-3}\right]\Bigg\}\Bigg] \tag{18}$$

Equation (14) was also used to obtain the limiting value of the boundary condition in eq. (18).

PARTICLE SIZE DISTRIBUTION

Characteristics of the solid phase in both types of reactors were expressed by the particle size distributions as a function of reactor height. A momentum balance for a spherical particle relative to the moving fluid is given as

$$\frac{dv}{dt} = \left(1 - \frac{\bar{\rho}_f}{\bar{\rho}_s}\right) g - \frac{C_D \bar{\rho}_f v^2 S}{2\,m} \tag{19}$$

in which v is the velocity of a particle relative to the moving fluid, m is the mass of a particle, S is the cross-sectional area of a particle, and C_D is the drag coefficient of a particle [5]. Then, for a spherical particle of diameter d_p, one has

$$S = \pi d_p^2/4 \qquad \text{and} \qquad m = (\pi d_p^2/6)\,\bar{\rho}_s \tag{20}$$

Then, eq. (19) becomes

$$\frac{dv}{dt} = \left(1 - \frac{\bar{\rho}_f}{\bar{\rho}_s}\right) \cdot g - \frac{3\,C_D v^2 \bar{\rho}_f}{4\,d_p \bar{\rho}_s} \tag{21}$$

For a steady state in a fluidized bed ($dv/dt = 0$), particles are stationary at the fluidizing level in a fixed-space coordinate system, and the terminal velocity of a particle with respect to a moving fluid is equal to the fluid velocity. Thus, one has

$$u_f = [4\,(\bar{\rho}_s - \bar{\rho}_f) d_p g / 3\,C_D \bar{\rho}_f]^{1/2} \tag{22}$$

At usual fluidization, the Reynolds number of a fluidized particle is generally in the range of $0.4 < (\mathrm{Re})_p < 500$. If the flow is too low, the bed does not fluidize. On the contrary, if the flow is too high, the bed may become entrained. Within this range of flow, an empirical correlation of the drag coefficient of flow around a single submerged sphere is given by

$$C_D = 10/(\mathrm{Re})_p^{1/2} \qquad \text{for} \qquad 0.4 < (\mathrm{Re})_p < 500 \tag{23}$$

in which $(\mathrm{Re})_p = d_p u_f \rho_f / \mu_f$ [4, 5]. An assumption was made that $\mu_f = \bar{\mu}_f$ (the local fluid viscosity, u_f, is the same as the material fluid viscosity, $\bar{\mu}_f$), which means that μ_f is independent of the volume fraction of fluid while u_f and ρ_f are dependent on the volume fraction of fluid ϵ_f.

Cylindrical Reactors

By use of eqs. (22) and (23), together with eqs. (3) and (9), an expression of particle diameter, d_p, in a cylindrical reactor at a steady state was obtained as

$$(d_p/D_0)_c = (^{15}\!/_2)^{2/3}\,\mathrm{Re}^{-1/3}\,\mathrm{Fr}_0^{2/3}\,A^{-2/3}\,(\epsilon_f)_c^{-1/3} \tag{24}$$

where D_0 is the diameter of the reactor inlet and

$$\mathrm{Re} = D_0 \bar{\rho}_f u_{z0}/\mu_f = \text{Reynolds number of fluid at reactor inlet} \tag{25a}$$

$$\mathrm{Fr}_0 = u_{z0}^2/gD_0 = \mathrm{Fr}\,(z_0/D_0) = \text{Froud number of fluid with respect to reactor inlet diameter} \tag{25b}$$

$$A = (\bar{\rho}_s - \bar{\rho}_f)/\bar{\rho}_f = \text{effective buoyant density} \tag{25c}$$

Tapered Reactors

By use of eqs. (22) and (23), together with eqs. (3), (4), (5), and (9), an expression of particle diameter, d_p, in a tapered bed reactor at a steady state was obtained as

$$(d_p/D_0)_T = (^{15}\!/_2)^{2/3}\,\mathrm{Re}^{-1/3}\,\mathrm{Fr}^{2/3}\,A^{-2/3}\,(\epsilon_f)_T^{-1/3}\,(1 + \zeta)^{-2} \tag{26}$$

QUANTITATIVE RESULTS

The volume fraction and the particle size distributions were calculated as a function of reduced reactor height with the following parameters:

$$\mathrm{Fr} = 0.01, 0.05, 0.1, 0.5, \text{ and } 1.0$$
$$F_\epsilon = 10^{-7} \text{ and } 10^{-4}$$

Equations (13) and (18) were used to calculate the volume fraction of fluid distribution for cylindrical and tapered geometry reactors, respectively. The results are shown in Figure 2. In a cylindrical geometry, the variation of the volume fraction with respect to the reactor height is almost linear, and the effect of the fluidization number, $F_\epsilon = \mathrm{Ar}/\mathrm{Re}^2$, on this relationship is very small, as can be seen from the case of $\mathrm{Fr} = 0.5$. If the Froude number decreases, the variation of the volume fraction of fluid with respect to the fluidization number becomes almost indistinguishable. The variation of the volume fraction with respect to the reactor height in a tapered geometry appears almost linear for low Froude numbers ($\mathrm{Fr} = 0.01$ and 0.05) and increases its downward curvature for the Froude numbers exceeding 0.1. The variations with respect to the fluidization number at $\mathrm{Fr} = 0.5$ and 1.0 are almost indistinguishable.

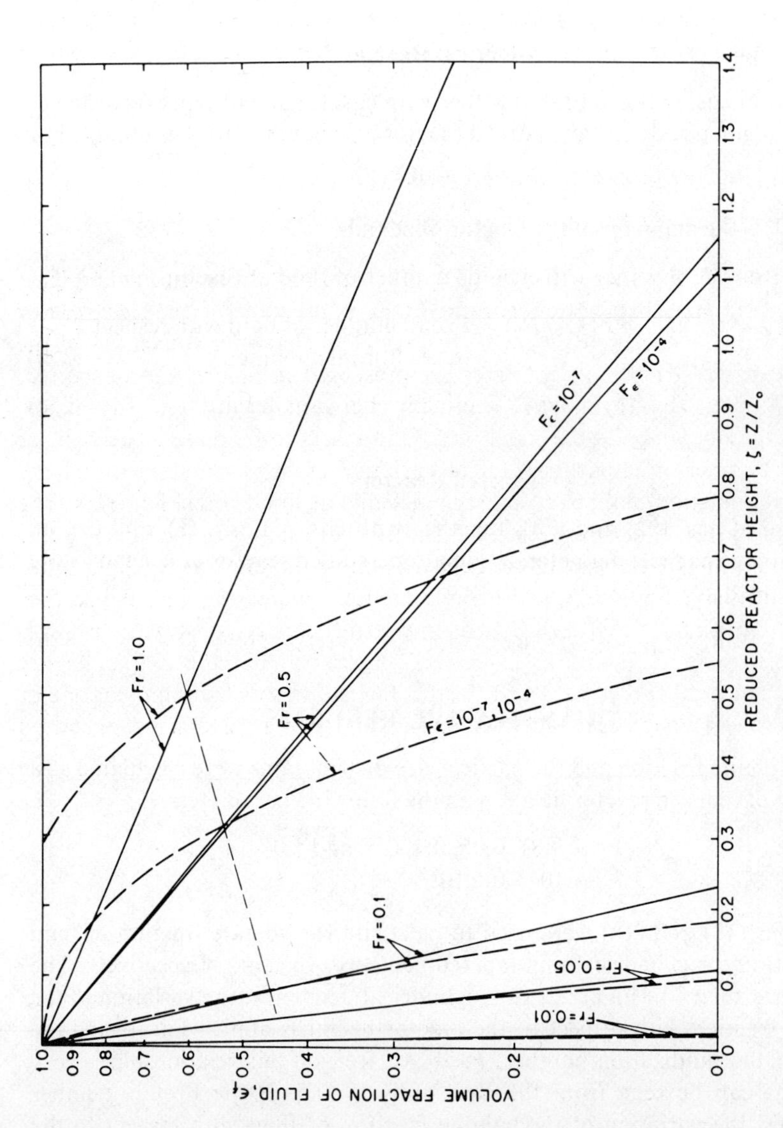

Fig. 2. Volume-fraction of fluid as a function of reduced reactor height in (——) cylindrical ($F_0 = 10^{-7}$) and (---) tapered ($F_\epsilon = 10^{-7}$) geometries.

The variation of particle size distributions with respect to the reactor height was calculated by eqs. (27) and (28), which were obtained by rearranging eqs. (24) and (26). Equations (27) and (28) define particle size distribution parameters for cylindrical $(Pd)_c$ and tapered $(Pd)_T$ geometries, respectively; these are analogous to the j_H and the j_D factors used in heat and mass transfer correlations [6]. For a cylindrical geometry, the particle size distribution parameter $(Pd)_c$ is

$$(Pd)_c = (d_p/D_0)\,\mathrm{Re}^{1/3}\,\mathrm{Fr}_0^{-2/3}\,A^{2/3} = (^{15}\!/_2)^{2/3}\,(\epsilon_f)_c^{-1/3} \tag{27}$$

and for a tapered geometry, the parameter $(Pd)_T$ is

$$(Pd)_T = (d_p/D_0)\,\mathrm{Re}^{1/3}\,\mathrm{Fr}_0^{-2/3}\,A^{2/3} = (^{15}\!/_2)^{2/3}\,(\epsilon_f)_T^{-1/3}\,(1+\zeta)^{-2}$$

Then, eq. (27), together with eqs. (13), (28), and (18), were used to calculate the particle size distribution parameter as a function of reduced reactor height with the specified Fr and F_ϵ values. The characteristic curves of the particle size distribution parameter are presented in Figure 3. As expected, the parameter in a cylindrical geometry increases almost linearily at low Froude numbers (Fr = 0.01 and 0.05) and starts to increase concavely for Froude numbers greater than 0.1. The variation of the parameter with respect to the fluidization number is indistinguishable at low Froude numbers. The variations at Fr = 0.5 for $F_\epsilon = 10^{-7}$ and 10^{-4} are shown in the figure. For a tapered geometry, the parameter behaves almost the same as that in a cylindrical geometry for a very low Froude number, increasing with the reactor height linearly, but the initiation of increasing concavity starts at a lower Froude number than that of cylindrical geometry. For the case of higher Froude numbers (Fr = 0.5 and 1.0), the particle size distribution parameter starts to show a decrease reaches a minima, and increase again as the reduced

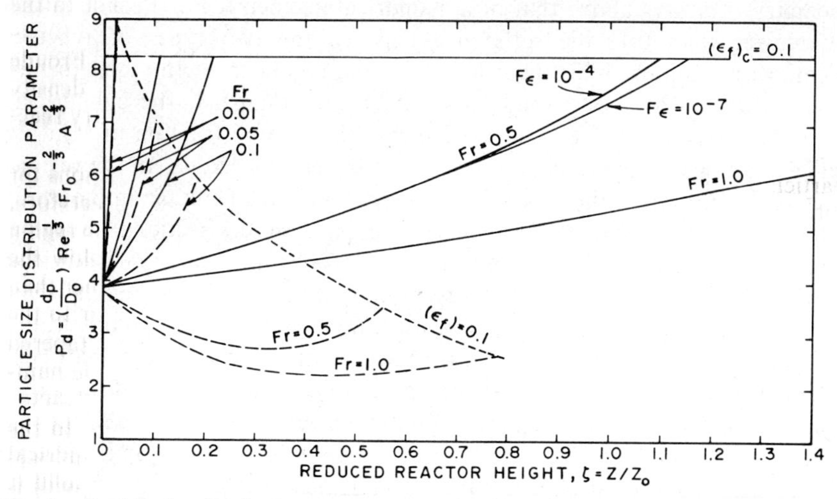

Fig. 3. Particle size distribution parameter as a function of reduced reactor height in cylindrical and tapered geometries. Symbols same as in Fig. 2.

reactor height increases. The variation of the parameter with respect to various fluidization numbers is almost negligible and much smaller than that for a cylindrical geometry.

DISCUSSION

The results presented here are based on the assumptions that the fluidized bed had reached the maximum bed expansion at a given steady-state condition, the solid particles were all spherical with a constant density, and the upward flow of liquid was a plug-flow. In a biological reactor, the adsorption or the attachment of biological material to the solid surface (immobilization of cells or enzyme) will cause changes in the density of particles, generally reducing the particle density. Thus, there will be a slight variation in the effective density which also appears in the expression for the Archimedes number, Ar, and thus for the fluidization number, F_ϵ. The variations of the volume fraction and of the particle size distribution parameter with respect to the fluidization number are very small and almost negligible for low Froude numbers for both geometries and indistinguishable in the range of Froude numbers investigated for a tapered geometry. Therefore, a constant density particle assumption is a valid one, in particular for a tapered geometry reactor.

As can be seen from Figure 2, curves of volume fraction distributions for both geometries intersect each other at a certain reactor height. Therefore, the advantageous characteristics of each type of reactor depend on the region where the characteristic curve of one type of reactor is located below the other and where the volume fraction of solid particles $(1 - \epsilon_f)$ is higher than the other (the higher biological activity). From the inlet of a reactor to the intersection height, the entrance zone, the volume fraction curve of a tapered geometry is always above that of a cylindrical geometry. For Froude numbers lower than 0.1, the difference between the two types of reactor geometries is almost indistinguishable for the region from the inlet to the insersection height. For Froude numbers higher than 0.1, the cylindrical geometry reactor shows the advantage that the volume fraction of solid is higher than that of the tapered geometry in the entrance zone. In the region after the intersection, the tapered geometry shows an advantage in the solid volume fraction. Even at the same reactor height, the cross-sectional area of a tapered geometry is always larger than that of a cylindrical geometry, thus providing more contact areas for a reaction, particularly in the region after the entrance zone. The entrance zone for a tapered geometry is biologically less active than that of a cylindrical geometry due to the smaller particle, volume fraction distribution. However the entrance zone can provide an environment for a biological substance to initiate a reaction as a time-lag zone. Therefore, less biological activity in this region is immaterial, and this is a characteristic that one desires to have in a biological reactor.

A static volume fraction of fluid in a packed bed is generally in the range between 0.4 and 0.6. This is the limiting value the volume fraction of fluid could reach. The volume fraction of fluid at the intersection height increases as the Froude number increases; that is, with an increase in the Froude number, the region favorable to the tapered geometry also increases. Thus, one may conclude that a tapered geometry reactor is a better reactor than that of a cylindrical geometry in a fluidized bioreaction, particularly with a high Froude number.

From Figure 3 one can see that the particle size distribution parameter Pd per unit reduced reactor height for a tapered geometry changes more widely than that of a cylindrical geometry for the case of higher Froude numbers (Fr = 0.5 and 1.0). It means that a tapered geometry has a higher particle size segregation power than does a cylindrical geometry for higher Froude numbers, while a cylindrical geometry shows the higher segregation power than that of a tapered geometry at smaller Froude numbers (Fr < 0.1). The variation of the distribution parameter with respect to the fluidization number F_ϵ is very small for high Froude numbers (Fr > 0.5) in a cylindrical geometry and is indistinguishable in a tapered geometry for the range of Froude numbers investigated. It is also indistinguishable in a cylindrical geometry for the cases of Fr < 0.1. With small Froude numbers, the segregation power of the two geometries becomes indistinguishable.

Therefore, a tapered geometry, fluidized-bed bioreactor is better than a cylindrical geometry reactor by allowing a wider range of fluidization velocities and by reducing instabilities such as plugging by biomass buildup.

The method of handling the nonspherical particles can be modified in the usual fashion by the introduction of appropriate shape factors in conjunction with the constant term in eqs. (27) and (28).

The author wishes to express his sincere appreciation to Don Broach, University of Tennessee Computing Center, for assistance in obtaining the numerical results, to S. H. Jury, University of Tennessee, Department of Chemical, Metallurgical, and Polymer Engineering, and to C. W. Hancher, W. W. Pitt, Jr., and R. K. Genung of Oak Ridge National Laboratory for their encouragement and many constructive discussions. This research was sponsored by the Division of Biomedical and Environmental Research, U.S. Dept. of Energy under contract No. W-7405-Eng-26 with the Union Carbide Corp.

References

[1] C. D. Scott and C. W. Hancher, *Biotechnol. Bioeng.*, *18*, 1393 (1976).
[2] C. D. Scott, C. W. Hancher, and S. E. Shumate II, in *Enzyme Engineering*, E. K. Pye and H. H. Weetall, Eds. (Plenum, New York, 1978), Vol. 3.
[3] W. W. Pitt, Jr., C. W. Hancher, C. D. Scott, and H. W. Hsu, *Chemical Engineering Progress*, to appear.
[4] D. Kunii and O. Levenspiel, *Fluidization Engineering* (Wiley, New York, 1969), p. 76.
[5] A. S. Foust, L. A. Wenzel, C. W. Clump, L. Maus, and L. B. Anderson, *Principles of Unit Operations* (Wiley, New York, 1960), p. 451.
[6] R. B. Bird, W. E. Stewart, and E. N. Lightfoot, *Transport Phenomena* (Wiley, New York, 1960), pp. 401 and 647.

Biological Removal of Metal Ions from Aqueous Process Streams

S. E. SHUMATE II, G. W. STRANDBERG, and J. R. PARROTT, JR.
Chemical Technology Division, Oak Ridge National Laboratory, Oak Ridge, Tennessee 37830

INTRODUCTION

Some aqueous effluent streams emanating from energy production processes contain dissolved heavy metals which, due to their chemical or radiological properties, can pose a hazard to the environment. The concentrations of such metals must be reduced to acceptable levels before the streams can be discharged to the environment. Processing steps in the nuclear fuel cycle generate wastewater streams containing a variety of dissolved heavy metals, including uranium. Conventional methods for removing heavy metals from aqueous streams include chemical precipitation, chemical oxidation or reduction, ion exchange, filtration, electrochemical treatment, and evaporative recovery. Such processes may be ineffective or extremely expensive when initial heavy-metal concentrations are in the range of 10–100 g/m^3 and discharge concentrations are required to be less than one g/m^3.

Another processing method that may be considered is the sorption and/or complexation of dissolved metal species by microorganisms. One of the earliest references to such a wastewater treatment concept for the disposal of radioactive metals was made by Ruchhoft [1] , who cited the removal of plutonium-239 from water by activated sludge. He observed that about 96% removal could be accomplished in a single stage of treatment and pointed out that a two-stage process could be used if a greater degree of decontamination was required. The decontamination process was described as the propagation of a microbial population "having gelatinous matrices with tremendous surface areas that are capable of adsorbing radioactive materials [1]." Other investigators [2] emphasized that the essential role of the biological population would be to function as an adsorbent. The concern expressed that "wastes can only be treated biologically if they are free from acids, alkalis, and toxic substances" [2] indicates that little thought was given to decoupling the two steps: propagation of a microbial sorbent, and contact of a metal-contaminated stream with the sorbent. In recent years,

Biotechnology and Bioengineering Symp. No. 8, 13–20 (1978)
Published by John Wiley & Sons, Inc.

however, some attention has been given to the use of dried, irreversibly degraded mycelium as an effective sorbent for the removal of metal radionuclides from aqueous streams [3].

Our approach has been to characterize a number of microbial species with respect to effectiveness in removing uranium from aqueous solution by resting (washed, resuspended) cells [4]. On the basis of this characterization, *Saccharomyces cerevisiae* and *Pseudomonas aeruginosa* were chosen for additional experiments to study the effects of initial metal concentration, temperature, and pH on the rate of uranium uptake in a well-mixed, single-stage contacting vessel. Our intent is to decouple the steps of biosorbent propagation and contact of an aqueous stream with the biosorbent for removal of dissolved heavy metals such as uranium. In this way, these steps can be optimized independently to yield a more effective overall process. The presence of growth-inhibiting substances in the wastewater stream would not have any detrimental effect on the biosorbent propagation step.

MATERIALS AND METHODS

Culture and Preparation of Cells

The yeast used in this study was *Saccharomyces cerevisiae* NRRL Y-2574 obtained from the Agricultural Research Service Culture Collection (Peoria, Ill.). The bacterium was a culture of *Pseudomonas aeruginosa* obtained from H. R. Meyer and J. Johnson, Colorado State University, which had been utilized in a study of plutonium uptake.

The general-purpose YM medium of Wickerham [5] was utilized for both culture maintenance and cell production. This medium contains glucose (1%), yeast extract (0.3%), malt extract (0.3%), and peptone (0.5%). All cultures were incubated at 28°C.

Sufficient cells for uranium uptake experiments were obtained by culturing the organisms for 24 hr in Fernbach flasks that contained 750 ml of YM medium and were shaken at 100 rpm (2-in. stroke). These flasks had been inoculated with 10 ml of a 24–72 hr shake culture started from an agar slant. Cells were recovered from the Fernbach cultures by centrifugation at 9000 *g* for 20 min and washed three times with deionized distilled water (Milli-Q Reagent-Grade Water System, Millipore Corp., Bedford, Mass.). The washed cell paste (refrigerated overnight) was resuspended in deionized distilled water to yield approximately 1 g wet cells/10 ml water. An aliquot of this cell suspension was dried at 110°C for determination of dry cell weight.

Uranium Uptake Experiments

Uranyl nitrate hexahydrate (J. T. Baker Chem. Co., Phillipsburg, N.J.) solutions were prepared with deionized distilled water so that the addition of a given volume of cell suspension provided the proper dilution to give the desired initial uranium concentration. Generally, 10 ml of the cell suspension were added to 40 ml of the uranium solution (preequilibrated at the required temperature) contained in a 250 ml Erlenmeyer flask. The mixture was shaken at 100 rpm at either 25 or 40°C in a water bath shaker (Aquatherm, model G-86, New Brunswick Scientific Co., Inc., New Brunswick, N.J.). At desired time intervals, samples were withdrawn, the cells were removed by centrifugation, and the uranium content of each supernatant fraction was determined. Distribution coefficients were calculated on the basis that the uranium removed from solution was sorbed onto biomass. Cell-free controls were run to ensure that the uranium removed from solution was associated with biomass.

Uranium Assay

A spectrophotometric assay based on the reaction of uranium with Arsenazo III was used. [6,7]. A 0.1% solution of Arsenazo III (Aldrich Chemical Co., Milwaukee, Wis.) was prepared by dissolving the dry powder in 0.1*N* NaOH. This solution was acidified to pH 1.5 with concentrated H_2SO_4 and finally diluted 1:10 with 0.1*N* H_2SO_4. The reagent was added at a ratio of 5:1 (v/v) to sample solutions that had been appropriately diluted to contain 1–20 ppm uranium. The solutions were mixed and allowed to stand approximately 10 min; then the absorbance of each was measured at 650 nm. Concentration was calculated from a standard curve.

RESULTS AND DISCUSSION

There are a number of mechanisms that could account for the association of uranium with the microbial cells recovered by centrifugation. Uranium could react to form a precipitate, discretely settled with the cells or as a fine colloid entrapped by extracellular polymers. Uranium may be transported into the cell and react to form a product that remains within the cell.

A number of investigators [8–12] have indicated that microbially synthesized polymers extending from the outer membrane of a cell are responsible for the binding of metal ions from solution. Metal cations may be complexed by negatively charged sugar units at the end of a polysaccharide chain. Rothstein and co-workers [8,13,14] have cited evidence that extracellular polyphosphate groups, associated with sugar metabolism, are responsible for the binding of uranium (uranyl ion) from aqueous solution. Polyphosphates complex metal ions by chelation through negatively charged oxygen atoms. [15]. The chelating power or capacity of a chain polyphosphate increases with increasing chain length. The nature of these suggested binding units and other potential uranium binding groups on the cell surface indicates that the binding might be dependent on the cell environment (i.e., the solution in intimate contact with the cell). The complex aqueous chemistry of uranium must be considered as well. However, from a practical standpoint, our initial concern has been focused on the identification of factors which would affect the association of uranium with biomass in a wastewater treatment process. Such factors might include soluble uranium concentration, pH, and temperature.

Figure 1 shows the rate and degree of uranium removal from solution by *S. cerevisiae* cells for a range of initial uranium concentrations from 10 to 500 g/m³. Even at an initial uranium concentration of 500 g/m³, the equilibrium solution concentration was just slightly higher than that which was reached with an initial uranium concentration of 10 g/m³.

The rate of uranium removal from solution was significantly affected by temperature at the higher initial uranium concentration as shown in Figure 2; the rate was considerably greater at 40 than at 25°C. This pronounced effect is seen again in Figure 3, which summarizes data from experiments wherein cells were sequentially contacted with solutions with initial uranium concentrations of 86 g/m³. The equilibrium uranium concentration in solution after the second exposure is just slightly higher than that reached after a single exposure to an initial uranium concentration of 100 g/m³ (Fig. 1).

Figures 4(a) and 4(b) illustrate the effect of pH on the rate of uranium removal from solution. The rate was greatest when the solution pH was

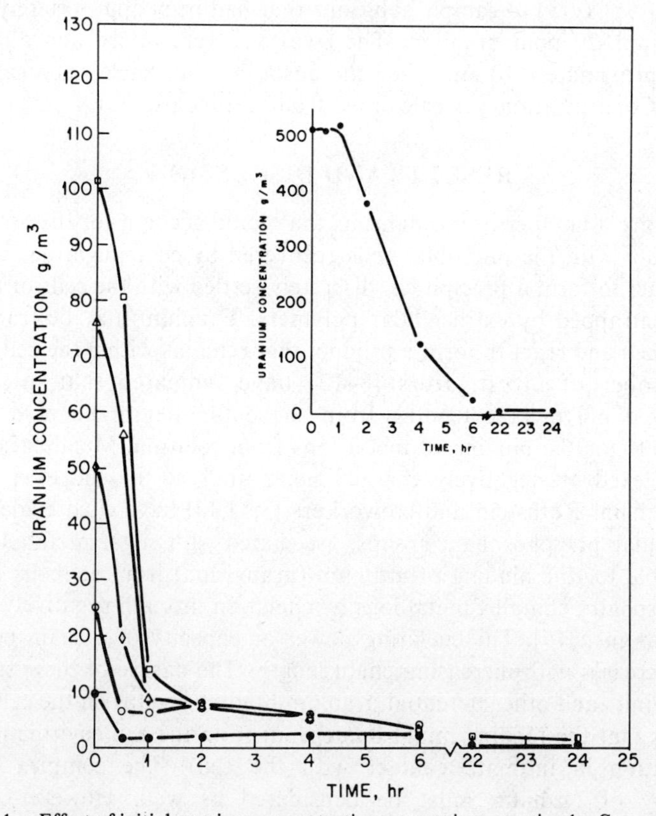

Fig. 1. Effect of initial uranium concentration on uranium sorption by *S. cerevisiae*.

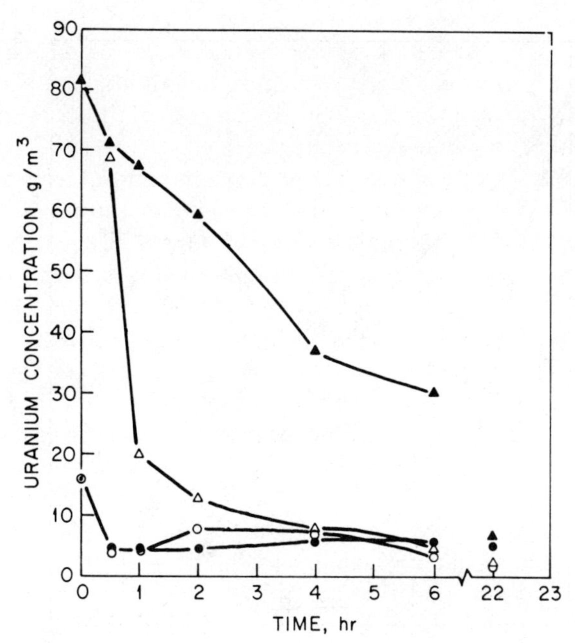

Fig. 2. Effect of temperature on uranium sorption by *S. cerevisiae*. (▲) 25°C; (△) 40°C; (○) 40°C; (●) 25°C.

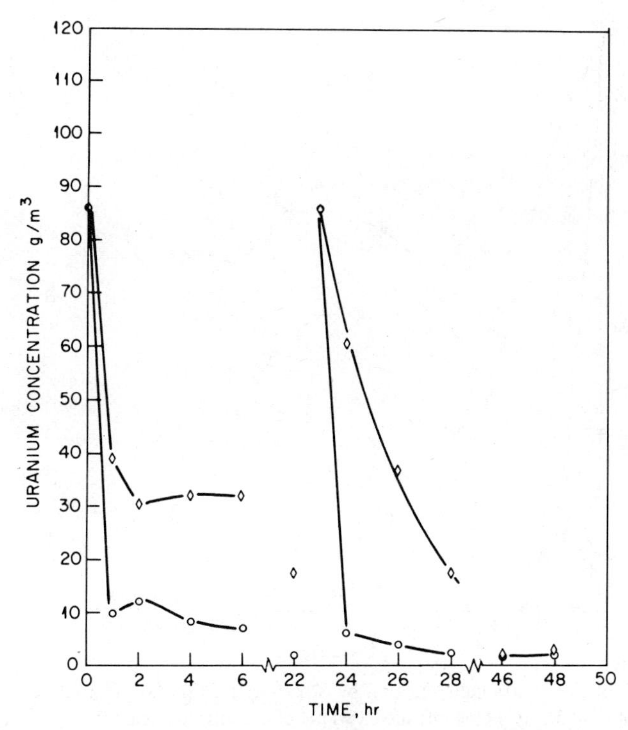

Fig. 3. Uranium sorption by *S. cerevisiae* during sequential exposure. (◇) 25°C; (○) 40°C.

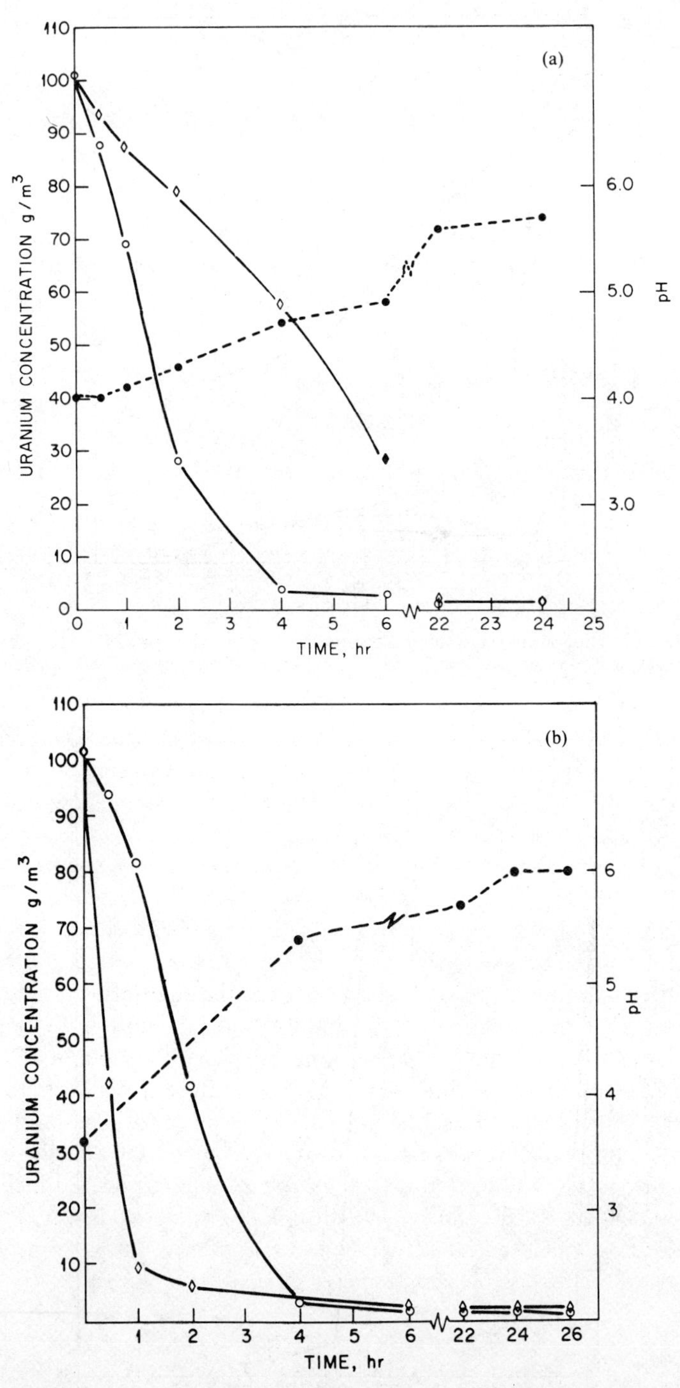

Fig. 4. Effect of pH on uranium sorption by *S. cerevisiae*: (a) (◇) pH controlled at 3.9; (b) (◇) pH controlled at 5.5. (○) Uncontrolled; (●) pH of uncontrolled reactor.

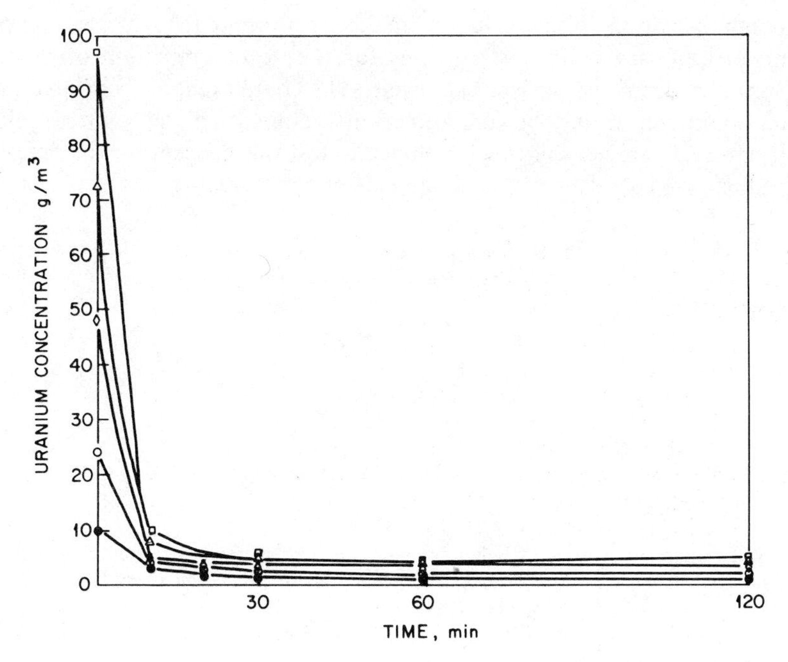

Fig. 5. Effect of initial uranium concentration on uranium sorption by *P. aeruginosa*.

maintained at 5.5; when no effort was made to control the solution pH, it gradually increased to a value between 5.5 and 6.0 as uranium was being sorbed or complexed by the yeast cells. The rate of uranium removal in this case was greater than when the pH was controlled at 3.9.

The rate and degree of uranium removal by *P. aeruginosa* cells for a range of initial uranium concentrations from 10 to 100 g/m^3 are shown in Figure 5. As is evident, the removal rate was much greater than that obtained using the yeast cells (Fig. 1). The data also indicate that the uranium concentration in solution had approached the equilibrium value after a contact time of only 10 min. This rapid uptake is similar to the effect observed by Zajic and Chiu [11] when using a growing *Penicillium* culture.

Soluble metal concentration, pH, and temperature are among the more readily determined and controlled parameters to be considered in the design of a process for the removal of a heavy metal from wastewaters by association with biomass. Additional factors that are being considered include the type, number, and reactivity of metal binding sites as affected by cell growth history and species differences, resistances to mass transfer, and other constitutents in the waste stream (e.g., competing ions).

CONCLUSIONS

Resting (washed, resuspended) microbial cells have been shown to effect rapid removal of uranium from aqueous solution. The rate and degree of uranium isolation from solution make this approach to heavy-metal removal quite promising as a means for the decontamination of process wastewaters from the nuclear fuel cycle. The complexed, concentrated uranium could conceivably be removed from the microbial cells and recycled. A bench-scale process is being developed to test this concept for the removal of radioactive heavy metals from aqueous process streams.

Research sponsored by the Divisions of Waste Management and of Biomedical and Environmental Research, U.S. Dept. of Energy, under contract No. W-7405-eng-26 with the Union Carbide Corp.

References

[1] C. C. Ruchhoft, *Sewage Works J.*, *21*(5), 877 (1949).
[2] G. E. Eden, in *Radioactive Wastes—Their Treatment and Disposal*, J. C. Collins, Ed. (E. & F. N. Spon Ltd., London, 1969), p. 141.
[3] R. Jilek et al., *Rudy* (*Prague*), *23*(5), 282 (1975).
[4] S. E. Shumate II et al., "Biological processes for environmental control of effluent streams in the nuclear fuel cycle," in *Proceedings Waste Management and Fuel Cycles '78* (University of Arizona Press, Tucson, in press).
[5] L. Wickerham, *Taxonomy of Yeasts*, Technical Bulletin No. 1029, U.S.D.A., Washington, D.C., 1951, p. 2.
[6] S. B. Savin, *Talanta*, *8*, 673 (1961).
[7] H. Onishe and K. Sekine, *Talanta*, *19*, 473 (1972).
[8] A. Rothstein and R. Meir, *J. Cell. Comp. Physiol.*, *34*, 97 (1949).
[9] P. R. Dugan et al., in *Advances in Water Pollution Research*, S. H. Jenkins, Ed. (Pergamon, New York, 1970), Vol. 2, p. III-20/1.
[10] P. R. Dugan and H. M. Pickrum, *Eng. Bull. Purdue Univ. Eng. Ext. Ser.*, *141*(2), 1019 (1972).
[11] J. E. Zajic and Y. S. Chiu, *Dev. Ind. Microbiol.*, *13*, 91 (1972).
[12] J. W. Costerton et al., *Sci. Am.*, *238*(1), 86 (1978).
[13] A. Rothsetin, A. Frenkel, and C. Larrabee, *J. Cell. Comput. Physiol.*, *32*, 261 (1948).
[14] A. Rothstein and C. Larrabee, *J. Cell. Comput. Physiol.*, *32*, 247 (1948).
[15] C. F. Bell, *Principles and Applications of Metal Chelation* (Clarendon, Oxford, 1977), p. 129.

Hysteresis Relationships between Product Rates and Cell Growth Rates in Fermentation Reactions

MICHAEL R. SMITH, ROBERT D. TANNER,*
KANZA H. BADAWI, and S. SHAHID HUSSAIN
Chemical Engineering Department, Vanderbilt University, Nashville, Tennessee 37235

INTRODUCTION

Luedeking and Piret [1] observed in 1959 that the rate of production of lactic acid by *Lactobacillus delbrueckii* is a linear function of cell growth rate and cell concentration:

$$\frac{d(P)}{dt} = \alpha \frac{d(X)}{dt} + \beta(X) \tag{1}$$

or

$$\frac{1}{(X)} \frac{d(P)}{dt} = \alpha \frac{1}{(X)} \frac{d(X)}{dt} + \beta \tag{1a}$$

Here, P is product concentration, X is cell concentration, and α and β are constants that are functions of pH and temperature. This relationship combines the concept of growth-associated product formation with the concept of non-growth-associated product formation and describes the rate of production of lactic acid in the growth phase of the fermentation. Aiyar and Luedeking [2] later observed that the formation of ethanol by *Saccharomyces cerevisiae* is also described by this equation.

While the Luedeking–Piret equation is an important correlation, its use in the fermentation industry has been limited due to its applicability to only a few systems. Rai's data [3] of the *Pseudomonas ovalis* NRRL B-1486 fermentation of glucose to gluconic acid are examples. A plot of specific product rate against specific cell growth rate for pH 5.8 and temperature 30°C (Fig. 1) is a straight line as predicted by the Luedeking–Piret model. However, for data collected at other temperatures, a plot of specific rates is not a straight line, but rather a hysteresis trajectory. Figure 2 exhibits this behavior.

Another system the Luedeking–Piret equation fails to describe is the production of intracellular free lysine by bakers' yeast [4]. Figure 3 shows a counterclockwise hysteresis trajectory for the plot of specific rates.

* To whom all correspondence should be addressed.

Biotechnology and Bioengineering Symp. No. 8, 21–35 (1978)

0572-6565/78/0008-0021$01.00

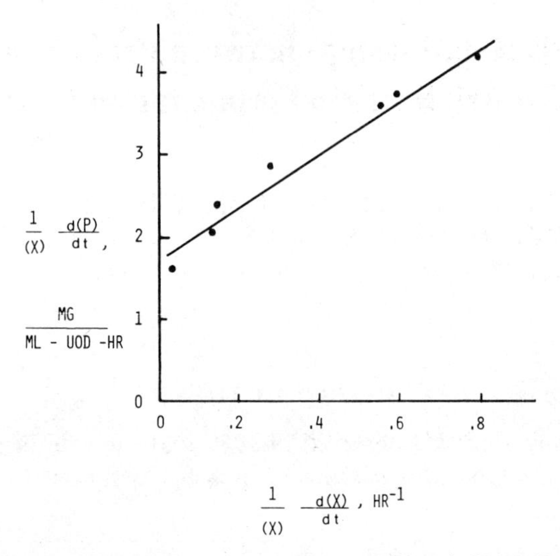

Fig. 1. Specific rates for gluconic acid production at 30°C and pH 5.8. Data from Rai [3].

These observations of hysteresis behavior encouraged us to examine the lactic acid and ethanol systems for deviations from linearity that would normally be attributed to experimental error and ignored. A plot of specific product rate against specific cell growth rate for the lactic acid fermentation at pH 6.0, an extreme pH for this system, shows signs of expanding from Luedeking and Piret's originally fitted straight line [1] into a hysteresis curve as shown in Figure 4. The behavior is obviously not very pronounced, but in light of other observations, it is worth noting. In addition to the lactic acid fermentation, Leudeking and Aiyar's ethanol data can exhibit hysteresis with a cusp (Fig. 5).

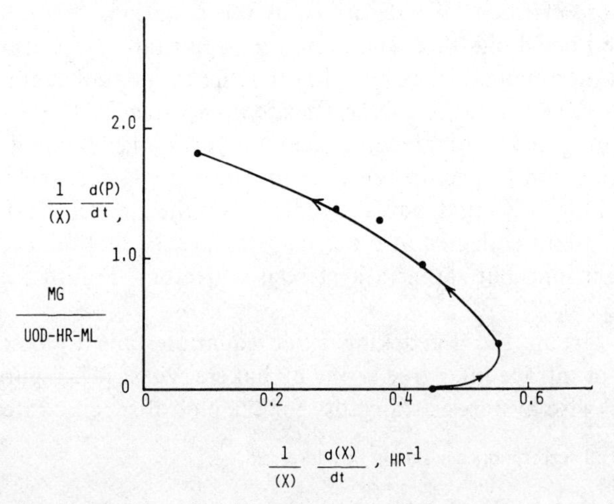

Fig. 2. Specific rates for gluconic acid production at 25°C and pH 5.8. Data from Rai [3].

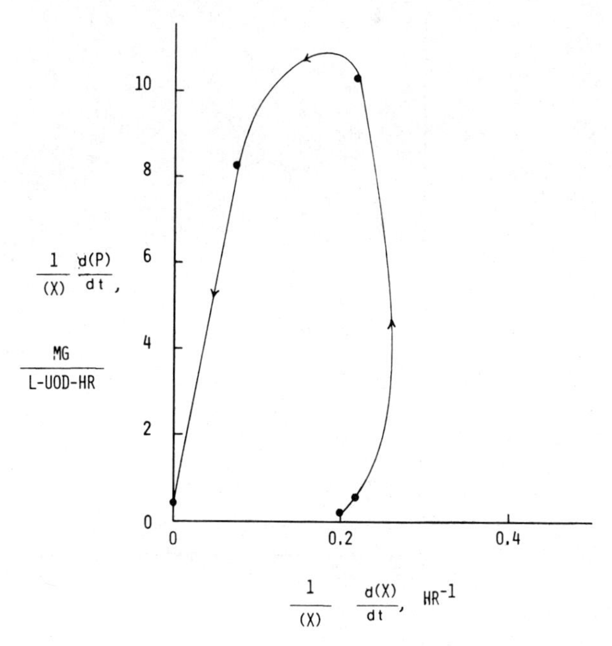

Fig. 3. Specific rates for intracellular free lysine production at 32°C and pH 5.0. Data from Ref. 4.

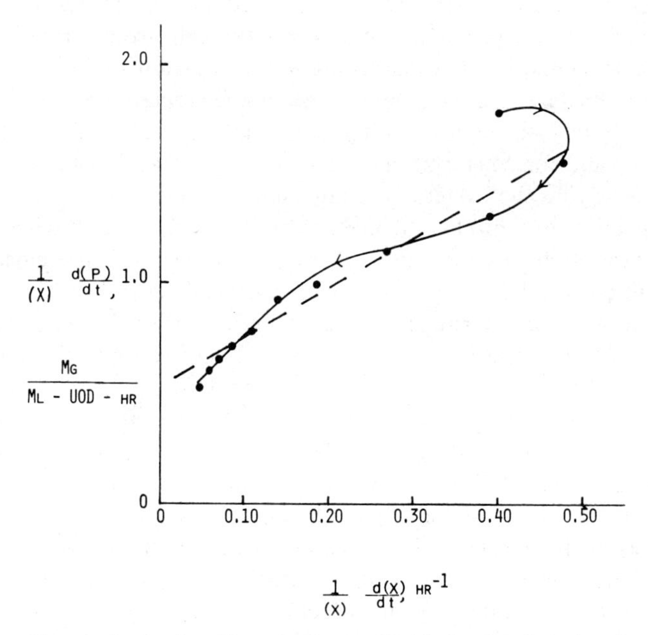

Fig. 4. Specific rates for lactic acid production at pH 6.0 showing deviations from linearity. Data from Luedeking and Piret [1].

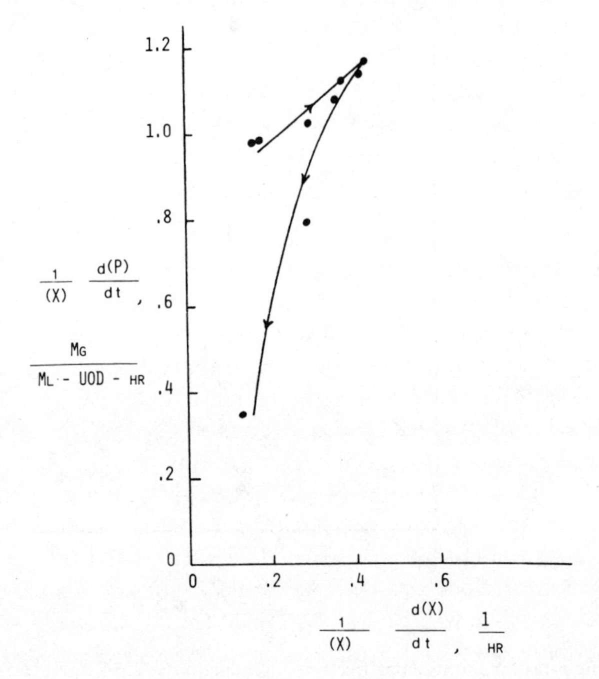

Fig. 5. Specific rates for ethanol production at pH 4.5 showing deviations from linearity. Data from Aiyar and Luedeking [2].

There are two primary motivations for studying the hysteresis relationship between specific product rate and specific cell growth rate. The first objective is to generalize the Luedeking–Piret equation to describe systems described by hysteresis curves. Such a generalized relationship may lead to sharpened control strategies for higher production rates of fermentation-derived fuels and for conservation of expensive substrates. The second motivation for studying the hysteresis relationship of product rate to cell growth rate is to gain insight into kinetic mechanisms of fermentation reactions.

Before examining product rate–cell rate relationships more closely, it is appropriate to recall previous work with hysteresis curves in fermentation reactions in order to determine the type of information we can hope to elucidate from the hysteresis behavior of product rate versus cell growth rate. It has been observed that Rai's data for the fermentation of glucose to gluconic acid describe hysteresis functions that have direction reversals as the temperature changes. Prior to Rai's study, only counterclockwise curves of product rate against the intermediate gluconolactone were observed. Subsequently, Tanner and Yunker [5] attributed this oscillation to possible relative changes in the concentration of two enzymes, a gluconolactonase and a protease. Furthermore, the presence of shunts in the biochemical pathway may be detected by examining the direction and convexity of hysteresis trajectories of product rate against intermediate concentrations [6,7]. This paper examines whether such information may be gathered when the more easily obtained cell concentration data are used in place of the intermediate concentration measurements. Such a substitution is often important in practice since intermediates are frequently difficult to monitor continuously.

HYSTERESIS ANALYSIS USING CONSTITUTIVE ENZYME FORMATION

Before attempting to explain hysteresis in plots of specific rates, we examined plots of product rate versus cell growth rate (also hysteresis curves). There are two motivations to this approach. The first is that dividing the rates by the cell concentration can suppress some trends that would be helpful to observe. The second motivation is that analysis was simplified and the behavior of product rate as a function of cell growth rate could be examined and explained using cell concentration.

Smoothed data describing the production of intracellular lysine in bakers' yeast [4] describe hysteresis relationships between product rate and cell growth rate. At the optimum conditions of pH 5.0 and temperature 32°C, the curve is counterclockwise (Fig. 6). Rates calculated from data collected when the temperature was held at 32°C and the pH allowed to vary uncontrolled describe a clockwise hysteresis curve (Fig. 7). Perhaps the explanation of this hysteresis reversal will give information about the fermentation mechanism that will lead to higher productivity of intracellular lysine.

Before attempting to analyze the production of intracellular lysine with a hysteresis diagnostic method, we decided to develop the method using a far less complex system. In view of its simple biochemical pathway (one intermediate, gluconolactone), the availability of data, and previous hysteresis analysis [5, 6] of the system, the fermentation of gluconic acid from glucose promised to be an excellent candidate.

A plot of rate of production of gluconic acid versus cell growth at 25°C and pH 5.8 is a wide, counterclockwise curve (Fig. 8). As the temperature is

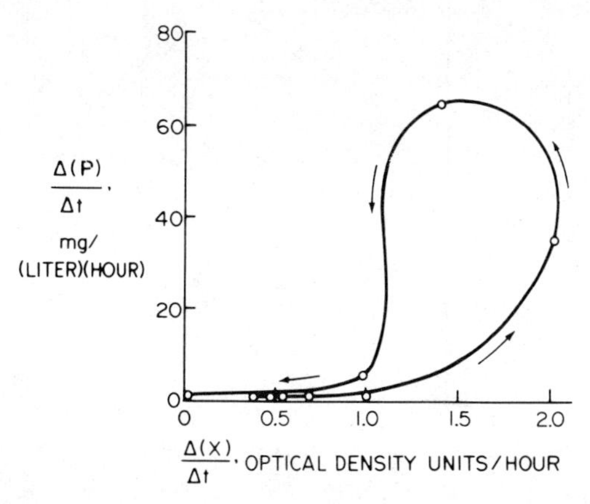

Fig. 6. Product and cell growth rates for intracellular free lysine production at 32°C and pH 5.0; counterclockwise hysteresis relationship. Data from Ref. 4.

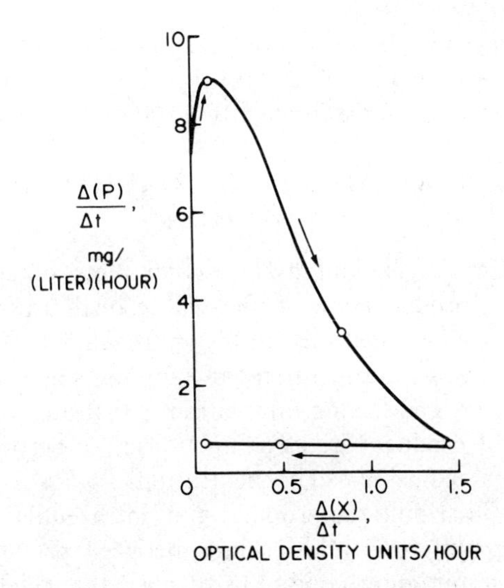

Fig. 7. Product and cell growth rates for intracellular free lysine production at 32°C and pH uncontrolled; clockwise hysteresis relationship. Data from Ref. 4.

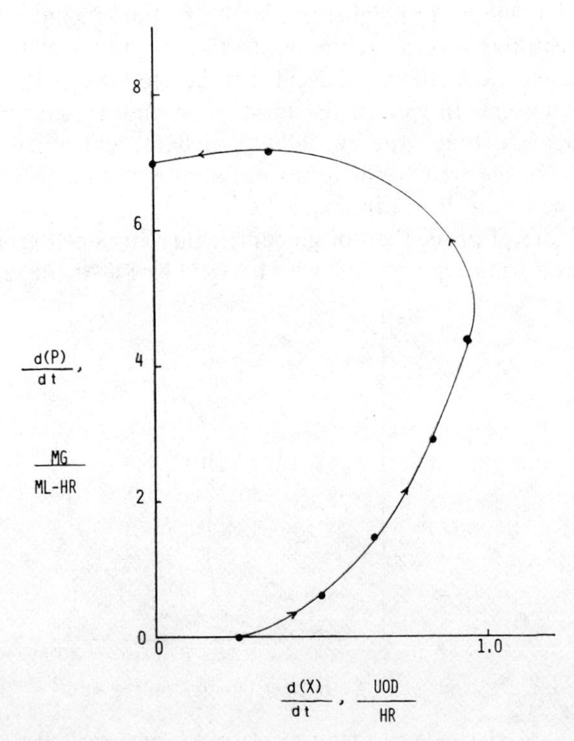

Fig. 8. Product and cell growth rates for gluconic acid production at 25°C. Data from Rai [3].

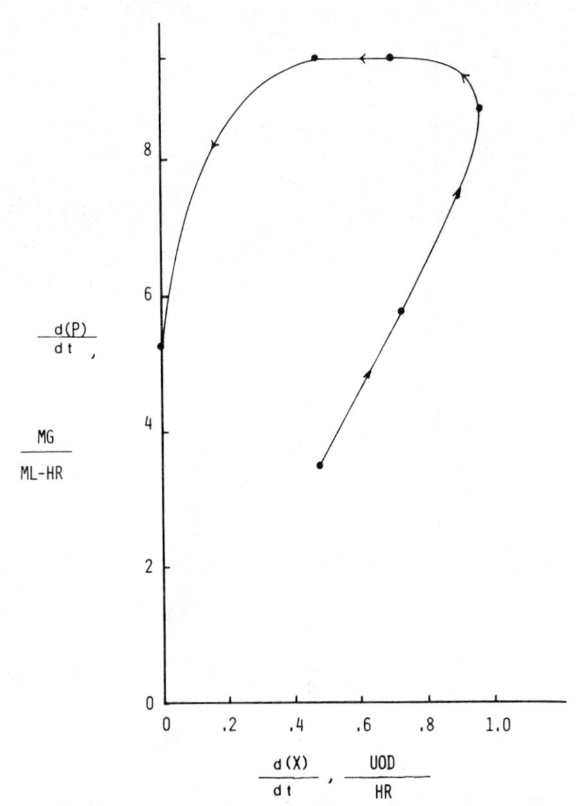

Fig. 9. Product and cell growth rates for gluconic acid production at 28°C. Data from Rai [3].

increased at constant pH, the curve begins to collapse on itself (Figs. 9–11). At 33.7°C the curve approaches the null configuration (Fig. 12). Then at 35.4°C the top portion of the curve has flipped over so that a "figure 8" is formed (Fig. 13). Previous work [7] shows that this curve can be an intermediate between clockwise and counterclockwise behavior; apparently, therefore, at some temperature higher than 35.4°C the rest of the curve will flip over, and a clockwise hysteresis trajectory will be observed.

To understand how the direction and slope of the curve is affected by the underlying structure of the system is important. The immediate goal is to simulate the phenomenon of the curve collapsing into a "figure 8" and reversing direction. From this simulation the foundation for the development of hysteresis analysis of product rate–cell growth rate relationships will be laid.

It is only appropriate to incorporate previously developed equations into our simulation. For a growth model the Monod equation is used; that is,

$$\mu = \frac{1}{(X)} \frac{d(X)}{dt} = \frac{\mu_{\max} (S)}{K_s + (S)} \tag{2}$$

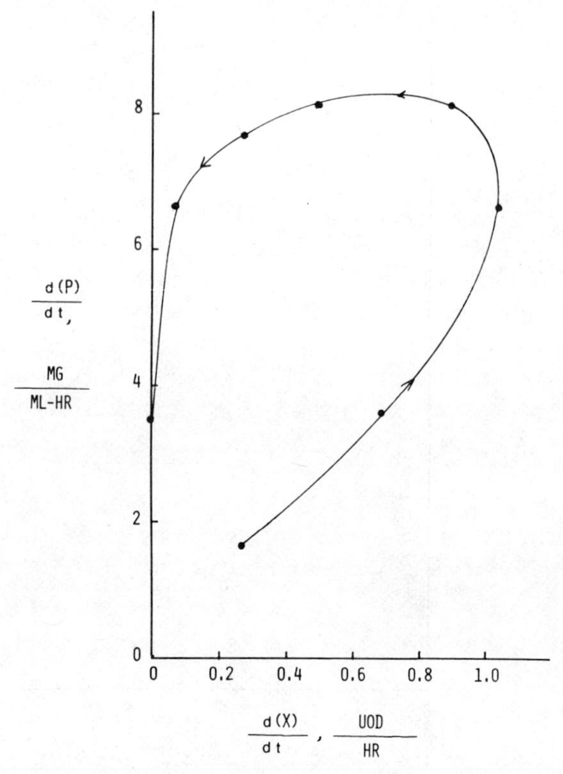

Fig. 10. Product and cell growth rates for gluconic acid production at 30°C. Data from Rai [3].

where the constant $\mu_{\max}$ is the maximum specific growth rate and K_s is the saturation constant. Along with the constant K_s, $\mu_{\max}$ may be determined from a reciprocal plot similar to the Lineweaver–Burk plot. Note that the Monod equation describes systems for which μ is zero when, and only when, substrate concentration is zero. Unfortunately, the gluconic acid system does not behave in this manner—cell growth stops before the substrate is completely gone. If S' is made the substrate concentration when cell growth stops, the Monod equation can be translated along the substrate axis to consider this non-Monod behavior:

$$\frac{1}{(X)} \frac{d(X)}{dt} = \frac{\mu_{\max} [(S) - S']}{K_s + [(S) - S']} \tag{3}$$

To simplify the simulation study, the Monod equation was not permitted to change with temperature; $\mu_{\max}$, K_s, and S' were held constant at 0.5 hr, 10 mg/ml, and 10 mg/ml, respectively, a reasonable first approximation for a narrow temperature range of the gluconic acid fermentation.

The Briggs–Haldane equation relates product rate to substrate concentration for an enzyme reaction. It is needed here to introduce the product rate to the simulation model in terms of another well-accepted equation:

$$\frac{dP}{dt} = \frac{V(S)}{K_m + (S)} \tag{4}$$

where

$$V = k(E)$$

V is the maximum product formation rate which is proportional to the total enzyme concentration in a closed enzyme system. The proportionality constant is the turnover number in a batch enzyme model. In a fermentation reaction, the concentration of enzyme changes as the cells grow and are restructured. In this analysis, E was allowed to vary linearly with cell concentration as for constitutive enzyme formation, so that

$$V = A + B(X) \tag{5}$$

where A and B are parameters.

For simplicity, K_m was held constant at 10 mg/ml in eq. (4) during the simulation study. To simulate the hysteresis behavior for gluconic acid, the parameters A and B were varied to reflect changes in the relationship

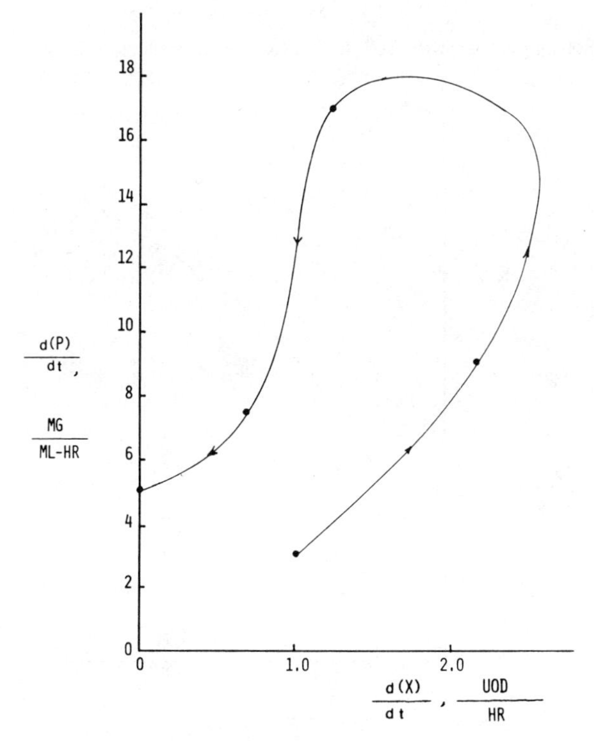

Fig. 11. Product and cell growth rates for gluconic acid production at 32°C. Data from Rai [3].

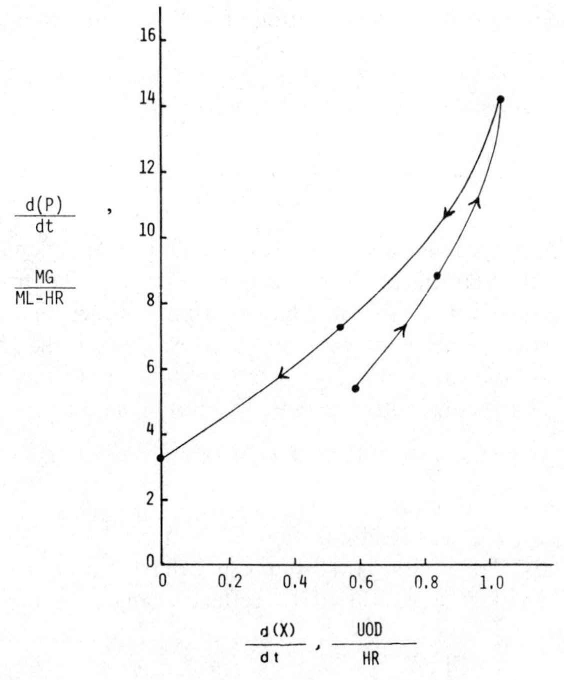

Fig. 12. Product and cell growth rates for gluconic acid production at 33.7°C. Data from Rai [3].

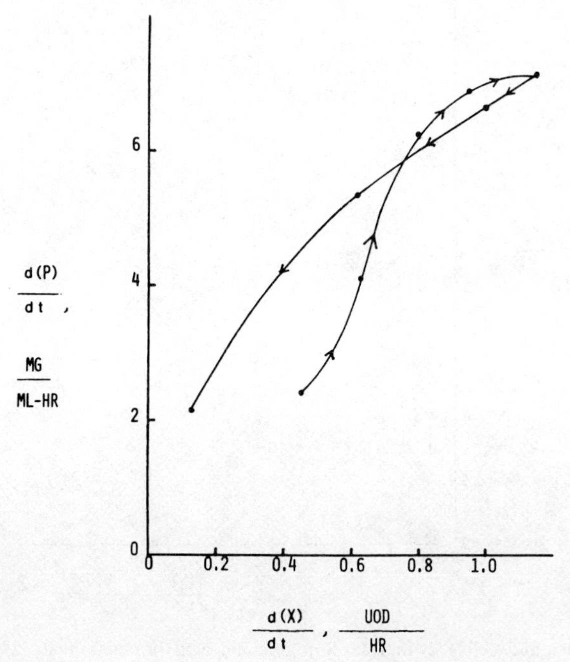

Fig. 13. Product and cell growth rates for gluconic acid production at 35.4°C. Data from Rai [3].

between cell concentration and the lumped enzyme. Here E represents both enzymes in the gluconic acid pathway. Owing to the inadequacies of the Monod and Briggs–Haldane equations and, more importantly, owing to the restrictions that were placed on the model by holding so may parameters constant, only the qualitative features of the curves could be reproduced, leaving the specific changes in shape and area to a more complete simulation study.

Figures 14–16 show hysteresis trajectories obtained from simulations of the described cell growth and product formation model when (S) and (X) are related using the data from the study reported in Figure 8. Since very early and very late time data were not reported, the corresponding trajectories were omitted in Figure 14–16. It is recognized, however, that these simulations must begin and end at the origin. Figure 14 shows a counterclockwise curve that results from allowing $B(X)$ to be much larger than A in eq. (5). This inequality is an indirect way of describing an increase in enzyme concentration during the course of fermentation. Figure 15 shows that the "figure 8" configuration is similar to that of the gluconic acid fermentation shown in Figure 13. This curve is the result of allowing $B(X)$ to be about the same magnitude as A (a moderate rate of enzyme manufacture

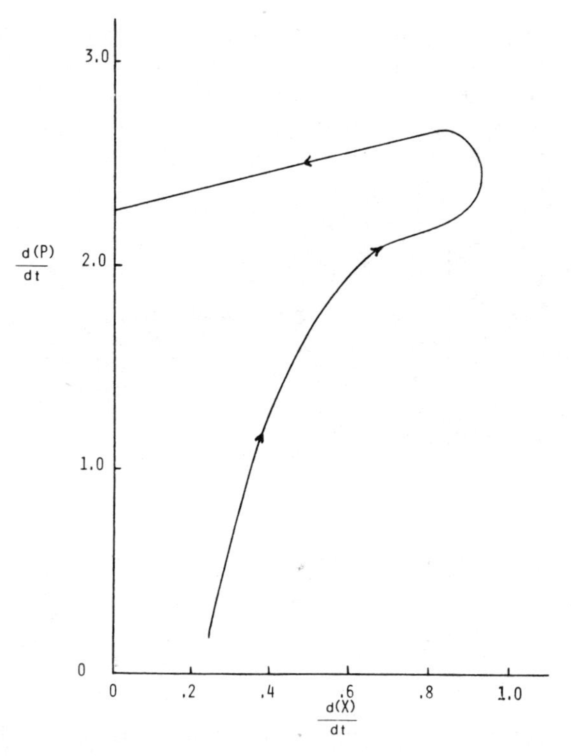

Fig. 14. Simulation of counterclockwise hysteresis relationship of product rate to cell growth rate. $V = A + B(X)$; $A = 0.001$; $B = 1$.

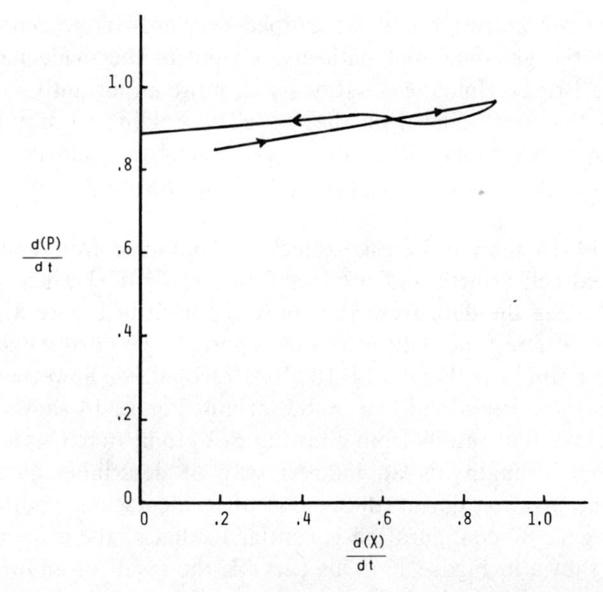

Fig. 15. Simulation of figure 8 relationship of product rate to cell growth rate. $V = A + B(X)$; $A = 1$; $B = 0.1$.

during the course of the fermentation). The third simulation, with $A = 1$ and $B = 0$, produced a clockwise hysteresis curve similar to that expected for temperatures higher than 35.4°C in the gluconic acid system. Because $B = 0$, this is the limiting case and would possibly correspond to a system in which a protease degrades the enzyme as soon as it is formed [5]. This

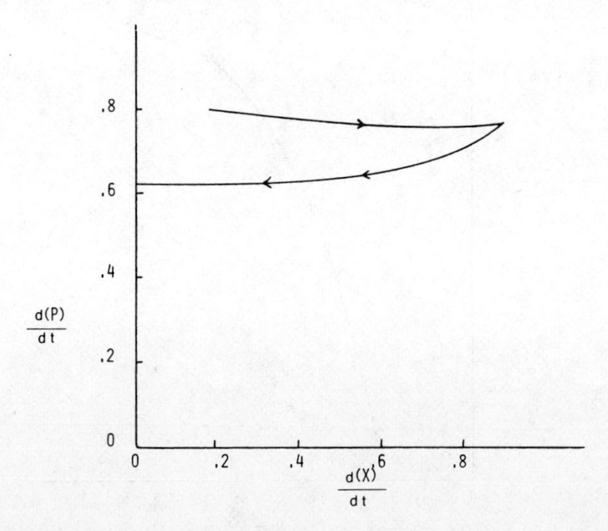

Fig. 16. Simulation of clockwise hysteresis relationship of product rate to cell growth rate. $V = A + B(X)$; $A = 1$; $B = 0$.

simulation of hysteresis reversal is important because it shows the feasibility of relating the concept of enzyme production in fermentation reactions with cell production.

DEVELOPMENT OF THE LUEDEKING–PIRET EQUATION FROM BIOCHEMICAL CONSTRUCTS

In the previous discussion, the feasibility of relating cell production to the synthesis of a key enzyme was shown. This demonstration was qualitative, however, and the argument for the two variables is now bolstered by a more rigorous material balance, where enzyme levels may appear implicitly.

An overall material balance of the enzymatic pathways can lead, in special cases, to the Luedeking–Piret equation for the growth phase. The material balance for a fermentation in which a product is enzymatically produced and in which cells are growing can be expressed in the following form:

$$(S) + \sum_{n=1}^{N} (I_n) + (P) + K(X) = S^* + KX^* \tag{6}$$

where (S) is the substrate concentration, typically in g/liter; I_n is the intermediate n concentration, typically in g/liter; (X) is the cell concentration, typically in units of optical density; K is the conversion constant from cell concentration in optical density units to cell concentration in units consistent with (S) and (I_n); and the asterisk indicates initial concentration.

Differentiating eq. (6) assuming the functions are continuous gives

$$\frac{d(P)}{dt} = -K\frac{d(X)}{dt} - \frac{d(S)}{dt} - \sum_{n=1}^{N}\frac{d(I_n)}{dt} \tag{7}$$

Defining a cell yield relationship between (S) and (X) (after Monod [8]) yields

$$Y \equiv \frac{-d(X)/dt}{d(S)/dt} \tag{8}$$

Combining eqs. (7) and (8) gives

$$\frac{d(P)}{dt} = \frac{1}{Y} - K\frac{d(X)}{dt} - \sum_{n=1}^{N}\frac{d(I_n)}{dt} \tag{9}$$

For the special case where Y is constant and $(1/Y - K) = \alpha > 0$ and where

$$\sum_{n=1}^{N}\frac{d(I_n)}{dt} = -\beta(X) \tag{10}$$

with β being another constant, the Luedeking–Piret equation is developed for the later growth phase of a fermentation process (where the intermediate concentrations are decreasing in value, and, hence, their rates are negative;

see eq. (1). The validity of eq. (10) seems to be enhanced as N increases. For the gluconic acid fermentation [3], where $N = 1$, the linearity implied by eq. (10) holds only approximately for the decaying branch of the single intermediate, gluconolactone, time curve and not at all for the rising portion of that curve.

Furthermore, the validity of eq. (10) as written is decreased if there are two or more products. When an intermediate such as ethanol becomes a desired product, then it may be best to write two product terms in eq. (6), while writing one intermediate term. Note that although we have defined I_n as an intermediate, $\sum_{n=1}^{N} (I_n)$ must include the concentrations of all species that are not substrate, the product P, or part of the cell mass. For instance, the lysine-enriched bakers' yeast process, ethanol and all its precursers must be included as I_n's. Therefore, eq. (10) used in conjunction with eq. (6) would not be expected to hold for the lysine product alone, which is consistent with the earlier observation that the Luedeking–Piret equation fails to apply to the enriched lysine yeast system.

If the restriction imposed by eq. (10) is relaxed but Y remains constant and $(1/Y - \mathrm{K}) = \alpha$, then eq. (9) describes a counterclockwise hysteresis relationship of $d(P)/dt$ to $d(X)/dt$ since $\sum_{n=1}^{N} d(I_n)/dt$ is positive initially and negative finally. This hysteresis notion is consistent with Aiyar and Luedeking's [2] original ethanol rate data for the pH's at 3, 4.5, and 6.8.

Another approach to this derivation is to let the $\sum_{n=1}^{N} d(I_n)/dt$ term be negligible and break out of the Y two terms:

$$\frac{1}{Y} = \frac{1}{Y_1} + \frac{\beta}{\mu}$$

where

$$\mu = \frac{1}{(X)} \frac{d(X)}{dt} \tag{11}$$

so that α then becomes $(1/Y_1 - K)$. Equation (11) is consistent with the observation of Senez and Belaïch [9] that Y increases as μ increases.

CONCLUSIONS AND RECOMMENDATIONS

By examples, the Luedeking–Piret equation has been shown to be a zero-area case of a more general hysteresis relationship. Initial attempts to exploit this relationship as a diagnostic tool for fermentation reactions indicate that useful information may be extracted from hysteresis curves of either product rate against cell growth rate or of specific product rate against specific cell growth rate.

The most promising approach to developing general models relating product rate to cell growth rate uses biochemical constructs with the general model. With this model, special conditions under which the Leudeking–Piret equation holds then become apparent. The restrictive conditions can then be identified for subsequent experimental verification.

It has been observed that product rate–cell growth rate relationships are often not as sensitive to changes in some engineering variables (such as temperature) as are the product rate–intermediate concentration relationships studied earlier [5]. But in the absence of intermediate concentrations, product rate–cell growth rate hysteresis curves still can serve as a diagnostic tool and, when carefully applied, as models for use in fermentation reactor design.

References

[1] R. Luedeking and E. L. Piret, *J. Biochem. Microbiol. Technol. Eng.*, *1*(4), 393 (1959).

[2] A. S. Aiyar and R. Luedeking, in *Chemical Engineering Progress Symposium Series, Bioengineering and Food Processing*, M. R. Sfat, Ed. (AIChE, New York, 1966), No. 69, Vol. 62, pp. 55–59.

[3] V. R. Rai, Ph.D. Thesis, Rutgers University, New Brunswick, N.J., 1973.

[4] R. D. Tanner, N. T. Souki, and R. M. Russell, *Biotechnol. Bioeng.*, *19*, 27 (1977).

[5] R. D. Tanner and J. M. Yunker, *J. Ferment. Technol.*, *55*, 143 (1977).

[6] R. D. Tanner, *AIChE J.*, *18*, 385 (1972).

[7] R. D. Tanner and L. H. DeAngelis, *Chem. Eng. J.*, *8*, 113 (1974).

[8] S. Aiba, A. E. Humphrey, and N. F. Millis, *Biochemical Engineering*, 2d ed. (Academic, New York, 1973), p. 141.

[9] J. C. Senez and J. P. Belaïch, in *Méchanismes de Régulation des Activities Cellulaires Chez Les Microorganisms* (Éditions du Centre National de la Recherche Scientifique, Paris, 1965), No. 124.

Membrane-Controlled Digestion: Effect of Ultrafiltration on Anaerobic Digestion of Glucose

T. W. JEFFRIES, D. R. OMSTEAD, R. R. CARDENAS,* and H. P. GREGOR

Department of Chemical Engineering and Applied Chemistry, Columbia University, New York, New York 10027

INTRODUCTION

The biological conversion of soluble organic materials to methane by anaerobic digestion is limited by the activity of methanogenic bacteria. In order to increase this rate, it is necessary to increase the supply of appropriate substrates, to remove inhibitory compounds, and to increase the methanogenic biomass. This study describes attempts to accomplish these goals by membrane-controlled anaerobic digestion.

The rate of methane production from H_2 and CO_2 is much greater than the rate of formation from organic acids. Bryant, McBride, and Wolfe [1] found that *Methanobacterium* strain MOH produces about 6 mmol/min or 16 liter/liter day of methane when grown on H_2 and CO_2. Augenstein, Costa, and Wise [2] have attained greater than 100 liter/liter day (specific productivity: 20 to 25 mmol CH_4/g cells hr) from H_2 and CO_2 by recycling the methanogenic biomass. By comparison, the rate of methane production from organic acids is slow [3] and generally does not exceed 1.0 liter/liter day in typical high-rate anaerobic digestion [4].

Hydrogen and CO_2 are present in anaerobic digesters as normal products of carbohydrate fermentations. As shown in Figure 1, cellulose is hydrolyzed by extracellular enzymes to form glucose. Glucose is then converted by a variety of acidogenic bacteria into acetic, propionic, butyric, valeric, and isovaleric acids and varying amounts of H_2 and CO_2. Although the initial hydrolysis of certain carbohydrates such as cellulose proceeds slowly, the formation of acids, H_2, and CO_2 from glucose is relatively rapid. The subsequent conversions of H_2 and CO_2 occurs rapidly, but the conversion of the organic acids formed to methane and CO_2 is slow. Acids higher than acetic are probably first converted to acetic acid and H_2 by acetogenic bacteria [5]. Acetogenesis is probably the rate-limiting step.

* Present address: Dept. of Civil and Envionmental Engineering, Polytechnic Institute of New York, Brooklyn, N.Y. 11201.

Biotechnology and Bioengineering Symp. No. 8, 37–49 (1978)

0572-6565/78/0008-0037$01.00

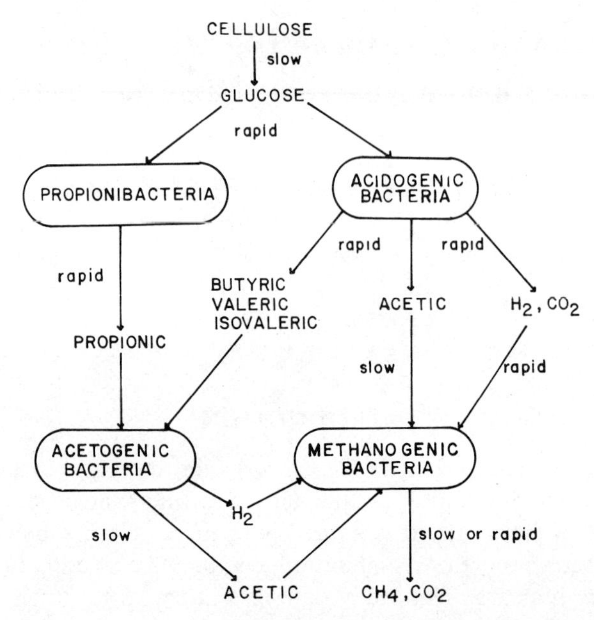

Fig. 1. Conversion of cellulose to acids, methane, and CO_2 in anaerobic digestion. Once cellulose is hydrolyzed, glucose is rapidly fermented to organic acids, H_2, and CO_2 by acidogenic bacteria. Hydrogen and CO_2 are rapidly converted to methane, but organic acids are converted slowly. Acetogenic bacteria degrade C-3 and higher organic acids to acetic acid and H_2 through a symbiotic relationship with methanogens.

The best studied acetogenic bacterium is the *S* organism isolated from Barker's cultures of *Methanobacillus omelianskii*. This organism oxidizes ethanol to acetic acid and H_2. The H_2 is consumed by a methanogenic bacterium [6, 7]. Hydrogen transfer between organisms has also been demonstrated in mixed cultures with *Selenomonas ruminatium* or *Ruminococcus flavefaciens* and *Methanobacterium ruminantium* [8, 9]. These mixed cultures produced large amounts of methane and acetate. In the absence of the methanogenic bacteria, greater amounts of higher volatile fatty acids and H_2 are formed which indicates that one of the functions of methanogenic bacteria is to minimize the accumulation of H_2 in the medium. If the acidogenic bacterium is able to dispose of reducing group activity as H_2 rather than through the reduction of acetyl CoA, it can derive an additional adenosine triphosphate (ATP) from the acetyl CoA [10]. Therefore, in the presence of methanogenic bacteria, a fermentation will tend toward the formation of methane, CO_2, and acetic acid.

The utilization of H_2 and CO_2 as a source for methane in anaerobic digesters is the rule rather than the exception. Large numbers of hydrogen-utilizing methanogenic bacteria can be cultivated from anaerobic digesters in contrast to the smaller numbers of acetate-utilizing methanogens [11, 12]. This could be attributable in part to the free energy available from these two pathways (i.e., −33 and −8.6 kcal, respectively). The cell yield/mol ATP utilized is relatively constant for all aerobic and anaerobic bacteria, and as the available free energy decreases, the cell yields and growth rates decrease as well [13, 14].

At present, only H_2 plus CO_2, formate, methanol, or acetate have been identified as substrates for methanogenic bacteria. Higher organic acids are probably degraded via symbiosis. Because the conversion of higher organic acids such as propionate or butyrate to H_2 and acetic acid is not energetically favorable, utilization of these products must be coupled with the formation of methane by methanogens. Only about −6 kcal is available per mole of propionate dissimilated by this pathway. Cell yields and growth rates of acetogenic and methanogenic bacteria are correspondingly lower. Thus, the conversion of volatile fatty acids to methane is rate limiting in anaerobic digestions.

Cell yields of methanogenic bacteria should increase with the supply of H_2. One way to accomplish this is to increase the feeding rate. Unfortunately, this also increases the rate of acid generation. As acetic acid accumulates, acidogenic bacteria shift to the formation of higher acids. This leads to even slower rates of methane production. At elevated hydraulic retention times, acidogenic and methanogenic bacteria are washed from the culture and digestion stops.

Hydrogen, formed during the dissimilation of glucose to acetic acid, can combine with CO_2 to form CH_4. The rate of this reaction greatly exceeds the rate of CH_4 production from acetic or higher organic acids. If the acids can be removed as they are formed, methanogenesis should proceed at a higher rate via this reaction. Moreover, production of methane by this pathway should not reduce the yield of acetic acid, so the acetic acid can be recovered as a by-product.

The objectives of this research were to enhance the rate of methanogenesis by removing acetic and higher organic acids by ultrafiltration as they were formed in an anaerobic digester and to concentrate and recover these acids by water-splitting electrodialysis and membrane-assisted solvent extraction.

EXPERIMENTAL

The overall process for membrane-controlled anaerobic digestion consists of four operations (Fig. 2). The first is a rapid, partial anaerobic digestion. This initial step produces CH_4, CO_2, and various organic acids. Ultrafiltration then separates the cells, extracellular enzymes and insoluble particulate material from the organic acids and nutrients. In this system, membranes of fixed-charge, sulfonic acid character are used to avoid membrane fouling. These membranes pass soluble materials of a molecular weight less than 1000. Water-splitting electrodialysis then separates and concentrates the salts into their component acids and bases. Ammonium hydroxide is

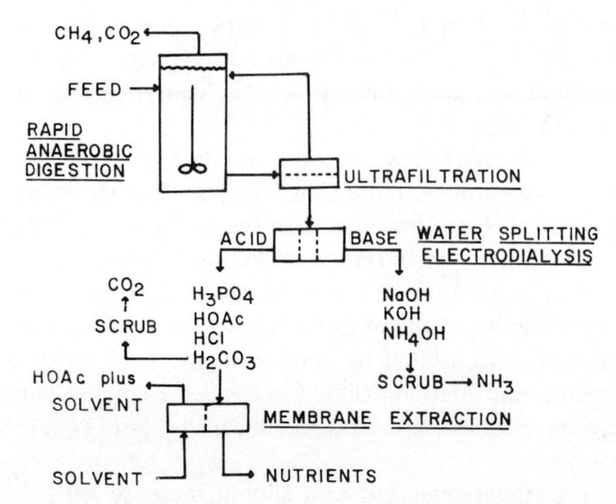

Fig. 2. Process diagram for membrane-controlled anaerobic digestion. Salts of organic acids are removed by ultrafiltration from anaerobic digestion as they are formed. Water-splitting electrodialysis separates these into their component acids and bases. Nitrogen is removed as anhydrous ammonia. Membrane-assisted solvent extraction concentrates and separates the organic acids from inorganic nutrients, which can then be recycled.

removed from the basic stream as gaseous ammonia by scrubbing and can be recovered as a by-product. The organic acids are removed from the acidic stream by membrane-assisted solvent extraction: nutrients, including phosporic acid, are recovered. In an integrated biofuels system, they could be returned to the growth cycle.

The details of the anaerobic digester and the ultrafiltration units are shown in Figure 3. The total volume of the experimental unit is about 3.0 liter. A peristaltic pump continuously feeds a 2.7-liter digester with synthetic medium containing 0.91 g $(NH_4)_2HPO_4$, 0.56 g NH_4Cl, 0.08 g KCl, 0.1 g $FeCl_3$, 0.2 g $MgCl_2 \cdot 6H_2O$, 0.022 g $AlCl_3 \cdot 6H_2O$, 0.02 g $CaCl_2 \cdot 2H_2O$, 5 mg $MnSO_4 \cdot H_2O$, 11.4 mg $CoCl_2 \cdot 6H_2O$, 0.2 mg $ZnCl_2$, 2 mg $(NH_4)_6MoO_{24} \cdot 4H_2O$, and 0.1 g yeast extract in 1.0 liter distilled water. Glucose was used as the carbon source; $NaHCO_3$ and Na_2CO_3 were added in varying amounts to keep the pH between 7.2 and 6.8.

Carbon dioxide in effluent digester gas was adsorbed in 1.0*N* KOH, and methane was measured by the difference between initial and final volumes in graduated gas analysis burettes.

Volatile fatty acids were determined by gas chromatography on a 6 ft × ¼ in. column of Chromosorb 101 coated with 2% phosphoric acid. The column was operated at 190°C with a helium carrier gas flow rate of about 100 ml/min. A flame ionization detector was used. Samples of digester fluid or permeate (5 ml) were acidified by the addition of 0.1 ml concentrated HCl. Digester samples were centrifuged to remove cells. Permeate samples were free of particulate matter and could be injected after acidification. The column required daily conditioning by the injection of about 50 μl water before use to obtain maximal sensitivity and minimal tailing of peaks. The lowest concentration of acetic acid that could be readily detected in a 1.0 μl sample was about 50 mg/liter. Determinations were accurate to within about 10% with repeated injections.

Cell biomass was determined as protein. The cell pellet from the volatile acid determination was washed in 5.0 ml of 1.0*N* HCl to remove residual metal sulfides, and the cells were taken up in distilled water. Cells were digested in 0.5*N* NaOH for 2 hr at 50°C, and cell protein was determined by the method of Lowry et al. [15]. Pepsin was used as a standard.

Ultrafiltration was performed at 55 psig. A diaphragm pressurizing pump provided a flow rate of approximately 100 ml/min. A magnetically driven centrifugal pump was used to recirculate the digestion fluid over the surface of the membrane of area 93 cm^2 at a rate of approximately 1 m/sec. Two membrane modules were connected in series. One used a commercially available polysulfone membrane having a pure water flux of 21.5 μm/sec-atm or 2.12 gfda (gal/ft^2-day-atm). The other was a nonfouling sulfonate membrane prepared in this laboratory [16].

The flux of the two membranes during operation is shown in Figure 4. The negatively charged sulfonate membranes maintained a flux near their initial rate, whereas the flux of the neutral polysulfone membranes declined to 5% of its water flux on contact with the digestion fluid and then fell to 0.4 μm/sec-atm. They could be used for about 15 hr before fouling had proceeded to the point where replacement was necessary.

Permeate from ultrafiltration was batch-fed into the water-splitting electrodialysis unit shown in Figure 5. In this process, anionic, cationic, and bipolar membranes in electrolytic cells are used to carry out the separation of salts into their respective acids and bases. For example, sodium acetate is converted into sodium hydroxide and acetic acid. The feed stream is introduced into the center chamber and the concentrated acidic and basic

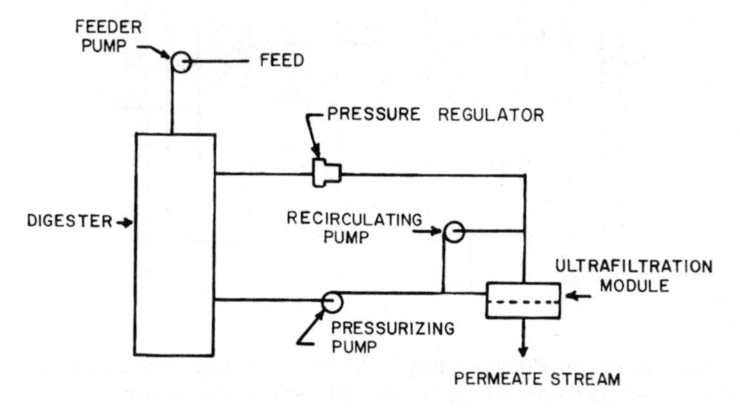

Fig. 3. Anaerobic digester and ultrafiltration cell.

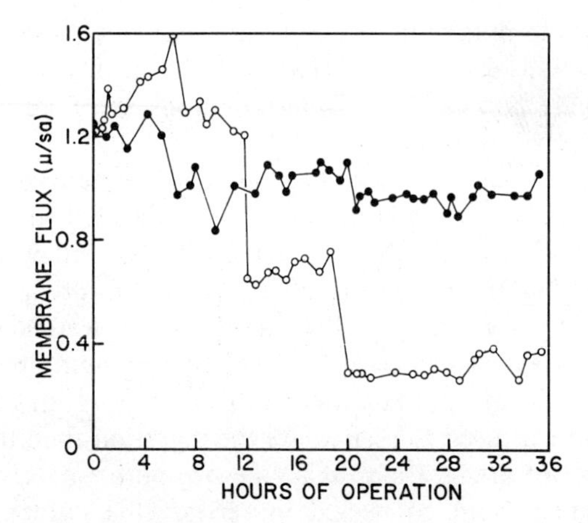

Fig. 4. Membrane flux of neutral and negatively charged membranes. (○) Neutral polysulfone membrane flux decayed with operation: flux of negatively charged (●) sulfonate membranes stayed near its initial rate.

streams are removed from the side chambers. The streams are circulated through the chambers in order to decrease the effects of concentration polarization at the membrane surfaces. The three compartments form a repeating unit cell. Generally, the voltage drop was about 3 V for each unit cell. Operation was carried out at a current density of 20 to 50 mA/cm^2, and current efficiencies approached 90%. Figure 6 shows typical results

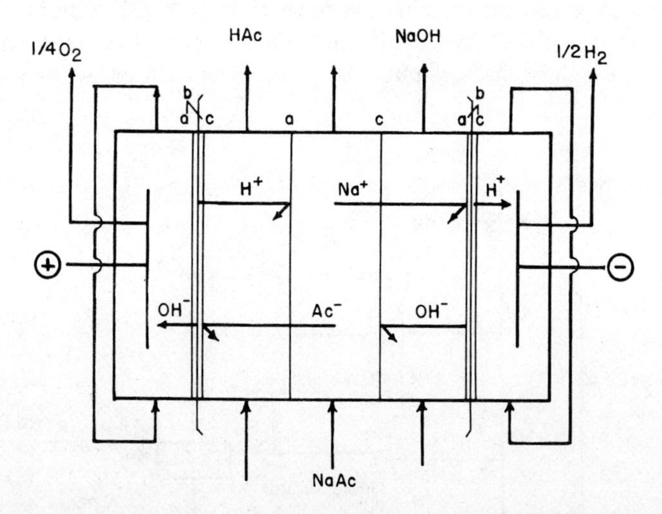

Fig. 5. Water-splitting electrodialysis module. Anion exchange membranes (a) will pass only anions: cation exchange membranes (c) will pass only cations. Bipolar membranes (b) consist of anion and cation exchange membranes back to back. The net reaction converts salts into their respective acids and bases.

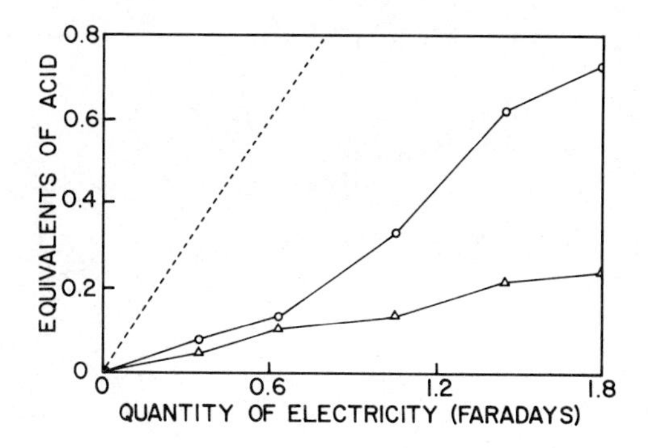

Fig. 6. Transport of organic acids by water-splitting electrodialysis. (---) 100% efficiency; (O) total organic acids; (Δ) acetic acid.

obtained with this system. Transport of organic acids as anions increased with time as their relative concentrations increased in the feed stream.

In membrane-assisted solvent extraction, the acid stream from water-splitting electrodialysis is passed across one face of the membrane while an alkaline solution is circulated across the opposite face. The membrane, which consists of an acetate-selective organic solvent immobilized in the polymeric gel membrane partitions the undissociated form of the acid strongly within it. Thus the solution on the feed side must be acidic to keep the acid in its undissociated form, and the process is driven by effecting a concentration gradient for the free acid by making the pH of the right-hand side solution high so that the acid is largely in the oil-insoluble anionic form on the opposite side.

RESULTS

In initial experiments, a conventional anaerobic digester was used without ultrafiltration, with the culture fed 0.5 g/liter of glucose each morning. Gas production and organic acids were followed for a 30-hr period. Results are shown in Figure 7. Within 1 hr after feeding, the gas production rate reached a peak and then declined rapidly. The concentration of volatile fatty acids also increased after feeding but did not reach a peak until after 2 hr and did not subsequently decrease at the same rate as gas production. The changes in gas production rate and CO_2 content were too transient for accurate observations. These experiments did indicate, however, that the gas production rate during glucose utilization was much higher than during subsequent utilization of the organic acids.

For the membrane-controlled experiments, a culture was started with effluent from active digesters that had maintained on glucose and a synthetic medium for 30 days. After an initial conditioning period in which low organic and hydraulic loading rates were employed, glucose loading rates were increased at a rate of 20% a day. Hydraulic loading rates were increased 20% a day from a detention time of 4.3 days (day zero) to 3 days

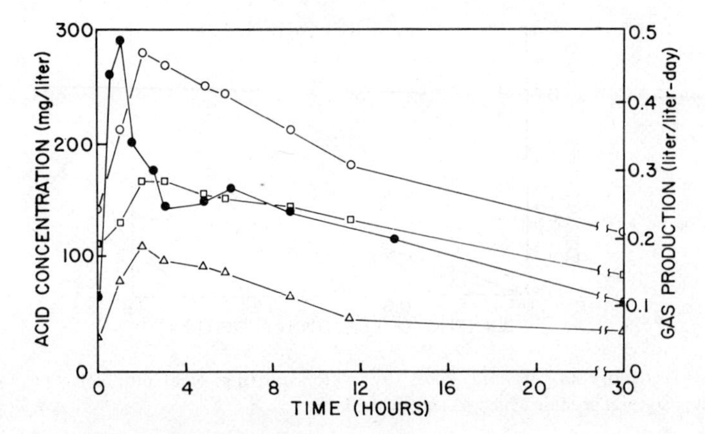

Fig. 7. Gas production and acid formation in a batch-loaded digester. (○) Total acids; (□) propionic; (●) total gas; (△) acetic.

at day 6. The hydraulic detention time was kept at 3 days until day 15 when it was decreased to 1.5 days (Fig. 8). Total gas production closely paralleled the rate of glucose additon. Initially methane comprised 70% of the evolved gas, but as glucose loading rates increased, the proportion fell. At day 11, methane was 50% of the effluent gas. The maximum rate of gas production (3.0 liter/liter-day) was attained on days 16 and 17. At this point, the methane production rate was 1.2 liter/liter-day or about 40% of the evolved gas. Approximately 85% of the methane and CO_2 were evolved during feeding and ultrafiltration. If feeding and ultrafiltration were continued on a 24-

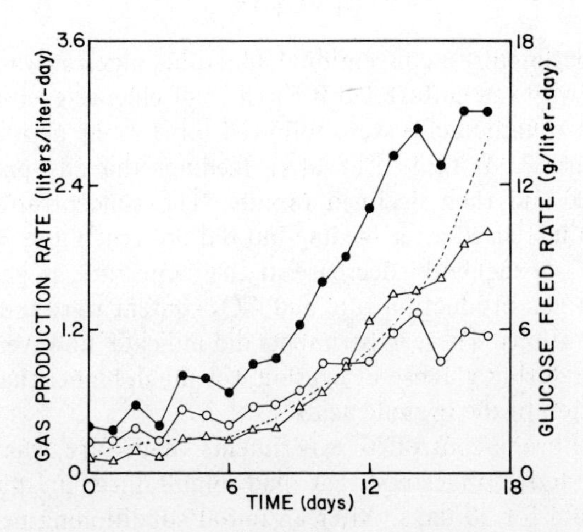

Fig. 8. Gas production in a membrane-controlled anaerobic digester. Digester volume was 3.0 liters. Incubation temperature was 35°C. Digester was continuously stirred. Glucose feed in synthetic medium was increased 20% a day. (●) Total gas; (– – –) glucose; (△) CO_2; (○) CH_4.

hr basis, total gas and methane production should approach 12.8 and 4.8 liter/liter-day, respectively.

Acid concentrations also increased with the glucose feeding rate as is shown in Figure 9. Initially, total acid concentrations were less than 100 mg/liter. From day 0 to day 14 acid concentrations increased during feeding and decreased overnight. This indicated net removal by methanogenesis. No significant accumulation of acetic acid occurred until day 8 when the feeding rate was 2.3 g/liter-day. On days 16 and 17, acetic acid decreased during ultrafiltration and increased overnight, indicating net formation by acetogenesis. Propionic acid began to accumulate on day 3 when the glucose feeding rate was 1.4 g/liter-day. Propionic acid concentrations also fluctuated, but to a lesser extent than acetic acid. Peak propionic acid concentrations were obtained on day 12. Butyric acid began to accumulate rapidly on day 13 when the glucose feeding rate was 5.0 g/liter-day. Overnight decreases in butyric acid concentration were observed on days 15 and 16. These decreases corresponded to the overnight increases in acetic acid and were therefore attributed to acetogenesis.

The moles of product formed per mole of glucose fed are shown in Figure 10. Carbon recoveries and redox balances were better for the period from day 9 on, and averaged 86% and 0.83, respectively, overall. At very low loading rates, methane production began to approach the value of 3.0 mol/mol fed expected from complete conversion of glucose to gas. At the highest loading rate the ratio was about 0.7 mol methane formed/mol glu-

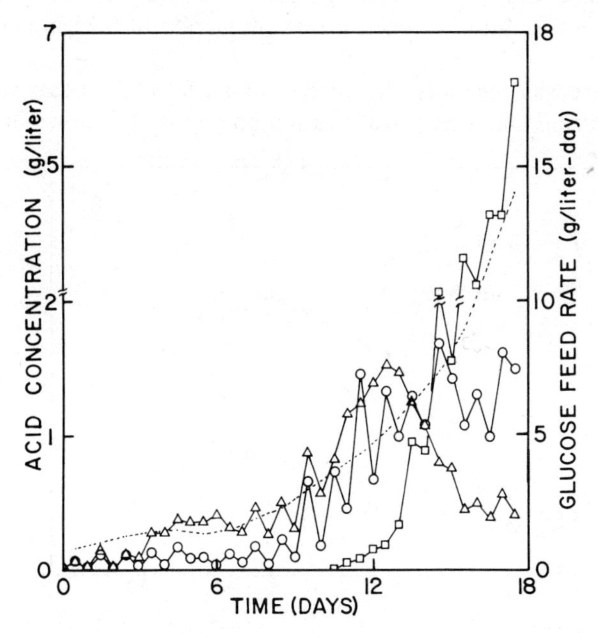

Fig. 9. Acid concentrations in membrane-controlled anaerobic digester. Conditions were the same as those listed in Figure 8. (□) Butyric; (---) Feed; (○) acetic; (△) propionic.

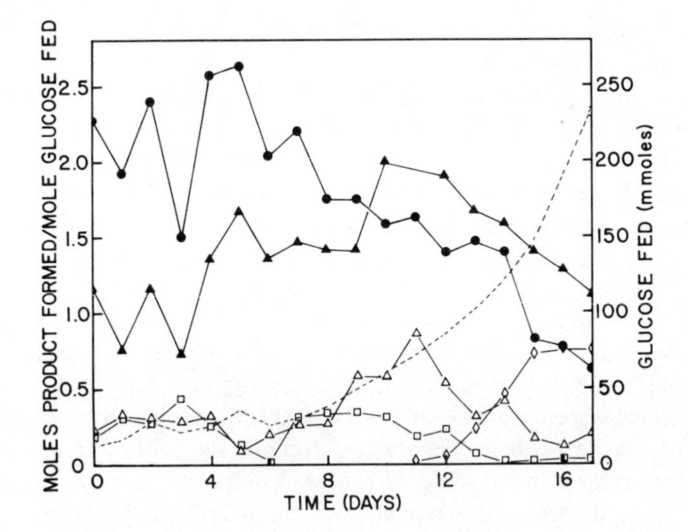

Fig. 10. Mol product formed/mol glucose fed in a membrane-controlled anaerobic digester. Conditions were the same as those listed in Figure 8. Acids were removed by ultrafiltration as the culture was fed. (●) CH_4; (---) glucose; (▲) CO_2; (△) acetate; (◇) butyrate; (□) propionate.

cose fed. Acetic acid production reached a peak rate of 0.86 mol/mol glucose fed on day 11. Propionic acid reached a peak concentration on day 12. The maximum rate of propionic acid formation was 0.44 mol/mol glucose fed. Butyrate formation reached a maximum rate of 0.75 mol/mol fed on day 17.

The culture was generally ultrafiltered about 6 to 8 hr a day. Gas production rates during and after ultrafiltration on day 10 are shown in Figure 11.

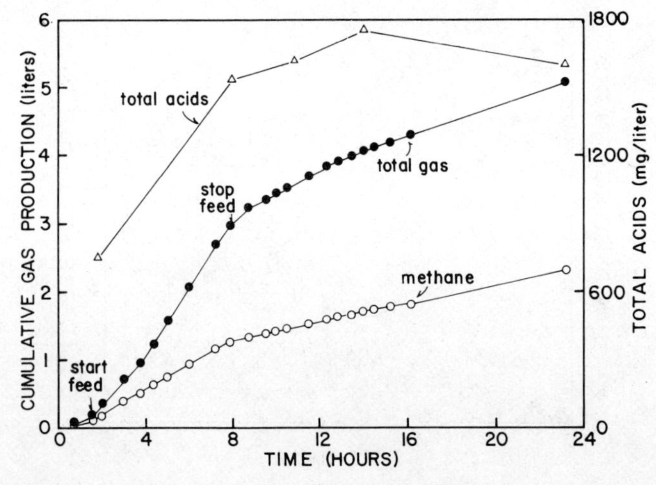

Fig. 11. Gas and acid production rates in anaerobic digestion before, during, and after feeding and ultrafiltration. Data are taken from day 10 of Figures 8, 9, and 10.

TABLE I

Gas Production as a Function of Glucose Addition

Glucose	CH_4	CO_2	CH_4	CO_2
	(mmol/liter-min)		(mol/mol fed)	
0	0.7	0.2	–	–
3.3	2.4	3.9	0.7	1.2
8.3	4.2	10.0	0.5	1.2
11.2	3.7	12.6	0.3	1.1

During addition of glucose (2 to 8 hr), the rate of total gas production was 28 mmol/hr (5 liter/liter-day). Immediately after glucose additon, the rate declined rapidly to 5 mmol/hr (1.0 liter/liter-day). The rates of methane production during and after glucose addition were 7.6 and 3.8 mmol/hr, respectively. The rate of gas formation during glucose addition is taken to be a composite of the rates of formation from organic acids and from H_2 yielded during the fermentation of glucose.

Within limits, the rate of gas formation was proportional to the rate at which glucose was added. Table I shows the rates of glucose addition and gas production on a single day. In the absence of glucose feeding, the gas was mostly methane. As the glucose feeding rate increased, the rate of methane production increased but the proportion of methane formed fell rapidly. Likewise, methane content was 70% in gas collected overnight but fell to between 25 and 35% of the gas collected during feeding as the experiment progressed.

One of the principal objectives of this work was to retain the cell biomass at high hydraulic loading rates. Biomass concentration increased from 430 mg protein/liter at day 0 to 2000 mg protein/liter at day 17. At that point the hydraulic detention time of the digester was 1.5 days.

DISCUSSION

This research has shown that the removal of organic acids from an anaerobic digester by ultrafiltration can proceed in a practical manner when using nonfouling membranes and that this treatment then allows the organic acids to be recovered by water-splitting electrodialysis and membrane-assisted solvent extraction. Moreover, CH_4 is produced at a high rate from H_2 and CO_2 during fermentation of the organic feedstock to acids.

The yield of methane obtained in rapid, partial anaerobic digestion depends upon the nature of the fermentation. The H_2 formed during conversion of glucose to acetyl CoA can either reduce acetyl CoA to higher organic acids (i.e., propionic, butyric) or reduce CO_2 to form CH_4. In the latter case, the acetyl CoA yields acetic acid. With a strict acetic acid fermentation, one would expect to observe 50% methane in the effluent gas and a 1/1 ratio of methane formed per glucose fed [eq. (1)]. This was approached on day 11 when acetic acid production reached a peak. With a butyric acid fermentation, one would expect to observe 25% methane in the effluent gas with a ratio of 2.0 mol of gas formed per mole of glucose fed [eq. (2)]. This was attained on days 16 and 17.

$$C_6H_{12}O_6 = 2CH_3COOH + CO_2 + CH_4 \quad (1)$$

$$6C_6H_{12}O_6 = 6CH_3CH_2CH_2COOH + 9CO_2 + 3CH_4 + 6H_2O \quad (2)$$

The actual concentration of methane on days 16 and 17 was higher than expected (40%). On days 15, 16, and 17 butyrate concentrations decreased overnight while acetate concentrations increased. These changes were attributed to acetogenesis. Acetogenesis should increase as cell recycle by ultrafiltration is extended.

With a propionic acid fermentation one would see the formation of propionic and acetic acids and CO_2 but no methane [eq. (3)]:

$$3C_6H_{12}O_6 = 4CH_3CH_2COOH + 2CH_3COOH + 2CO_2 + 2H_2O \quad (3)$$

The removal of organic acids by ultrafiltration enhances the gas production rate. Normal loading rates for first-stage, high-rate digesters were between 1.6 and 8 g/liter-day of volatile solids [18]. Detention times are 10 to 15 days. Higher loading rates lead to accumulation of acids and ultimate digester failure. In the experiments described herein, we fed up to 14 g/liter-day with a hydraulic detention time of 1.5 days. Acids were removed by ultrafiltration. Cultures were fed and ultrafiltered only 6 hr a day, so loading rates of up to 56 g/liter-day and detention times of 0.38 days were attainable with the present system. Higher rates of ultrafiltration are readily achieved by using larger membranes, higher pressures, and/or higher flux membranes. The sulfonic acid membranes used in this study had a porosity which cut off at about MW 1000. Higher flux membranes would cut off at about 10,000 and should be entirely adequate for this application, because so few solutes are in the range between 1000 and 10,000. The membranes that foul show a sharply reduced MW cutoff. The polysulfone membranes have a nominal cutoff of MW 20,000 at 21 μm/sec-atm but, after fouling, appear to cut off at less than MW 1000.

No cells are lost in the permeate, so the carbon loading rate can be varied independently of the hydraulic loading rate to attain the optimal volatile acid concentrations and methane production rates. The ultrafiltration process can readily function at several percent solids (of these kinds), so it should be possible to attain 20 g/liter (dry wt) or more of biomass in the recirculating fluid. The principal problem in ultrafiltration is the viscosity of the digester contents. Ultrafiltration can retain molecules with a molecular weight greater than 20,000, so extracellular enzymes would not be lost in the permeate. Insoluble materials such as cellulose are also retained. Slower-growing organisms are, of course, retained, so acetogenesis and methanogenesis rates should increase relative to acidogenesis. If anaerobic filters are used in conjunction with ultrafiltration, even higher concentrations of actively digesting biomass should be obtained.

An acetic acid and methane fermentation with a standing methanogenic biomass of 10 g/liter and a specific activity of 20 mmol CH_4/g-hr [2] would yield about 110 liter of CH_4 and 580 g of acetic acid per liter of reactor volume per day from 780 g of cellulose per liter per day. A compact system with this design could be readily transported and could be used to convert agricultural wastes to fuel and chemical feedstocks. Future studies in this program will be directed toward the degradation of cellulosic and algal biomass.

This study was supported by The Fuels from Biomass Systems Branch, Division of Solar Energy, Department of Energy under Contract Eg-77-S-02-4292; the authors express their appreciation of this support. They also acknowledge with thanks the assistance of Y. Chang and Y. Kuo who set up the electrodialytic and solvent extraction devices. We also thank Professor Henry R. Bungay III for his suggestions and advice, which were invaluable in the early stages of this study.

References

[1] M. P. Bryant, B. C. McBride, and R. S. Wolfe, *J. Bacteriol.*, *95*, 1118 (1968).

[2] D. C. Augenstein, J. Costa, and D. L. Wise, *Investigation of Converting the Product of Coal Gasification to Methane by the Action of Microorganisms* (NTIS, FE-2203-12, 1976).

[3] J. C. Zeikus, P. S. Weimer, D. R. Nelson, and L. Daniels, *Arch. Mikrobiol.*, *104*, 129 (1975).

[4] A. W. Lawrence and P. L. McCarty, *J. Water Pollut. Control Fed.*, *41*, R1 (1969).

[5] M. P. Bryant, in *Microbial Energy Conversion*, H. G. Schlegel and J. Barnea, Eds. (Pergamon, New York, 1977).

[6] C. A. Reddy, M. P. Bryant, and M. J. Wolin, *J. Bacteriol.*, *109*, 539 (1972).

[7] M. P. Bryant, E. A. Wolin, and R. S. Wolfe, *Arch. Mikrobiol.*, *59*, 20 (1967).

[8] M. Chen, and M. J. Wolin, *Appl. Environ. Microbiol.*, *34*, 756 (1977).

[9] M. J. Latham and M. J. Wolin, *Appl. Environ. Microbiol.*, *34*, 297 (1977).

[10] E. L. Iannotti, D. Kafkewitz, M. J. Wolin, and M. P. Bryant, *J. Bacteriol.*, *114*, 1231 (1973).

[11] R. S. Wolfe, *Adv. Microb. Physiol.*, *6*, 107 (1971).

[12] J. G. Zeikus, *Bacteriol. Rev.*, *41*, 514 (1977).

[13] R. K. Thauer, K. Jungermann, and K. Decker, *Bacteriol. Rev.*, *41*, 100 (1977).

[14] J. C. Senez, *J. Bacteriol.*, *26*, 95 (1962).

[15] O. H. Lowry, N. J. Roseborough, A. L. Farr, and R. J. Randall, *J. Biol. Chem.*, *193*, 265 (1951).

[16] C. C. Gryte and H. P. Gregor, *J. Polym. Sci., Polym. Phys. Ed.*, *14*, 1839 (1976).

[17] C. C. Gryte and H. P. Gregor, *J. Polym. Sci., Polym. Phys. Ed.*, *14*, 1855 (1976).

[18] E. J. Kirsch and R. M. Sykes, *Prog. Ind. Microbiol.*, *9*, 155 (1971).

Potential of Large-Scale Algal Culture for Biomass and Lipid Production in Arid Lands

Z. DUBINSKY,* T. BERNER,† and S. AARONSON
Biology Department, Queens College, City University of New York, Flushing, New York 11367

INTRODUCTION

Increasing awareness that the earth is limited in its ability to adequately support the runaway human population is evident in scientific [1] and political [2] quarters and on local [3] and international [4] levels. The major areas of concern are food-protein shortages [5]; fossil fuel price increases and anticipated depletion [6, 7]; and the pollution of waters [8], soils [9], and the atmosphere [8–12] by direct and indirect human activities. We shall attempt to examine the applicability of large-scale algal culture as a contribution to the alleviation of some of the above-mentioned problems.

Whittaker and Likens [13] classified 18×10^6 km^2 (12% of total land area) as desert and semidesert scrub, an additional 24×10^6 km^2 (16%) as extreme deserts, and 8.7×10^6 km^2 (6%) as dry, hot deserts. The primary productivity of these three areas, which ranges between 0 and 0.7 g/m^2/day, may be compared with a range of 0.3–1.1 g/m^2/day for the 14×10^6 km^2 (9.6%), which is the area of cultivated land. The deserts' barrenness appears even more striking against the 86 g/m^2/day theoretical maximal productivity attainable at 6.6% solar radiation utilization efficiency values given by Bassham [14] or even the measured 38 and 52 g/m^2/day for sugar cane and corn, respectively, during the peak of their growth season (Table I). In the vast dry, hot desert belt, mostly north and south of the equator, primary productivity is limited by the lack of suitable water, as the plentiful light and elevated temperatures could otherwise sustain near maximal photosynthesis.

The culture of marine, brackish-water, or sewage-tolerant microalgae in desert areas, where land is virtually valueless, could yield crops of the same

* Permanent address: Life Sciences Dept. Bar-Ilan University, Ramat-Gan, Israel.
† Permanent address: School of Education "Oranim" Dept. of Biology, Haifa University, Tivon, Israel.

Biotechnology and Bioengineering Symp. No. 8, 51–68 (1978)

0572-6565/78/0008-0051$01.00

TABLE I

Measured Yields and Maximal Photosynthetic Productivities of Some Higher Plants and Algae[a]

	Assumed Radiation Kcal cm^{-2} y^{-1}	Yield $g\ m^{-2}\ d^{-1}$	Yield $t\ ha^{-1}\ y^{-1}$	Efficiency (%) Total	Efficiency (%) Ph A.R.
Theoretical maximum					
High Solar Desert (annual)	200	86	313	6.6	15.3
U.S. Average (annual)	144	61	224	6.6	15.3
Higher Plants					
Sugar Cane (max.)	247	38		2.4	5.6
" " (average)	168	31	112	2.8	2.3
Corn (max.)	247	52		3.2	7.4
" (average)	168	4	13	0.4	0.9
Alfalfa	247	23		1.4	3.3
Sugar Beet (max.)	247	31	33	1.9	4.4
" " (average)	168	9		0.8	1.9
Algae (averages)					
<u>Scenedesmus</u> <u>acutus</u> (Dortmund)	86*	10(15)**	25	1.2	2.8
" " (Bangkok)	146*	15(15)	55	1.5	3.5
" <u>quadricauda</u> (Trebon)	86*	4.7(16)	17	0.8	1.8
" <u>acutus</u> (Rome)	140*	15(17)	55	1.6	3.7
Green Algae (Israel)	180*	25(18)	92	2.3	4.6
Diatoms (Woods Hole)	120*	7(19)	25	0.8	1.8
<u>Porphyra</u> <u>spp</u>. (Japan)	110*	0.23(20)	0.84	0.04	0.08
<u>Eucheuma</u> <u>striata</u> (Philippines)	147*	3.64(20)	13	0.35	0.8
<u>Gracilaria</u> <u>spp</u>. (Taiwan)	100*	4.17(21)	15	0.6	1.4
<u>Gracilaria</u> <u>foliifera</u> (U.S. Mass.)	120*	16.9(22)	62	2.1	4.9
<u>Hypnea</u> <u>musciformis</u> (U.S. Mass.)	120*	17.6(22)	65	2.2	5.1

[a] Mainly after Bussham [14].

* Radiation data from Ehricke [23].

** See references.

order of magnitude as the total present human agricultural product, even at efficiencies presently achieved in conventional agriculture. At present, algal sewage oxidation ponds of pilot plant and full-field scale yield as much as 25–45 g/m^2/day on a continuous basis in Israel (Table II). However, such intense algal growth cannot be sustained on unenriched ocean waters. The low concentration of essential nutrients, such as nitrogen and phosphorus, in the photic zone of the tropical seas causes their notorious transparency and, less welcome, associated virtual sterility. This problem could be overcome with the conventional, though undesirable, energy intensive, fertilizer-based type of approach [27–29]. Alternatively, temperature gradients across the ocean thermocline could be used to pump nutrient-rich, abyssal waters, an approach being considered at present for offshore mariculture [30]. A third approach would be an integrated system using environmentally objectionable animal and human wastes from cities and feedlots (to be built in the future) as the valuable nutrient mines to be tapped by the algae [31–35]. The most obvious macronutrients thus obtained would be nitrogen, phosphorus, and carbon. Large-scale production of flue gas (power plants, industry, etc.) could also be coupled to the algal ponds and serve as a source of CO_2 [36] and energy for temperature regulation and stirring, thus adding assets to algal culture and reducing the level of atmospheric pollution [37].

The advantage of marine and brackish-water microalgal cultures over conventional agriculture in such regions resides in their large surface to

TABLE II

Some Maximal Algal Yields in Large-Scale Mass Culture or in Nature

Organism	Location	Max Yield ($g\ m^{-2}d^{-1}$)	Ref.
Blue-green bacteria			
Spirulina platensis	Dortmund	17	15
Spirulina platensis	Bangkok	20	15
Spirulina platensis	Mexico	18	15
Spirulina platensis	Florence	36	17
Eukaryotic algae			
Chlorella pyrenoidosa	Trebon	19	24
Diatoms	Woods Hole (Mass.)	13	19
"	Ft. Pierce (Florida)	23	19
Green algae, wastewater	Israel	38	18
Nannochloris sp.	Trebon	13	24
Scenedesmus acutus	Dortmund	28	15
Scenedesmus acutus	Bangkok	35	15
Scenedesmus acutus	Rome	38	17
Scenedesmus obliquus	Rupite (Bulgaria)	45	15
Scenedesmus obtusiusculus	Florence	39	17
Scenedesmus quadricauda	Trebon	15	24
Stigeoclonium sp.	Trebon	24	24
Ulothrix acuminata	Trebon	23	24
Natural algal blooms			
Loch Leven	England	4	25
Lake George	Uganda	94	26

volume ratios, resulting in fast growth, their lack of nonphotosynthesizing organs or tissues, high nutritional value, and their ability to use seawater. An extra bonus of growing aquatic plants would be the evaporative cooling of the ponds which eliminates the need for air conditioning, which would be required in controlled-environment agriculture based on growing land plants in greenhouses. It remains to be seen, however, what the potential uses of algae so grown would be, as only the economic analysis of algal culture systems can determine whether they hold promise for the future.

USES OF ALGAE

Seaweeds have been used extensively in the Orient since antiquity and to some extent as animal feed elsewhere and as a source of alginates and various stabilizers and emulsifiers in modern food industries [38, 39]. Until now the use of microalgae has been rather rare, as in the indigenous collection of bloom algae (*Spirulina*) in Lake Chad and Lake Texcoco for human consumption [37, 40]. Nevertheless, since the late forties, scientific attention has been drawn towards the potential in mass cultivation of microalgae. These early efforts culminated in the concerted Carnegie Institution of Washington project [41].

In the course of these and subsequent studies it became evident that microalgae were rich in proteins (Table III) of good, essential, amino acid composition (Table IV) and that they were a good source of vitamins (Table V). The algae were grown on defined mineral media, and the nutritional and environmental requirements of a number of species were determined. However, from the beginning, it became apparent that the competitive advantages resulting from small size are inherently linked with difficult and costly harvesting methods. The cost of production was estimated at the time (1957) at about 50 cents/lb [95]. This price (considered high in 1957), coupled with traditional human nutritional habits, rendered the project economically unattractive in competition with conventional agriculture for any of the following: land, water, fertilizer, and the required energy and capital inputs [27–29]. Microalgae also share the nutritional shortcomings of other single-cell protein sources for humans: high content of nucleic acids with its potential for causing kidney and liver problems [55] and a high content of polyunsaturated fatty acids causing digestive tract problems, such as diarrhea [56].

Algal culture becomes increasingly profitable as the extent of integration of the algal "farms" with other, juxtaposed, complementary units increases [57]. Moreover, algal culture systems should be designed as adaptive entities deploying both the inherent physiological adaptability of the algae and the flexible engineering approaches to optimize product properties according to changing market trends and national policies. Algal culture facilities integrated with sewage treatment plants or feedlots would offset a major

TABLE III

Gross Organic Chemical Composition of Microorganisms, as Compared with Conventional Protein Sources

	% of Cell Dry Wt.				
	Protein	Carbohydrate	Total Lipid	Total Nucleic Acid	Ref. Source
Prokaryota					
Bacteria	36-80	2-36	1-39	4-34	
Blue-green bacteria	36-65	8-20	2-13	3-8	
Eukaryota					
Algae	9-62	3-58	1-76	3-6	
Fungi	3-55	36-39	1-70	5-13	
Protein sources*					
Egg	49	3	45		
Meat muscle	57	2	37	1	
Fish	55		38		
Milk	27	38	30		
Corn	10	85	4		
Wheat	14	84	2		
Soy flour	47	41	7		
Eukaryota, Algae					
Agmenellum quadruplicatum	36	32	13		43
Amphidinium carteri	28	31	18		43
Chaetoceros sp.	35	7	7		43
Chlamydomonas specifera	37	58	6		44
Chlorella pyrenoidosa	57	26	2		43
Chlorella specifera	50	10-25	20		44
Coscinodiscus sp.	17	4	2		43
Dictyopteris delicatula	9	25	0.8		44
Dunaliella salina	57	32	6		43
Exuviella sp.	31	37	15		43
Gonyaulox polyedra	28	31	5		43
Gracilaria domingensis	8	47	0.2		44
Hypnea musciformis	11	34	0.4		44
Macrocystis pyrifera	41	58	1		44
Monochrysis lutheri	49	31	12		43
Ochromonas danica	26-44	6-50	39-71		43
Phaeodactylum tricornutum	33	24	7		43
Protosiphon botryoides	47	10	37		43
Scenedesmus obliquus	48-52	24-27	7-13		43
Skeletonema costatum	37	21	5		43
Syracosphaera carterae	56	18	5		43
Tetraselmis maculata	52	15	3		43
Ulva fasciata	10	47	0.4		44
Ulva specifera	30	70	trace		44

* Reference 42.

part of the initial investment, as environmental protection requires sewage disposal and treatment, even if algae are not produced [37, 58].

If algae are to be grown on sewage alone, the sewage has to be either domestic or from feedlots, as the inclusion of industrial wastes (unless treated at the originating factory) might adversely affect the algae [59] or result in the accumulation of heavy metals [60] or organic chemicals [61, 62] that would render the algae unsafe for use as either food or feed. In algal culture systems operating on sewage, the effluent is a valuable product—water, unfit for drinking but suitable for cooling in refineries, power plants, and similar industrial installations. Such effluent is being used for the irrigation of crops not meant for human consumption, such as cot-

TABLE IV

Algal Content of Essential Amino Acids

Organism	FAO standard amino acids										Ref
	Cys	Ile	Leu	Lys	Met	Phe	Thr	Trp	Tyr	Val	
Prokaryota											
Blue-green bacteria and bacteria											
Anabaena cylindrica	±	+	+	+	+	+	+	+	+	+	44
Anabaena flos-aquae		+		+	+	+	+	+	+	+	44
Calothrix sp.	±	+	+	+	+	+	+	+	+	+	45
Hydrogenomonas sp.	+	+	+	+	±	+	+	±	+	+	17
Nostoc commune	±	+	+	+	+	+	+	+	+	+	45
Phormidium uncinatum	+	+	+	+	+	+	+	+	+	+	44
Pseudomonas sp.	-	+	+	+	+	+	+		+	+	17
Rhodopseudomonas gelatinosa		+	+	+	+	+	+			+	46
Rhodopseudomonas sp.	±	+	+	+	+	+	+	±	+	+	17
Spirulina maxima	+	+	+	+	+	+	+	+	+	+	47
Spirulina platensis	+	+	+	+	+	+	+	+	+	+	44
Tolypothrix tenuis	±	+	+	+	+	+	+	+	+	+	45
Eukaryota,											
Algae											
Chlorella ellipsoidea	+	+	+	+	+	+	+		+	+	44
Chlorella pyrenoidosa		+	+	+	+	+	+	-	+	+	44
Chlorella sorokiniana	+	+	+	+	+	+	+	+	+	+	44
Chlorella specifera		+		+	+	+	+	+	+	+	44
Chlorella vulgaris	±	+	+	+	±	+	+	-	+	+	44
Cylindrocystis brebissonii	+	+	+	±	+	+	+	±	+	+	48
Gigartina acicularis	±	±	+	±	±	±	±	±	±	±	49
Gonyaulax polyedra	+	+	+	+	+	+	+	+	+	+	50
Gracilaria compressa	±	±	±	±	±	±	±	±	±	±	49
Gracilaria confervoides	±	+	+	+	±	+	+	±	±	+	49
Hypnea musciformis	±	±	+	±	±	±	±	±	±	±	49
Maesotaenium kramstai	+	+	+	+	+	+	+	±	+	+	48
Ochromonas malhamensis		+	+	+		+	+		+	+	48
Pithophora sp.	+	+	+	+	+	+	+	±	+	+	48
Scenedesmus acutus	±	+	+	+	+	+	+	+	+	+	45
Scenedesmus obtusiusculus	+	+	+	+	+	+	+	+	+	+	45
Staurastrum cristatum	+	+	+	+	+	+	+	±	+	+	48

+ = Amino acid present in significant amount.
± = Amino acid present in trace: 1% of total amino acids.
– = Not present.

ton, or for drip irrigation of fruits like dates and citrus relatively isolated from the soil. The residual phosphorus and nitrogen [63, 64] in the effluent increases its agricultural value while saving on energy at present invested in denitrification of wastes on one hand and the production of ammonia and nitrate for fertilizer on the other. Where seawater would be used, it could be partly recirculated and enriched by addition of a suitable proportion of sewage to meet the algal nutritional requirements.

In principle, the algae, once harvested, could also be used as a source of biomass for fuel, a proposal that seems unattractive to us at the current market price of crude oil, even after raising the energy value of the biomass by fermenting it to ethanol or methane [57]. The extra step of fermentation, although increasing the energy content of the product and facilitating its transportation and distribution, would incur additional costs and dissipate some of the photosynthetically stored energy, thus further reducing the ultimate economic potential.

As animal feed, algae would have to be able to compete with such protein sources as soy meal which currently sells at 170 dollars/metric ton [58, 65] or with fish meal whose price fluctuates along with the Humboldt current between 200–400 dollars/ton. Nutritional studies have shown that fish [66,

TABLE V

Vitamin Content of Algae[a]

Organism	Vitamin (ng mg^{-1} cell dry wt.) A	D	E	Ascorbate C	Biotin	B6	B12	Folates	Nico-tinate	Panto-thenate	Ribo-flavin	Thiamine	Ref
Chlorophyceae													
Caulerpa racemosa				1000-3200									51
Chlamydomonas reinhardii	105	-	4000	2000	0.26	nd	nd	9	nd	nd	nd	nd	52
Chlorella ellipsoidea	5*				0.2	0.3-3	.04-.09	22-47	112-125	4-9	23-37	10-23	51
Chlorella vulgaris	nd		2000	2000	0.45	11	nd	11	56	79	68	nd	52
Enteromorpha linza	0.003*			100-2570	0.2		0.1	0.3	28		1	2	51
Monostroma nitidum		3		750-800	0.1		0.01	0.4	10	4	9	1	51
Scenedesmus obliquus	nd		1000	15000	nd	nd	nd	6	nd	nd	46	nd	52
Ulva pertusa				270-410	0.2		0.1	0.1	8	2	3	1	51
Chrysophyceae													
Ochromonas danica	137	-	2170	830		23	0	9	89	37	35	9	53
Dinophyceae													
Peridinium cinctum fa. westii	-	-	-	-	0.2-0.3	1-3	0.2-0.2	0.4-0.7	9-18	7	26-61	2-9	54
Phaeophyceae													
Colpomenia sinuosa					0.1		0.08	0.05	5	3	5	0.3	51
Dictyopteris prolifera				10000	0.2		0.02	0.2	18	4	4	0.5	51
Dictyota dichotoma					0.2		0.01	0.5	15	0.7	6	0.8	51
Ecklonia cava									19	0.5	3	1	51
Hizikia fusiforme	0.005*	16		0-920	0.2		0.006	0.2	7	2	3	0.3	51
Hydroclathrus clathratus					0.2		0.07	0.9	4	3	3	0.3	51
Laminaria sp.	0.004*			30-910		0.3	0.003		30		0.2	0.9	51
Myelophycus caespitosus					0.2		0.01	0.1	14	3	3	0.3	51
Padina arborescens					0.2		0.004	0.5	9	2	1	0.3	51
Sargassum fulvellum					0.1		0.03	0.6	9	2	5	0.3	51
Sargassum nigrifolium					0.2		0.02	0.3	17	0.5	6	0.4	51
Sargassum thunbergii					0.3		0.05	0.3	5	9	5	0.4	51
Spathoglossum pacificum					0.2		0.01	0.6	25	0.3	0.8	0.4	51
Rhodophyceae													
Chondrococcos japanicas					0.04		0.2	0.1	8	1	11	1	51
Chondrus ocellatus				160	0.07		0.09		30	7	15	2	51
Gelidium amansii					0.1		0.04	0.8	20	1	18	2	51
Gloiopeltis tenax					0.04		0.02	0.7	24	6	15	3	51
Gracilaria gigas	0.008*				0.02		0.2	0.3	8	2	1	2	51
Gracilaria textorii					0.2		0.08	0.7	44	10	7	5	51
Grateloupia ramosissima					0.08		0.03	0.7	25	2	6	1	51
Hypnea charoides					0.09		0.03	0.5	22	7	3	1	51
Laurencia okamurai				40	0.1		0.1	0.8	39	9	10	0.5	51
Lomentaria catenata					0.09		0.03	0.2	24	12	3	1	51
Porphyra tenera	0.5*			100-8310	0.3		0.3	0.09	68	10	23	2	51

[a] Key: * International units, I.U./mg; nd not detected; – not found.

67], poultry [37, 66–68], and swine [37, 68, 69] can be raised on diets where part of the protein in the feed is replaced by algal meal, with no adverse affects on growth or nutritional characteristics of the meal.

ALGAL LIPIDS

The main effort of our research was to examine the potential use of microalgae as a source of lipids. Our work has confirmed many previous observations made of the high lipid content of some algal species, as well as the ability of others to divert their metabolism toward lipid synthesis. We have so far examined the lipid content of over 24 algal species isolated from as diverse environments as fresh water, seawater, hypersaline brines, desert soils, and sewage oxidation ponds. The results of these analyses are summarized in Table VI.

These algae were grown by us in batch or continuous 10-liter cultures, where cell concentrations ranged from 100 to 1500 mg/liter. We have also

TABLE VI

Lipid Content of Algae and Blue-Green Bacteria

Organism	Total Lipid % of dry Weight	% of total lipid			Hydrocarbon % dry weight
		Neutral lipid (NL)	Glyco-lipid (GL)	Phospho-lipid (PL)	
Prokaryota					
Desert Blue-green bacterium #4F	8.9	68	12	16	0.13
Desert Blue-green bacterium #92	6.5	11	41	50	
Eukaryota					
Bacillariophyceae					
Cylindrotheca closterium	26.7	42	13	45	
Cylindrotheca fusi formis	28.1	54	14	31	
Fragilaria construens	16.0	52	21	26	
Phaeodactylum tricornutum	31	14	39	47	
Pennate marine diatom #9	28.8	60	20	21	0.62
Pennate marine diatom #PB48	35.2	46	44	10	
Chlorophyceae					
Botryococcus braunii	52.9	89	4	7	19
Clorococcum oleofaciens	21.6	62	34	4	
Chlorosarcinopsis negevensis	32.2	66	13	20	
Dunaliella salina[1]	47.2	21	62	17	
Dunaliella primolecta	53.8	43	38	20	
Radiosphaera negevensis	43.0	53	20	26	
Scenedesmus sp.[3]	24.7	35	24	41	
Stichococcus bacillaris	31.9	50	7	43	
Desert #103	33.7	41	6	53	
Xanthophyceae					
Botrydium granulatum	15.3	44	17	39	2.8
Algae from sewage[2]					
Chlorella-Euglena	22.6	48	11	41	
Euglena	11.0	81	7	11	
Micractinium	17.4	35	43	21	
Oocystis	19.9	68	15	17	0.13
Scenedesmus	22.2	47	42	11	

[1] Pilot-plant scale, unialgal culture, Israel.

[2] High rate sewage oxidation pond with the species shown predominant on harvesting, Israel.

[3] Pilot-plant scale, unialgal culture, Dortmund, West Germany.

TABLE VII

Examples of Environmental Manipulation of the Lipid Content of Algae and Blue-Green Bacteria

Environmental Variable	Organism	Variable and Lipid Ranges (%)				Reference
Light intensity	*Spirulina platensis*	10 klux 4.2	-	40 klux 6.2		70
	Nitzschia closterium	200 f.c. 20.8	-	2000 f.c. 22.8		71
Temperature	*Ochromonas danica*	15°C 39	20°C 44	25°C 49	30°C 53	72
	Cyanidium caldarium		20°C 4.5	45°C 6.0		73
Nitrogen depletion	*Chlorella pyrenoidosa*	With N 10		Without N 70		69
	Monodus subterraneus	22		40		69
	Tribonema sp.	10		22		69
Salinity (NaCl%)	*Botryococcus braunii*	0 36	5 49	6 51		*
Senescence	*Chlorella vulgaris*	Young 22		Old 28		74
	Scotiella sp.	27.5		32		74
	Euglena gracilis	14.5		19.0		74
	Tribonema aequale	11.5		16		74
	Navicula pelliculosa	24.0		29.0		74
	Phaeodactylum tricornutum	19		32		*
	Desert Blue-Green #92	9		26		*
Combination of factors	*Stichococcus bacillaris*	11.9		38.9		75
	Chlorella pyrenoidosa	4.5		85.6		75

* Results from this laboratory.

analyzed samples grown in a pilot plant algal culture facility of GSF in Dortmund, Germany, and in Haifa, Israel, as well as from experimental, pilot plant, and full-scale sewage oxidation ponds in Israel.

Table VII shows changes in the lipid content of some algal species when grown under various environmental conditions or at different points along their growth curve. Such shifts have been shown to occur in nature and in laboratory experiments in response to various, seemingly unrelated environmental factors such as nutrient limitation (in particular nitrogen starvation), irradiance levels, temperature, osmotic pressure, as well as senescence. The mechanism relating the external triggering with the biochemical changes in the cells have for the most cases not yet been clearly elucidated.

We have also further analyzed the algal lipids into three major classes: neutral lipids, glycolipids, and phospholipids (Table VI) showing that most algae are rather high (30–60% of total lipids) in neutral lipids. The total, free and esterified, fatty acid content is likely to be above 50%, another fact worth remembering, as fatty acids have a distinct market value [76].

In some instances certain algae produce more than 50% of their dry biomass as specialized lipids. *Dunaliella* spp. regulate their osmotic potential by producing up to 80% of their dry weight as glycerol [77–79], a finding that is at present being applied on a pilot-plant scale under an Israeli patent.

Botryococcus braunii is associated with fossil fuel deposits of the type known as boghead coal [80], and under late-bloom conditions it was found to contain up to 86% hydrocarbons [81]. We have been able to find media and conditions suitable for growing this organism and also found it to be unusually high (for an alga) in lipids in general (53%) and in hydrocarbons in particular (19% of the total lipids, 10% of the dry biomass), as shown on Table VI. Even at the lipid levels already obtained by us, *Botryococcus braunii* compares very favorably with other plant species currently being considered as hydrocarbon sources [82]. *Botryococcus braunii*, originally described mainly from fresh waters, has been cultured by us on media with osmotic potentials, up to those of seawater, with no adverse effects.

ECONOMICS OF ALGAL MASS PRODUCTION

From the above data some economic projections can be made. A corn grower in the United States can at present expect to gross 500 dollars/ha (ha = hectare)/year (if the crop averages 250 bushels or 6.25 ton/ha/year [57]. This figure is useful for comparison, although many farmers around the globe have to be content with a small fraction of the above mentioned income. Even at 25 g/m^2/day average algal production annual yield would be 91.25 tons/ha/year.

Estimating the algae to be roughly 75% carbohydrate and protein at caloric values of 15 GJ (GJ)/metric ton and 25% lipids (see Tables III and VI) at 40 GJ/ton [57], the total energy yield per hectare per year would be

$$0.75 \times 91.25 = 68.4 \text{ ton/ha/year}$$
$$68.4 \times 15 = 1026.6 \text{ GJ/ha/year}$$
$$0.25 \times 91.25 = 22.8 \text{ ton/ha/year}$$
$$22.8 \times 40 = 912.5 \text{ GJ/ha/year}$$

Summing these values yields 1939 GJ/ha/year.

The biomass-cost target at present is set at 0.95–1.40 dollars/GJ [85], which would bring 1842–2714.7 dollars/ha/year well above either the expected price for corn or the projected 400–560 dollars/ha/year revenue from lumber-based energy farms [57]. The quite striking difference of over threefold gross revenue per unit area per year (see Table VIII) is mainly due to the difference in projected yields between the 25 tons/ha/year in lumber farms [57] and 6.25 tons/ha/year corn [57] as against over 90 tons of dry algal biomass, and to a lesser degree due to the higher oil content of the algae.

TABLE VIII

Estimated Gross Income from Some Proposed "Biomass for Energy" Systems

System	Production: Dry Biomass $g\ m^{-2}\ d^{-1}$	$t\ ha^{-1}\ y^{-1}$	$lb\ acre^{-1}\ y^{-1}$	Energy $GJ\ t^{-1}$	$GJ\ ha^{-1}\ y^{-1}$	$Mbtu\ acre^{-1}\ y^{-1}$	Gross Income $\$\ ha^{-1}\ y^{-1}$ at (83) \$0.95 GJ	\$1.40GJ
"Hydrocarbon plants" containing 10% oil (79) $0.1 \times 40\ GJ\ t^{-1} + 0.9 \times 15\ GJ\ t^{-1}$ (57)	2.7	10	8.922	17.5	175	67.1	166 230*	233 296.5
Energy Farm for wood (57)	6.9	25	22.305	17	425	162.9	403	595
Lumber, average for U.S.A., at 57% softwood (84, 85)	0.42	1.5	1,379	(86) 19.2	28.8	11.0	27.4	40.3
Algal mass culture	10	36.5	32,565	21.25	776	297	737 1,316	1,086 1,500
containing 25% "oil"	25	91	81,190	"	1,934	741	1,837 3,282	2,708 3,742
$0.25 \times 40GJ\ t^{-1}$ $0.75 \times 15\ GJ\ t^{-1}$	50	183	163,272	"	3,889	1,490	3,695 6,600	5,445 7,526

* Underlined values calculated assuming oil is extracted and sold separately at current crude price, 14.5 dollars/barrel, 7 barrels/ton.

Nielsen et al. [82] estimate that at 10% "acetone and benzene extractables," high latex and hydrocarbon desert shrubs yielding 24 tons/ha/year would produce oil at near current world crude oil prices (14.5 dollars/barrel) [87] even if the extracted biomass is discarded as useless.

If the algae would be used as a substitute for soy meal, selling currently for 170 dollars/ton [65], the economic forecast would look even better, allowing a gross income of 15,513 dollars/ha/year or better—if the algae are first extracted for their oil. This oil alone, if *not* used as fuel, could amount to 22.8 tons/ha/year, when soy oil prices are at present over 20 cents/lb or 440 dollars/ton [65], which would equal $22.8 \times 440 = 10{,}053$ dollars/ha/year, still leaving the high-protein residue (Table IX). The vast world fatty-acid market for industry (Table X) [76] is willing to pay somewhat less than the price realized by oils meant for human consumption. Coconut oil sells at 16 cents/lb (although it has occasionally gone as high as 60 cents/lb) [76]. The extraction of the oil will obviously incur some extra cost although technologies are currently available for such processes.

Large-scale, worldwide algal culture based on nutrient imput from human and farm wastewater seems feasible, when algal growth potential (AGP) of sewage and its global production [89, 90] are considered. It has to be remembered, however, that where most sewage is at present produced, either land would not be available for extensive algal culture or its price would be prohibitive. Moreover, in many such areas radiation levels and temperatures are too low. In arid regions algal culture systems should be built in conjunction with the establishment of new population and husbandry concentrations or be linked with existing urban centers like those in

TABLE IX

Estimated Income from Algal Culture[a]

Proposed Use of Algae	Gross Income from algae alone			Net Income, deducting $80 per ton for harvesting and drying (88)			Net Income, adding income from sewage treatment and effluent sale**			Net Income, after deducting $43,248 production costs ha^{-1} y^{-1} (58)			Net Income, adding income from sewage treatment $13,690 and effluent sale $47,906 (58)		
	I*	II*	III*	I	II	III	I	II	III	I	II	III	I	II	III
For combustion at $1.175 GJ^{-1}	912	2,272	4,570	-2,008	-5,008	-10,070	182 3,102	452 7,732	910 15,550	-42,336	-40,976	-38,678	19,260	20,620	22,918
If oil sold separately at $14.5 barrel (87)	1,610	4,013	8,071	-1,309	-3,267	- 6,569	880 3,800	2,193 9,473	4,111 19,051	-41,638	-39,235	-35,177	19,958	22,361	26,419
For animal feed, to replace soy meal at $170 per ton (65)	6,205	15,470	31,110	3,285	8,190	16,470	5,475 8,395	13,650 20,930	27,450 42,090	-37,043	-27,778	-12,138	24,553	33,818	49,458
If oil sold separately at 20¢ per lb., while residue sold as feed at $170 per ton	8,669	21,613	43,463	5,749	14,333	28,823	7,939 10,859	19,793 27,073	39,803 54,443	-34,579	-21,635	215	27,017	39,961	61,811

[a] Income = dollars/ha/year.

* I, daily production of 10 g/m^2; II, 25 g/m^2; III, 50 g/m^2.

** Assuming 10^6 gal sewage treated per 1 ton of algae produced. Upper figure of each two based on 50 dollars/10^6 gal sewage treated and 10 dollars from the sale of the effluent. Lower, underlined figures based on 125 dollars/10^6 gal sewage treated and 15 dollars for 10^6 gal of effluent [88].

TABLE X

Typical Disposition of 1973 Production of Fatty Acids[a]

Industry	Percentage	In 10^6 of lbs
Surfactants - soaps	33	392
Fatty nitrogens	18	214
Rubber Industry	10	119
Surface coatings	10	119
Grease - heavy metal soaps	5	60
Textile Industry	5	60
Plasticizers	4	48
Food additives	1	12
Cosmetic	1	12
Pharmaceuticals, export and other	13	154
	100%	1,190

[a] End-use consumption of fatty acids in the United States. From Johanson [76].

the southwestern United States, Cairo, or Teheran, in tightly coupled material and energy-conserving networks [33]. In the vicinity of population centers and feedlots, the extra bonus of sewage treatment will offset much of the considerable initial investment in the establishment of the algal ponds and ancillary equipment. This investment has recently been estimated at 80,000–100,000 dollars/ha [58] including capital investment in pond construction, as well as equipment for algal separation and drying.

If we regard algal culture facilities as sewage treatment plants, the economics of the capital investment are affected most favorably (Table IX). Oswald and Golueke [88] suggest that each ton of algae grown is equivalent to the treatment of 10^6 gal of sewage at 50–125 dollars, depending on the method of treatment, while in Israel current (1977) costs to treat sewage to a degree making it acceptable for marine discharge are estimated at 142 dollars for 10^6 gal [58], a figure which was used for the preparation of Table IX.

Whenever such systems would be operating without the admixture of seawater, the effluent could become a valuable product partially replacing other water sources selling in the United States at 10–15 dollars/10^6 gal (1967) [88]. In arid lands the price of water is considerably more. For example, the current price of water for agriculture in Israel is 240 [91]–500 [58] dollars/10^6 gal. When mixed with seawater, the effluent could either be discharged with minimal adverse environmental effects or be used for salt tolerating crops, such as dates or other plants currently being explored [92].

Taking into account the capital investment (excluding land cost) and processing costs totaling 43,248 dollars/ha/year [58] the estimated income from algal production will not come close to showing a worthwhile profit (Table IX). This fact remains even at optimistic (50 g/m^2/day) annual average yields and maximal prices for the product. If coupled to sewage treatment, which must be done in any case, at great expense, algal culture is still not expected to show great profit, but this procedure could certainly reduce the costs incurred. Only when the potential value of the effluent is added to the income from the algal product and the sewage treatment can algal mass culture systems show a comfortable profit margin (Table IX). Increase in scale [32], harnessing of wind energy and industrial flue gases to pond circulation and reduction in pond construction costs may further improve the economics of the proposed system.

MAJOR PROBLEM AREAS

Many problems remain to be solved before the potential of algal mass culture can begin to be realized. The screening of the vast majority of algal species for growth rate, protein quantity and quality, oil production and temperature, and salinity resistance has yet to be done. The best algae (some of which we have possibly presented here) for upscaling, singled out by such screening, should then be studied in detail. The purposes of such a study would be as follows:

a) The determination of the environmental conditions required for sustained dominance of the selected algae over "wild" species under nonsterile field conditions [93].

b) Improved understanding of the biochemistry and physiology of lipid and hydrocarbon production in algae, in order to facilitate optimization of lipid or protein production at different points along the algal-culture pond system, or at various points in time, in response to economic realities.

In addition, studies in the following areas should also be conducted:

c) A joint biological-engineering effort to find the best sewage-effluent and sewage–seawater mixture ratios, the right pond depths at different seasons, optimal algal densities for maximal production (or for cheapest separation!), horizontal flow rates required for good nutrient supply, and to prevent sinking and means to avoid objectionable anaerobic conditions at night.

d) Finding the most suitable method for harvesting the algae. The separation of the algae remains at present the most energy-intensive step in the entire process. Research in progress in Germany, in Israel, and in the United States is aimed at choosing the best method. Alum and polyelectrolyte flocculation, electroflotation, centrifugation as well as microstraining and other filtration methods are among those being considered [93, 94].

e) The processing of the algae once separated. This problem, however, does not seem to be of comparable magnitude to those mentioned above. The slurry obtained after separation would have to be treated according to the designated use. It could be sun-dried, fermented as obtained, steam-dried, or spray-dried and later pelletized before or after lipid extraction. These kind of technologies for the processing are commonplace in food technology.

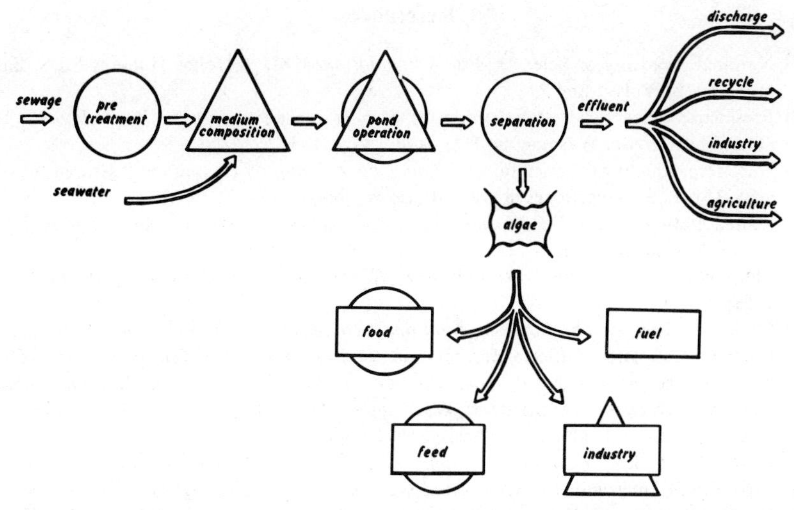

Fig. 1. (△) Major scientific, (○) technological, and (▭) economic problem areas in the exploitation of algal mass culture.

Figure 1 points out the major scientific and technological bottlenecks remaining to be overcome before large-scale algal culture becomes an economical reality.

CONCLUSIONS

i) Algal mass culture may be economically competitive with other land uses in hot arid lands.

ii) Algae can be grown on human and animal wastes as nutrient sources, or on mixtures of such wastes with seawater.

iii) The integration of algal mass cultures with municipal and agricultural sewage treatment installations and with industrial plants favorably affects the economics, energetics, and environmental impact of such systems.

iv) Algal culture systems can be designed to flexibly adapt to maximize biomass, lipid, or effluent production, in response to changing needs.

v) Present market values favor the use of algae for animal feed supplement, after being first extracted for their lipids.

vi) In arid lands, the effluent obtained from sewage, after algal separation, may prove as valuable as the algae themselves.

This work was supported in part by the Petroleum Research Fund, administered by the American Chemical Society No. 9085 AC1, and by the National Institutes of Health, Institutional grant 5-S05-RR-07064-10, to Queens College of the City University of New York. We also thank Professor G. Shelef and Dr. R. Moraine of the Sherman Environmental Engineering Research Center, Technion, Haifa, Israel for algal samples and for a preprint of their paper in press, Dr. E. Stengel of Gesellschaft fur Strahlen and Umweltforschung, Dortmund, Federal Republic of Germany, and Dr. A. Ben Amotz, Israel Oceanographic and Limnological Research Ltd., Haifa, Israel for algal samples.

References

[1] National Academy of Sciences, *Rapid Populations Growth* (Johns Hopkins Press, Baltimore, 1971), Vols. 1 and 2.

[2] President's Science Advisory Committee, *Report of the Panel on the World Food Supply* (The White House, Washington, D.C., 1967) Vols. 1–3.

[3] U.S. Department of Agriculture, *National Food Situation* (Economic Research Service, NSF-151, U.S. Department of Agriculture, Washington, D.C., 1975).

[4] United Nations World Food Conference, *Assessment of the World Food Situation* (FAO, Rome, November, 1974).

[5] Third World Food Survey, *Freedom from Hunger Campaign Basic Study* (FAO, Rome, 1963), Vol. 11.

[6] M. K. Hubbert, in *The Environmental and Ecological Forum*, 1970–1971, U.S. Atomic Energy Commission, Office of Information Services, Oak Ridge, Tenn., 1972, pp. 1–50.

[7] National Research Council, *Guayule: An Alternative Source of Natural Rubber*, National Academy of Sciences, Washington, D.C., March 1977. (Available from National Technical Information Service Springfield, Va.)

[8] I. H. Suffet, Ed., *Fate of Pollutants in the Air and Water Environments, Part 2, Advances in Environmental Science and Technology* (Wiley, New York, 1977), Vol. 8.

[9] R. Hague and V. H. Freed, *Environmental Dynamics of Pesticides* (Plenum, New York, 1975).

[10] M. Stuiver, *Science*, *199*, 53 (1978).

[11] M. Siegenthaler and H. Oescheger, *Science*, *199*, 388 (1978).

[12] I. J. Higgins and R. G. Burey, *The Chemistry and Microbiology of Pollution* (Academic, New York, 1975).

[13] R. H. Whittaker and G. E. Likens, in *Primary Productivity of the Biosphere*, H. Lieth and R. H. Whittaker, Eds. (Springer Verlag, New York, 1975).

[14] J. A. Bassham, *Science*, *197*, 630 (1977).

[15] C. J. Soeder, *Naturwissenchaften*, *63*, 131 (1976).

[16] T. Simmer, in *Studies in Phycology*, B. Fott, Ed. (E. Schweizerbartische Verlagsbuchhandlung, Stuttgart, 1969).

[17] G. Florenzano, R. Materassi, and C. Paoletti, *Acqua Aria*, *8*, 640 (1975).

[18] G. Shelef, R. Moraine, T. Berner, A. Levi, and G. Oron, in *Photosynthesis 77: Proceedings of the Fourth International Congress on Photosynthesis*, D. O. Hall, J. Combs, and T. W. Goodman, Eds. (The Biochemical Society, London, 1978).

[19] J. C. Goldman, J. H. Ryther, and L. D. Williams, *Nature*, *254*, 594 (1975).

[20] H. S. Parker, *Aquaculture*, *3*(4), 425 (1974).

[21] Y. C. Shang, *Aquaculture*, *8*(1), 1 (1976).

[22] B. E. Lapointe, L. D. Williams, J. C. Goldman, and J. H. Ryther, *Aquaculture*, *8*(1), 9 (1976).

[23] A. Ehricke, in *McGraw-Hill Encyclopedia of Energy*, Daniel N. Lapedas, Ed. (McGraw-Hill, New York, 1976), pp. 626–633.

[24] F. Hindak, *Algological Studies* (*Trebon*), *1*, 77 (1970).

[25] E. D. LeCren, *Philos. Trans. R. Soc. London*, *274*, 359 (1976).

[26] P. H. Greenwood, *Philos. Trans. R. Soc. London*, *274B*, 375 (1976).

[27] M. J. Perelman, *Environment*, *14*(8), 8 (1972).

[28] D. Pimentel, L. E. Hurd, A. C. Bellotti, M. J. Forster, I. N. Oka, O. D. Sholes, and R. J. Whitman, *Science*, *182*, 443 (1973).

[29] D. Pimentel, W. Dritschilo, J. Krummel, and J. Kutzman, *Science*, *190*, 754 (1975).

[30] K. M. Rodde, J. B. Sunderlin, and O. A. Roels, *Aquaculture*, *9*, 203 (1976).

[31] W. Schmitt, *Ann. N.Y. Acad. Sci.*, *118*, 645 (1965).

[32] J. Oswald and C. G. Golueke, in *Single Cell Protein*, R. I. Mateles and S. R. Tannenbaum, Eds. (MIT. Press, Cambridge, Mass., 1968).

[33] H. T. Odum, *Environment, Power and Society* (Wiley-Interscience, New York, 1970), 331 p.
[34] J. H. Ryther, W. M. Dunstan, K. R. Tenore, and J. E. Hugnenin, *BioScience*, *22*, 144 (1972).
[35] G. Shelef, *Combined Systems for Algal Wastewater Treatment and Reclamation and Protein Production* (Technion-Israel Institute of Technology, Haifa, 1975) (annual report).
[36] E. Stengel, *Ber. Dtsch. Bot. Ges.*, *83*, 589 (1970).
[37] G. Clement, in *Single Cell Protein II*, S. R. Tannenbaum and D.I.C. Wang, Eds. (MIT Press, Cambridge, Mass., 1975).
[38] V. J. Chapman, *Seaweeds and Their Uses* (Methuen, London, 1970).
[39] S. Bonotto, in *Marine Ecology*, O. Kinne, Ed. (Wiley, New York, 1976), Vol. III, Part 1, pp. 467–501.
[40] M. Schwimmer and D. Schwimmer, *The Role of Algae and Plankton in Medicine* (Grune and Stratton, New York, 1955).
[41] J. S. Burlew, Ed., *Algal Culture from Laboratory to Pilot Plant* (Carnegie Institution of Washington, Washington, D.C., 1953).
[42] S. A. Miller, in *Single Cell Protein*, R. Mateles and S. R. Tannenbaum, Eds. (MIT. Press, Cambridge, Mass., 1968), Vol. I.
[43] S. Aaronson, *Arch. Hydrobiol.* (Suppl.), *41*, 108 (1972).
[44] L. C. Lokar, *Ind. Aliment.*, *13*, 64 (1974).
[45] C. Paoletti, G. Florenzano, R. Materassi, and G. Caldini, *Sci. Tecnol. Alimenti*, *3*, 170 (1973).
[46] R. H. Shipman, I. C. Kao, and L. T. Fan, *Biotechnol. Bioeng.*, *17*, 1561 (1975).
[47] G. Clement, M. Rebeller, and P. Trambouze, *Rev. Inst. Francais Petrole*, *23*, 702 (1968).
[48] H. E. Schlichting, *J. Econ. Bot.*, *25*, 317 (1970).
[49] G. Bruni and B. Stancher, *Rass. Chim.*, *26* (3), 751 (1974).
[50] S. Patton, P. T. Chandler, E. B. Kalon, A. R. Loeblich, G. Fuller, and A. A. Benson, *Science*, *158*, 789 (1967).
[51] A. Kanazawa. *Bull. Jpn. Soc. Sci. Fisher.*, *29*, 713 (1963).
[52] S. Aaronson, S. W. Dhawale, N. J. Patni, B. DeAngelis, O. Frank, and H. Baker, *Arch. Microbiol.*, *112*, 57 (1977).
[53] S. Aaronson, B. DeAngelis, O. Frank, and H. Baker, *J. Phycol.*, *7*, 215 (1971).
[54] S. Aaronson, T. Berman, and D. Wynne (Unpublished).
[55] A. J. Sinskey and S. R. Tannenbaum, in *Single Cell Protein II*, S. R. Tannenbaum and D. I. C. Wang, Eds. (MIT Press, Cambridge, Mass., 1975).
[56] N. S. Scrimshaw, in *Single Cell Protein II*, S. R. Tannenbaum and D. I. C. Wang, Eds. (MIT Press, Cambridge, Mass., 1975).
[57] E. S. Lipinsky, *Science*, *199*, 644 (1978).
[58] R. Moraine, G. Shelef, A. Meydan, and A. Levin, "Algal SCP from wastewater treatment and renovation process," *Biotechnol. Bioeng.* (in press).
[59] E. J. Luard, *Phycologia*, *12*, 29 (1973).
[60] V. Gerhards and H. Weller, *Z. Pflanzenphysiol.*, *82*, 292 (1977).
[61] A. Södergren, *Vatten*, *2*, 90 (1973).
[62] M. J. Suess, *Sci. Total Environ.*, *6*, 239 (1976).
[63] Z. Dubinsky and T. Berner, *Combined Systems for Algal Wastewater Treatment and Reclamation and Protein Production* (Bar Ilan University, Ramat-Gan, Israel, 1975).
[64] G. Shelef, *Combined Systems for Algal Wastewater Treatment and Reclamation and Protein Production* (Technion-Israel Institute of Technology, Haifa, 1976) (annual report).
[65] *The New York Times*, Friday, March 10, 1978).
[66] Z. Berk, S. Mokady, B. Hepher, and E. Sanbank, *Proc. Biochem.* (in press).

[67] T. G. Bahr, D. L. King, H. E. Johnson, and C. L. Kerns, *Dev. Industr. Microbiol.*, *18*, 121 (1977).
[68] O. P. Walz, F. Koch, and H. Bruen, *Z. Tierphysiol. Tierernahr. Futtermittelkde.*, *35*, 55 (1975).
[69] G. E. Fogg and D. M. Collyer, in *Algal Culture from Laboratory to Pilot Plant*, J. S. Burlew, Ed. (Carnegie Inst., Washington, D.C., 1953).
[70] O. N. Albitskaya, G. N. Zaitseva, M. V. Pakhomova, O. I. Goronkova, G. S. Silakova, and T. M. Ermokhina, *Mikrobiologiya*, *43*, 649 (1974).
[71] D. M. Orcutt and G. W. Patterson, *Lipids*, *9*, 1000 (1974).
[72] S. Aaronson, *J. Phycology*, *9*, 111 (1973).
[73] B. L. Adams, V. McMahon, and J. Seckbach, *Biochem. Biophys. Res. Commun.*, *42*, 359 (1971).
[74] D. M. Collyer and G. E. Fogg, *J. Exp. Bot.*, *6*, 256 (1955).
[75] N. W. Milner, in *Algal Culture from Laboratory to Pilot Plant*, J. S. Burlew, Ed. (Carnegie Inst., Washington, D.C., 1953).
[76] A. G. Johanson, *J. Am. Oil Chem. Soc.*, *54*, 848A (1977).
[77] L. J. Borointzka and A. D. Brown, *Arch. Microbiol.*, *96*, 37 (1974).
[78] A. Ben Amotz and M. Avron, *FEBS Lett.*, *29*, 153 (1973).
[79] A. Ben Amotz and M. Avron, *Plant Physiol.*, *51*, 875 (1973).
[80] K. B. Blackburn, *Trans. R. Soc. Edinburgh*, *58*, 841 (1936).
[81] A. C. Brown and B. A. Knights, *Phytochemistry*, *8*, 543 (1969).
[82] P. E. Nielsen, H. Nishimura, J. W. Otvos, and M. Calvin, *Science*, *198*, 942 (1977).
[83] R. F. Ward, in *Proceedings of Fuels from Biomass Conference at the University of Illinois*, J. T. Pfeffer, Ed. (Univ. of Illinois Press, Champaigne-Urbana, 1977).
[84] U.S. Forest Service Forestry Resources Report 1974.
[85] C. C. Burwell, *Science*, *199*, 1041 (1978).
[86] O. D. Lorenzi, Ed., *Combustion Engineering* (Riverside Press, Cambridge, Mass., 1952).
[87] *Crude Oil Price*, Bulletin No. 105, Amoco, Chicago, 1978.
[88] W. J. Oswald and C. G. Golueke, in *Algae, Man and the Environment*, D. F. Jackson, Ed. (Syracuse U. P., Syracuse, N.Y., 1968).
[89] W. J. Oswald, in *Prediction and Measurement of Photosynthetic Productivity*, IBP/PP Technical Meeting, Trebon, 1969, Wageningen, 1970.
[90] C. J. Soeder, *Wasser/Abwasser*, *113*, 583 (1972).
[91] Agricultural Attache, Israel Embassy, Washington, D.C., personal communication, 1978.
[92] E. Epstein and J. D. Norlyn, *Science*, *197*, 249 (1977).
[93] W. J. Oswald and J. R. Benemann, in *Biological Solar Energy Conversion*, A. Mitsui, S. Miyachi, A. San Pietro, and S. Tamura, Eds. (Academic, New York, 1977).
[94] J. R. Benemann, J. C. Weissman, B. L. Koopman, and W. J. Oswald, *Nature*, *268*, 19 (1977).
[95] D. R. Thacker and H. Babcock, *J. Solar Energy Sci. Eng.*, *1*, 37 (1957).

Alternatives for Energy Savings at Plant Level for the Production of Alcohol for Use as Automotive Fuel

RODOLFO ESPINOSA, VINICIO COJULUN, and FIDIAS MARROQUIN
Destiladora de Alcoholes y Rones, S.A., Santa Lucia Cotz., Guatemala

INTRODUCTION

Over the past ten years, the growing concern about the energy crisis, which was originated by the fossil oil shortage, and the calculated speculation with the existing reserves have directed investigative efforts toward a proper and efficient utilization of other nonconventional sources of energy.

Some countries have developed extensive projects for the massive production of alcohol from carbohydrates to be utilized as a partial substitute for gasoline. The classical example is Brazil, where the vast land extension for the sugar cane plantations represent an invaluable source of renewable energy.

Central American countries like Guatemala base their economy on agriculture, and the sugar production has more than doubled within the past four years [1]; as a consequence the sugar mills' facilities have been expanded considerably, but the amount of by-products has also increased in proportion. These facts and the already available technology for the use of alcohol in automotive combustion represent a good opportunity for developing countries to modify advantageously their economies.

GENERAL CHANGE OF SCOPE

The market trend in the sugar cane industry will set the pace for the introduction of the various technological advances in alcohol production, but a new image seems to be developing against the traditional, outdated, and inefficient production systems of the past.

A comparison of Figures 1 and 2 illustrates the overall change in the general scope of alcohol production. Obviously, lower energy consumption, higher process yields, and maximum reutilization of wastes and by-products are sought.

As previously stated, depending on the sugar international market the main change will occur on the sugar cane processing itself, that is bypass the sugar manufacturing and ferment directly the cane juice. The implicit advantage is the steam generation by the onsite-producted bagasse. At this

Biotechnology and Bioengineering Symp. No. 8, 69–74 (1978)

0572-6565/78/0008-0069$01.00

point, needless to say, the sugar farming, sugar mills, and the alcohol industry must be integrated as a whole, making it even more efficient, because the transportation costs of raw materials will be drastically reduced.

Steam consumption reduction is the best-selling product of engineers and equipment manufacturers these days. Some European firms allow themselves to guarantee a steam consumption as low as 3.5 lb steam/lb alcohol, against the 4.8–5 lb steam/lb product for plants before the general concern about energy savings.

Those steam savings are claimed to be achieved by means of innovations such as indirect steam heating of the distillation columns. These processes allow for steam recovery and the thermal or mechanical recompression of the steam exhausts of the several side processes, such as the evaporation of slops and the drying of yeast.

Some other process shortcuts have been visualized; Maldonado et al. [2] proposed airlift fermentors for both a continuous fermentation stage and air-stripping alcohol recovering. There have been also some efforts to introduce the concept of direct sugar cane fermentation, combined with water extraction of the sugar; since the sugar in solution is constantly being converted to alcohol, the concentration gradient between the solution and the cane never reduces, given the possibility of a mechanically less complicated sugar extraction process.

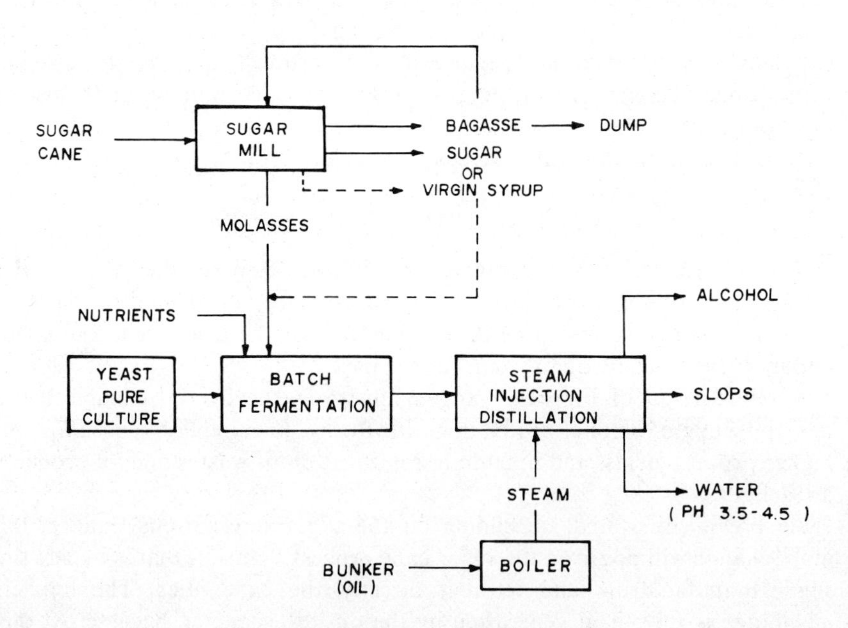

Fig. 1. Present alcohol production process.

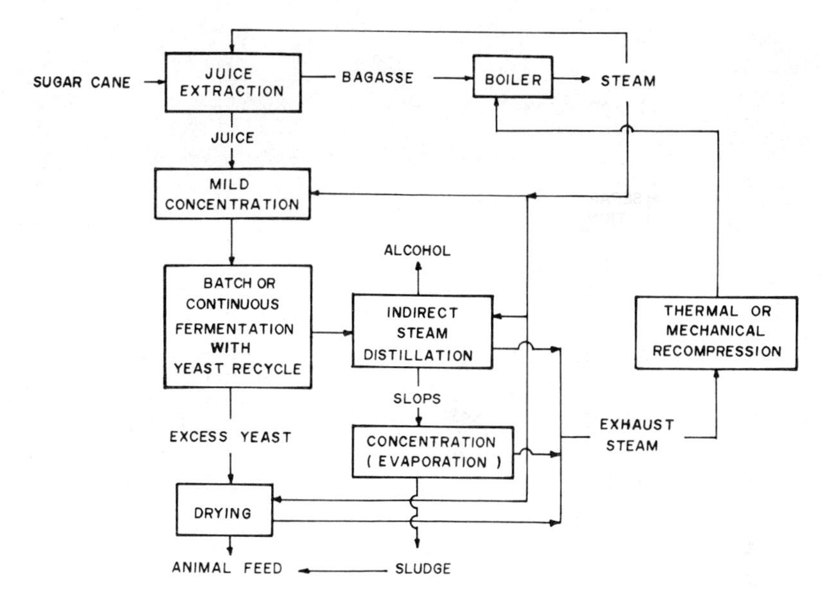

Fig. 2. Modified alcohol production process.

IN-PLANT IMPROVEMENTS

The evaluation of a given fermentation system allows the operator to optimize his yields and productivity. Biotechnology has been of great help in that matter. The search for better ways to convert the sugar into the alcohol at Destilería de Alcoholes y Rones, S.A., in Guatemala, has led to a series of experiments and tests of both continuous and yeast recycle (Melle–Boinot) fermentation systems.

Although the Melle–Boinot process has been tried since 1939, its use has not been widely adopted, mainly because the kinetics of the process are apparently related to the characteristics of a given yeast strain. Hodge and Hildebrant [3], describing the process in 1954, reported that 2 or 3% of the sugar fermented in a normal process is converted to cell mass. In subsequent cycles, assuming there is no cell growth, the alcohol conversion would increase by 1–1.5%. In 1956, Palacios [4] reported a yield increase of 4–7% under the same assumptions but in efficiency ranges of 62–67% of the theoretical conversion.

In early tests at DARSA, a 7% rise was obtained for isolated full-scale fermentations with efficiency ranges of 82–88%, not taking into account the sugar losses in the clarification of molasses—which during proper operation should be nil. Figure 3 illustrates a typical layout of the Melle–Boinot process.

The rate of product formation in the alcohol fermentation depends not only on the cell mass concentrations, X, but also on the cell growth rate

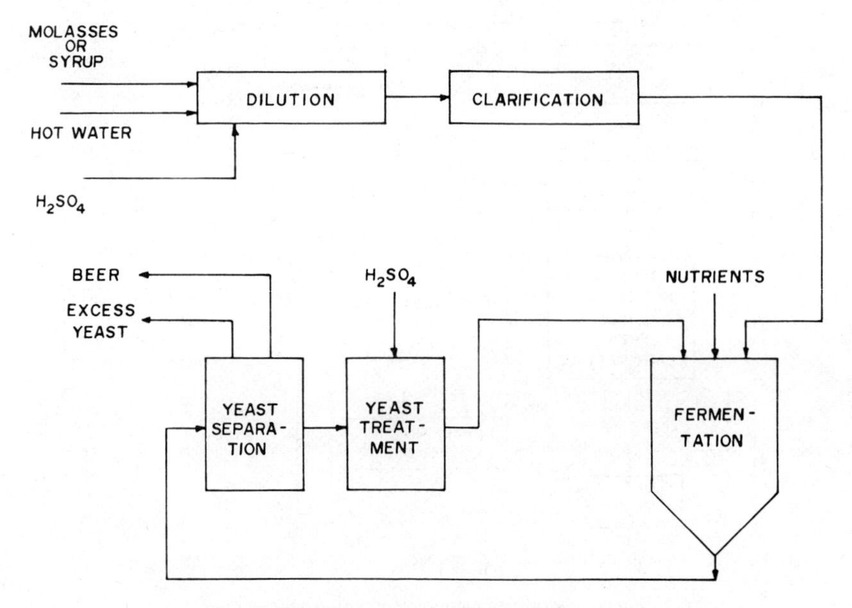

Fig. 3. Batch fermentation with recycle of yeast.

according to the model [5]:

$$\frac{dP}{dt} = a\,\frac{dX}{dt} + bX$$

where P is the product concentration, X is the cell mass concentration, t is time, and a, b are constants.

It is clear then, that there is a yeast cell mass buildup along with the

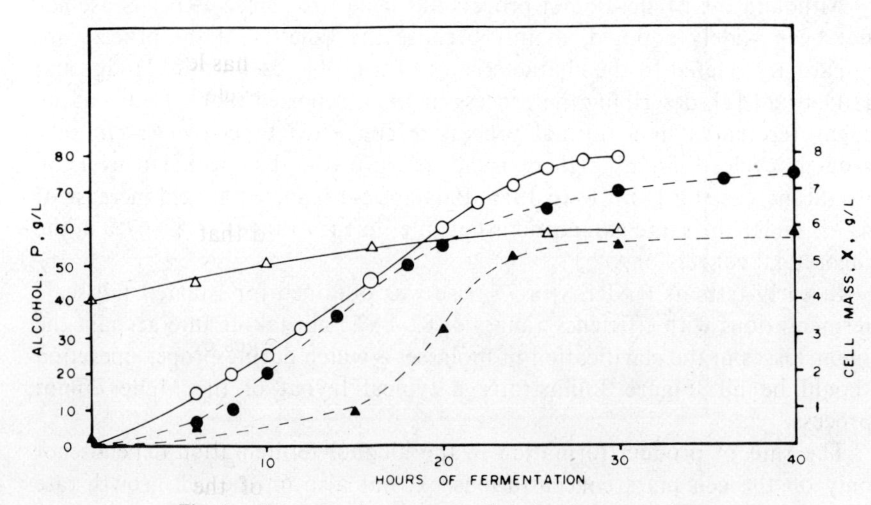

Fig. 4. Yeast recycle vs. nonrecycle in batch fermentation.

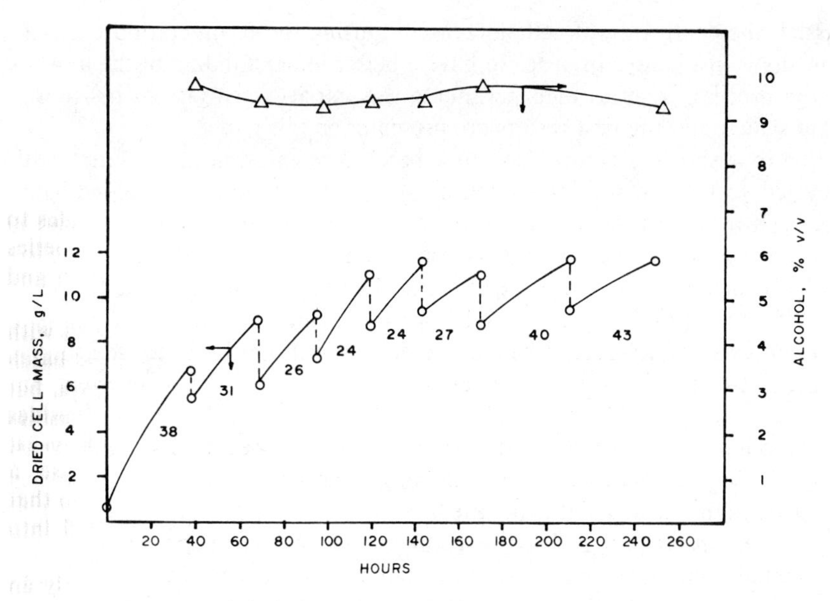

Fig. 5. Yeast recycle in fermentations with initial sugar concentration of 170 g/liter.

alcohol production, although the existing literature on the process does not foresee or discuss this aspect. Once the optimum population of yeast is sur-

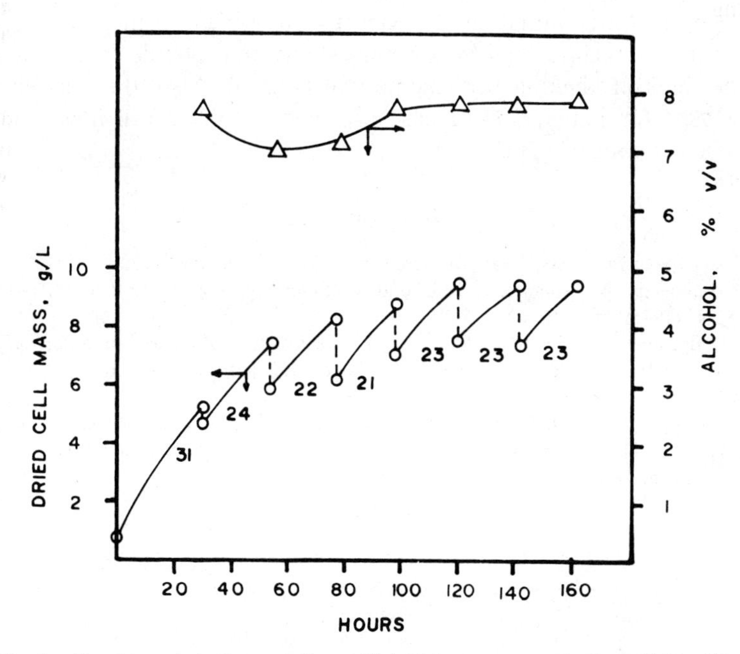

Fig. 6. Yeast recycle in fermentations with initial sugar concentration of 130 g/liter.

passed, the death rate of cells increases, causing the fermentation cycles to slow down gradually. In order to have a better understanding of the kinetics of the process, a series of experiments was carried out both at bench and plant scales, and the first results are presented in this paper.

Figure 4 shows a comparison of a batch fermentation and a batch with recycled yeast fermentation. It can be observed that, since the second batch has been massively inoculated, little sugar is converted into cell mass, but instead, it is being used to increase the alcohol content in the beer; besides the yield increase, the main attractive is the time saving. Because the yeast must be removed from the beer before it reaches the stationary phase, a continuous second stage in the storage or beer well has been devised so that the last 10 or 15 g/liter of remaining sugar can be also converted into alcohol.

The point at which the fermentation must be stopped shows regularly an 80% viability at the yeast cells; therefore, about 20% of the yeast was removed in the subsequent experiments.

The *Saccharomyces cerevisiae* strain had shown a rather high resistance to alcohol concentrations of 12–14%, but in-plant experience showed that working with reuse of the yeast at those levels did not allow this to work for more then three or four cycles. A deactivation seems to occur due to a prolonged exposure to high concentations of alcohol. This process is better observed in Figure 5, which represents repetitive cycles of a single fermentor with recycled cell mass; after two successful and very short cycles the fermentation starts to slow down, until it eventually stops.

Working at lower initial sugar concentations which result in 13% lower final alcohol concentrations shows more stable and reliable fermentations, with the same production per volume unit (Fig. 6). Also, the yeast cream can be used for many more cycles, provided the dried cell mass can be maintained at about 7 g/liter.

References

[1] *Revista Gerencia*, 4, Assn. Nac. De Gerentes De Guatemala, April 1978.

[2] O. Maldonado, R. Espinosa, C. Rolz, and A. Humphrey, *Ann. Technol. Agri.*, *24* (3–4), 335–342 (1975).

[3] H. Hodge and F. Hildebrandt, "Alcoholic Fermentation of Molasses," in *Industrial Fermentations*, L. A. Underkofler and R. J. Hickey, Eds (Chemical Publishing, Inc., New York, 1954).

[4] H. Palacio, *Fabricación del Alcohol* (Salvat Editores, S.A., Madred, España, 1956).

[5] S. Aiba, A. Humphrey, and N. Millis, *Biochemical Engineering* (University of Tokyo Press, Japan, 1972).

Selective Hydrolysis of Hardwood Hemicellulose by Acids

Y. Y. LEE, C. M. LIN, T. JOHNSON, and R. P. CHAMBERS
Department of Chemical Engineering, Auburn University, Auburn, Alabama 36830

INTRODUCTION

The hemicellulose fraction of biomass has not received much attention as a source of fuel or chemicals, although hemicellulose accounts for 10–40% of the carbohydrate and lignin content of various forestry and agricultural residues [1]. In enzymatic hydrolysis of cellulosic biomass, both hemicellulose and cellulose are in fact hydrolyzed to give their respective sugar products, for example, xylose and glucose, because of hemicellulase activity of common cellulase. However, xylose is always mixed with glucose. Unless it is separated from the main product, it is not likely that the xylose fraction will be utilized in further processing. Moreover, enzymatic hydrolysis of natural cellulosic substances such as wood or agricultural residues has been found inefficient without excessive grinding or substantial chemical pretreatment [2–4]. An adverse situation is also encountered in acid hydrolysis of cellulosic biomass insofar as hemicellulose is concerned. During an acid hydrolysis process aiming at glucose as the main product, hemicellulose is hydrolyzed more rapidly than cellulose. Consequently, when the reaction is carried out under the conditions that favor cellulose hydrolysis, most of the xylose is decomposed to furfural, and very little of it is actually recovered [5].

The fact that the hemicellulose is considerably easier to hydrolyze to monomeric sugars than is the cellulose [6] provides a possibility that the hemicellulosic fraction can be selectively hydrolyzed from biomass. Also, extraction of hemicellulose from the cellulosic biomass could increase the value of the biomass because the residue remaining after hemicellulose extraction is more susceptible to subsequent cellulose hydrolysis by enzyme or acid. From the standpoint of biomass utilization, hemicellulose utilization can therefore be considered both as a recovery and a pretreatment process.

In this study, low-level acid hydrolysis was investigated as a method of selectively recovering hemicellulose sugars from hardwood residue. Both strong and weak acids (sulfuric, sulfurous, and acetic) were used at various concentrations, seeking for an optimum processing condition. The spent

Biotechnology and Bioengineering Symp. No. 8, 75–88 (1978)

0572-6565/78/0008-0075$01.00

residue was further hydrolyzed by *Trichoderma viride* enzyme to evaluate the hemicellulose hydrolysis step as a pretreatment process in glucose production. Finally, this paper discusses the process economics, projecting our laboratory data to a commercial-scale production of xylose from hardwood residue.

EXPERIMENTAL

Materials

The hardwood residue (southern red oak) was supplied to us in the form of sawdust by a local lumber mill (Buchanan Lumber Co., Montgomery, Ala.). The sawdust was screened, and the median fraction (12–40 mesh) was used in this study. The cellulase enzyme was a gift from the U.S. Army Research and Development Center at Natick, Mass. According to the supplier, the enzyme source is *Trichoderma viride* (QM 9414), and the cell-free extract was freeze-dried before shipment (batch No. 11). Other characteristics of the enzyme are as follows: FP activity = 0.125 IUB/mg powder, soluble protein = 0.24 mg/mg powder, CMCase activity = 6.66 unit/mg powder.

Hemicellulose Hydrolysis

Acid hydrolysis of the hardwood residue was conducted in a 2-liter pressure reactor (Parr Instrument Co., model 4501-4522). The stainless steel reactor was equipped with a pressure-seal stirrer, an electrical heater, and an on–off temperature controller. With these provisions the temperature was controlled to within 2°C of the set-point. Eighty g prescreened and oven-dried sawdust were weighed into the reactor and mixed with 800 ml acid solutions. The reactor was then heated to and maintained at the desired temperature. About 30 min of preheating were required to reach the set temperature. In this report, the term "reaction time" excludes such preheating time. The reactor was agitated at a constant speed of 45 rpm throughout the preheating and reaction period. In the hydrolysis experiment using SO_2, the pure gas was initially blown into the reactor to purge out the air. With the gas outlet closed, the SO_2 pressure in the reactor was then raised to the desired level and left at that pressure for 2–3 hr at room temperature to saturate the water content in the reactor. Because of desorption and thermal expansion, the reactor pressure rose to between 170 and 220 psi after heating. The SO_2 pressure specified in this report indicates the initial pressure before heating. The processed hydrolysate samples were taken through a porous stainless-steel filter.

Enzyme Hydrolysis of Spent Residue

The treated and untreated hardwood sawdust samples were ground by a laboratory ball mill (Norton) and screened. The fraction that passed through 170 Tyler mesh was used as the substrate. The substrate solution containing 5 wt % solid 0.1*M* phosphate buffer (pH 4.7) was prewarmed to 50°C in a shaker bath. The dry cellulase powder (300 mg) was then weighed into the substrate solution (25 ml buffer, 1.25 g solid) to initiate the reaction. Liquid samples were withdrawn through a syringe attached with a Millipore filter and stored for sugar analysis.

Sugar Analysis

Liquid samples taken from acid hydrolysis were placed in a hot water bath at 90°C for 3 hr for complete hydrolysis of oligosaccharides into monomers. Samples from SO_2 experiments were placed in a sonic bath for an additional 30 min to drive off residual SO_2 gas. The acid and enzyme hydrolysates were first analyzed for glucose using a Beckman Glucose Analyzer (model EAR-2001). The acid hydrolysates were further analyzed for determination of total reducing sugar using dinitrosalicylic acid [7]. A few samples of acid hydrolysates were also analyzed for the individual sugar contents using a high pressure liquid chromatograph (HPLC) (Waters Associate, μ-carbohydrate column, RI detector). The liquid samples were neutralized with $Ba(OH)_2$ and filtered before injection. The HPLC results (a sample chromatogram shown in Fig. 1) indicated that xylose and glucose

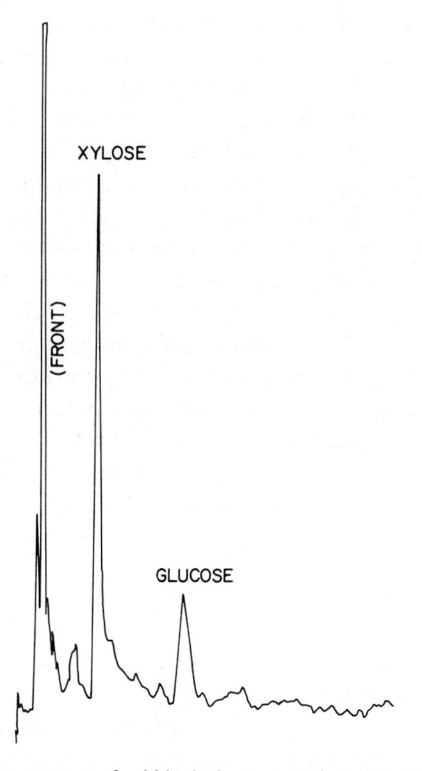

Fig. 1. HPLC chromatogram of acid hydrolysate sample (0.2%, H_2SO_4, 170°C, 2 hr).

were the major constituents in the hydrolysate. The sugar content from hemicellulose (predominately xylose) was determined from the difference between total reducing sugar and predetermined glucose. The hemicellulose sugar thus determined is expressed as xylose+ in this paper since it equals approximately the xylose content.

RESULTS AND DISCUSSION

Acid Hydrolysis of Hemicellulose

Hydrolysis of hemicellulose in general requires milder reaction conditions than those found in cellulose hydrolysis. Since the objective of this work is to recover xylose selectively from biomass complex and at the same time enhance the susceptibility of the remaining residue, it is particularly important to adjust the reaction condition so that the hemicellulose fraction is hydrolyzed in a reasonably short time and yet the cellulose content is undestroyed. It has been reported that the hemicellulose in wood or agricultural residues could be hydrolyzed using sulfuric acid [1, 8–11]. However, pretreatment has been totally disregarded in these studies. In the present work, sulfuric, acetic, and sulfurous acids were employed as the hydrolytic catalysts. In the studies involving sulfuric acid, three levels of acid concentration (0.1, 0.2, 0.4 wt %) were employed at a fixed reaction temperature of 170°C. The variation of xylose and glucose components during batch hydrolysis of the hardwood residue is shown in Figure 2. The major component is xylose rather than glucose, meeting our first objective of selectivity in hydrolysis. The glucose component released from cellulose did not exceed 12% of the total cellulose matter originally present in the biomass. The xylose+ content (as determined by total reducing sugar minus glucose) reached maximum within 2 hr of reaction time for 0.1 and 0.2% H_2SO_4 and gradually decreased due to decomposition. HPLC analysis of the product showed that about 90% of the nonglucose sugar (xylose+) was xylose. On the other hand, the curve for 0.4% H_2SO_4 showed a steady decrease of xylose+, which indicates that the maximum sugar concentration is obtained during the preheating period. In other words, the reaction conditions of 0.4% H_2SO_4, 170°C were too stringent to recover xylose effectively. The maximum xylose+ concentration obtained for 0.1 and 0.2% H_2SO_4 was about 160 mmol. With a solid-to-liquid ratio of 1:10, this figure represents the xylose yield of 24 g/100 g dry wood. Taking the hemicellulose content of southern red oak as 22.74% [12], the yield of xylose+ is calculated to be 94% of the theoretical maximum (accounting water addition in hydrolysis). A near quantitative recovery of hemicellulose sugar is therefore achieved using either 0.1% or 0.2% H_2SO_4. For better selectivity in hydrolysis (or low glucose production), however, the lower acid level (0.1%) is considered a preferable reaction condition.

The results obtained from hydrolysis experiments using sulfur dioxide are shown in Figure 3. In this case, the acid concentration was fixed (saturated

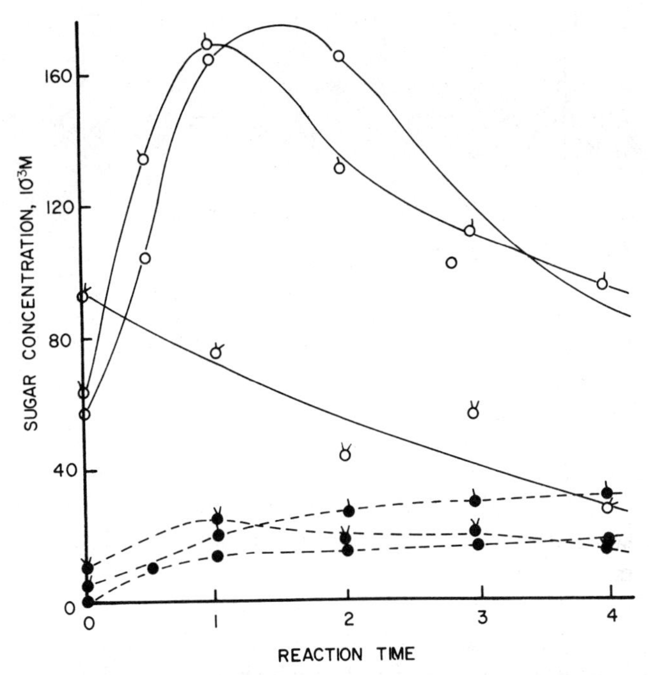

Fig. 2. Variation of sugar concentration during hydrolysis of hemicellulose in hardwood by sulfuric acid, 1:10 solid–liquid ratio, 170°C. (—) Xylose; % H_2SO_4: (○) 0.1%, (○) 0.2%, (○) 0.4%. (---) Glucose; % H_2SO_4: (●) 0.2%; (●) 0.1%; (●) 0.4%.

with SO_2 at 10 psig, 22°C) while the temperature was varied. Selective hydrolysis is again achieved at 150°C. (Note that the glucose level is near zero.) However, the molar yield of xylose+ reached only 40% of theoretical maximum at this temperature. The reaction pattern at 170°C was quite different from that at 150°C in that hemicellulose hydrolysis was completed during the preheating period. Consequently, only the decomposition of sugar was shown in the figure, and the maximum xylose+ concentration appeared at zero reaction time. The corresponding molar yield was 83%. The glucose concentration at that point reached 28 mmol, which is equivalent to 11.3% of the total cellulose content originally present in the biomass. It seems obvious that xylose yield increases with the reaction temperature, while the selectivity of hydrolysis decreases with it, and that 170°C is near the optimum temperature when both aspects are considered. The data also indicate that the yield and selectivity can be improved by close control of reaction time and temperature, for example, by stopping the reaction at some point during the preheating period or taking an equivalent strategy. The chemical requirement of 16 g SO_2/100 g water under the present conditions is extremely high compared to 0.1–0.2% for the H_2SO_4 method. From an economic standpoint this high requirement would be a significant disadvantage. The SO_2 requirement was high because SO_2 is not

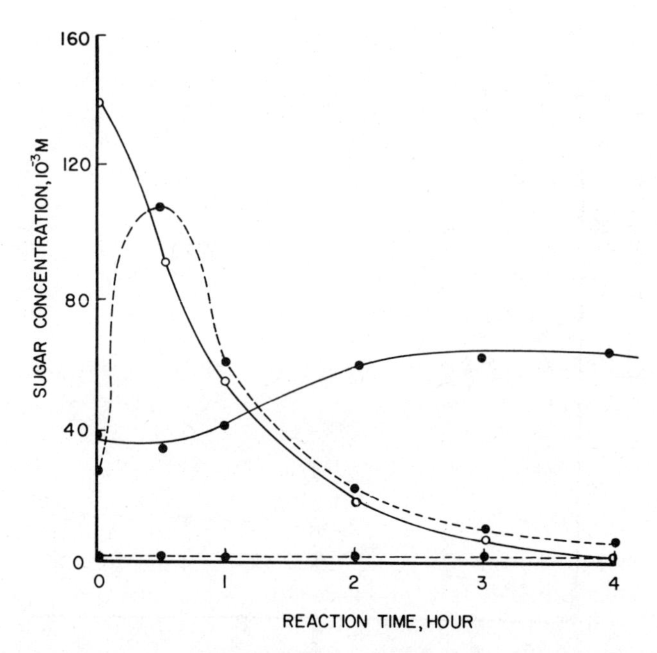

Fig. 3. Variation of sugar concentration during hydrolysis of hemicellulose in hardwood by SO_2, 1:10 solid–liquid ratio. (—) Xylose: (●) SO_2, 10 psig, 150°C; (O) SO_2, 10 psig, 170°C. (- - -) Glucose: (●) SO_2, 10 psig, 170°C; (●) SO_2, 10 psig, 150°C.

only acting as a catalyst but also is reacting with lignin to form lignosulfite. The present treatment condition is, in fact, similar to that of the sulfite pulping process. Delingnification in the SO_2 treatment was also indicated by the high weight loss of 54% after processing as compared to 32% in the sulfuric acid and 24% in the acetic acid processing. The unreacted SO_2 gas, however, can be recovered at the end of the process and reused.

The hydrolysis results with acetic acid when two levels of acid concentration (1% and 10%) were applied at 170°C are shown in Figure 4. As indicated in the figure, the catalytic activity of acetic acid was far inferior to that of either sulfuric acid or SO_2. After 4 hr reaction, the yield of hemicellulose sugars reached only 35% of the theoretical maximum for 1% acid treatment. The yield was even lower when 10% acid was used, perhaps because of faster decomposition.

Enzymatic Hydrolysis of Spent Residues

The hardwood residue that remains after hemicellulose hydrolysis still contains most of its cellulose. As was pointed out earlier, the spent residue would presumably be made more susceptible to subsequent enzymatic hydrolysis after undergoing an acid treatment process. To verify this point experimentally, we ground the spent residues and two additional reference substrates (untreated hardwood and filter paper) and then hydrolyzed them by *T. viride* cellulase. The experimental conditions were kept identical for all the substrates: 5% slurry, 50°C, pH 4.8, 170-mesh particle size. The hydrolysis results are shown (Figs. 5–7) by the plots of glucose accumulation versus reaction time (cellobiose excluded).

Significant increases in both reaction rate and glucose yield are shown for the pretreated residues over untreated ones. The initial reaction rates up to 5 hr of reaction time were almost identical for all substrates. Marked differences, however, emerged after that period. The relative reaction rates estimated by the slopes in the figure during a 10- to 30-hr span and the relative glucose yield in 50 hr are listed in Table I. The spent residues from sulfuric acid treatment have the highest enhancement, about a sixfold increase over the untreated control, approaching the value of 6.9 for filter paper. A somewhat lower but still quite significant enhancement ranging from 2.4 to 4.5 was observed for the residues treated with SO_2 or acetic acid. A similar pattern was shown in glucose yield, the sulfuric acid–treated residue again having the highest enhancement (2.6 times that for the untreated control). The relative yield reported here is not based on the cellulose content but on the bulk initial weight of the substrate.

It has been speculated that the susceptibility of cellulosic biomass to enzymatic hydrolysis is closely related with its crystallinity [13]. We have accordingly measured an index of relative crystallinity of the various spent

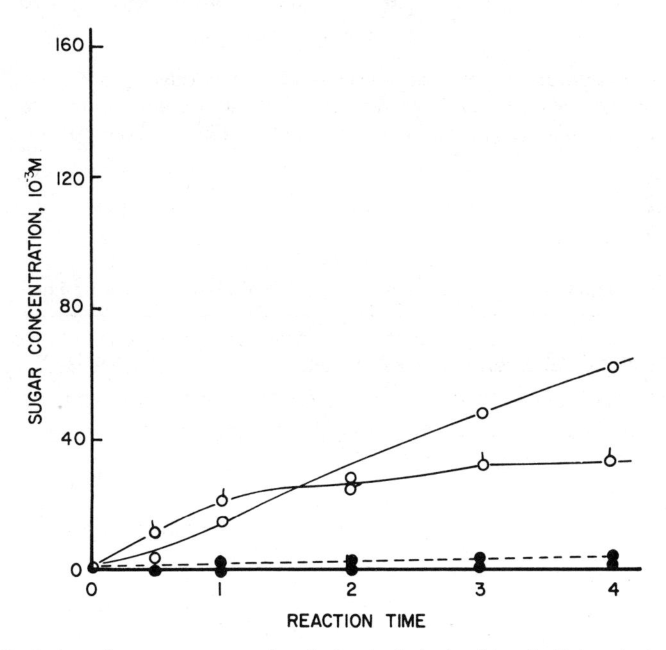

Fig. 4. Variation of sugar concentration during hydrolysis of hemicellulose in hardwood by acetic acid, 1:10 solid–liquid ratio. (—) Xylose; (- - -) glucose. % HAC: (O) 1%, 170°C, (O̍) 10%, 170°C.

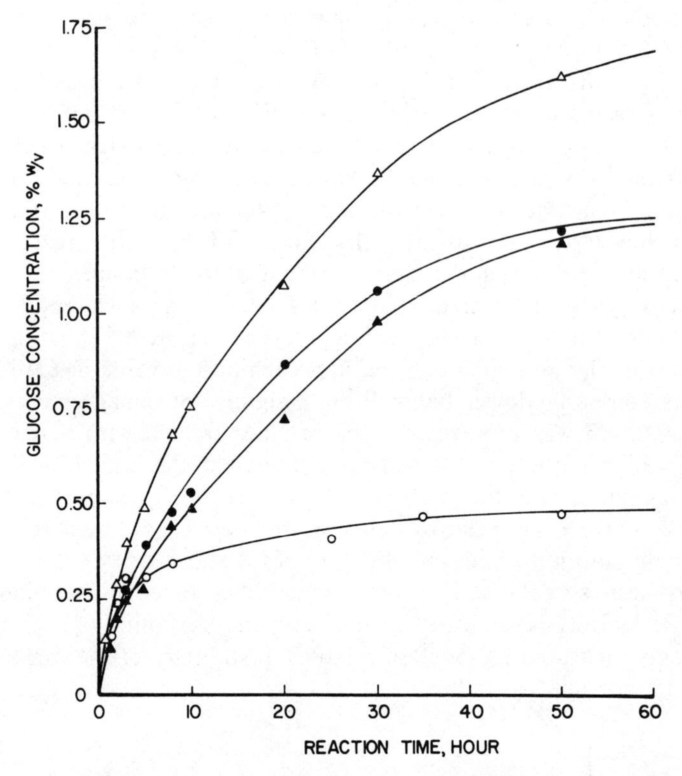

Fig. 5. Enzymatic hydrolysis of spent residue, substrate pretreated by sulfuric acid (reaction conditions: pH 4.8, 50°C, 5% solid, 300 mg enzyme/25 ml). (Δ) Filter paper; (●) pretreated with 0.2% H_2SO_4, 170°C; (Δ) pretreated with 0.1% H_2SO_4, 170°C, (O) untreated.

TABLE I

Comparative Enhancement in Enzymatic Hydrolysis of Spent Residue

Prehydrolysis Condition	Relative Reaction Rate During 10-30 hr.	Relative Glucose Production in 50 hr.
Untreated (Control)	1.0	1.0
H_2SO_4, 0.1%, 170°C	5.6	2.5
0.2%, 170°C	6.3	2.6
SO_2, 10 psig, 170°C	3.5	2.3
10 psig, 150°C	2.4	1.1
HA_c, 1%, 170°C	3.4	1.7
10%, 170°C	4.5	2.2
Filter Paper	6.9	3.4

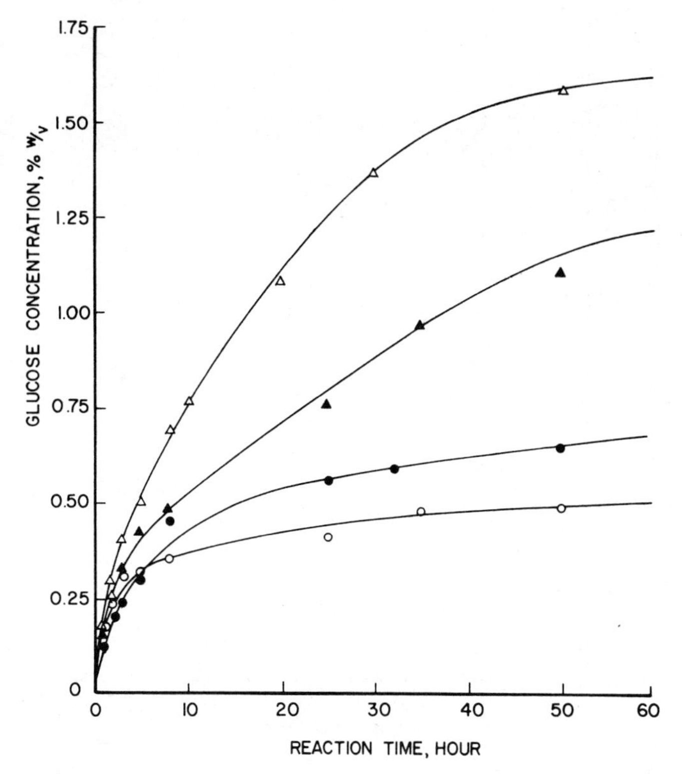

Fig. 6. Enzymatic hydrolysis of spent residue, substrate pretreated by SO_2 (reaction conditions: pH 4.8, 50°C, 5% solid, 300 mg enzyme/25 ml). (△) Filter paper; (▲) pretreated with SO_2, 10 psig, 170°C; (●) pretreated with SO_2, 10 psig, 150°C, (○) untreated.

residues from the grinding time needed to achieve same degree of size reduction. Shown in Table II is the required milling time needed to have 50% of each substrate pass through a 170-mesh sieve. Judging from the milling time, the crystallinity of the hardwood residue is tremendously reduced by acid treatments. This effect alone would be a significant advantage for hemicellulose extraction because a certain degree of size reduction is a necessary step in various types of biomass processing. The most noticeable reduction was the residue treated with SO_2, for which the milling time was reduced from 480 min to 6–15 min. Comparing Tables I and II shows that the susceptibility is indeed related to crystallinity in the broad sense that each pretreatment substrate has higher reaction rate and shorter milling time. Closer observation, however, reveals a complication in the relationship because the order in crystallinity is not in accordance with the degree of enhancement. This complication in turn led us to consider another important factor directly related with the susceptibility—the surface area. Although the substrates are ground to equal average sizes, the microstructures of the particles can still make significant differences in surface

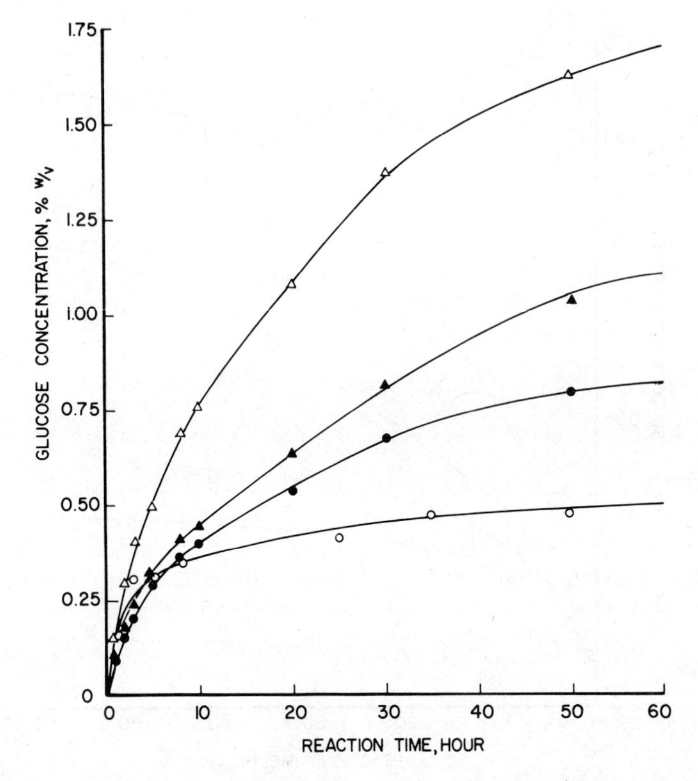

Fig. 7. Enzymatic hydrolysis of spend residue, substrate pretreated by acetic acid (reaction conditions: pH 4.8, 50°C, 5% solid, 300 mg enzyme/25 ml). (△) Filter paper; (▲) pretreated with 10% HAC, 170°C; (●) pretreated with 1% HAC, 170°C; (○) untreated.

TABLE II

Grinding Efficiency of Spent Hardwood Residue

Hydrolysis Condition	Milling Time to Collect 50% in 170 Mesh
Untreated	480 min.
H_2SO_4, 0.1%, 170°C	180
H_2SO_4, 0.2%, 170°C	45
SO_2, 10 psig, 170°C	6
SO_2, 10 psig, 150°C	15
HAC, 1%, 170°C	110
HAC, 10%, 170°C	55

areas. In support of this proposition are the electron micrographs (Fig. 8) of treated and untreated residues taken in our previous study [5]. Comparison of the electron micrographs definitely indicates that the residue treated with sulfuric acid has much greater surface area than the one treated with SO_2. The micrographs clearly show that the wood structure is made more porous and the cellulosic fibers are exposed as the result of hemicellulose extraction. From these observations it seems obvious that both crystallinity and surface area are related to the susceptibility and that the susceptibility enhancement in sulfuric acid treatment is mainly a result of the increase in surface area, whereas in SO_2 treatment the benefit comes mainly from reduced crystallinity.

Process Economics in Pentose Production from Hardwood

The foregoing experiments show that the hemicellulose sugars (predominately xylose) can be extracted selectively by acids. On the basis of our laboratory data, we therefore made an economic analysis for producing hemicellulose sugars from cellulosic biomass. A process using 0.1% sulfuric acid was chosen, as it was proven to be most efficient in hemicellulose hydrolysis. Figure 9 shows the simplified flow diagram for the process.

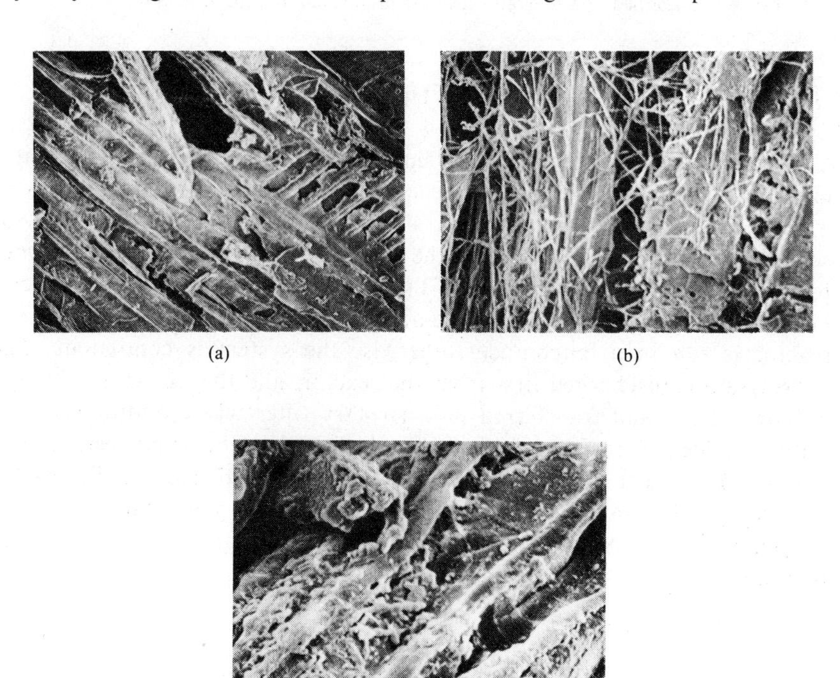

(a) (b) (c)

Fig. 8. Electron micrograph of hardwood residue (southern red oak): (a) Untreated (500×); (b) treated with H_2SO_4 (500×), and (c) treated with SO_2 (500×).

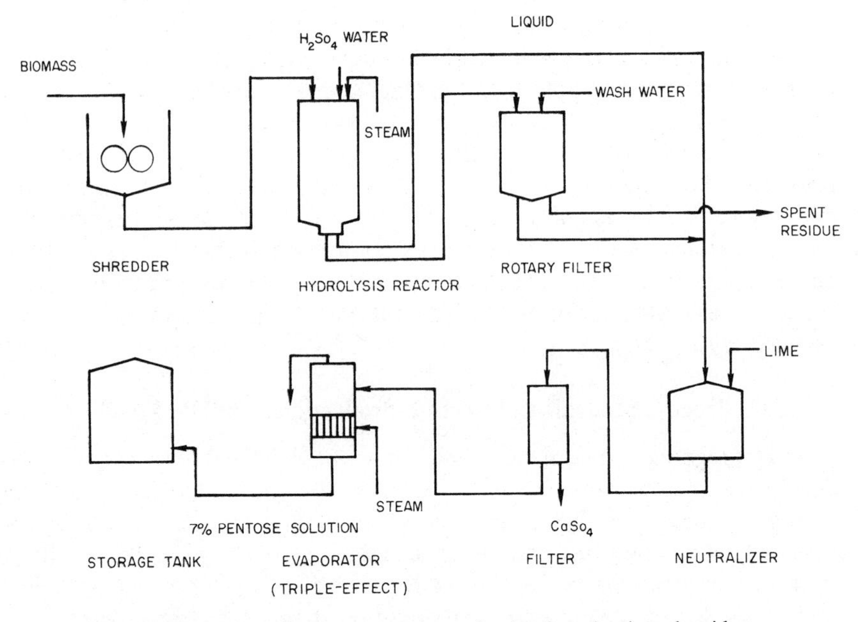

Fig. 9. Simplified flow diagram—pentose production from hardwood residue.

Process Description

The biomass is passed through a shredder, reducing the size into about 4 mesh before it is charged into the reactor. The solid feed is then mixed with water and sulfuric acid to form a slurry. Readjustment of the water-to-solids ratio was made to 4:1 from the 10:1 of our laboratory study in order to reduce the steam consumption in the evaporation stage. Steam is directly injected into the reactor to raise and maintain the reaction temperature. The reactor is run in a batchmode; otherwise the system is continuous. The hydrolysate is discharged first from the reactor, and the wet spent biomass is blown down and transferred into a rotary filter where additional wash water is added to recover residual sugar adhered in the spent residue. The liquid product is then neutralized with lime and put through additional an filtering process for $CaSO_4$ removal. At the final stage the product is concentrated by a triple-effect evaporator to a 7% sugar level, a concentration suitable for subsequent fermentation.

Production Cost

Table III presents the economics for the process. For a plant processing 1000 tons of biomass per day the total capital investment is $6.65 million, for which 38% is for the equipments. For yearly production of 66,000 tons of sugar (solid basis), the total production cost amounted to $3.829 million.

TABLE III

Cost Analysis of Pentose Production from Hardwood[a]

A. CAPITAL INVESTMENT		
Purchased Equipment		$ 2.55 MM
Reactors	$ 0.81 MM	
Shredder	0.21	
Wash Tank	0.09	
Spent Residue Filter	0.28	
Neutralizer	0.08	
$CaSo_4$ Filter	0.01	
Evaporator	0.75	
Storage Tanks	0.20	
Pumps	0.12	
Installation (0.80 PE)		$ 1.15
Piping (0.31 PE)		0.79
Instrumentation (0.13 PE)		0.33
Buildings and Service Facilities (0.20 PE)		0.51
Engineering, Construction (0.50 PE)		1.28
Contingency		0.55
	Total:	$ 6.65 MM

B. PRODUCTION COST	Annual Cost	% Breakdown
Raw Material	$ 0	0
Chemicals	0.228 MM	6.0%
Labor & Maint.	0.135	3.5
Utilities	2.635	68.8
Steam (2.451)		
Electricity (0.138)		
Water (0.046)		
Depreciation	0.665	17.4
Taxes and Insurance	0.166	4.3
TOTAL PRODUCTION COST/YR	$ 3.829 MM	
BY-PRODUCT CREDIT	0	
NET PRODUCTION COST/YR	3.829 MM	
PRODUCT COST/LB PENTOSE IN 7% SOLUTION	¢ 2.90	

[a] 1000 ton/day dry feed.

This figure is translated into 2.9 cent/lb sugar. One item particularly notable in the processing cost is the cost of steam that accounts for 64% of the total. The analysis has been made under the assumption that the process is an integral part of a complete processing plant in which the cellulose is also utilized. The treated solid biomass (a by-product), which is now more susceptible to cellulose hydrolysis, is in fact increased in value. However, no by-product credit was taken for pretreatment, putting an equal value for both raw material and by-product (note that zeroes were put in both places in Table II). Without claiming by-product credit, the production cost of 2.9 cent/lb for sugars from hemicellulose is still considered very attractive in view of the recent cost figures for glucose from cellulosic biomass in enzymatic process that ranges from 6 to 20 cent/lb [14, 15]. The low cost of the sugar is attributed to low catalyst cost and fast reaction in comparison to enzymatic hydrolysis.

References

[1] J. W. Dunning and E. L. Lathrop, *Ind. Eng. Chem.*, *37*, 24 (1945).

[2] M. Mandels, L. Hontz, and J. Nystrom, *Biotechnol. Bioeng.*, *16*, 1471 (1974).

[3] T. K. Ghose and J. A. Kostick, *Adv. Chem. Ser.*, *95*, 415 (1979).

[4] C. R. Wilke, R. D. Yang, A. S. Sciamanna, and R. Freitas, "Studies on enzymatic hydrolysis of lignocellulosic materials," paper presented at the 157th ACS National Meeting, Anaheim, Calif., 1978.

[5] Y. Y. Lee, T. Yue, and A. R. Tarrer, "Acid hydrolysis of oak sawdust," paper presented at AIChE National Meeting, Kansas City, Mo., 1976.

[6] G. V. Shultz and E. Huseman, *Z. Phys. Chem.*, *B52*, 42 (1942).

[7] G. L. Miller, *Anal. Chem.*, *31*, 426 (1959).

[8] J. F. Harris, G. J. Hajny, M. Hannan, and S. C. Rogers, *Ind. Eng. Chem.*, *38*, 896 (1946).

[9] J. F. Harris, J. F. Saeman, and E. G. Locke, *For. Prod. J.*, *8*,(9), 248 (1958).

[10] M. S. Dudkin and V. E. Starichkova, *Zh. Prikl. Khim.*, *4H*(12), 2711 (1968).

[11] E. Maekawa and E. Kitao, *Makuzai Kenkyu*, *37*, 6 (1966).

[12] J. F. Harris, Forest Products Laboratory, USDA, Madison, Wis., personal communication.

[13] J. A. Howell and J. D. Stuck, *Biotechnol. Bioeng.*, *17*, 873 (1975).

[14] J. M. Nystrom, R. K. Andren, and A. L. Allen, "Enzymic hydrolysis of cellulosic wastes—The status of the process technology and economic assessment," paper presented at 81st AIChE Meeting, Kansas City, Mo., 1976.

[15] D Hsu, M. Ladisch, and G. T. Tsao, "Pretreatment of cellulose as a means to increase yield of glucose from enzyme hydrolysis," paper presented at the 157th ACS National Meeting, Anaheim, Calif. 1978.

Cellulase Production by a New Mutant Strain of *Trichoderma reesei* MCG 77

BENEDICT J. GALLO, RAYMOND ANDREOTTI, CHARLES ROCHE, DEWEY RYU,* and MARY MANDELS

Food Sciences Laboratory, U.S. Army Natick Research and Development Command, Natick, Massachusetts 01760

INTRODUCTION

There is an increasing interest in the bioconversion of biomass to produce chemical and liquid fuels thereby conserving petroleum. Cellulose is the most abundant renewable natural product, and its hydrolysis to glucose by acid or enzyme is a good approach to its more effective utilization. At Natick, we are trying to develop a process for hydrolysis of waste cellulose to glucose with cellulolytic enzymes produced by the fungus *Trichoderma reesei*. The technical feasibility of this process has been demonstrated, but further improvements are required to make it economically attractive [1]. Because production of the enzyme is a major cost factor, we are trying to improve yields and productivity of the enzyme by screening for improved mutant strains and by optimizing fermentation conditions. We would like to report in this paper the results of our studies using our new mutant strain of *T. reesei* MCG 77† [2]. The performance of the strain in 15-liter fermentations in terms of its productivity of cellulase is compared to its parental strain QM 9414 and another mutant strain NG 14 obtained from Rutgers [3, 4].

Cellulase in *Trichoderma* is an inducible enzyme, and 0.75% cellulose flask cultures typically give a cellulase of low specific activity about 0.6 filter paper units/mg soluble protein in the broth. Enzyme yields may be increased by using a medium containing a high concentration of cellulose. Since the consumption of high levels of cellulose tends to develop acidic conditions that reduce growth and inactivate enzymes, experiments using shaken flasks are restricted to the use of low levels of cellulose or otherwise require high salt concentrations to supply nitrogen and to control pH. It is difficult to evaluate cellulase production of new mutant strains under such conditions. In fermentors larger than 5-liter size, pH can be controlled with ammonia to supply nitrogen as needed, and high levels of cellulose can be

* Present address: The Korea Advanced Institute of Science, Seoul, Korea.

† On deposit at the Northern Regional Research Center, U.S.D.A., Peoria, Illinois; RRL 11,236.

Biotechnology and Bioengineering Symp. No. 8, 89–101 (1978)
Published by John Wiley & Sons, Inc.

used with low salt media to provide increased levels of cellulase [5, 6]. In this work, the cellulase productivity of the new mutant strain at concentrations of 2% and 6% cellulose was studied. The enhanced rate of enzyme production of strain MCG 77 was also studied and confirmed.

MATERIAL AND METHODS

Organism

Three mutant strains of *T. reesei* Simmons [2–4] were used for this study (see Fig. 1). These strains were formerly assigned to the species *T. viride* Persoon *ex* Fries [7].

Strain MCG 77 was selected based on its ability to clear cellulose on an agar plate containing 8% glycerol and its near freedom from glycerol repression in submerged culture. Its immediate parental strain, TKO, 41, survived both the ultraviolet light and the Kabicidin (polyene antibiotic) treatments. Strains QM 9414 and NG 14 were maintained on potato dextrose agar (PDA); MCG 77 was maintained on agar slants containing Vogels salts (1964) with biotin and cellulose [1.25 or 2.25% (w/v)] because this strain grows slowly and fails to conidiate on PDA.

Fermentation

Experiments with 10-liter submerged-culture working volume were carried out in Magnaferm Fermentors, model MA 114 (New Brunswick Scien-

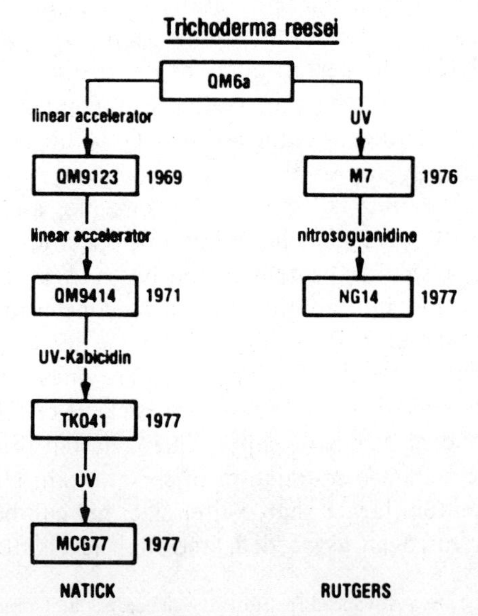

Fig. 1. Genetic lineage of current cellulase mutant strains of *T. reesei*: a summary showing the genetic relationship of reported cellulase-enhanced strains to the wild-type strain of *T. reesei* QM 6a. Years that the strains were isolated are indicated, and the type of mutagen used is given with the exception of Kabicidin which is a fungicide.

tific Company). This fermentor utilizes a 15-liter glass vessel equipped with a magnetically driven, triple-propeller stirrer and a mechanical foam breaker. Temperature, dissolved oxygen, and pH were controlled and were continuously recorded during the experiments. The pH was initially allowed to fall from 5.0 to 3.0 and controlled at 3.0 by using 2*N* NH_4OH. The temperature of the fermentation was controlled at 27°C; aeration was set at 2 liter/min flow (0.2 v/v/m) at 7 psig pressure; the agitation speed was varied between 300 to 500 rpm to maintain positive dissolved oxygen and good mixing. The basic salt medium used in these experiments was as described previously [8] except that urea was omitted. Proteose peptone (0.1%) and Tween 80 (0.1%) were added to fermentation media. Fermentors were inoculated at 10% and 20% (v/v) with 3-day-old inoculum cultures that were started with conidia from slants. Cellulose (0.75%) and/or glucose (0.75%) were used for inoculum medium or as indicated in the text. Glucose was autoclaved separately as a solution or in the form of dry granules and added to the medium when cool. After samples were removed from the fermentor, the solids and broth were separated by centrifugation or filtration by suction through glass fiber filters, and the samples were refrigerated or frozen until analysis.

Substrates

Pure cellulose substrates used included: SW40 (Brown Co., Berlin, N.H.) hammer-milled 40-mesh spruce wood pulp. BW200 (Brown Co., Berlin, N.H.) ball-milled 200-mesh SW40. FB cotton: two-roll milled cotton was prepared in this laboratory by passing absorbent cotton through a Farrel Birmingham Mill for 1 min at 10-mil gap [9].

FBWS40: Two-roll milled SW40 prepared in this laboratory by passing hammer-milled 40-mesh spruce wood pulp through a Farrel Birmingham Mill for 1 min at 20-mil gap [9].

Analyses

Dry weight measurements represent the mycelial weight and residual cellulose. The solid fraction was dried at 80°C overnight.

Mycelial protein was determined by extracting three times in 1*N* NaOH at 50°C and measuring the protein (by a microbiuret procedure against bovine serum albumin as the standard).

Soluble protein in the filtrate was measured by the Lowry [10] procedure after precipitation with 10% trichloracetic acid and bovine serum albumin as the standard.

Reducing sugar was measured as glucose by a dinitrosalicylic acid procedure [11].

Residual cellulose was estimated by methods previously described [12].

Cellulase activity is expressed in international units (μmol glucose released/min) of enzyme activity measured with carboxymethylcellulose

(CMC) or with filter paper (FP) [13]. In some tests cellulase activity was also measured as *shaken* filter paper cellulase units (× 0.56 = unshaken units) or a shaken disk assay [14] (× 1.6 = unshaken filter paper cellulase units). Cotton activity is equivalent to milligrams of reducing sugar (as glucose) produced when 1 ml culture filtrate acts on 50 mg cotton for 24 hr at pH 4.8, 50°C, in an unshaken test tube. β-Glucosidase was measured in international units using salicin [12].

RESULTS AND DISCUSSION

Fermentation of 2% Cellulose by Mutant Strains

When the three mutants of *T. reesei*, QM 9414, NG 14, and MCG 77, were grown on 2% cellulose in batch fermentation for seven days, high yields of cellulase (approximately 4 filter paper cellulase units/ml) were produced by each of them. The MCG 77 strain shows the fastest rate of enzyme production and the highest productivity (over 40 units/liter/hr at three days) (Fig. 2 and Table I). As noted previously for the strain QM

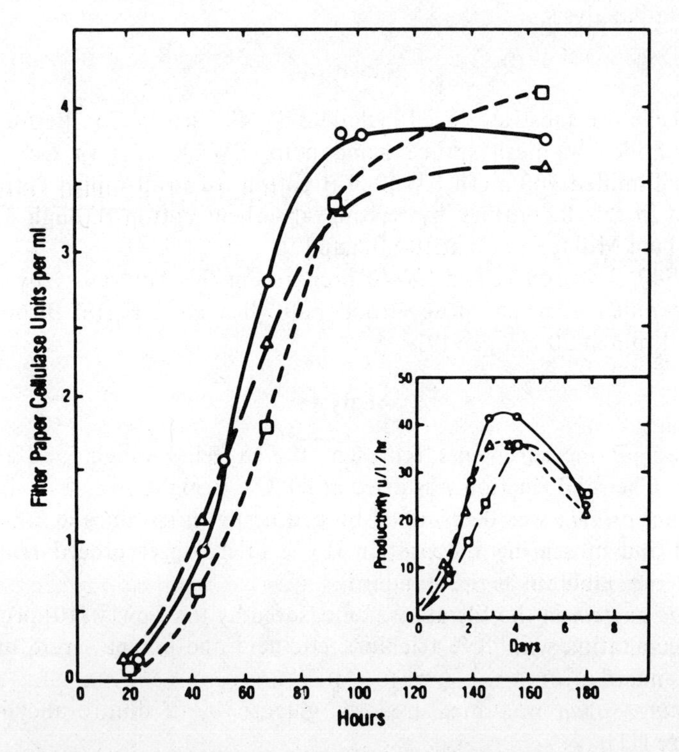

Fig. 2. Cellulase synthesis by three strains of *T. reesei*: (○) MCG 77; (△) NG 14; and (□) QM 9414. Conditions: pH control, ≥3; substrate, 2% ball-milled pulp (BW 200); inoculum, 10% (v/v) mycelial; cellulase, filter paper cellulase units/ml; and productivity, filter paper cellulase units/liter/hr.

TABLE I

Filter Paper Cellulase Units and Cellulase Productivity Obtained in 2% Cellulose Fermentations Run at pH 3.0 and pH 5.0[a]

pH Control	Strains of *T. reesei*							
	MCG 77				NG14			
	FPU/ml		Productivity $U/l \cdot hr^{-1}$		FPU/ml		Productivity $U/l \cdot hr^{-1}$	
	51 hrs	Max (time)	51 hrs	Max (time)	51 hrs	Max (time)	51 hrs	Max (time)
≥5	1.9	2.9 (92 hrs)	36.3	36.3 (51 hrs)	1.1	2.8 (164 hrs)	22	25 (67 hrs)
≥3	1.6	3.9 (92 hrs)	31	41.8 (92 hrs)	1.3	3.6 (164 hrs)	26	36 (67 hrs)

[a] Table showing filter paper cellulase units/ml broth (FPU/ml) and the cellulase productivity (filter paper units/liter/hr) at the end of 51 hr and at their maximum value. Fermentor residence time at which the maximum values were attained is also given. Conditions in which these fermentations were run are described in Figure 2.

9414, the optimal pH for growth was 5.0 and that for enzyme yield was 3.0 [13].

Fermentation of 6% Two-Rolled Milled Cotton by Mutant Strains

In a typical *Trichoderma* cellulose fermentation of strain QM 9414, the first 50 hr is the period of active growth [13]. The pH is allowed to fall rapidly to about 3.0, which is the control point. By the time most of the cellulose is consumed, NH_4OH is also consumed rapidly and biomass reaches its maximum, but little enzyme has been produced. Most of the enzyme in a batch is produced during the second 50-hr period as residual cellulose is consumed, little or no reducing sugar is detected, ammonia consumption eases, biomass falls slowly, and finally the pH begins to rise. Increasing the initial cellulose concentration is a good way of increasing yield [5], but attempts to prolong the enzyme production phase by adding only cellulose later in the fermentation have been only modestly successful. However, the enzyme production phase could be prolonged somewhat by using a substrate that is more slowly consumed. Because of its high crystallinity, fibrous cotton is the most resistant pure cellulose, but it is a poor substrate for growth. Recent work in our laboratory has shown that two-roll milling is an excellent pretreatment of cellulose for saccharification, increasing both reactivity and bulk density of the cellulose [9]. When cotton is two-roll milled, it becomes much more susceptible to cellulase, but it is still much more resistant than cellulose pulps. When the three mutant strains were grown on 6% two-roll milled cotton, growth was excellent, but even after 15 days some residual cellulose remained in the culture, enzyme was still being produced, the pH was still at the control level of 3.0, and NH_4OH was still being consumed. The enzyme yields of 10–15 filter paper units/ml (14–21 mg soluble protein/ml) were the highest we have seen so far (Table II and Fig. 3). In parallel with excellent growth the major fermentation product was extracellular protein, which was five to nine times higher than the mycelial protein (not shown). High cotton activity was achieved early in the fermentation by strain MCG 77 as shown in the inset of Fig. 3. The high cotton activities reached by the three strains suggest that high levels of the exo-β-glucanase and cellobiosyl hydrolase were being synthesized and, thus, a potent cellulase produced.

Repression Effect of Glucose on Cellulase Synthesis

The effect of glucose on the synthesis of several enzymes is well documented [15–17]. Repression of cellulase synthesis by glucose in a number of other fungi including *T. viride*, *Myrothecium verrucaria*, and *Neurospora crassa* have also been reported [18–20]. Glucose also appears to repress cellulase synthesis in *T. reesei.* To demonstrate the effect that glucose has on cellulase production, strains QM 9414 and MCG 77 were grown on 2% cellulose plus 1% glucose and cellulase synthesis followed (Fig. 4). The induction and synthesis of cellulase occurred after free glucose was completely consumed. However, the rate of cellulase synthesis in strain QM 9414 was not as high as in strain MCG 77, and the cellulase yields were also lower. The new mutant behaved quite differently: the level of mycelial protein reached its peak at a higher level and earlier than strain QM 9414. In addition, cellulase biosynthesis was fully recovered as soon as glucose was exhausted and a good yield of cellulase was achieved (Fig. 5). The final yields and rates of soluble protein synthesis achieved by strains QM 9414 and MCG 77 in these fermentations are compared to those in which 2% cellulose alone was used (Fig. 6). In the case of the new mutant, the final

TABLE II

Fermentation of Two Roll-Milled Cotton by Mutant Strains[a]

Strain	QM9414		NG14		MCG77	
1N NH_4OH ml/liter	105		174		128	
	Soluble Protein mg/ml	Productivity U/l · hr^{-1}	Soluble Protein mg/ml	Productivity U/l · hr^{-1}	Soluble Protein mg/ml	Productivity U/l · hr^{-1}
44 hr	0.77	8.	1.0	10.	0.53	7.
74 hr	---	18.	---	26.	---	19.
92 hr	3.3	24.	5.4	30.	4.3	29.
164 hr	9.2	33.	12.8	38.	10.9	38.
212 hr	11.2	39.	16.4	56.	12.8	47.
332 hr	13.6	30.	21.2	45.	16.2	32.

[a] Table shows synthesis of extracellular soluble protein by 3 mutant strains of *T. reesei* and the filter paper cellulase productivity (units/liter/hr) achieved. Fermentation conditions are described in Figure 3.

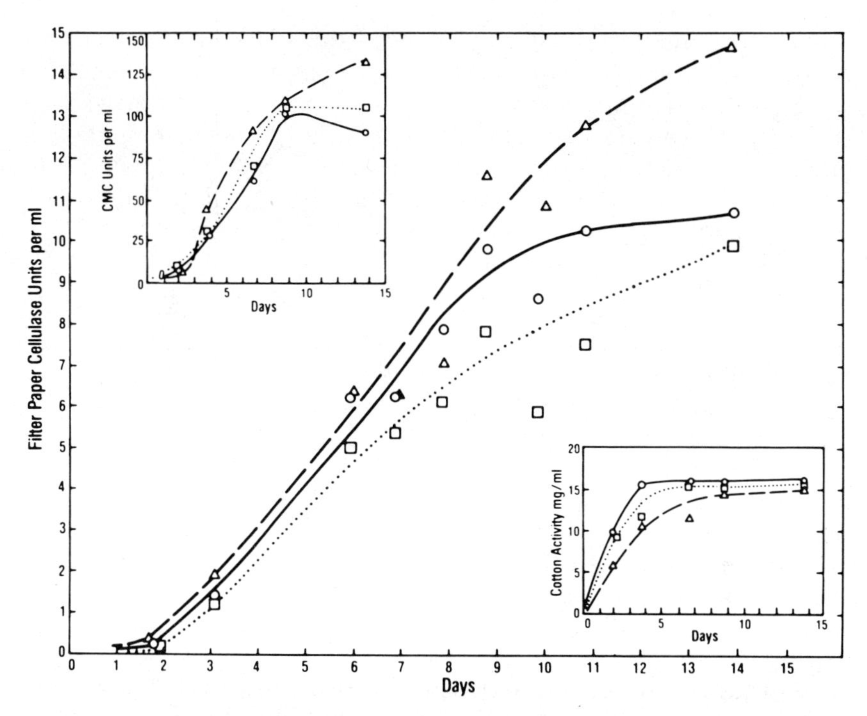

Fig. 3. Cellulase synthesis by three strains of *T. reesei* in a 6% cellulose fermentation: (○) MCG 77; (△) NG 14, and (□) QM 9414. Conditions: pH control, ≥3; substrate, 6% two-roll milled cotton; and inoculum, 20% (v/v) mycelial; Cellulase, CMC and filter paper cellulase units/ml and cotton activity as mg reducing sugar/ml.

yield of soluble protein and its rate of synthesis were almost the same in both fermentations whereas the soluble protein produced by strain QM 9414 when grown on the glucose-supplemented medium was appreciably less than in the fermentation on cellulose alone. The reduced cellulase yield and somewhat lower rate of synthesis in strain QM 9414 when grown on glucose supplemented media were previously reported by Mandels [21]. Similar effects are seen on the cellulase produced by other fungi [5, 22, 23].

The repression effect of free glucose on cellulase synthesis is again seen when 2% cellulose fermentations of strains QM 9414 and MCG 77 were fed a pulse of 1% glucose in the early stages of fermentation. Immediately after the injection of the glucose, the synthesis of soluble protein and cellulase was temporarily repressed in both fermentations. After the glucose was exhausted, strain MCG 77 resumed synthesis of soluble protein and cellulase at the same rate as before the addition of glucose and achieved about the same soluble protein and cellulase yields as the control fermentation which was not pulsed with glucose (Fig. 7). Strain QM 9414, however, showed a significantly lower rate of soluble protein synthesis and cellulase after the glucose was exhausted.

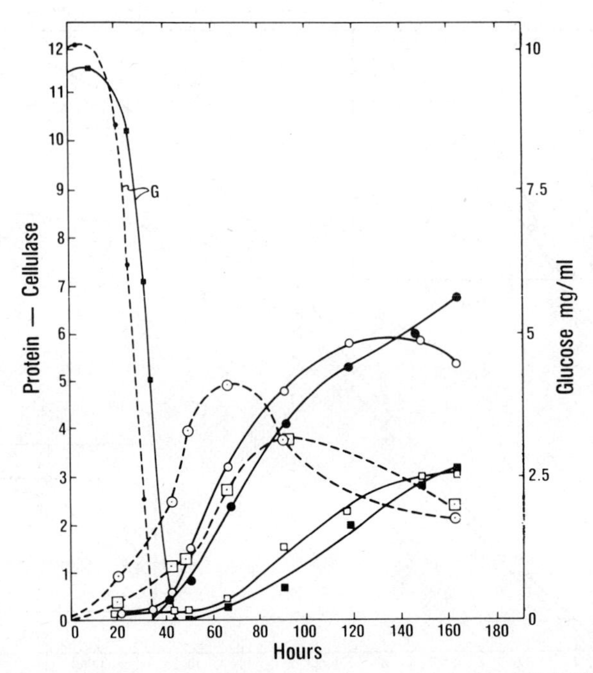

Fig. 4. Effect of glucose on cellulase synthesis in *T. reesei* strains MCG 77 and QM 9414. Conditions: pH control, ≥3; substrates, 2% ball-milled pulp (BW 200) plus 1% glucose; inoculum, 10% (v, v) mycelial; glucose, (G), MCG 77 (●) and QM 9414 (■); Cellulase: (●) filter paper cellulase units/ml MCG 77 and (■) and QM 9414; Protein: soluble protein as mg/ml: (○) MCG 77 and (□) QM 9414, and mycelial protein as mg/ml: (⊙) MCG 77 and (⊡) QM 9414.

These data indicate that cellulase synthesis in strain MCG 77 is repressed by free glucose and apparently is not affected by the repression (partial) of cellulase synthesis in the subsequent period as seen in strain QM 9414. The results also indicate that glucose-grown mycelium can be used to inoculate cellulose fermentations without any loss in cellulase yields.

Use of Glucose-Grown Mycelium to Increase Cellulase Production

To capitalize on the faster rate of cellulase synthesis of strain MCG 77 and the reduced initial lag period, several cellulose fermentations were carried out using susceptible cellulose (roll-milled and ball-milled) at 4% and 50% inocula, which were grown on glucose. The results show that very rapid growth and high enzyme productivity were achieved (Fig. 8). The average productivity of 72 filter paper units/liter/hr on the 84-hr batch cycle basis attained in this experiment is one of our best results to date. A very high level of cotton activity (17 mg/ml at 84 hr) was also achieved early in the fermentation.

Approximately half of the soluble protein and cellulase was synthesized during the growth phase of the fermentation. The mycelium grown on glu-

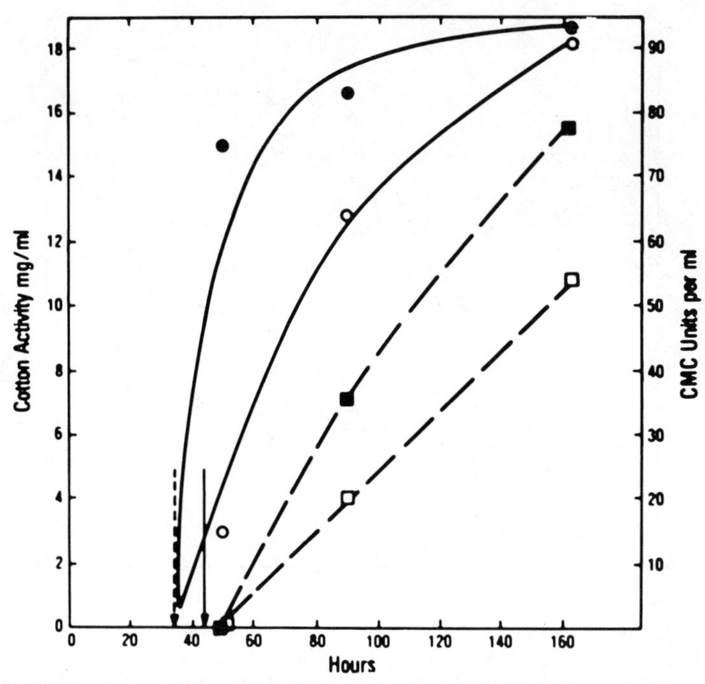

Fig. 5. Cellulase synthesis by strains MCG 77 and QM 9414 after their repression by glucose. Fermentation conditions are described in Fig. 4. Other conditions: cotton activity (mg reducing sugar/ml): (●) MCG 77 and (■) QM 9414; CMC activity (units of CMC/ml): (○) MCG 77 and (□) QM 9414; and glucose exhaustion, (¦) MCG 77 and (↓) QM 9414.

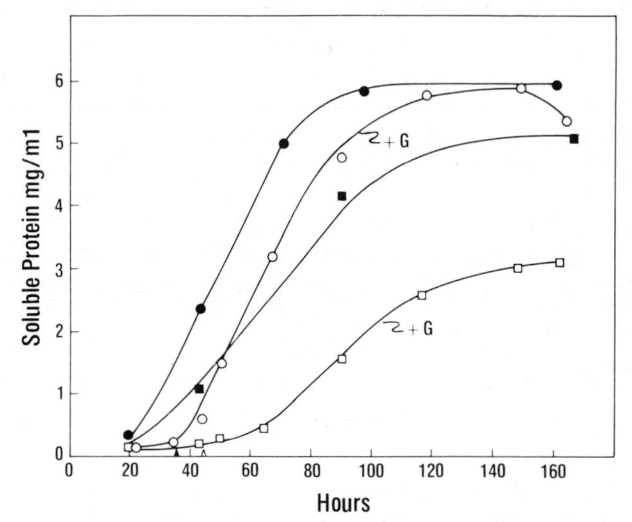

Fig. 6. Soluble protein synthesized in batch fermentations using 2% cellulose with and without 1% glucose supplement. Conditions: pH control, ≥3; strains: (■, □) QM 9414 and (○, ●) MCG 77; fermentor substrates: 2% ball-milled pulp (solid) and 2% ball-milled pulp + 1% glucose (open) (+G); inoculum, 10% (v/v) mycelial; and glucose exhaustion, (△) QM 9414 and (▲) MCG 77.

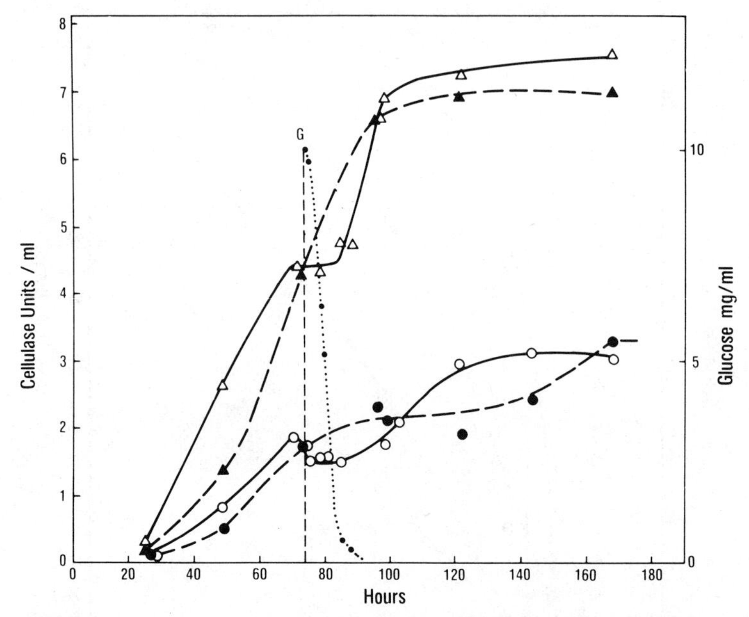

Fig. 7. Effect of a 1% glucose pulse on a 2% cellulose fermentation of strain MCG 77. Conditions: pH control, ≥3; growth fermentor substrate, 1% glucose (w/v); cellulose fermentor substrate, 1.5% two-roll milled pulp + 0.5% two-roll milled cotton (w/v); inoculum for growth fermentor, 10% (v/v) glucose-grown mycelium started with 2.8×10^9 conidia; inoculum for cellulose fermentors, 10% (v/v) glucose and fermentor-grown mycelium; glucose, G (· · ● · ·) pulsed at 74 hr; and cellulase units: Control cellulose fermentor FPU (●); CMC (▲); glucose-pulsed cellulose fermentor FPU (○), CMC (△).

TABLE III

Productivity of Soluble Protein Synthesis in Several Cellulose Fermentations[a]

Fermentor substrate and % (w/v)	pH Control	*T. reesei* Strains		
		MCG77	NG14	QM9414
		P. (Time)	P. (Time)	P. (Time)
BW 200 (2%)	≥5	60 (64 hrs)	39 (92 hrs)	-
BW 200 (2%)	≥3	63 (92 hrs)	43 (167 hrs)	36 (163 hrs)
Glucose (1%) + BW 200 (2%)	≥3	50 (117 hrs)	-	20 (148 hrs)
FB Cotton (6%)	≥3	48 (332 hrs)	63 (332 hrs)	40 (332 hrs)
FBSW 40 (4%)	≥3	110 (84 hrs)	-	-

[a] Summary of the productivities attained in cellulose fermentations using the three mutant strains of *T. reesei* in mg soluble protein/liter broth/hr and values were based on the highest amount of soluble protein produced in the fermentation. Fermentor residence time at which these values were attained is also given. Fermentation conditions for these data were previously described in the text and/or figures. Dash indicates that fermentations under these conditions were not run.

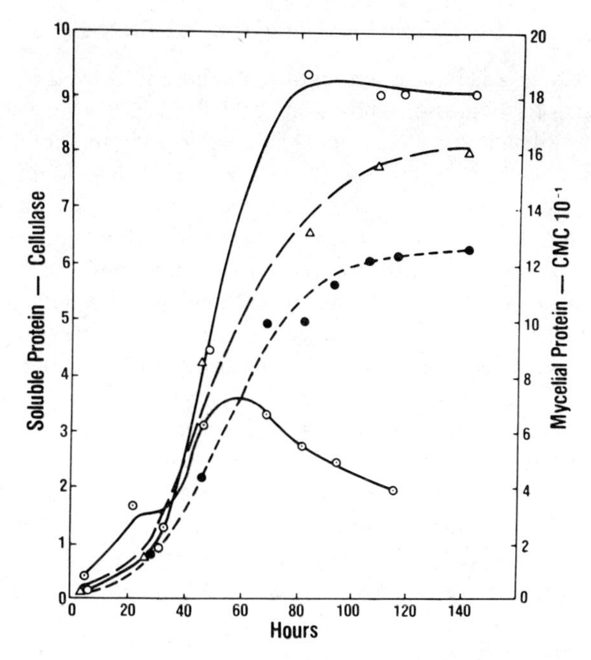

Fig. 8. Cellulase synthesis in a cellulose fermentation of strain MCG 77 started with a 50% mycelial inoculum. Conditions: pH control, ≥3.0; growth fermentor substrate, 2% glucose plus 0.5% two-roll milled pulp (w/v); cellulose fermentor substrate, 400 g two-roll milled pulp; inoculum for growth fermentor, 10% (v/v) three-day mycelium started with 5.03×10^8 conidia; inoculum for cellulose fermentor, 50% (v/v) glucose, fermentor-grown mycelium; cellulase, filter paper cellulase units/ml broth (●) and CMC units/ml (△); soluble protein mg/ml (○); and mycelial protein, mg/ml (⊙).

cose appears to be as good as that grown on a mixture of glucose and cellulose although the latter type of inoculum gave the higher cellulase productivity when 4% two-roll milled pulp was used (Fig. 8). The synthesis of soluble protein began within 3 hr after inoculation and continued at a relatively high rate for the first four days. Three of the four fermentations conducted in this study showed very similar profiles of soluble protein synthesis and achieved an average soluble protein productivity of 79 mg/liter/hr. The maximum productivity achieved was 110 mg soluble protein/liter/hr. Productivities based on soluble protein (the most reliable measure of extracellular cellulase in these experiments) for these and other fermentations are shown (Table III).

These results show that heavy mycelial inoculum of strain MCG 77 grown on glucose or a similar strain of *T. reesei* can effectively eliminate or reduce the initial lag in cellulase synthesis encountered in cellulose fermentations and significantly increase the productivity of the fermentation. Furthermore, higher productivities of cellulase could be maintained for a prolonged period by using heavier mycelium inoculum and higher concentrations of cellulose.

SUMMARY AND CONCLUSIONS

Productivity in a cellulase fermentation can be improved by optimization of media and fermentation conditions and by strain improvement, which enables us to alleviate or circumvent the genetic and metabolic controls that reduce or limit enzyme synthesis. So far, we found that *T. reesei* is the best organism to use for cellulase production because it produces high levels of a stable and complete cellulase complex. The wild strain produces little cellulase because of the restrictions placed on cellulase synthesis by genetic and metabolic controls. The mutant strains with enhanced cellulase productivity QM 9414 and NG 14 are also subject to catabolite repression. With these mutants the best yields are attained when growth is good but restrained by substrate limitation or low pH. In this paper, we discussed the results obtained in using new improved strains of *T. reesei*. Using these mutants we were able to increase productivity, also, by using higher concentrations of cellulose. For instance, high cellulase productivity was attained by using 6% two-roll milled cotton which contains about 80% fairly susceptible cellulose and about 20% residue that is very resistant to cellulase. The mutant strains grow rapidly on the susceptible portions of the cellulose, producing mostly biomass, and then continue hydrolyzing the resistant fraction of the cellulose with active enzyme production over a 14-day period, which results finally in broths containing up to 2% extracellular protein.

Strain MCG 77 gives yields similar to the other enhanced mutant strains QM 9414 and NG 14 in regular cellulase fermentations. However, this strain grows rapidly on soluble substrates without subsequent restrictions in cellulase synthesis and has a faster rate of metabolism and enzyme production. Therefore it is a good candidate for practical application. We should, however, continue to look for higher yielding mutants with faster rates of metabolism and thus higher cellulase productivities.

We thank Charles Macy and Thomas Tassinari of NARADCOM for two-roll milling several cellulose substrates used in this study. Also thanks are due Dr. A. S. Sussman, University of Michigan, Ann Arbor, for his editorial suggestions and comments. Dr. Gallo and Dr. Ryu were supported by the National Research Council Associateship Program at NARADCOM. Financial support for this research was also provided under Interagency Agreement No. E (49–28)-1007 with the U.S. Department of Energy. Dr. Montenecourt and Dr. Eveleigh kindly provided the *T. reesei* NG-14 strain for our comparative study.

References

[1] J. M. Nystrom and A. L. Allen, *Biotechnol. Bioeng. Symp.*, *6*, 55 (1976).
[2] B. J. Gallo, "A Cellulase Regulatory Mutant of *Trichoderma reesei*," in preparation.
[3] M. Mandels, J. Weber, and R. Patrizek, *Appl. Microbiol.*, *21*, 152 (1971).
[4] B. S. Montenecourt and D. E. Eveleigh, *Appl. Environ. Microbiol.*, *34*(6), 777 (1977).
[5] D. S. Sternberg, *Biotechnol. Bioeng.*, *18*, 1751 (1976).
[6] D. S. Sternberg and S. Dorval, *Biotechnol. Bioeng.*, in press.
[7] E. G. Simmon, *Abstracts of 2nd Internat. Mycol Congress*, Tampa, Fla., 1977, p. 618.
[8] M. Mandels, *Biotechnol. Bioeng. Symp.*, *5*, 681 (1975).

[9] T. Tassinari and C. Macy, *Biotechnol. Bioeng.*, *18*, 1321 (1970).
[10] O. H. Lowry, N. J. Rosebrough, A. L. Farr, R. J. Randall, *J. Biochem.*, *193*, 265 (1951).
[11] G. L. Miller, *Anal. Chem.*, *31*, 426 (1959).
[12] M. Mandels, R. Andreotti, and C. Roche, *Biotechnol. Bioeng. Symp.*, *6*, 21 (1976).
[13] R. E. Andreotti, M. H. Mandels, and C. Roche, *Proceedings*, *Bioconversion Symp.* (I.I.T., Delhi, 1977), p. 249.
[14] B. S. Montenecourt, D. E. Eveleigh, G. K. Elmund, and J. Parcells, *Biotechnol. Bioeng.*, *20*(2), 297 (1978).
[15] J. Mandelstam, *Biochem. J.*, *82*, 489 (1962).
[16] N. H. Horowitz and R. L. Metzenberg, *Ann. Rev. Biochem.*, *34*, 527 (1962).
[17] D. L. Hanks and A. S. Sussman, *Am. J. Bot.*, *56*(10), 1160 (1969).
[18] T. Nisizawa, H. Suzuki, and K. Nisizawa, *J. Biochem.*, *71*, 99 (1972).
[19] M. A. Hulme and D. W. Stranks, *J. Gen. Microbiol.*, *69*, 145 (1971).
[20] B. J. Gallo, "Cellulase System of *Neurospora crassa*," Ph.D. thesis, University of Michigan, Ann Arbor, 1977.
[21] M. Mandels and R. E. Andreotti, *Biochemistry*, in press.
[22] D. Boothby, *Phytochemistry*, *9*, 127 (1970).
[23] H. C. DeMenezes, T. J. B. DeMenezes, and H. V. Boa, Jr., *Biotechnol. Bioeng.*, *15*, 1123 (1973).
[24] H. J. Vogel, *Amer. Nat.*, *98*, 435 (1964).

Simultaneous Cellulose Hydrolysis and Ethanol Production by a Cellulolytic Anaerobic Bacterium

CHARLES L. COONEY, DANIEL I. C. WANG, SY-DAR WANG, JENNIFER GORDON, and MARGARITA JIMINEZ

Department of Nutrition and Food Science, Massachusetts Institute of Technology, Cambridge, Massachusetts 02139

INTRODUCTION

While there is much disagreement on when the availability of fossil fuels, especially oil, will begin to decline, there is little disagreement on the fact that the decline is inevitable [1]. In anticipation of this event, considerable attention is being directed toward the utilization of renewable resources for the eventual large-scale production of fuels and chemicals [2, 3].

Biomass in the form of cellulose, hemicellulose, and lignin provides a means of collecting and storing solar energy and hence represents an important energy and material resource. However, prior to the use of this resource, it is necessary to convert it to a usable form such as a gaseous (methane, synthesis gas) or liquid (ethanol, acetic acid, methanol) feedstock that will fit into our chemical economy. For this reason, a variety of strategies have been proposed and explored ranging from direct chemical methods such as pyrolysis to biological methods such as enzyme hydrolysis and fermentation. However, the high moisture content of biomass makes it less suitable for most chemical treatments and more suitable for biological processing.

One of the most publicized approaches toward biomass utilization is enzyme-catalyzed hydrolysis of cellulose and hemicellulose to low-molecular-weight components that can serve as substrates for fermentation to fuels or chemicals. Much work has focused on the use of enzymes produced from the fungus *Trichoderma reesei* [4, 5] grown aerobically on cellulose. Cellulases are isolated and used to effect cellulose hydrolysis in separate saccharification reactors. We present here an alternative approach in which cellulase production, cellulose hydrolysis, and fermentation are carried out simultaneously in a single operation. To achieve this, we utilize the anaerobic, thermophilic bacterium *Clostridium thermocellum*.

This direct approach to biomass utilization eliminates the need for separate enzyme production and hydrolysis reactors. Furthermore, it takes advantage of the fact that during enzyme production, cellulose can be converted directly to useful products such as ethanol and acetic acid. In

Biotechnology and Bioengineering Symp. No. 8, 103–114 (1978)

0572-6565/78/0008-0103$01.00

addition, since the microbial production of many other chemicals and fuels is achieved anaerobically, there is the possibility of using mixed-culture fermentations to achieve direct conversion of cellulose to a variety of products. Results presented here show the direct fermentation of cellulose to ethanol and acetic acid with the simultaneous accumulation of fermentable sugars.

MATERIALS AND METHODS

Microorganism

Clostridium thermocellum ATCC 27405 was used in all of these studies. Except where noted, it has been grown in the following CM3 medium (g/liter) [6]: KH_2PO_4, 1.5; K_2HPO_4,2.9; $(NH_4)_2SO_4$, 1.3; $MgCl_2$, 1.0; $CaCl_2$, 1.5; $FeSO_4$ (5% solution), 25 μl/liter; resazurin (0.2% solution), 1 ml/liter; yeast extract, 2.0; cysteine hydrochloride, and cellulose as Solka Floc SW-40 (Brown Co., New Hampshire) or ball-milled corn residue, or cellobiose; concentrations are noted in the text. Medium pH was initially adjusted to 7.6, except where noted, with NaOH before autoclaving and was reduced by gassing with prepurified N_2 or CO_2.

Culture Growth

Stock cultures are grown in Hungate anaerobic tubes (Bellco, Inc.) containing 9 ml CM3 medium plus 1 ml inoculum until cellulose is visibly degraded and turbidity appears. When refrigerated at 4°C, they can be stored up to five weeks. Transfers are made every three weeks.

Flask cultures are carried out in 500 ml Erlenmeyer flasks with 250 ml medium. The flasks are converted for anaerobic use [7].

Bench-scale fermentors (2 or 5 liter) were employed with agitation speed maintained at 100 rpm. Anaerobiosis was achieved and maintained by sparging N_2 or CO_2 during the fermentation. Ammonium hydroxide was used for pH control. The age of the culture inoculum was critical and was selected to be 18–30 hours to provide an actively growing culture.

Analytical Methods

Cell growth was monitored by turbidity (red filter on a Klett–Summersion Colorimeter) during growth on cellobiose and Solka Floc. During growth on Solka Floc, the broth was filtered on Whatman No. 1 filter paper; *C. thermocellum* would pass through the filter, and cell concentration could be measured turbidimetrically. Corrections were made for cells that adsorbed onto the cellulose by measuring in separate experiments cell adsorption as a function of concentration.

Residual solids are measured using 5 ml of the fermentation broth after centrifugation at 20,000 *g* for 15 min. Supernatant is withdrawn with a Pasteur pipette and the pellet is resuspended in 5 ml of 8% formic acid to achieve cell lysis. The residual solids are filtered through preweighed metricel GA-6 Nuclepore filters (diameter = 25 mm, pore size = 0.45 μm) and allowed to dry overnight at 75°C. The dried filter plus solids are weighed and the residual solids are determined as the difference.

Broth-reducing sugars were determined on supernatants using a colorometric dinitrosalysilic acid (DNS) reagent as described by Miller [8] with glucose as a standard.

Cellulolytic activity in the broth is measured in 0.2 ml supernatant added to 1 ml of a 2% carboxymethylcellulose (Sigma) dissolved in 0.05*M* sodium citrate buffer, pH 4.8. Supernatant (0.2 ml) is added to a third tube containing 1 ml of the buffer alone as a control. After 1 hr of incubation at 60°C, DNS reagent is added to measure reducing sugars.

Hydrolysis products from biomass are measured by either high-performance liquid chromatography (HPLC) or thin-layer chromatography (TLC). The HPLC was performed on a Waters Associates (Milford, Mass.) system with a Waters μBondapak column or a Whatman Partisil 10-PAC column. The operating conditions are flow rate, 1.5 ml/min; operating pressure, 800 psi; attenuation, 1×; chart speed, 10 mm/min; temperature, ambient; injection volume, 25 μl; solvent, acetonitrile; water, 80:20.

Xylose, glucose, and cellobiose were used to calibrate the column, and raffinose was included in each sample as an internal standard. When the Waters μBondapak carbohydrate column was used, the procedure was as follows: Samples are prepared by centrifugation at 12,000 rpm for 20 min. Two ml supernatant are transferred to tubes containing 500 mg of a mixed-bed ion exchange resin (Bio-Rad AG501-X8 consisting of equivalent amounts of AG5OW-X8, H^+ form cation exchanger, and AG1-X8 OH-form anion exchanger). To this mixture, 0.2 ml of an 11 mg/ml solution of raffinose is added such that the final concentration in the sample is 1 mg/ml. The sample and resin are contacted for 30 min at room temperature and then transferred to an Amicon Centriflo membrane cone, with a molecular weight cutoff of 25,000. The membrane cone and its support are centrifuged at 900 *g* for 30 min. The ultrafiltration serves two purposes: it removes large-molecular-weight compounds, especially proteins, that can interfere with column performance and produces a particulate-free ultrafiltrate that is ready for injection into the HPLC column.

As an alternative to the use of mixed-bed ion exchange resins for sample cleanup, a guard column packed with Whatman HC-Pellosil has been used in series with the Whatman Partisil 10-PAC analytical column. Samples are again centrifuged through Amicon ultrafiltration membrane cones for the removal of large-molecular-weight compounds. The filtrates are then directly injected into the HPLC system.

Ascending paper thin-layer chromatography was done with Whatman No. 1 filter paper. The solvent was *n*-butanol:acetic acid:water in the ratio 4:2:1. Aniline hydrogen phthalate (1%) was used as the developing agent.

Ethanol and acetic acid were measured by gas chromatography. Ethanol assay was measured in a stainless-steel column (5 ft × 1/8 in. o.d.) containing Chromosorb 101 (mesh 80/100). The temperature was isothermal (110°C) with injector and detector temperatures at 200 and 250°C, respectively. The flow rate of carrier N_2 was 40 ml/mm.

Acetic acid was measured in samples acidified with 2*N* HCl. A Teflon column (8 ft × 1/8 in. o.d.) containing Chromosorb 101 was used. Temperature was isothermal (170°C) with injector and detector temperatures and carrier gas flow as noted above.

RESULTS

Direct Fermentation of Cellulose

Clostridium thermocellum is an obligately anaerobic bacterium with the ability to utilize cellulose [6]. Results shown in Figure 1(a) illustrate the kinetics of growth for *C. thermocellum* on ball-milled corn residue. Cell growth in this experiment was monitored by carbon dioxide production using the empirically derived correlation. This method of cell mass estimation was found to be particularly useful for growth on solid substrates. In

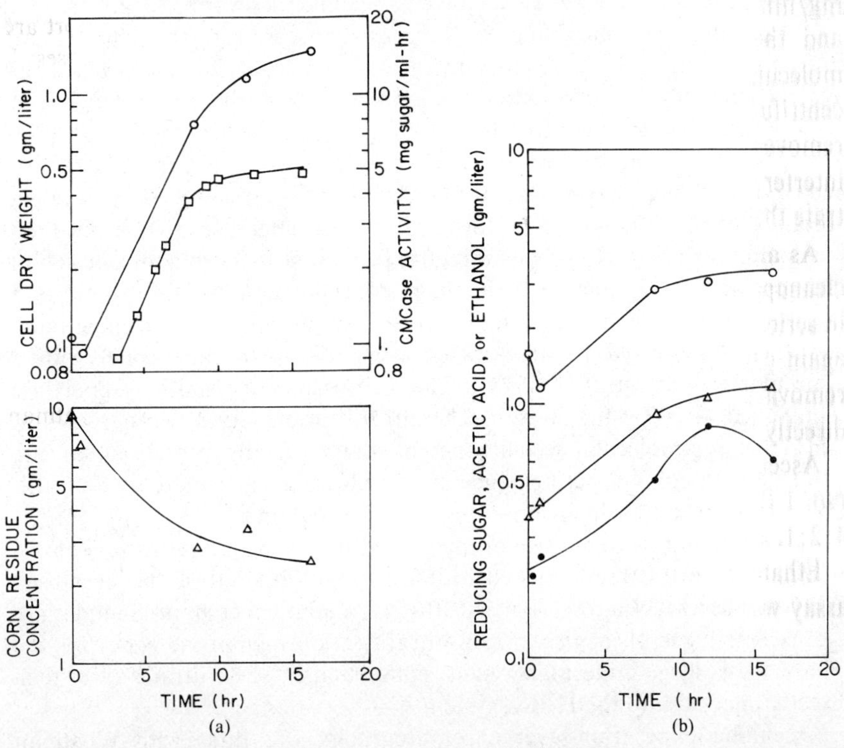

Fig. 1. Kinetics of growth and product formation of *C. thermocellum* growing on corn residue at 60°C and pH 6.9. (a) Cell growth, enzyme production, and substrate consumption, (O) CMCase, (□) cell dry weight, (Δ) corn residue solids. (b) Reducing sugar, acetic acid, and ethanol accumulation during the fermentation. (O) Reducing sugar, (Δ) acetic acid, (●) ethanol.

TABLE I

Specific Rate of Product Formation by *Clostridium thermocellum* on Different Cellulose Substrates and Cellobiose

TIME OF FERMENTATION (hr)	SPECIFIC RATE OF PRODUCT FORMATION		
	REDUCING SUGAR (g /g cell-hr)	ETHANOL (g /g cell-hr)	ACETIC ACID (g /g cell-hr)
CORN RESIDUE			
3	0.53	0.23	0.34
4	1.03	0.20	0.32
6	0.98	0.19	0.36
8	0.96	0.23	0.38
10	0.52	0.23	0.16
SOLKA FLOC			
4	0.35	0.11	0.07
7	0.36	0.13	0.05
10	-	0.16	0.06
CELLOBIOSE			
4		0.26	0.16
6		0.27	0.20
8		0.20	0.12
10		0.12	0.10
12		0	0.11

this fermentation, the initial concentration of corn residue was 9.5 g/liter and 2.5 g/liter remained at the end, which corresponds to a net degradation of 74%. The maximum specific growth rate was 0.3 hr^{-1} with the pH and temperature controlled at 6.9 and 60°C, respectively. Also shown in Figure 1(a) is the cellulase activity in the broth measured as CMCase. The enzyme activity assayed illustrates the growth-associated accumulation of extracellular and unbound cellulase. However, this measured enzyme activity is not as meaningful as the *in vivo* rate of cellulose hydrolysis.

It is important to note here that during growth of *C. thermocellum* on cellulosic biomass at pH values less than 7.1, there is substantial accumulation of reducing sugars, as well as fermentation products resulting from cellulose utilization. This fact can be seen from the results in Figure 1(b); at the end of the fermentation, there were 3.4 g/liter reducing sugars, 1.2 g/liter acetic acid, and 0.5 g/liter ethanol (0.8 g/liter at its maximum).

The sugars accumulated in the broth during cellulose (Solka Floc) utilization are shown to be primarily glucose, cellobiose with tracer of xylose (not shown); the kinetics of sugar accumulation are shown in Figures 2(a) and 2(b). The sum of glucose and cellobiose do not equal the total reducing sugars, which is due to the presence of some higher-molecular-weight products in the broth and to some loss of sugars in preparation for separation on HPLC.

The important point to be made here is that it is possible to convert, in a single-step fermentation, biomass to sugar and chemicals, for example, ethanol and acetic acid. Some typical results for the utilization of corn residue, Solka Floc, and cellobiose are compared in Table I. These calcula-

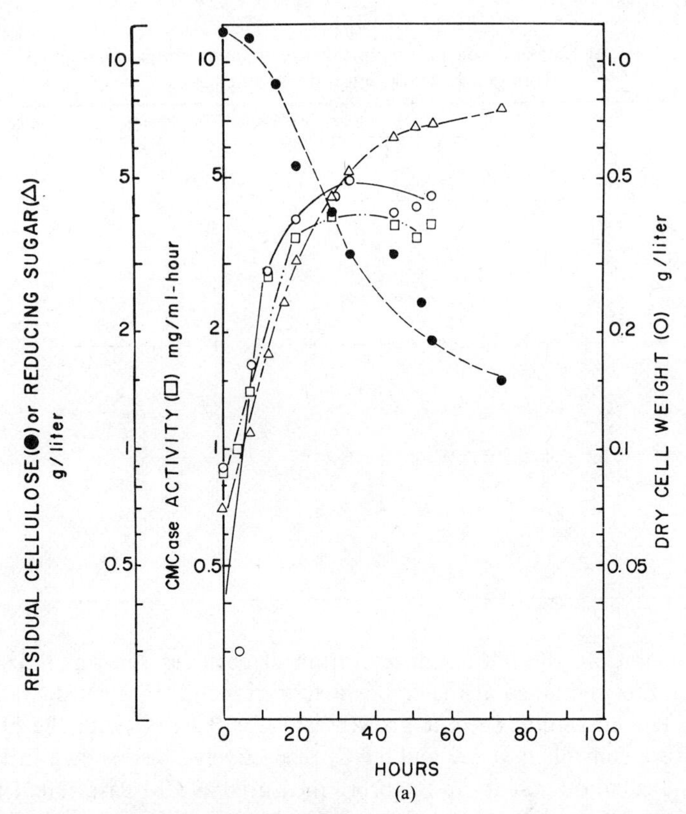

Fig. 2. Kinetics of growth and product formation for *C. thermocellum* growing on Solka Floc. (a) Cell mass, cellulase, as CMCase and reducing sugar accumulation (△) reducing sugar, (○) dry cell weight; (□) CMCase, (●) residual cellulose. (b) Results of HPLC analysis of the sugars accumulated in the broth. (○) Total reducing sugar; (▲) glucose, (□) cellobiose. Analysis was done with the Whatman column.

tions are the specific rates of sugar, ethanol, and acetic acid accumulation. Interestingly, the results show that with respect to cellulose degradation and product formation corn residue shows the highest rates.

Ethanol Production

Our objective here is to develop further the direct production of ethanol from cellulose. However, we used in these studies Solka Floc as the substrate which permitted an easier examination of the kinetics of the fermentation. In order to increase the concentration of end products, a fed-batch fermentation was performed and the results are shown in Figure 3. Solka Floc was added at intervals during the 60-hr fermentation. The ethanol and acetic acid accumulation reached 4 g/liter each. A total of 47.5 g cellulose was fed and 16 g/liter remained at the end. This represents 66% degradation of the cellulose. The reducing sugar concentration reached 8.2 g/liter.

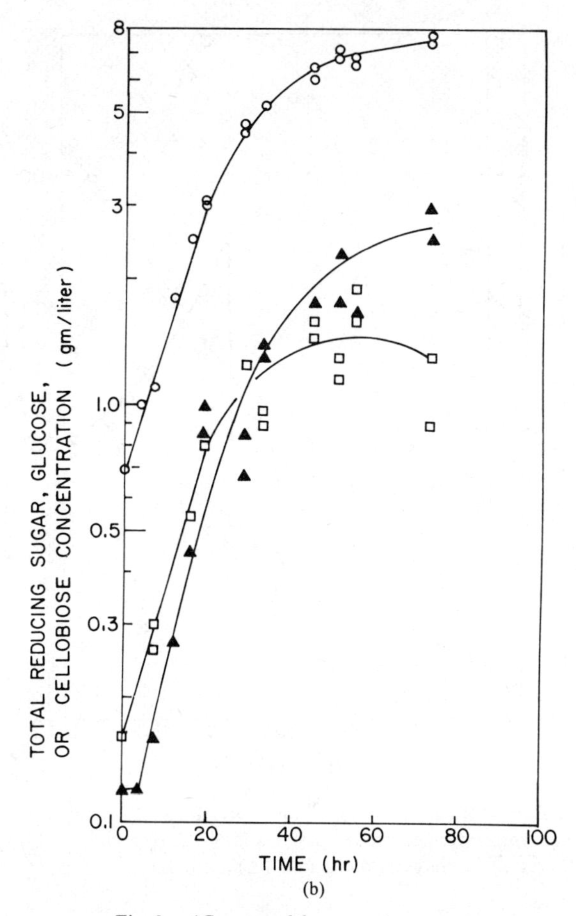

(b)

Fig. 2. (*Continued from previous page.*)

This fermentation was controlled at a pH of 7.1 and, during the active growth period, paper chromatographs of the broth showed that most of the sugars accumulated were pentoses (about 10% of the Solka Floc is xylan). However, when growth slowed down (at about 40 hr) cellobiose begins to accumulate. From the results it appears that cellulose hydrolysis can proceed even after growth has stopped and the production of ethanol and acetic acid continues although at a reduced rate.

A summary of three fermentations using *C. thermocellum* on varying amounts of cellulose (as Solka Floc) is presented in Table II. From these results, it is seen that a yield of 0.2 to 0.25 g each of ethanol and acetic acid per gram of cellulose degraded is obtained as well as substantial amounts of reducing sugars. The yield of total products is typically 40 to 50%.

If this direct cellulose fermentation is to become a technically viable alternative, it is essential that the amount of accumulated ethanol be increased. A major problem at this time is the alcohol inhibition on growth.

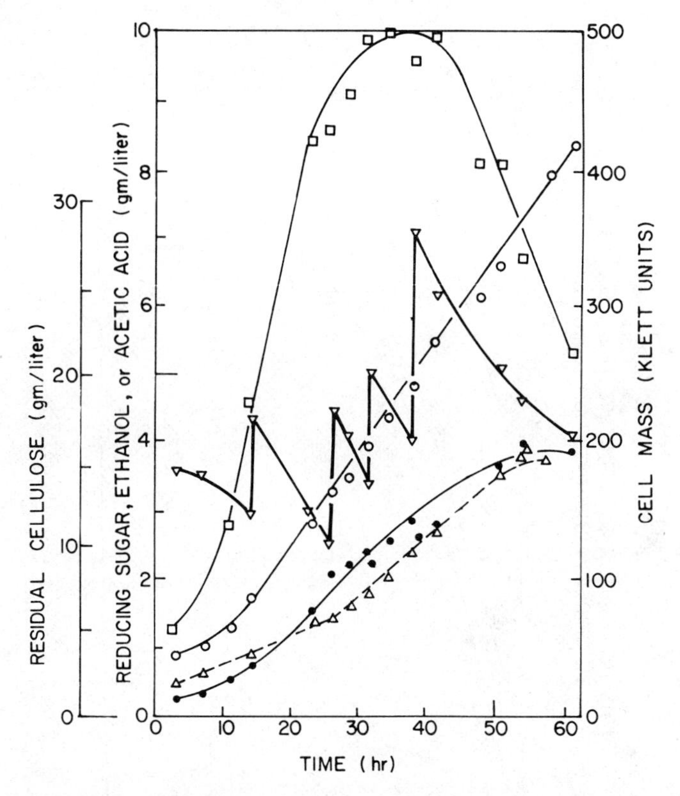

Fig. 3. Growth and product formation by *C. thermocellum* in a fed-batch fermentation with Solka Floc; pH and temperature controlled at 7.1 and 60°C, respectively. (□) Cell mass; (○) reducing sugar, (▽) residual cellulose; (●) ethanol, (△) acetic acid.

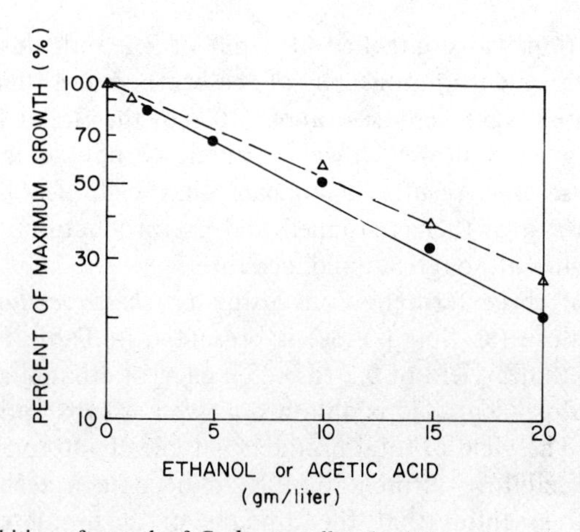

Fig. 4. Inhibition of growth of *C. thermocellum* by (△) acetic acid and (●) ethanol.

TABLE II

Summary of Results from *C. Thermocellum* Grown on Varying Concentrations of Solka Floc

RUN	A	B	C
Cellulose Degraded (g/ℓ)	7.2	9.6	31.6
Cell Dry Weight (g/ℓ)	0.8	1.1	1.5
Reducing Sugar on Accumulated (g/ℓ)	1.9	1.9	8.3
Yield of Reducing Sugars (g/g)	0.26	0.20	0.26
Ethanol Produced (g/ℓ)	1.1	2.0	4.8
Yield of Ethanol (g/g)[1]	0.21	0.26	0.21
Acetic Acid Produced (g/ℓ)	1.3	1.9	3.8
Yield of Acetic Acid (g/g)[1]	0.25	0.25	0.16
Yield of Acetic Acid[1] and Ethanol (g/g)	0.46	0.51	0.37

[1] Yields are calculated by dividing the ethanol or acetic acid produced by the differences between the cellulose degraded and the reducing sugar accumulated.

Shown in Figure 4 are the results of the effect of alcohol and acetic acid concentration on the inhibition of growth of this organism. To overcome this problem, *C. thermocellum* was carried through a serial transfer in medium containing 10 g-liter cellobiose in which the ethanol concentration was gradually increased from 10 to 40 g/liter. After 40 transfers, an isolate was obtained that would grow in the presence of 40 g/liter ethanol. This culture, S-2, was then transferred 15 more times on medium with 10 g/liter cellulose and a single-colony isolate was selected, S-4. This organism is tolerant to more than 5 vol % ethanol. To test its ability to produce ethanol, this new isolate (S-4) was grown on cellobiose, added incrementally to an anaerobic flask with 250 ml medium. pH was controlled manually, one or two times per day. The results are shown in Figure 5. After 100 hr, 32 of the total 45 g/liter of cellobiose were utilized, and 9 g/liter of ethanol (corrected for dilution) were produced; a yield of 0.28 g ethanol/g glucose was achieved. Although not shown, substantial amounts of acetic acid are also accumulated. Even after 100 hr, ethanol production was continuing at about 0.1 g ethanol/liter hr.

DISCUSSION

Results presented here focus on the direct fermentation to degrade cellulose to useful products. This approach avoids separated processes for enzyme production and subsequent cellulose hydrolysis. As a consequence of this simplified scheme, there is the possibility of reducing both the capital investment and operating cost when compared to a two-step process. This

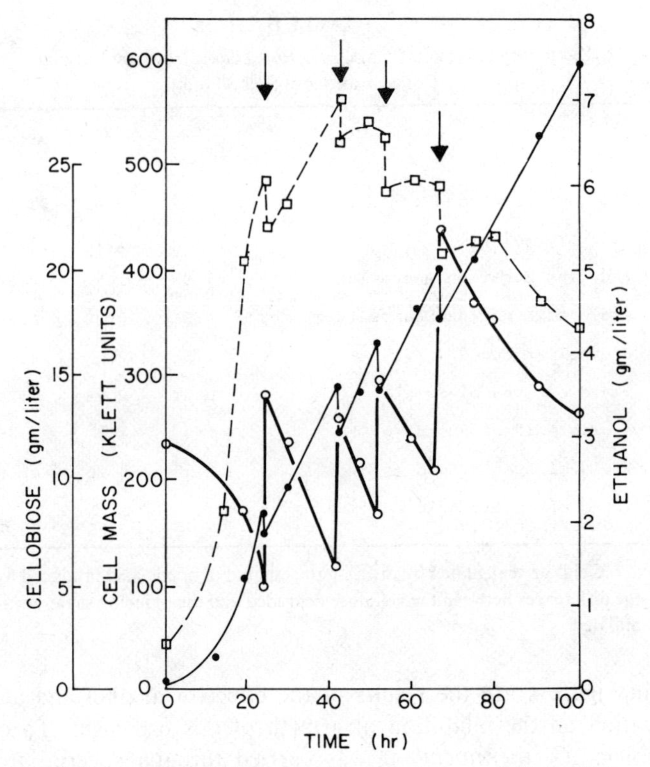

Fig. 5. Fed-batch fermentation of *C. thermocellum* growing on cellobiose which was added at the time marked by arrows. (●) Ethanol; (□) cell mass; (○) cellobiose.

direct route is summarized by the block diagram in Figure 6 for a hypothetical example of 200×10^6 kg/year of corn residue.

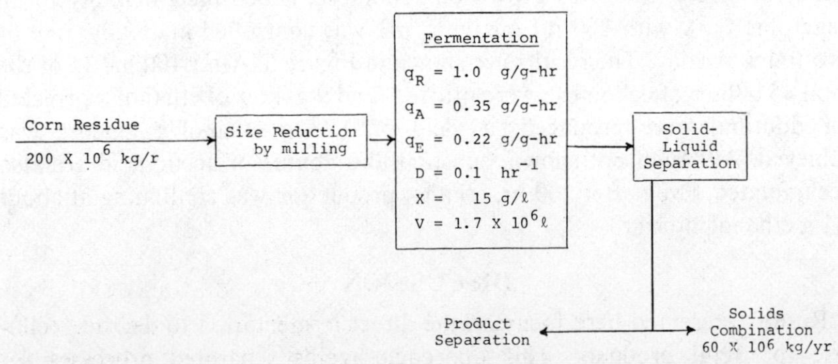

Fig. 6. Block diagram of a direct cellulose fermentation process scheme for a facility handling 200×10^6 kg/corn residue fed/year with a fermentable polysaccharide content of 70% on a dry weight basis. Ethanol: 16×10^6 kg/yr; yield = 0.11 g/g. Acetic acid: 25×10^6 kg/yr; yield = 0.18 g/g. Reducing sugar: 71×10^6 kg/yr; yield = 0.51 g/g.

The concept of a single-step cellulose fermentation has received little attention for the production of materials such as ethanol, although it has been investigated for methane production. For example, Pfeffer [9] and Cooney and Wise [10] used single-stage, continuous, thermophilic anaerobic digestors to convert cellulose to methane. Interestingly, it has been noted [10, 11] that substantial amounts of organic acids can be accumulated in these digestors. In spite of these and many other studies, the most common approach for cellulose utilization has been to separate the enzyme production and cellulose hydrolysis stages.

The production of ethanol and acetic acid is coupled with growth and culture maintenance since they are catabolic products of cellulose, while reducing sugar accumulation is not coupled with growth. Material balances for product formation may be written as

$$\frac{dP_A}{dt} = q_A X - DP_A \qquad \text{(acetic acid)} \tag{1}$$

$$\frac{dP_E}{dt} = q_E X - DP_E \qquad \text{(ethanol)} \tag{2}$$

$$\frac{dP_R}{dt} = q_R X - DP_R \qquad \text{(reducing sugars)} \tag{3}$$

where P_A, P_E, and P_R are the concentrations of acetic acid, ethanol, and reducing sugar; q_A, q_E, and q_R are the corresponding specific productivities (g product/g cell hr); D is the dilution rate; and X is the cell concentration.

Average values (Table I) for the specific productivities observed during growth on corn residue are $q_A = 0.35$, $q_E = 0.22$, and $q_R = 1.0$ g/g cell hr. From knowledge of the maximum specific growth rate for *C. thermocellum*, which is about 0.3 hr^{-1} on cellulose, we can examine the steady-state dependence of product concentration on cell concentration. Typically, one might operate at a dilution rate $D = 0.10$ hr^{-1}. Therefore, to achieve an ethanol concentration of 50 g/liter, a cell density of 23 g/liter is required. This is far above the value of 6 g/liter observed by Wang and Fleischaker with *C. thermoaceticum* (unpublished results) and 8 g/liter observed by Wise et al. [12] for a mixed culture of anaerobes in a continuous culture system with recycle. In the example in Figure 6, we choose an achievable value of 5 g/liter for cell mass.

There are, however, several alternative approaches to the problem. One method is to increase cell densities by cell recycle; because many of the cells are adsorbed to the residual solids in a cellulose fermentation, this is a viable approach. However, of greater impact is the control of cellular metabolities such that ethanol is formed rather than acetic acid and hydrogen gas and such that the reducing sugars are converted to ethanol rather than allowed to accumulate. This additonal sugar utilization could be achieved in part by improved metabolism of *C. thermocellum* or the addition of a second culture. The latter route is attractive to achieve utilization of sugars such as glucose and xylose that are not metabolized by *C. thermocellum*. Continuing work on this direct fermentation is focused on both of these approaches toward process improvement.

Examining further the example in Figure 6, it is seen that the total product yield on degraded solids is 29% for acetic acid and ethanol. If the ethanol were produced instead of acetic acid and if the sugars could be converted to ethanol at a conservative yield of 0.3 g/g, then the overall yield of ethanol would be 35% of the solids degraded or 25% of the total solids fed. These values are felt to be the goals for this approach to the direct utilization of cellulose.

The authors wish to acknowledge the support of Department of Energy Grant No. Eg-77-S-02-4198 and the Corning Glass Works Fellowship for support of J. Gordon.

References

[1] C. L. Wilson, *Energy: Global Prospects 1985–2000* (McGraw-Hill, New York, 1977).
[2] E. S. Lipinsky, *Science, 199*, 644 (1978).
[3] E. L. Gaden, Jr., M. H. Mandels, E. T. Reese, and L. B. Sparo, Eds., *Enzymatic Conversion of Cellulosic Materials, Biotechnol. Bioeng. Symp. 6* (1976).
[4] M. Mandels, L. Hontz, and J. Nystrom, *Biotechnol. Bioeng., 16*, 1471 (1974).
[5] D. Sternberg, *Biotechnol. Bioeng. Symp., 6*, 35 (1976).
[6] R. J. Weimer and J. G. Zeikus, *Appl. Environ. Microbiol., 33*, 289 (1977).
[7] L. Daniels and J. G. Zeikus, *Appl. Environ. Microbiol., 29*, 710 (1975).
[8] G. L. Miller, *Anal. Chem., 31*, 426 (1959).
[9] J. T. Pfeffer, *Biotechnol. Bioeng., 16*, 771 (1974).
[10] C. L. Cooney and D. L. Wise, *Biotechnol. Bioeng., 17*, 1119 (1975).
[11] C. L. Cooney and R. B. Ackerman, *Eur. J. Appl. Microbiol., 2*, 65 (1975).
[12] D. L. Wise, C. L. Cooney, and D. C. Augenstein, *Biotechnol. Bioeng.*, in press.

Solar Bioconversion Systems Based on Algal Glycerol Production

L. A. WILLIAMS,* E. L. FOO, A. S. FOO, I. KÜHN, and C.-G. HEDÉN

Microbiological Engineering Research Group, Department of Bacteriology, Karolinska Institute, S-104 01 Stockholm, Sweden

INTRODUCTION

The basic human needs of any society at any level of development are often summarized by the so-called five *F*s: food, fodder, fuel, fertilizer, and fiber. Any society that wishes to live within its energy income, which we all shall eventually be forced to do, must fulfill these basic needs through the utilization of only solar energy, in its various forms. In principle, all these human needs can be satisfied by photosynthetic energy capture and carbon and nitrogen fixation of higher plants in conjunction with some associated microorganisms. Although such agrarian systems have in the past supported all mankind and currently support a large fraction of it, it seems likely that the demands of a very large human population and rising expectations regarding the standard of living will preclude any voluntary and smooth return to an idyllic agrarian past.

What will be required is the purposeful application of microbiological and technical knowledge accumulated only within the last 100 years in order to devise and develop systems for solar energy bioconversions which are more useful to modern man, due to intrinsically higher rates or efficiencies or to the diversity and utility of products produced. In such bioconversion systems it is desirable that the primary photosynthetic energy-capturing process be coupled as closely and efficiently as possible to the further bioconversion processes. In addition, it is necessary that such bioconversion processes be operated in the form of integrated process systems, where the by-products and wastes of one process serve as the raw materials and nutrients of another—in other words, maximum possible recycling of nutrients in all their forms. Finally, it is desirable that such bioconversion systems be sufficiently simple and flexible to allow their use at all levels of technology, ranging from highly industrial settings to the village level.

Bioconversion systems based on lignocellulosic materials have been much discussed and studied in recent times. Perhaps their greatest attraction is their ability to utilize readily available wastes and to yield a wide variety of

* Present address: Dept. of Viticulture and Enology, University of California, Davis, CA 95616.

Biotechnology and Bioengineering Symp. No. 8, 115–130 (1978)
 0572-6565/78/0008-0115$01.00

products, including SCP, liquid fuels, solvents, and chemicals, through fermentations of the primary bioconversion products, that is, glucose and other carbohydrates. However, there are also many associated problems such as collection, pretreatment, and hydrolysis, which are the subject of continuing work. On the other hand, bioconversion systems based on micro- and macro-algae have been aimed mostly at only a few products: crude algal SCP, alginate and other hydrocolloids, methane, and perhaps nitrogen-rich biofertilizers (in conjunction with wastewater treatment in many cases).

However, there does exist an algal photosynthetic system that fixes the major proportion of its carbon in the form of a low-molecular-weight, soluble, polyhydroxyalcohol (glycerol), which is very well suited as substrate for a variety of bioconversion processes. (This system based on the halophilic algae *Dunaliella* is described in more detail below.)

This glycerol-producing system is most attractive because it combines many advantages of both lignocellulose and algal systems and permits great flexibility in further bioconversions. In addition to the obvious products such as glycerol itself and crude algal protein, other possible products of bioconversion include yeast SCP, mannitol, fructose, ethanol, and various other organic acids and solvents. In addition, the crude algal glycerol may serve as the energy source for biological nitrogen fixation or as the electron donor for photosynthetic hydrogen production. These various hypothetical conversion routes are summarized in Figure 1.

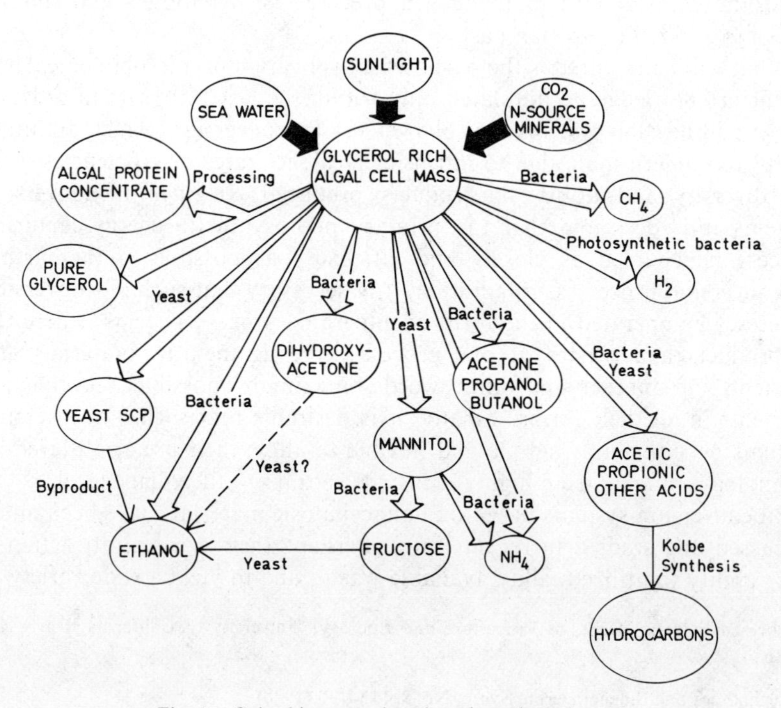

Fig. 1. Solar bioconversions based on algal glycerol.

TABLE I

Summary of Literature on *Dunaliella* Growth Characteristics

Ref.	Species	NaCl M[a]	Temperature °C	pH	Doubling Time hours	Maximum Population 10^6 cells/ml
(1)	D. bioculata	1.0	33.5+	-	5.7	7.2
	D. primolecta	0.5	33.5+	-	5.4	4.0
(2)	D. viridis	1.5+	30+	7-9	-	39.0
	D. salina	1.5+	30+	7-9	-	1.9
(3)	D. tertiolecta	0.02-1.0+	18	-	24[b]	5.0
(4)	D. tertiolecta	0.5-0.75	33+	-	4.8	-
(5)	D. tertiolecta	0.5	25+	-	10	-
(6)	D. tertiolecta	-	20	-	9	-
(7)	D. viridis	1.0-2.0+	25	7.5	19[b]	-
(8)	D. parva	1.5	32	7.4	15	2.0[b]
(9)	D. tertiolecta	1.0	20+	6.0+	17[b]	-
(10)	D. tertiolecta	1.2+	20	6.0+	-	-
(11)	D. tertiolecta	0.5-2.0+	28	8.2	10[b]	1.0[b]
(12)	D. sp.	2.75+	32+	6.2	10	-
(13)	D. parva	1.5	25	7.4	15	4.0
(14)	D. tertiolecta	0.2+	26	7.5	10	-
	D. viridis	2.0+	26	7.5	20	-
(15)	D. tertiolecta	1.4+	-	6.0	20	28.0
(16)	D. salina	2.0+	33+	7.5	24	100.0
(17)	D. sp.	1.4-2.9+	-	-	-	2.0
(18)	D. parva	1.5	25	7.5	15	-

+ Designates "optimum" conditons found by authors.

[a] NaCl molarities are only approximate due to conversion from other units.

[b] Indicates approximate values estimated from reported results.

ALGAL SUBSYSTEM

Algae of the genus *Dunaliella* are found in a wide range of environments, varying from nearly fresh water to saturated salt solutions. Some species are euryhaline while others have a definite halophilic character. Over the years, the growth characteristics of many species have been investigated for various reasons; much of this information is summarized in Table I [1–18].

Nutrition and Environment

Most species grow well on an inorganic salts medium, with no organic growth factors apparently required. Heterotrophic growth is not common. Table I clearly shows the wide range of salt concentrations over which growth occurs; there is some disagreement with regard to optima, even within the same species. It has been shown for several species that seawater (approximately 0.5*M* NaCl), at concentration factors of up to 10 times the original, provides an excellent growth medium, when supplemented with a nitrogen source and phosphate, and it has been suggested that the high osmotic pressure of seawater concentrates becomes inhibitory before any of the individual ions reach toxic levels [2, 3]. It has also been shown that *Dunaliella* supernatant brines may be reused, after nutrient augmentation, for further cultivations with no signs of autoinhibition [1, 19].

Ammonium, nitrate, and urea can serve as good nitrogen sources, with ammonium seemingly the best under most circumstances [2, 20, 21]. Uric acid and some amino acids are also utilized to lesser extents by some species [2]. The half-saturation constants for uptake of nitrate and ammonium have been reported to be $1.4 \mu M$ and $0.1 \mu M$, respectively, for *Dunaliella tertiolecta* [22].

The form of inorganic carbon utilized seems to be the dissolved gas species [23]. Evidence has been presented indicating a major role for the enzyme carbonic anhydrase for making dissolved carbon dioxide available from bicarbonate, especially under conditions of low carbon dioxide solubility, such as high salinity and high temperature [11, 23].

Several studies on temperature relationships have been indicated optima of 30–33°C and at least some tolerance for higher temperatures [5, 12]. The pH values used by most workers lie in the range 6–8. The optimal pH is somewhat uncertain. The response of *Dunaliella* species to light intensity is very similar to other green algae, though they can perhaps withstand extremely high intensities better [5, 12]. As seen in Table I, the best doubling times reported have been 5–6 hr, although reports of 10–15 hr have been much more usual. Maximum populations reported are usually in the range of $1\text{–}10 \times 10^6$ cells/ml, corresponding approximately to 0.1 to 1.0 g dry weight/liter [1, 15]. However, much higher populations have also been achieved in some instances [2, 15, 16].

Glycerol Production

The fact that glycerol is an important photosynthetic product in *D. tertiolecta* was first discovered in 1964 by Craigie and McLachlan [24]. They also showed a strong dependence of glycerol production on the salt

TABLE II

Glycerol Production by *Dunaliella*

Ref.	Species	grams intracellular glycerol / liter algal culture	Molar glycerol in cell	Molar NaCl in medium
(24)	D. tertiolecta	0.02 - 0.2[a]	-	-
(25)	D. parva	0.08 - 0.12[a]	2.1	1.5
(14)	D. tertiolecta	0.01 - 0.1[a]	1.4	1.4
	D. viridis	0.1 - 1.0[a]	4.4	4.2
(15)	D. tertiolecta	0.09 - 1.1	2.0[a]	2.7
(18)	D. parva	0.05 - 0.1[a]	2.1	2.1

[a] These values are very approximate estimates based on insufficient information given.

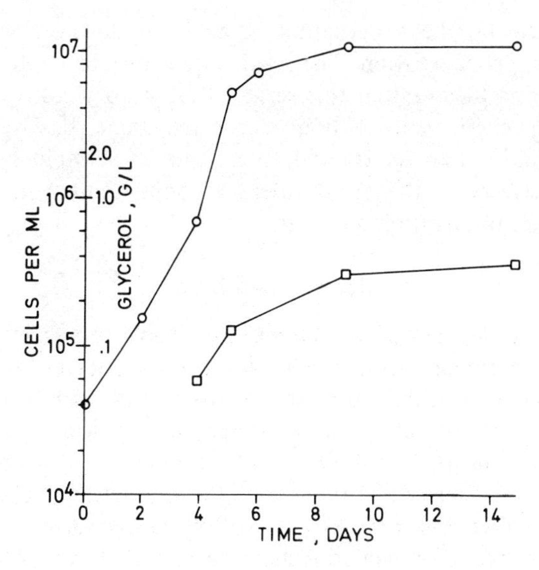

Fig. 2. Growth and glycerol production by *D. parva:* mineral medium with 2.5*M* NaCl [13], 5% CO_2 in air at 0.1 v/v/m, 30°C, approximately 8000 lx, 5 cm culture depth. (○) Cell count, (□) intracellular (pellet) glycerol.

concentration of the medium. Since then, the role of photosynthetically produced glycerol in the osmotic regulation of *Dunaliella* species growing at high salinities has been further elucidated by groups in Germany [10, 15], Israel [18, 25], and Australia [14].

From a practical viewpoint, the important results of this work are the amounts of intracellular (i.e., harvestable) glycerol attainable, expressed as intracellular grams per liter of cell suspension and also as absolute concentration inside the cells. Some of this information, reported or estimated from the literature, is shown in Table II. In very dense cultures, the intracellular glycerol can approach 1 g/liter cell suspension. For *D. tertiolecta* at 16% NaCl, glycerol constitutes 82% of the "dry weight," and the cell contents are estimated to be a 15% (v/v) glycerol solution [15]. Internal concentrations in extremely halophilic species may be even higher [14].

Preliminary Experimental Results

To date, our work has been limited to simple, small-scale indoor experiments with one strain of *D. parva*. The results of an early cultivation in a simple sparged bottle are shown in Figure 2. As expected, the pellet (intracellular) glycerol content follows the increase in cell number, reaching a value of 0.32 g glycerol/liter cell suspension. (These preliminary results have not been corrected for an unknown concentration factor caused by evaporation.) In addition, a 30-liter scale cultivation has been carried out successively reusing the supernatant brine from the previous cultivation, with nutrient supplementation. The results are shown in Table III.

It is interesting to note that this series of laboratory cultivations in 2.5*M* NaCl medium, carried out without sterile technique, has resulted in low-level contamination by bacteria and at least one as yet unidentified mycelial fungus. The effects of these halotolerant contaminants on yield are not certain but seem to be minimal so far.

Harvest and Lysis

Ease of harvesting is one of the most desirable properties of any algae intended for biomass production. *Dunaliella* species are unicellular flagellates which probably remain in suspension mostly due to their motility. They are usually ovoid in shape, with a width of 6–7 μm and length of 10–12 μm [6, 7, 26]. *Dunaliella tertiolecta*, grown in seawater, is reported to have a density of 1.08 g/cm^3 [26]. The settling velocity of *D. tertiolecta* in seawater has been experimentally determined to be 1.6 cm/hr and the corresponding calculated density, to be 1.15 g/cm^3 [27].

Cultures at higher salinities would have to contend with higher densities of the suspending medium, but the cells would also increase in density due to the higher content of glycerol (ρ = 1.26 g/cm^3). Neutral buoyancy might occur at the same point, which is very undesirable. The relationship between salinity and cell density must be investigated very thoroughly for several species.

Any method that inhibits the natural motility of the cells to allow gravity sedimentation is probably the simplest and cheapest method, at least for the initial concentration of cultures. The fact that cultures sediment at low temperatures has been mentioned [28]. Our own experience has shown that cultures in 2.5*M* NaCl medium sometimes sediment to a great extent if placed in the dark at room temperature.

Other possible methods of initial harvest include flotation, chemical flocculation, filtration, and phototactic accumulation. These methods are generally more complicated and have not as yet been investigated fully by us.

TABLE III

30 Liter Cultivations with Brine Reuse

Batch Number	Cell count at Harvest cells/ml	Intracellular glycerol at harvest g/l cell suspension
1	3.7×10^6	.23
2	5.9×10^6	.27
3	4.2×10^6	.18

Once harvested and dewatered to remove as much of the brine as possible, the algal cells must be lysed to release the intracellular glycerol. The remarkable advantage of *Dunaliella* in this respect is that it lacks any rigid cell wall; the cells are natural naked protoplasts [1, 16, 25, 28].

Complete lysis can be obtained by resuspending brine-grown cells into the original volume of distilled water, although this results in a very dilute glycerol solution. Resuspension in some small fraction of the original volume can result in a considerable concentration factor, although there are practical limitations set by residual salt levels and incomplete lysis. Some combination of this osmotic shock technique and higher temperatures might be beneficial.

Since the cells are essentially small bags of glycerol solution surrounded by a naked membrane, presumably phospholipid in nature, lysis with minimal dilution is in principle also achievable by using detergents or phospholipase enzymes. Perhaps the phospholipases could even be produced by the organisms used in the next bioconversion step.

To produce a crude algal lysate suitable for further bioconversion by fermentation, it will be necessary to achieve a concentration factor of at least 100 times over the intracellular glycerol concentration in the original culture. A great fraction of this concentration can be achieved due to the beneficial sedimentation and easy lysis characteristics of the organism. In achieving these concentrations, residual salt is probably a problem which must be dealt with. Perhaps halotolerant organisms for the further bioconversion steps could be found; then a high salt content would be beneficial in diminishing contamination problems.

BIOCONVERSION SUBSYSTEMS

Most of the bioconversions suggested in Figure 1 should be considered hypothetical at present. However, in most cases there is some evidence in the literature suggesting that these routes are technically feasible using pure glycerol, and so they seem likely candidates for utilization of glycerol-rich algal lysates. Information on fermentations using glycerol as substrate was found to be relatively scarce, probably due to the high cost of conventional glycerol and, thus, to its undesirability as a fermentation substrate.

Glycerol

Probably the most valuable chemical product recoverable from the algal subsystem described above is glycerol itself, and this is apparently the focus of patent applications and pilot-scale work being undertaken industrially in Israel [29]. Glycerol prices in the United States are currently about 50–52 cents/lb [30].

The price of glycerol, both natural and synthetic, is very dependent on the price of petroleum. Glycerol prices more than doubled in 1974, coinciding with huge increases in crude oil prices [30]. Very recent estimates of this price sensitivity indicate a glycerol price increase of 2.9 cents/lb for every dollar/barrel increase in petroleum prices [31]. It was also estimated that glycerol production by fermentation of carbohydrates could become competitive with conventional sources when petroleum prices exceed 15 dollars/barrel [31].

In the production of glycerol from natural sources, one of the big costs is due to the required purification from crude dilute aqueous solutions. For fat-derived glycerol, the older alkaline saponification technologies yielded lye solutions containing 2–15% glycerol, while more modern autoclave hydrolysis technology results in "sweet water" of up to 20% glycerol [32]. Glycerol production by carbohydrate fermentation also yields a broth of 5–10% glycerol in the best cases [32]. These concentrations of starting material compare well with the reported intracellular glycerol concentrations in *Dunaliella* cell mass (Table II) of 18–36% (2–4*M*). In order to produce pure glycerol from such crude solutions, multieffect evaporation and steam-vacuum distillation are normally used, and the total steam requirement has been reported to be in the range of 5–10 lb steam/lb pure glycerol [32]; thus the purification energy is about the same as the energy content of the glycerol produced. Various solvent extraction processes have also been proposed for purification from crude, high-salt-content solutions [33]. Some of these may be applicable directly to the algal cell mass to bring about lysis and extraction in one step.

The production of pure glycerol from *Dunaliella* cell mass seems well worth investigating in countries with well-defined chemical market requirements, the required level of chemical technology, and the necessary climatic conditions and saline water resources.

Starch

Another remarkable feature of the *Dunaliella* osmoregulatory system is its ability to rapidly convert the intracellular glycerol into insoluble starch storage products. This conversion occurs when cells from a high osmotic pressure medium are shifted to a lower osmotic level. If the shift is not large enough to cause immediate osmotic lysis, then the organism quickly recovers osmotic balance by converting the excess intracellular glycerol to osmotically inactive starch. The conversion process is reported to occur within about 90 min [25].

The starch-enriched algal biomass produced by such a method could then obviously be used as a substrate for many conventional and well-known fermentations normally based on plant starch. This technique for algal starch production might be useful for producing special isotope-labeled carbohydrates for applications such as enzyme mechanism studies or biomedical techniques such as emission tomography.

Algal Protein

The utilization of *Dunaliella* as a source of algal SCP was suggested over 20 years ago [1] and has recently been reemphasized [28]. The crude protein content is around 50% [1, 28], and the amino acid profile compares well with other types of SCP [28]. The great advantage of *Dunaliella* is the lack of an indigestible cell wall [1, 28], thus allowing its use as food for all types of animals, including fish, poultry (carotene content also beneficial), and swine, as well as ruminants. However, the high residual salt content of harvested algae is a problem requiring special attention [28].

It may also be possible to produce some higher quality protein concentrates from *Dunaliella*, while at the same time recovering a glycerol-rich product for fermentation. In principle, there is very little difference between the lysates of *Dunaliella* discussed above and the crude juice extracted from plant leaves in various leaf protein concentrate (LPC) processes [34]. Thus it should be possible to apply similar methods of thermal protein coagulation to produce an algal protein concentrate (APC) analogous to LPC, leaving behind a glycerol-rich, low chlorophyll solution for fermentation applications.

Yeast SCP

One of the most interesting and immediately available methods of utilizing glycerol-rich algal lysates is the aerobic cultivation of yeast for SCP production. If the protein separation methods suggested above should result in a large reduction of the chlorophyll content of lysates, it may even be possible to produce a relatively clean yeast SCP for human consumption. Although many yeasts will utilize glycerol aerobically, probably not all species will. A preliminary screening carried out by us on a variety of randomly chosen species showed utilization by about half the strains tested (Table IV). It is likely that permeability plays a large role in glycerol utilization, and perhaps more widespread glycerol utilization could be detected under other conditions.

The batch growth of *Candida utilis* in 5% glycerol medium is shown in Figure 3. The results are similar to yeast growth on carbohydrates, with a final yield factor $Y_{x/s}$ of 0.4 g dry weight/g glycerol. The transient production of ethanol is also of interest (see below), probably indicating oxygen limitation.

The ability of glycerol-rich algal lysate to support yeast growth has also been investigated. Lysates concentrated by vacuum evaporation were found to support excellent growth by a *Pichia* sp., the only yeast so far tested. The results of one experiment are shown in Table V. This experiment resulted in a fictitiously high yield factor of 0.86, based only on the glycerol used. Obviously the lysate contained other soluble compounds which served as good substrates also, probably algal carbohydrates, amino acids,

TABLE IV

Aerobic Glycerol Utilization by Yeasts

Yeast	Utilization of glycerol[a]
C. intermedia	+
C. utilis	+
Hansenula anomala	+
Kloeckera apiculata	-
Pichia sp.	+
Rhodotorula ruba	-
Saccharomyces sp.	-
S. carlsbergensis	-
S. rosei	+
Torulopsis sp.	+
Torula sp.	+
Zygosaccharomyces mellis	-
Z. rouxii	-

[a] Growth on agar medium containing yeast nitrogen base (BBL) plus 5% glycerol.

peptides, etc. Obviously, these crude algal lysates are excellent substrates for yeast cultivation.

Ethanol

The production of ethanol from glycerol-rich *Dunaliella* biomass offers a unique system for liquid fuel and chemical production which is not achievable by other algal systems.

The obvious organism of choice for alcohol production would be yeast. However, there seems to be no evidence in the literature that any yeast can anaerobically metabolize glycerol to yield ethanol. It appears that yeasts as a group do not share the various mechanisms that allow certain bacteria to

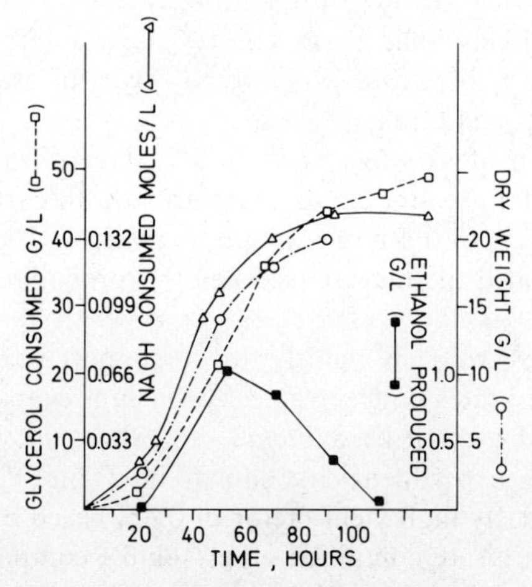

Fig. 3. Batch growth of *C. utilis:* medium: 50 g/liter glycerol, 0.01*M* phosphate, yeast nitrogen base (BBL), pH 5.8, 28°C, surface aeration by stirring only.

TABLE V

Yeast Growth on *Dunaliella* Lysate

	Before growth	After growth
Total dry weight (g/liter)	6 (algal debris)	12
Yeast dry weight (g/liter)	-	6 (by difference)
Glycerol (g/liter)	8	1
Other soluble compounds	?	?

perform a glycerol fermentation (see below). However, as shown above (Fig. 3), *Candida utilis* growing aerobically with oxygen limitation shows transitory production of small amounts of ethanol. It might be possible to substantially increase this production by altering environmental conditions and to utilize a two-stage culture system (with the first stage as an oxygen-limited vacuum fermentation and the second stage well aerated) to simultaneously produce ethanol and yeast cell mass. Other indirect routes for yeast ethanol fermentation could be postulated if the glycerol is first transformed to some other substrate such as dihydroxyacetone, mannitol, or fructose by preliminary fermentations with other organisms (see below).

However, there are at least two interesting bacterial fermentations of glycerol that produce ethanol as a major product. The first occurs in gram negative bacteria of the so-called enteric group. Strains of *Klebsiella* (*Aerobacter*) were shown to ferment 1 mol glycerol to 1 mol ethanol and 1 mol formic acid almost quantitatively, which results in perfect redox balance [35, 36]. The theoretical yield of ethanol is 0.5 g/g glycerol, and the actual yield obtained in manometric studies was 0.43 [35]. The theoretical energy recovered in the ethanol is 82.5% of that of the glycerol.

We have begun to investigate this fermentation under more realistic conditions and substrate levels. Figure 4 shows results obtained for this fermentation using 20 g/liter glycerol in a mineral salts medium. The acid

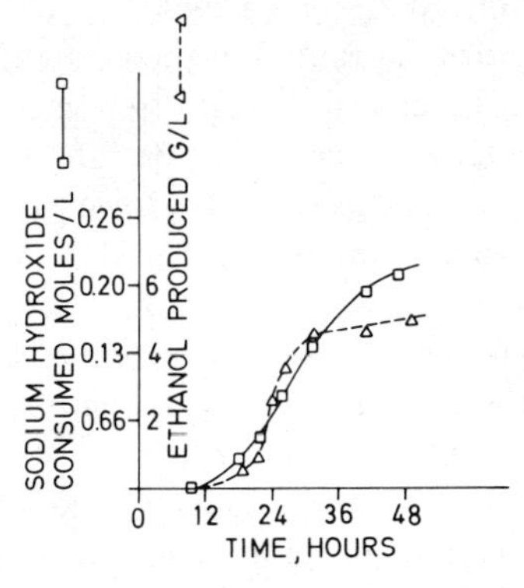

Fig. 4. Fermentation of glycerol to ethanol by *Klebsiella aerogenes*, strain 1033/2103 [36]: mineral medium with 20 g/liter glycerol, pH 7.0, 28°C.

production was fairly close to the expected theoretical value, and the major acid has been shown to be formic as expected. However, the ethanol yield was only half the theoretical, 0.25 g/g glycerol. It should be emphasized that these are very preliminary results and can probably be greatly improved after further optimization.

Other related gram-negative organisms such as *Escherichia*, *Citrobacter*, *Serratia*, and *Hafnia* also probably carry out forms of glycerol fermentation, although *Citrobacter* has been shown to form large amounts of trimethylene glycol (1,3-propanediol), a viscous liquid with very few uses. Products of this *Citrobacter* fermentation were reported to be trimethylene glycol ($Y_{x/s}$ = 0.2–0.3), ethanol ($Y_{x/s}$ = 0.1–0.24), formic, acetic and lactic acids, and hydrogen [37].

The other interesting bacterial fermentation for ethanol production from glycerol uses *Bacillus macerans*. In World War I, this and related organisms (*B. acetoethylicum*, considered identical with *B. macerans*) were investigated as methods of producing acetone and ethanol from starchy materials [38]. Although yields on carbohydrates were not very attractive, it was coincidentally discovered that the yield of ethanol from glycerol was high ($Y_{x/s}$ = 0.40–0.43) with no acetone formation. Other work has shown that H_2, acetic, and formic acids are formed, and an even higher yield of ethanol was reported ($Y_{x/s}$ = 0.6) [39]. The big disadvantage with these *Bacillus* fermentations is that they are apparently very slow (10–15 days with 2% glycerol), although modern research might greatly improve on these 60-year-old results.

Solvents and Organic Acids

The production of such chemicals as acetone, butanol, propanol, and several organic acids from glycerol could also be of interest. Unfortunately, there is not a large amount of readily available information on glycerol fermentations by anaerobes likely to make such transformations. According to *Bergey's Manual* [40] only 3 of 11 species in group 1, 8 of 22 species in group 2, and 3 of 19 species in group 3 of the genus *Clostridium* are capable of fermenting glycerol. The characteristic often varies from strain to strain within a species. *Clostridium butyricum* is designated as variable, while *Cl. acetobutylicum* is listed as negative [40]. However, other reports indicate that *Cl. acetobutylicum* can ferment glycerol [41]. Although *Cl. glycolicum* should be of interest, because it converts 2 mol propylene glycol quantitatively into *n*-propanol and propionic acid, it reportedly cannot ferment glycerol [42]. Strict rumen anaerobes, such as *Selenomonas*, are also known to produce several organic acids from glycerol [43].

Glycerol can be fermented to propionic acid and acetic acid by *Propionobacter* sp., with good yields of propionate ($Y_{x/s}$ = 0.76) [44]. In addition, lactic acid can be produced from glycerol by *Streptococcus faecalis* [45].

Other Chemicals

Dihydroxyacetone is apparently the only fermentation product currently produced from glycerol. It is produced oxidatively by *Gluconobacter suboxydans*. It is a specialty item used in cosmetics and suntan preparations and is of little interest here since it probably requires pure glycerol as substrate [46].

Mannitol can be produced from glycerol by yeasts of the genus *Torulopsis* [47]. This aerobic process can result in yield factors ($Y_{x/s}$) of around 0.5, and the mannitol can then be converted almost quantitatively to fructose by *Gluconobacter* oxidation [47]. This mannitol-producing process might have applications in mixed cultures or two-stage systems to extend the versatility of glycerol as a substrate.

Nitrogen Fixation

Biological nitrogen fixation, unless photosynthetically driven, requires large amounts of cheap substrate, and it would seem that the glycerol-rich algal lysates discussed here might be applicable. Several of the organisms mentioned above, such as *Klebsiella* and *Clostridium*, might be exploited for this purpose. In addition, we have been examining several *Azotobacter* strains. Our preliminary results indicate that, although some strains will grow on glycerol in the presence of fixed nitrogen, no substantial growth occurs under conditions requiring fixation.

The combination of nitrogen fixation with yeast SCP production in mixed culture is also of interest. The culture supernatant of a *Pichia* cultivation on glyerol was found to support good growth of *Azotobacter*, presumably due to acids produced and partial autolysis products. In this regard, the mannitol-producing yeast mentioned above is a prime candidate for mixed culture since mannitol is the usual substrate for *Azotobacter* cultivations.

Methane and Hydrogen

The anaerobic digestion of algae to methane is an obvious and well-known bioconversion process needing little comment. It may be possible that the *Dunaliella* cell mass is more easily or rapidly digested than algae having cell walls, but no information is available.

The possible use of crude glycerol as the election donor for photoproduction of hydrogen by photosynthetic bacteria is another long-range possibility. Of 11 species of *Rhodospirillum* and *Rhodopseudomonas* listed in *Bergey*'s *Manual* [40], only one, *Rhodopseudomonas sphaeroides*, is reported to photoassimilate glycerol. Four are listed as not utilizing glycerol, and the rest are unknowns. A mutant of *Rhodopseudomonas capsulata* able to photoassimilate glycerol was also recently reported [48].

DISCUSSION

It is not our contention that all of the bioconversion alternatives outlined here are economically or even technically feasible, given present knowledge. It is only desired to point out the numerous possibilities based on algal glycerol production. It must be realized that economic feasibility should not be judged only in the context of industrialized countries, but consideration must also be given to applications in less developed countries, where such bioconversion industries utilizing "appropriate biotechnology" might contribute significantly to self-reliance and economic development.

In the context of the United States, it is possible to make some rough estimates for the cost of crude algal glycerol for use as a fermentation substrate. Two recent studies carried out for the Department of Energy have estimated the cost of producing microalgae biomass with land-based systems of enormous scale, utilizing CO_2 from combustion processes, and incorporating almost complete recycle of nitrogen nutrients. These highly engineered systems probably represent the ultimate in cost reduction for microalgae systems. For a reasonable yield of 20 tons dry weight per acre-year, the Dynatech study [49] projects a cost of 4.6 cents/lb ash-free dry weight in the form of a fermentable microalgae slurry. For the same basic conditions, the CSO-California study [50] projects a cost of 2 cents/lb. Since it should be possible to produce *Dunaliella* in saline water systems at comparable yields, and with about 50% of the "dry weight" as glycerol, the cost of fermentable glycerol would be twice these figures, in the range of 4–10 cents/lb. This price for a fermentable substrate lies in the same range as many estimates of the price of glucose from waste cellulose hydrolysis, although it does not compare well with conventional carbohydrates at current price levels.

One of the major technical drawbacks associated with solar energy conversion by means of halophilic algal systems is the fact that the high salinity probably precludes simultaneous waste treatment in the sense of conventional sewage oxidation ponds. This means that high salinity systems may have severe problems with obtaining an adequate supply of nitrogen and carbon. Although nitrogen may possibly be recycled and augmented via biological fixation, any algal-based system producing an end product other than hydrogen, electricity, or heat immediately runs into the carbon balance problem. Therefore, it is particularly important that the algal cultivation subsystem be integrated with the further bioconversion fermentations to allow maximum recycle of CO_2. A hypothetical integrated system for ethanol, SCP, and methane production is shown in Figure 5. Even such a highly integrated system will require carbon inputs, and it would probably be necessary to operate it in conjunction with conventional land-based agriculture. Such integration with several products produced will probably greatly improve the economics of any system proposed.

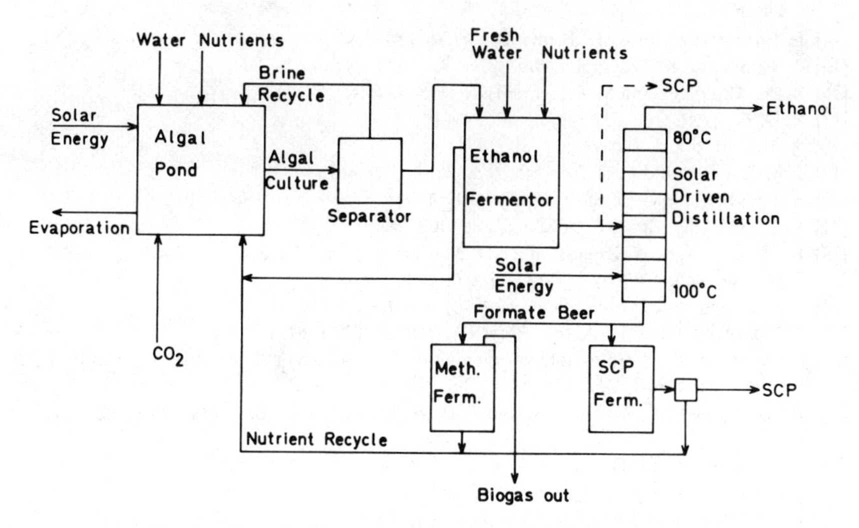

Fig. 5. Hypothetical integrated system for ethanol production.

CONCLUSION

The capture of solar energy in the form of glycerol-rich cell mass of the halophilic algae *Dunaliella* offers some unique possibilities for the production of food, fuels, and chemicals. Especially interesting is the possibility for an all-microbial route to liquid fuels such as ethanol starting with CO_2, although the most immediate applications will probably be the production of glycerol itself.

This work has been supported entirely by the Swedish Board for Energy Source Development.

References

[1] B. P. Eddy, *J. Exp. Bot.*, *7*, 372 (1956).
[2] A. Gibor, *Biol. Bull.*, *111*, 223 (1956).
[3] J. McLachlan, *Can. J. Microbiol.*, *6*, 367 (1960).
[4] R. W. Eppley and F. M. Macias, *Am. J. Bot.*, *50*, 629 (1963).
[5] H. R. Jitts, C. D. Mc Allister, K. Stephens, and J. D. H. Strickland, *J. Fish. Res. Board Can.*, *21*, 139 (1964).
[6] R. W. Eppley and P. R. Sloan, *Physiol. Plant.*, *19*, 47 (1966).
[7] M. K. Johnson, E. J. Johnson, R. D. Mac Elroy, H. L. Speer, and B. S. Bruff, *J. Bacteriol.*, *95*, 1461 (1968).
[8] A. Ben-Amotz and B. Z. Ginzburg, *Biochim. Biophys. Acta*, *183*, 144 (1969).
[9] K. Wegmann and H. Metzner, *Arch. Mikrobiol.*, *78*, 360 (1971).
[10] K. Wegmann, *Biochim. Biophys. Acta*, *234*, 317 (1971).
[11] A. Latorella and R. L. Vadas, *J. Phycol.*, *9*, 273 (1973).
[12] O. Van Auken and I. B. McNulty, *Biol. Bull.*, *145*, 210 (1973).
[13] A. Ben-Amotz and M. Avron, *Plant Physiol.*, *53*, 628 (1974).

[14] L. Borowitzka and A. D. Brown, *Arch. Microbiol.*, *96*, 37 (1974).
[15] G. Frank and K. Wegmann, *Biol. Zbl.*, *93*, 707 (1974).
[16] A. A. Abdullaev and V. E. Semenenko, *Fiziol. Rast.*, *21*, 1145 (1974).
[17] T. D. Brock, *J. Gen. Microbiol.*, *89*, 285 (1975).
[18] A. Ben-Amotz, *J. Phycol.*, *11*, 50 (1975).
[19] J. McLachlan and C. S. Yentsch, *Biol. Bull.*, *116*, 461 (1959).
[20] E. Paasche, *Plant. Physiol.*, *25*, 294 (1971).
[21] P. K. Bienfang, *Limnol. Oceanogr.*, *20*, 402 (1975).
[22] R. W. Eppley, J. N. Rogers, and J. J. McCarthy, *Limnol. Oceanogr.*, *14*, 912 (1969).
[23] L. A. Loeblich, *J. Phycol. Suppl.*, *6*, 9 (1970).
[24] J. S. Craigie and J. McLachlan, *Can. J. Bot.*, *42*, 777 (1964).
[25] A. Ben-Amotz and M. Avron, *Plant Physiol.*, *51*, 875 (1973).
[26] C. A. Price, L. R. Mendiola-Morgenthaler, M. Goldstein, E. N. Breden, and R. R. L. Guillard, *Biol. Bull.*, *147*, 136 (1974).
[27] R. W. Eppley, R. W. Holmes, and J. D. H. Strickland, *J. Exp. Mar. Biol. Ecol.*, *1*, 191 (1967).
[28] N. Gibbs and C. M. Duffus, *Appl. Env. Microbiol.*, *31*, 602 (1976).
[29] "Science/technology concentrates," *Chem. Eng. News*, 18 July, 1977.
[30] Anderson, E. V., *Chem. Eng. News*, 20 June, 11 (1977).
[31] G. E. Tong, *Chem. Eng. Prog.*, *74(4)*, 70 (1978); see also *Chem. Eng. News*, 28 Nov., 15 (1977).
[32] T. M. Godfrey, in *Glycerol*, C. S. Miner and N. N. Dalton, Eds. (ACS Monograph Series, Reinhold, New York, 1953).
[33] J. C. Elgin, US patent 2,479,041, Aug. 16, 1949.
[34] N. W. Pirie, Ed., *Leaf Protein*, IBP Handbook No. 20 (Blackwell Scientific, Oxford, 1971).
[35] B. Magasanik, M. S. Brooke, and D. Karibian, *J. Bacteriol.*, *66*, 611 (1953).
[36] E. C. C. Lin, *Annu. Rev. Microbiol.*, *30*, 535 (1976).
[37] M. N. Mickelson and C. H. Werkman, *J. Bacteriol.*, *39*, 709 (1940).
[38] J. H. Northrop, L. H. Ashe, and J. K. Senior, *J. Biol. Chem*, *39*, 1 (1919).
[39] H. B. Speakman, *J. Biol. Chem.*, *64*, 41 (1925).
[40] R. E. Buchanan and N. E. Gibbons, Eds., *Bergey's Manual of Determinative Bacteriology*, 8th ed. (The Williams and Wilkins Co., Baltimore, 1974).
[41] E. Simon, *Nature*, *152*, 626 (1943).
[42] L. W. Gaston and E. R. Stadtman, *J. Bacteriol.*, *85*, 356 (1963).
[43] P. N. Hobson and S. O. Mann, *J. Gen. Microbiol.*, *25*, 227 (1961).
[44] A. S. Phelps, M. J. Johnson, and W. H. Peterson, *Biochem. J.*, *33*, 726 (1939).
[45] I. C. Gunsalus, *J. Bacteriol.*, *54*, 239 (1947).
[46] S. R. Green, in *Microbial Technology*, H. J. Peppler, Ed. (Reinhold, New York, 1967).
[47] H. Onishi and T. Suzuki, *Biotechnol. Bioeng.*, *12*, 913 (1970).
[48] D. Lueking, L. Pike, and G. Sojka, *J. Bacteriol.*, *125*, 750 (1976).
[49] Dynatech R/D Company, presented at The Workshop on the Economics and Engineering of Large-Scale Algal Biomass Systems, MIT, Jan. 24–25, 1978.
[50] J. R. Benemann, P. Pursoff, and W. J. Oswald, *Engineering Design and Cost Analysis of a Large Scale Microalgae Biomass System* (CSO International, Concord, Calif., 1978).

Organic Chemicals and Liquid Fuels from A[illegible] Biomass

J. E. SANDERSON, D. L. WISE, and D. C. AUGENSTEIN
Dynatech R/D Company, Cambridge, Massachusetts 02139

1. INTRODUCTION

The recent energy crisis has precipitated an interest in finding sources of clean energy other than petroleum. This interest has led to a number of programs to convert our vast coal reserves into clean-burning easily transported fuels. However, even if these programs are commercially successful, they will provide only a temporary solution to the energy problem because our coal reserves, although vast, are finite. Ultimately, we must depend on renewable energy sources with minimum environmental impact.

A ubiquitous renewable energy source is the biomass created by the photosynthetic reduction of carbon by vegetable matter. As a result, interest is beginning to concentrate on using cellulose from vegetable matter with high yields per acre as a source of energy, either by direct combustion, or conversion into methane or other fuels by fermentation processes.

The composition of organic matter varies in many respects, depending on its source; two of these respects are lignin content and moisture content. Organic matter high in lignin and low in moisture content is particularly well suited for conversion to energy by direct combustion. Lignin is considerably higher in heating value than cellulose (~12,700 BTU/lb [1] vs. 7500 BTU/lb), but this heating value is not available to fermentation processes. Lignin is quite resistant to biological attack unless fairly rigorous pretreatments are employed. Water may be considered to have a negative heating value in direct combustion. To take water at, say, 68°F, vaporize it, and heat it to flue gas temperature (600°F) requires about 1300 BTU/lb [2]. A comparison of two hypothetical examples of biomass on potential energy yield, both from fermentation and from direct combustion, is shown in Table I.

By fermentation 50% more energy is potentially available from the high moisture biomass, whereas it is completely useless as a fuel for direct combustion without predrying. A further consideration is that in the woody substrate the cellulose* fraction consists of substantial amounts of

* Including materials of similar oxidation state.

Biotechnology and Bioengineering Symp. No. 8, 131–151 (1978)
 0572-6565/78/0008-0131$01.00

TABLE I

Comparison of Hypothetical Examples of Biomass on Potential Energy Yield

	Composition %				BTU/lb (DAF)	
Substrate	Cellulose*	Lignin	H_2O	Ash	Fermentation	Combustion
Woody Biomass	50	25	20	5	$\frac{3750 + 0 + 0}{.75} = 5000$	$\frac{3750 + 3175 - 260}{.75} = 8900$
High Moisture Biomass	12	0	85	3	$\frac{900 + 0 + 0}{.12} = 7500$	$\frac{900 + 0 - 1100}{.12} = -1700$

* Including materials of similar oxidation states.

crystalline cellulose and pentosans, which are difficult to ferment to liquid fuels.

2. PROCESS DESCRIPTION

The process which is the subject of this paper is shown in Figure 1. The first step is to ferment a suitable form of biomass (work to date has been primarily with marine algae) to dilute organic acids in a nonsterile, fixed packed-bed fermentor. The organic acids are then removed from the fermentor and concentrated by extraction. This extraction procedure may be a conventional liquid–liquid extraction, a membrane process, or an ion exchange technique. The concentrated acid solution is then either distilled to give acetic acid or electrolyzed to give olefins or aliphatic hydrocarbons. If olefins are the desired product, the olefins are then fractionally distilled to give individual olefins. The fermentation conditions and extraction procedure are selected based on the products desired. If acetic acid is the desired product, the fermentation is run to give primarily acetic acid as its product. If olefins are desired, the fermentation is run to give acetic, propionic, butyric, and valeric acids, and the extraction is carried out so as to remove the higher acids preferentially. If liquid hydrocarbons are the desired product, the fermentation is run to give primarily butyric, valeric, and caproic acids, and the extraction is run to remove these acids preferentially to the lower acids. Each of these steps will be described in turn.

3. MARINE ALGAE

3.1. Marine Algae as a Source of Organic Matter

The primary types of macrocellular marine algae are the brown algae (e.g., *Macrocystis*, *Laminaria*), which make up the bulk of the matter usually referred to as kelp, the green algae (e.g., *Ulva*), which constitute the sea-lettuce-type materials that are eaten in the Orient as well as the smaller varieties often responsible for fouling the hulls of ships, and the red algae (e.g., *Chondrus*, *Gelidium*, *Gracilaria*), which produce the hydrocolloids carrageenan and agar that are used commercially as thickeners.

Macrocellular marine algae, unlike most other forms of vegetable matter, do not contain appreciable quantities of cellulose. Their biodegradable fraction consists primarily of carboxylated or sulfonated galactose and/or mannose chains. Anaerobic fermentation of marine algae has shown that these materials are rapidly and thoroughly converted to organic acids.

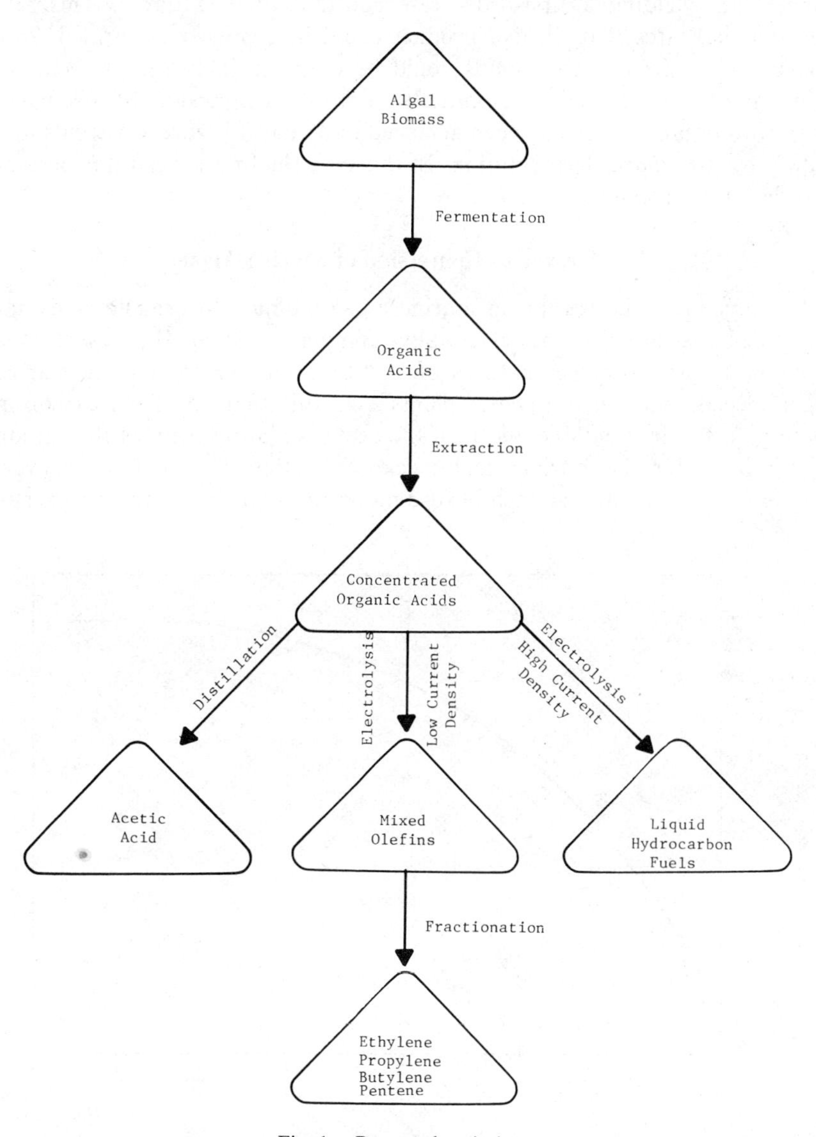

Fig. 1. Process description.

3.2. Fermentation Rates for Marine Algae

The rate of conversion of marine algae to organic acids is very rapid. Figure 2 is a plot of total organic acid concentration (as acetic acid) versus time for *Chondrus crispus*, a marine algal species. The rate decreases with increasing acid concentration, indicating either product inhibition, pH inhibition, or substrate depletion. Subsequent experiments indicated that the effect is caused largely by pH inhibition. Figure 3 is a semilog plot of the same data, yielding a first-order rate constant of 0.77 day^{-1}. This rate constant indicates that, if acid product could be removed as formed, 98% conversion of fermentable solids could be obtained in five days. Without removal of product acids, essentially complete conversion of *Chondrus crispus* to organic acids has been achieved in 35 days. Figure 4 presents rate data for a *Gracilaria* fermentation. In this case the first-order rate constant was determined to be 0.62 day^{-1}.

3.3. Degree of Conversion of Marine Algae

Conversion of the carbon in marine algae biomass to organic acids has been demonstrated to be essentially complete. Table II presents the maximum conversion percentages on a total solids basis of three marine algal species and two freshwater plants assuming that all of the carbon in these plants is fermentable and is in the same oxidation state as the carbon in cellulose. For the marine algal species the estimated maximum conversion is 43–45%, assuming each hexose molecule is converted to three acetic

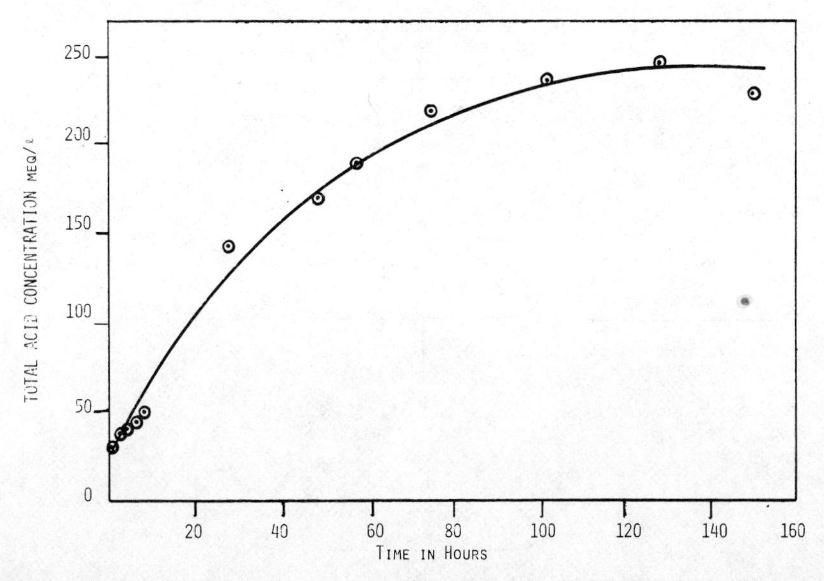

Fig. 2. Total acid concentration vs. time; 15% solids, 90% *Chondrus crispus*, 10% digester effluents (DE).

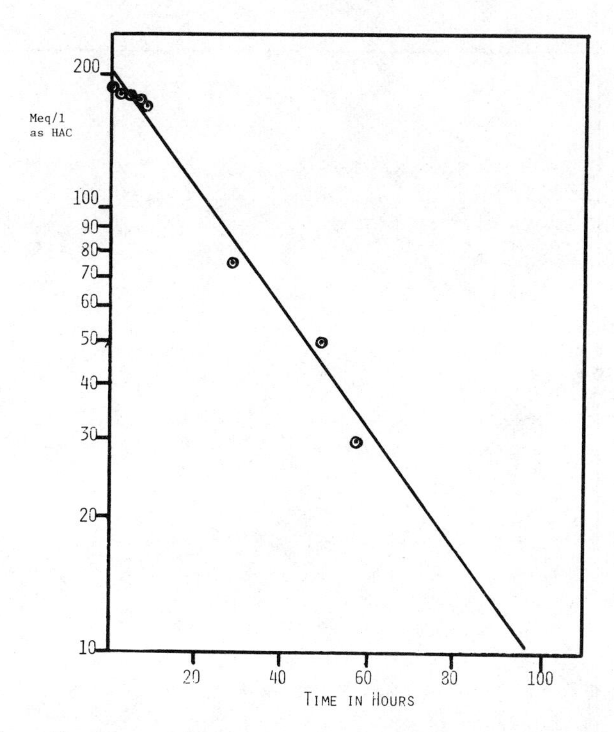

Fig. 3. ln $C_\infty - C$ vs. time; 15% solids, 90% *Chondrus crispus*, 10% DE. $K = 0.032/\text{hr} = 0.77/\text{day}$. $(\delta C/\delta T)_0 = 168$ mequiv/liter day.

acid molecules. Table III presents the results of an unstirred fermentor containing 5% total solids of *Chondrus crispus* inoculated with sewage sludge and run at a controlled pH of 5.5–6.0. The pH was measured at each sampling and adjusted with either calcium carbonate or phosphoric acid. There is a considerable amount of scatter in the data. In order to obviate

TABLE II

Estimated Conversion Percentages

Substrate	% C*	Maximum % Conversion
Hydrilla	33	83
Gracilaria	17	43
Duckweed	34	85
Ulva Lactuca	17	43
Chondrus Crispus	18	45

* Galbraith Labs, Knoxville, Tennessee.

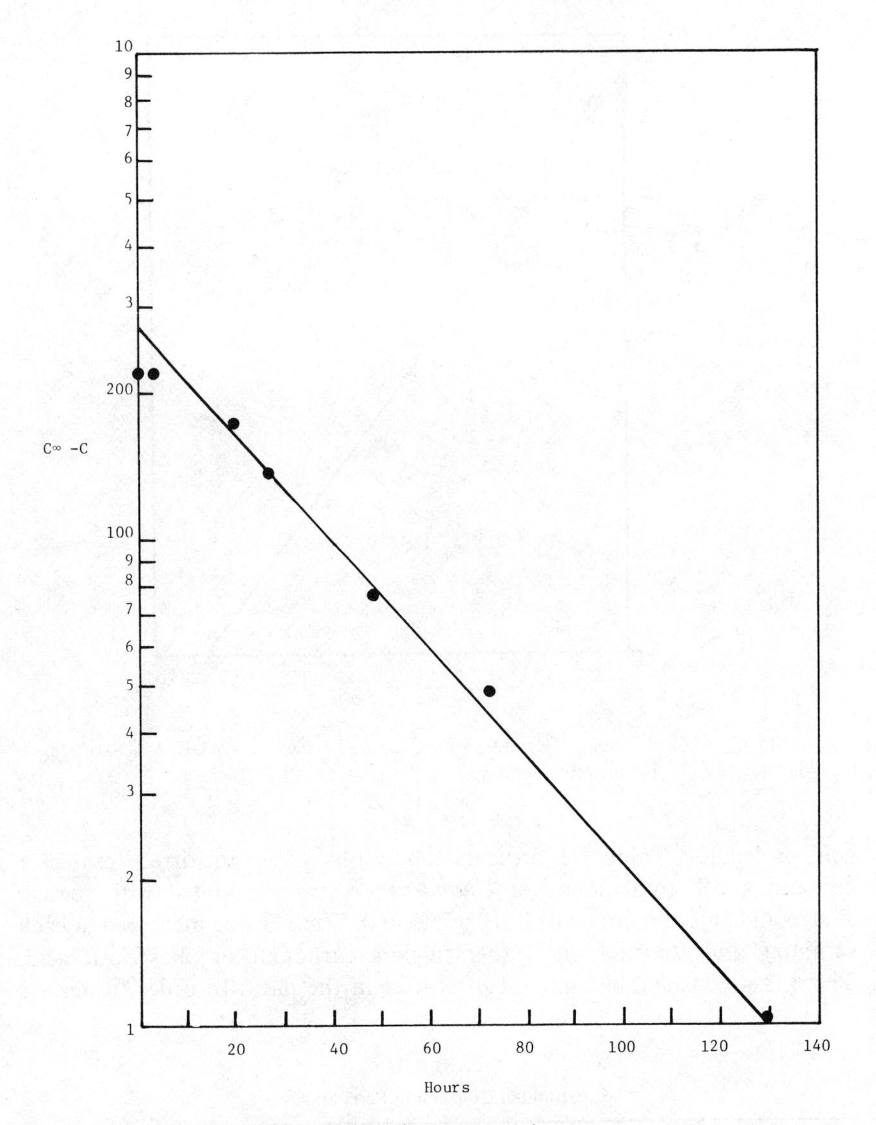

Fig. 4. Run No. 30517-4: 15% solids, 90% *Gracilaria*/10% DE. C_{∞} = 225 mequiv/liter; K = 0.026 hr^{-1} = 0.62 day^{-1}; $(dC/dt)_0$ = 140 mequiv/liter day.

some of the difficulties associated with interpreting these data, concentration of each acid versus time is plotted in Figure 5, and a smooth curve is drawn through each set. From these plots the average final concentrations are calculated and included in Table III. On this basis the conversion based on total solids is 40%, indicating essentially complete conversion of carbon to organic acids on a three acetic acid equivalents per hexose unit basis.

The anaerobic fermentation of cellulosic material to acetic acid by sewage sludge flora has been studied by Jeris and McCarty [3] who showed by radioactive tracer studies that the process is best described as

$$C_6H_{12}O_6 + 2H_2O \rightarrow 2CH_3COOH + CO_2 + 4H_2$$

presumably proceeding through pyruvate via the Embden–Myerhoff pathway. It was shown in these tracer studies that approximately 70% of the methane formed in anaerobic digestion processes was derived from acetic acid.

Anderson et al. [4] have shown that *Clostridium thermoaceticum* can produce three acetates from each glucose molecule, two by the route just described and the third by the reduction of the CO_2 formed to formate, which is further reduced by intermediates of the tetrahydrofolate pathway. Ultimately, carboxylation occurs to give acetate. The overall reaction is shown as

$$\text{glucose} + 3\text{ADP} + 3\text{Pi} \rightarrow 3\text{acetate} + 3\text{ATP}$$

These workers were able to obtain an actual yield of 2.5 acetates per glucose molecule.

TABLE III

Organic Acid Concentration vs. Time[a]

Time, Day	pH	A[1]	P	B	V	C	Total	A/T	R[2]	%[3] Conv.
0	---	0	0	0	0	0	0	---	0	0
1	---	25	10.5	6	0	0	41.5	60%	58	7%
2	4.5	14.5	8.5	7.5	6	0	36.5	40%	68	8%
5	---	29	4	17.5	3	0	53.5	54%	90	11%
6	---	33	14.5	24	15	~1	81.5	38%	171	20%
7	---	25.5	8	20	6.5	0	60	43%	111	13%
8	5.3	28	7	27.5	4	5	71.5	39%	142	17%
12	5.2	70	15	26	5	0	116	60%	178	21%
19	5.2	46	8.5	13	6	30	103.5	44%	233	28%
26	5.9	45	6	28	4	31	114	39%	263	32%
33	5.9	68	9	29.5	6.5	32.5	145.5	47%	309	37%
41	6.0	51	7	24	4	13	99	52%	188	23%
47	5.9	124	25	33.5	3	3	188.5	66%	273	33%
54	5.9	53	7	23.5	5.5	25.5	116.5	45%	249	30%
61	---	5.5	11	28	7	33	134	41%	299	36%
68	6.1	58	14	29	10	34	145	40%	324	39%
75	5.8	59	5	28	7	34	133	44%	297	36%
81	6.2	60.5	9	29	12	43	154	39%	360	43%
89	5.7	64	4	28	10	47	153	42%	362	43%
96	5.8	60	12	45	19	42	178	34%	422	51%
103	5.7	56	6.5	33	12.5	35	143	39%	331	40%
		Average final concentrations								
∞	---	60	10	32	12	35	149	40%	336.5	40%

[a] Experiment No. 00354-1-5b—5% *Chondrus crispus*, controlled pH.

[1] A = acetic, P = propionic; B = butyric; V = valeric, C = caproic acid concentration in mequiv/liter.

[2] R = Total reducing equivalents = A + 1.75P + 2.5B + 3.25 V + 4.0C.

[3] % Conv., percent conversion based on total solids.

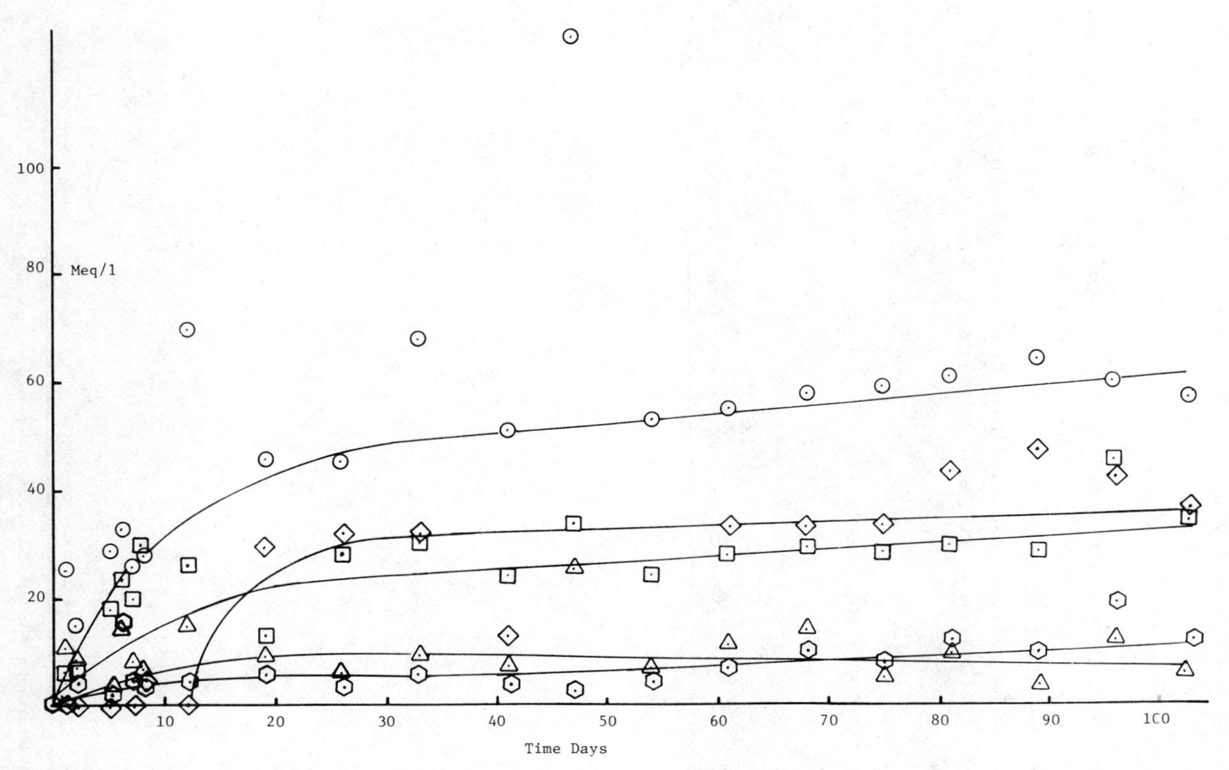

Fig. 5. Organic acids concentration vs. time. Experiment No. 02354-1; 5% *Chondrus crispus*. (⊙) Acetic; (△) propionic; (⊡) butyric; (⊕) valeric; (◇) caproic.

Other organisms have been shown to be able to synthesize acetate from CO_2 as well [5], some of which are apparently present in sewage sludge since it has been shown at Dynatech R/D Company that acetate is formed in the sewage sludge catalyzed anaerobic fermentation of CO_2 and hydrogen [5]. It therefore appeared likely that the fermentation of biomass by sewage sludge flora would produce more than two acetic acids molecules from each hexose. These acetic acid molecules could then be converted to the observed product mix according to the following set of disproportionation reactions:

$$7CH_3COOH \rightarrow 4CH_3CH_2COOH + 2CO_2 + 2H_2O$$

$$5CH_3COOH \rightarrow 2CH_3CH_2CH_2COOH + 2CO_2 + 2H_2O$$

$$13CH_3COOH \rightarrow 4CH_3CH_2CH_2CH_2COOH + 6CO_2 + 6H_2O$$

$$4CH_3COOH \rightarrow CH_3CH_2CH_2CH_2CH_2COOH + 2CO_2 + 2H_2O$$

It is also possible that hydrolyzed cellulose is converted to only two acetic acid molecules and that the hydrogen produced is utilized to form the higher acids, perhaps in the following manner:

$$3CH_3COOH + 2H_2 \rightarrow 2CH_3CH_2COOH + 2H_2O$$

$$2CH_3COOH + 2H_2 \rightarrow CH_3CH_2CH_2COOH + 2H_2O$$

$$5CH_3COOH + 6H_2 \rightarrow 2CH_3CH_2CH_2CH_2COOH + 6H_2O$$

$$3CH_3COOH + 4H_2 \rightarrow CH_3CH_2CH_2CH_2CH_2COOH + 4H_2O$$

In either case, if mixed acids are the desired product, high conversions are obtained from marine algae.

It is also interesting to note from the data in Table III that the acids with even numbers of total carbon atoms predominate. This phenomenon is also observed in naturally occurring fats and oils and suggests that fatty acids are synthesized biologically from acetate via two-carbon additions.

The column in Table III labeled A/T is the ratio of acetic to total acid. In this fermentation these ratios run from 40% to 60%, with the higher values at lower total acid concentrations. This effect can be exaggerated by proper choice of conditions so that A/T ratios of greater than 80% can be obtained at low conversions. However, if acetic acid is the desired product, preliminary results indicate that thermophilic conditions (~55°C) favor acetic acid formation. Table IV presents the results to date of a thermophilic fermentation *Chondrus crispus* at 5% total solids including 1% sewage sludge (bots) as an inoculum. Other possible ways of maximizing the A/T ratio may be to use chlorinated analogs of the higher acids as inhibitors. This approach has not yet been investigated.

4. FERMENTATION

The results presented in Sec. 3 were obtained in static fermentors. These experiments were run to determine the total available reducing equivalents

TABLE IV

Thermophilic Fermentation of *Chondrus crispus*[a]

Time, Day	A	P	B	V	C	T	A/T	R	% Conv.
0	0	0	0	0	0	0	–	0	0
1	5	0	0	0.5	0	6	91	7	1%
2	19	0	∼1	0	0	20	95	22	3
4	67	0	0	0	0	67	100	67	8
5	34	0	0	0	0	34	100	34	4%
7	37	0	0	0	0	37	100	37	4
8	29	0	0	0	0	29	100	29	3
13	66	0	<1	0	0	67	99	67	8
17	65	0	15	0	0	80	81	103	12
24	63	3	5	0	0	71	89	81	10

[a] Experiment No. 00398-1–5% solids.

and to estimate conversion rates from marine algal biomass. In the process described in Sec. 2, the fermentation step is envisioned to take place in a fixed packed-bed fermentor.

Engineering at Dynatech [6] has shown that a number of constraints imposed by conventional digestion systems are removed and that operating parameters may be improved in a reactor system termed the fixed packed-bed fermentor. It has been found that the important function of mass transfer ordinarily accomplished by stirring in conventional digesters can be equally well fulfilled by a moving stream of liquid of the proper composition and moving at the required rate through the waste. With proper arrangement of piping inlets and outlets, this same liquid stream may be used to control the temperature, add nutrients, control pH, and remove reaction inhibiting products. It has also been determined that toxic components may be removed from the liquid phase by suitable means, such as activated carbon in an external circulation loop, without the necessity of treating the digesting material directly. In addition, the piping arrangement involved may be utilized to dewater the undigested material when the digestion stage is completed.

The solid substrate loading in the fixed packed-bed fermentor can be much higher than in a conventional digester when the necessity for stirring is imposed. Furthermore, without the disruptive effect of stirring, the solid substrate itself conveniently serves as a support for the microorganisms that attack it.

In Figure 6(a) is shown one possible arrangement for the fixed packed-bed fermentor. In this case it may be used as a batch reactor for organic

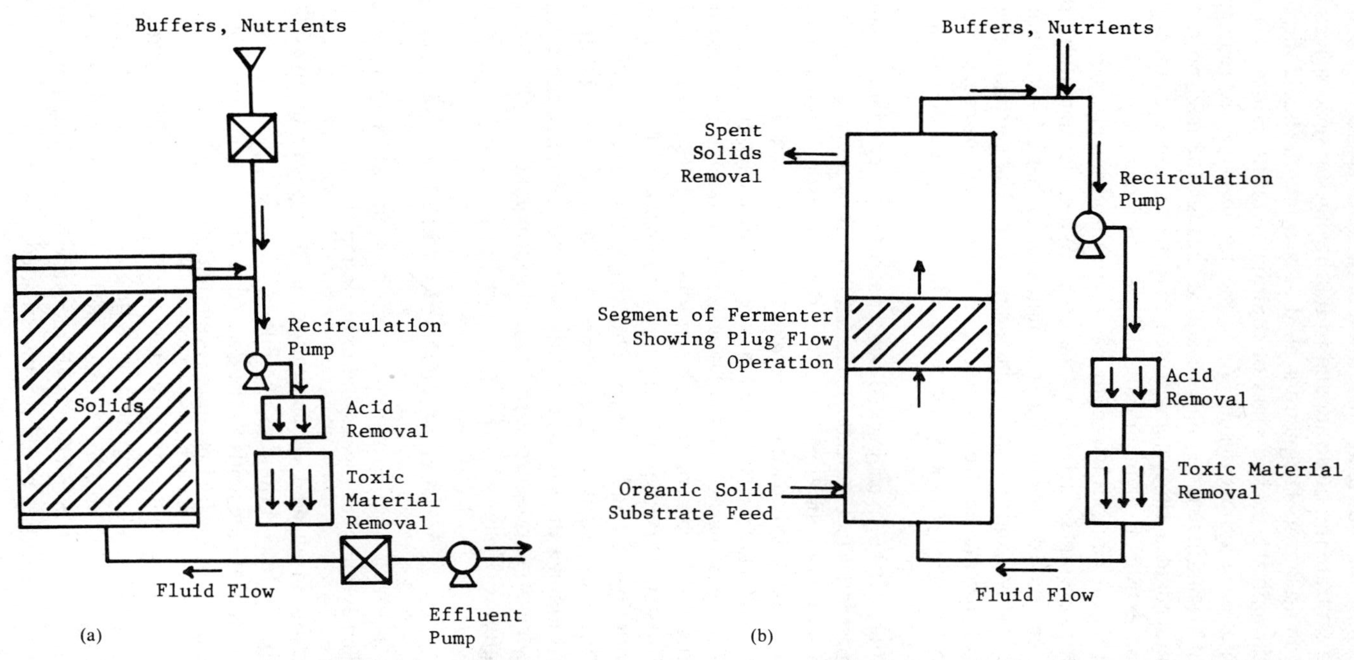

Fig. 6. Possible arrangements for the fixed packed-bed fermentor: (a) schematic diagram of fixed packed-bed fermentor; (b) continuous-flow packed-bed fermentor.

acid production. The fermentor is filled with solid substrate. An aqueous medium containing nutrients and microorganisms is added in sufficient quantity to saturate the substrate and to fill the digester head space and circulation loop as shown. The pump is used to circulate the liquid through the loop and through the packed mass of comminuted solid substrate, containing fermentation microorganisms. As fermentation proceeds, required buffers and additional nutrients may be added through the inlet port, and the acid is removed in a subsystem, which may be an ion exchange column, liquid–liquid extractor, or membrane device. Toxic compounds may also be removed from the aqueous phase, by passing it through a cleaner which may contain, for example, activated charcoal. When fermentation is economically completed, the undigested material is partially dewatered by opening the valve and pumping the liquid out. The partially dewatered material is then removed, fresh substrate is added, and the cycle is repeated. Clearly, this batch system is the most simple one and serves largely to describe the concept.

Application of the fixed packed-bed concept to a continuous plug-flow system is presented in Figure 6(b). Here solid substrate is fed into the bottom of the fermentor and removed from the top of the fermentor. As with the batch-type packed-bed system, fermentor liquid is circulated through the bed and organic acids are recovered in an external subsystem.

Figure 7 is a photograph of the Dynatech Lab, pilot-scale, fixed packed-bed fermentor. It is constructed of 12-in.-diameter glass tubing and is 15 ft in height. It is supported by a triangular framework of 1.25-in. pipe and is enclosed with plywood and Plexigas. The temperature is controlled at 100°F by circulating warm air inside the enclosure. A flow rate of 3 gph (gal/hr) is sufficient to fluidize the substrate, initially at 10% solids, in the reactor to reduce the operating pressure at the pump (approximately 7 ft below the top of the column) to 3 psig.

5. EXTRACTION PROCESSES

Concurrent with the fermentation process an extraction process is required. The extraction process should be designed

1) To maintain an acid concentration in the digester to give a desired product mix at a rate as rapid as possible.
2) To concentrate the acids to the desired level for futher processing.
3) To remove the desired products preferentially.

It is clear that the choice of extraction method as well as the fermentation condition depends on the product mix desired. Some possible extraction methods are listed in Table V. Of these methods, ion-exchange with conventional elution, solvent extraction with recovery by re-extraction, and diffusion membrane processes have been investigated.

Fig. 7. Large fixed packed-bed fermentor.

The use of a conventional strong anion exchange resin with quarternary ammonium sites was considered for use as an acid removal technique. The

TABLE V
Extraction Methods

1. ION EXCHANGE
   a. Conventional Elution
   b. Methanol Elution
2. SOLVENT EXTRACTION
   a. Recovery by Distillation (Hydrosciences process)
   b. Recovery by Re-Extraction
3. MEMBRANES
   a. Diffusion
   b. Electrodialysis

hydroxylated resin is inserted in the recirculation loop of the fixed packed-bed reactor as in Figure 6(a). The hydroxyl ions are displaced by the organic acid anions. When the resin is spent (as indicated by no pH change across the resin), it is removed and replaced by fresh resin. The spent resin is then eluted with an equivalent amount of mineral acid to remove the organic acids collected. The resin is then regenerated by eluting with alkali.

Use of strong anion-exchange resins is unsuitable in this application for several reasons. First, replacement of anions in the fermentation broth such as acetate, butyrate, and bicarbonate tends to increase the pH of the fermentor to the p*K* of the resin. The fermentation does not function at pHs above about 8.0. Second, the resin will reduce the phosphate level and the levels of other nutritive anions to very low levels. Third, one equivalent of mineral acid and more than one equivalent of alkali are required to generate each equivalent of organic acid. More than one equivalent of alkali is required because the equilibrium constant for the reaction

$$NR_4^+X^- + OH^- \rightleftarrows NR_4^+OH^- + X^-$$

is less than one. However, weak anion exchange resins with equilibrium constants greater than one are available and will be investigated shortly for their utility in this application.

Solvent extraction of organic acids from aqueous solution is a technique that has been under consideration for some years. In fact, Goering is credited for suggesting ethyl acetate as an extractant for acetic acid in 1833 [7]. Ethyl acetate, along with diethyl ether, continued to be the most efficient extraction medium at low acid concentrations [8] until recently when novel solvent extraction media were developed. This new technology is based on the use of trioctylphosphine oxide in combination with other solvents, the details of which are proprietary to Hydrosciences, Inc. [9].

An alternate approach to conventional solvent extraction followed by distillation is solvent extraction followed by re-extraction into aqueous base. The organic acids are accumulated in the aqueous base as salts, and when their concentration reaches about 25% by weight they are neutralized with mineral acid and distilled. This approach requires only one equivalent each of acid and base to concentrate the organic acid product from 1–3% in the fermentor to 20 or 25%.

Although some difficulty has been experienced in the use of this technique in the laboratory (such as emulsion formation) is appears that this approach is the most suitable for use on a large scale. The problems are not fundamental in nature but rather are those which are generally amenable to engineering solutions and, in fact, may have already been solved by Hydrosciences, Inc.

It was found that an extraction medium of this type could be incorporated into a membrane consisting of a cellulose acetate gel supported by stainless-steel screen. Membranes under investigation for some time are prepared by dissolving 6% by weight of cellulose triacetate in acetic acid. The screen is then impregnated with this solution and immersed in water to gel the cellulose triacetate. The water is changed periodically to remove as much of the acetic acid as possible. The membrane is then checked for leaks, and if any are present they are plugged with more of the cellulose triacetate solution and re-equilibrated with water. The membrane is then exchanged with isopropanol to remove the water, then with Freon TF to remove the isopropanol. The final step is to equilibrate the membrane with Freon TF containing 10% by weight of trioctylphosphine oxide.

Membranes prepared in this way diffuse organic acids at approximately the same rate as would be expected for an equal thickness of the Freon–TOPO mixture itself and do not pass appreciable quantitites of ionic species under osmotic pressures of 1100 $lb/in.^2$. It appears that, for experimental purposes, this membrane extraction procedure will prove more useful than solvent extraction although, for commercial purposes, potential stability problems in the presence of base and anaerobic bacteria may limit the useful lifetime of the membrane to an extent which will make this approach unattractive economically.

More recently, another membrane system has been developed which appears to have potential on a commercial scale. It is a hollow fiber membrane device consisting of silicone rubber tubing swollen with Freon TF and soaked in a 20% TOPO in Freon solution. A membrane of this composition consisting of a 40-ft length of 3/8-in i.d., 1/2-in. o.d. silicone rubber tubing is currently in operation with the lab/pilot, fixed packed-bed fermentor and is performing well although at present there is no evidence that the TOPO was actually picked up by the membrane. A membrane extractor consisting of 64 four-foot lengths of 1/16-in. i.d., 1/8-in. o.d. silicone rubber tubing is currently under construction.

In principle, a membrane of this type could be constructed from any polymeric, preferably partially crosslinked, tubing which may be swollen by a solvent for TOPO or which has a favorable partition coefficient for organic acids against water. These systems have the promise of being sufficiently resistant to alkali and biological attack as well as being sufficiently durable against mechanical fatigue to be suitable for use in a commercial process.

None of the processes described will remove acetic acid preferentially from a dilute organic acid mixture. Strong anion-exchange resins will remove organic acids indiscriminately. Some selectivity for acetic acid may be obtained with weak ion exchange resins based on acid strength,* but the difference in acid strengths of the product acids is small.

Extraction with organic solvents will generally remove higher acids in preference to lower ones because the partition coefficients of aliphatic organic acids are proportional to chain length, as expected.

A possible exception to this statement is trioctylphosphine oxide, which appears to remove organic acids relatively indiscriminately. However, this compound is a solid at mesophilic temperatures and has therefore been used in solution in organic solvents. The organic solvents used in this work, aliphatic and fluorinated hydrocarbons (selected on the basis of their compatibility with the fermentation process), are solvents for organic acids themselves; therefore, these systems remove higher acids preferentially. If TOPO were dissolved in a nonsolvent for all of the product acids, indiscriminate acid extraction might be obtained, but no such solvent comes immediately to mind. Other possibilities are 1) to operate the extraction at elevated temperatures with pure TOPO, which should be compatible with thermophilic fermentations and 2) to use a lower molecular-weight phosphine oxide or a mixture of two phosphine oxides that are liquid at mesophilic temperatures.

The membrane processes investigated give the same selectivity as the extraction processes because they are driven by the same partition coefficients. In principle, however, it should be possible to construct a membrane which favors the lower acids based on higher diffusion coefficients of smaller molecules. However, as this effect is proportional to the cube root of molecular weight, the selectivity would be expected to be minor.

In summary, if acetic acid is the desired product, it is best to run the fermentation to favor this product. The preferred conditions appear to be low acid concentrations and thermophilic operation. The preferred extraction method is to use one extraction which favors the higher acids, followed by an indiscriminate extraction. The product of the second extraction may then be distilled to yield glacial acetic acid. The product of the first extraction may be 1)fractionally distilled to give higher organic acids; 2) electrolyzed to give olefins or aliphatic hydrocarbons; or 3) fed back into the digester.

If olefins are the desired product, it is assumed that ethylene, propylene, and 1-butene are preferred, based on the size of the market and the value of

* This phenomenon is expected to favor acetic acid.

these products. It is necessary, therefore, to operate the fermentor at sufficiently low total organic acid concentrations to minimize caproic acid production, which gives 1-pentene as an electrolysis product, and to use an extraction method which removes higher acids preferentially. Either liquid–liquid extraction or the membrane techniques discussed are suitable. The primary criterion for selection is a high partition coefficient for propionic acid, which is electrolyzed to ethylene, the most valuable olefin product.

For aliphatic hydrocarbon products, the preferred fermentation conditions are higher total acid concentrations to maximize caproic acid formation and extraction conditions to favor preferential removal of the highest acids. Liquid–liquid extraction or the membrane processes described previously, each without the use of trioctylphosphine oxide, would be suitable. The higher acids are preferred products because these have the most favorable energy balance in the electrolysis.

6. ELECTROLYSIS

Electrolysis of monobasic aliphatic organic acids generates paraffins olefins, esters, alcohol, and carbon dioxide at the anode, and hydrogen gas at the cathode. Anode reactions occur via different mechanisms. The Kolbe reaction at an anode consists of the reduction of acid anion at the electrode surface to form an acid radical, which subsequently decomposes to form CO_2 and the alkyl radical as in Eq. (1). Two alkyl radicals then combine to form the Kolbe product; see Eq. (2).

$$RCOO^- \xrightarrow{-e^\ominus} RCOO\cdot \rightarrow CO_2 + R\cdot \tag{1}$$

$$2R\cdot \longrightarrow R\text{–}R \tag{2}$$

As the anode reaction occurs, hydrogen ions are reduced at the cathode, forming hydrogen gas,

$$2H^+ \xrightarrow{+2e^\ominus} H_2$$

Because the Kolbe product results from a bimolecular reaction, high concentration of the alkyl radical will favor this dimerization. For an electrolytic process, the requirement of high-alkyl radical concentration at the electrode surface means a high current density. A current density of 0.25 A/cm^2 is considered the lower limit for aliphatic hydrocarbon formation [10]. This current density is well above the current density anticipated from diffusion. The current density estimated from Fick's law is

$$I/A = 0.02C$$

where C is in mol/liter. However, stirring has been shown to reduce the effective boundary layer thickness and thereby to increase the current density by up to a factor of 20, which will give a current density in the right range at a concentration of about 1 mol/liter.

Experimental results have shown that electrolysis proceeds at very low electrolysis potentials. A half-wave potential of less than 0.5 V was obtained when current was measured as a function of potential in a stagnant 10% (w/v) valeric acid solution, buffered to pH 6 in a solution containing a much higher concentration of depolarizer than electroactive species, so that diffusion of acid anions is not affected by electrode–ion attractions (Fig. 8). The electrolysis current displayed a plateau from 0.7 to 1.3 V, corresponding to the diffusion current, but increased as the electrolysis potential increased above 1.3 V. The increased current was due to electrolysis of water. However, no oxygen evolved as would be expected from water electrolysis, when potential was increased to 23 V.

A mechanism that might explain the experimental lack of oxygen generation in high-voltage electrolysis was suggested by Glasstone and Hickling and is referred to as the "Hydrogen Perioxide Theory" [11]. These investigators suggest that hydrogen peroxide, formed at the anode from hydroxyl

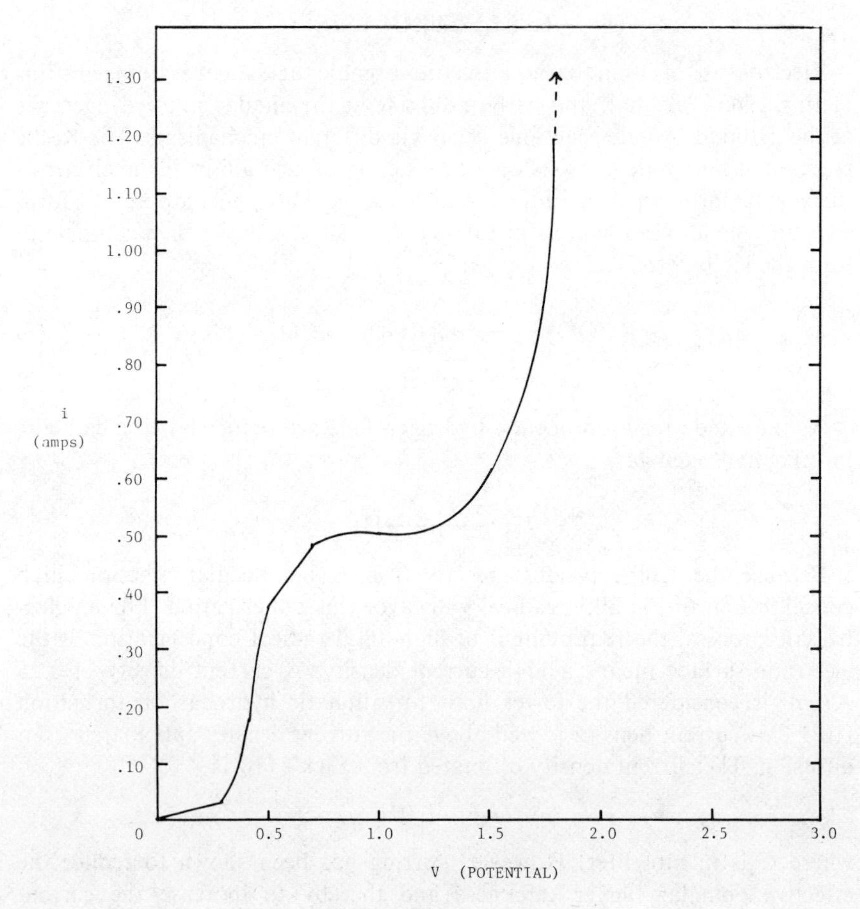

Fig. 8. Volts vs. current.

radicals, or the hydroxyl radicals themselves prior to dimerization, react with organic acid anions to form organic acid radicals. Radicals then combine as in eqs. (1) and (2) to yield alkane and carbon dioxide.

$$RCOO^{\ominus} + \cdot OH \rightarrow RCOO\cdot + OH^{\ominus}$$

$$2RCOO^{\ominus} + H_2O_2 \rightarrow 2RCOO\cdot + 2OH^{\ominus}$$

It is significant to note that anodes consisting of substances which are catalysts or which are oxidized to substances which are catalysts for hydrogen peroxide decomposition do not produce the Kolbe product.

Electrolysis of organic acids has been found to give a mixture of products, including olefins, alcohols, and esters as well as aliphatic hydrocarbons presumably from the following set of reactions:

$$H_2O \xrightarrow{-e^-} H^+ + OH\cdot \xrightarrow[-OH^-]{RCOO^-} RCOO\cdot \xrightarrow{CO_2} R\cdot \rightleftarrows R'CH=CH_2 + H\cdot$$

$RCOO^-$ $\xrightarrow{-e^-}$ $RCOO\cdot$; $R\cdot \rightarrow \frac{1}{2}R_2$; $RCOO\cdot + R\cdot \rightarrow RCOOR$; $OH\cdot + R\cdot \rightarrow ROH$; $OH\cdot + H\cdot \rightarrow H_2O$

Table VI lists the product composition from the electrolysis of a number of organic acids [10].

One may speculate from this reaction scheme that to maximize aliphatic hydrocarbon the preferred conditions are the following:

(1) High temperature to increase the rates of low activation energy processes, that is, diffusion to increase diffusion-controlled current density and decarboxylation to minimize ester formation.

(2) Low voltage to minimize hydroxyl ion formation, thus minimizing alcohol formation and perhaps reducing olefin formation.

(3) High current density to maximize $R\cdot$ concentration, favoring bimolecular product. To obtain high current densities at low voltages it is necessary to have high stirring rates. Use of depolarizers (supporting electrolytes) is counterproductive.

In order to maximize olefin production, on the other hand, it is necessary to run at low current densities to avoid dimerization. This is accomplished most economically by run at extremely low working voltages and large electrode areas. If voltages are kept below that required to electrolyze water, then platinum electrodes should not be necessary. Furthermore, sufficient hydrogen is produced by these processes to supply the required current at a potential of 0.7 V from a state-of-the-art fuel cell. As this potential is on the plateau region of Figure 8 it is likely that when optimized the olefin electrolysis and perhaps the paraffin electrolysis as well may be run with no external energy requirement. As shown in Table VI, propionic and butyric acids tend to favor olefin formation, that is, ethylene and propylene,

TABLE VI

Kolbe Electrolysis of Aliphatic Monobasic Carboxylates

ACID	STRUCTURE	PERCENTAGE YIELD		
		PARAFFIN	OLEFIN	ESTER
Acetic	CH_3COOH	85	2	2
Propionic	C_2H_5COOH	8	66	5
n-Butyric	C_3H_7COOH	14.5	53	10
iso-Butyric	$(CH_3)_2CHCOOH$	trace	62	10
n-Valeric	C_4H_9COOH	50	18	4
iso-Valeric	$(CH_3)_2CHCH_2COOH$	43	42	5
Methyl Ethyl acetic	$(CH_3)(C_2H_5)CHCOOH$	10	42	10
Trimethyl acetic	$(CH_3)_3C\ COOH$	13	52	0
Caproic	$C_5H_{11}COOH$	75	7	1.5
Lauric	$C_{11}H_{23}COOH$	45	—	—
Myristic	$C_{13}H_{27}COOH$	33.7	—	—
Palmitic	$C_{15}H_{31}COOH$	30	—	—
Stearic	$C_{17}H_{35}COOH$	27.6	—	—

whereas valeric and caproic acids tend to give alkanes; the unoptimized electrolysis process tends to give the desired products.

This is to acknowledge with great appreciation the assistance of Dr. John H. Ryther, Woods Hole Oceanographic Institution, Woods Hole, Massachusetts, for supplying sufficient amounts of the freshwater weeds *Hydrilla* and Duckweed, and the marine algae *Ulva lactuca* (green algae) and *Gracilaria* (red algae). The *Chondrus crispus* (Irish moss) was purchased from Marine Colloids, Inc., Rockland, Maine. The work was primarily supported by contract No. EG-77-C-02-4388 from the U.S. Dept. of Energy.

References

[1] S. I. Falkehaz, *J. Appl. Polym. Sci., Appl. Polym. Symp.*, *28*, 247–257 (1975).
[2] B. F. Hoanel, *Peat Lignite and Coal* (Government Printing Bureau, Ottawa, 1914), p. 55.
[3] J. S. Jeris and P. L. McCarty, *J. Water Pollut. Control Fed.*, *37*, 178 (1965).
[4] J. R. Anderson, A. Schaupp, C. Neustater, A. Brown, and L. O. Ljungdahl, *J. Bacteriol.*, *114*, 143 (1973).
[5] D. C. Augenstein and D. L. Wise, *Investigation of Converting the Products of Coal Gasification to Methane by the Action of Microorganisms*, Quarterly Report on ERDA Contract No. E(48-18)-2203, Report No. FE-2203-8, May 15, 1976. (Available from NTIS, P.O. Box 62, Oak Ridge, Tenn.)
[6] D. C. Augenstein, D. L. Wise, and C. L. Cooney, *Resour. Recovery Cons.*, *2*, 257 (1976/77).

[7] E. L. Jones, *Chem. Ind.*, 1590 (1967).
[8] W. V. Brown, *Chem. Eng. Progr.*, *59*, 10 (1963).
[9] R. W. Helsel, *Chem. Eng. Progr.*, *73* (1977).
[10] M. J. Allen, *Organic Electrode Processes* (Reinhold, New York, 1958).
[11] S. Glasstone and A. Hickling, *J. Chem. Soc.*, 1878 (1934).

Assessment of *Nif*-Derepressed Microorganisms for Commercial Nitrogen Fixation*

CARL J. WALLACE and BARRY O. STOKES
Jet Propulsion Laboratory, California Institute of Technology, Pasadena, California 91103

INTRODUCTION

Ammonia is a key chemical in the U.S. economy and is used in the manufacture of plastics, explosives, synthetic organic chemicals, and fertilizers. Fertilizer constitutes the major use at 74% of the total 16.5×10^6 tons produced in 1975[1]. The application of nitrogen fertilizer is a critical factor in maintaining U.S. agricultural productivity. Hardy concludes that nitrogen fertilization is the single most important nonbiological factor in increasing cereal grain yields [2, 3].

The majority of the commercially fixed ammonia is prepared by the Haber-Bosch process [4], which is well engineered and quite efficient. While current production is largely based on natural gas [5], the process can be easily adapted to other fossil fuels at higher cost [6, 7] so castastrophic dislocations in production are unlikely. Hardy [2] estimates that 500–600 ammonia plants could supply projected world-wide needs through the year 2000. While the need for alternative technologies for nitrogen fixation is not acute, conventional ammonia plants are expensive, costing as much as 10^8 dollars per 1000 tons per day capacity [2]. This fact together with recent interest in developing non-fossil-fuel-based technologies has focused attention on biological systems as an alternative means of fixing nitrogen. Global biological nitrogen fixation currently accounts for 175×10^6 metric tons of nitrogen fixed per year which exceeds the amount fixed by nonbiological processes [8]. In addition, 45×10^6 metric tons are produced biologically in agriculturally important settings [8]. Obviously, efforts to enhance biological nitrogen fixation have considerable potential for reducing our fossil-fuel-based fertilizer usage.

One approach to enhancing nitrogen fixation in free-living microorganisms is the development of *nif*-derepressed strains. Free-living microbes produce ammonia only when fixed nitrogen is unavailable in the

* This paper presents the results of one phase of research carried out at the Jet Propulsion Laboratory, California Institute of Technology, and supported by the National Science Foundation, Grant No. AER76-09093, by agreement with the National Aeronautics and Space Administration.

Biotechnology and Bioengineering Symp. No. 8, 153–174 (1978)

0572-6565/78/0008-0153$01.00

environment. [9–14]. *Nif*-derepressed strains are insensitive to repression by ammonia and are, therefore, genetically enhanced for nitrogen fixation. Two general types of *nif*-derepressed microbes have been otained during the last few years, namely, control mutants and auxotrophs (glutamate or glutamine). Control mutants were obtained by selecting for *nif*-derepressed strains among revertants from mutant strains of *Azotobacter vinelandii* that could not fix nitrogen [15]. Nitrogen fixation in these strains is not repressed by ammonia and the organisms grow normally on N_2 gas as the sole nitrogen source [15]. *Nif*-derepressed strains of *Klebsiella pneumoniae* [16–18] and *Spirillum lipoferum* [19] have been obtained by genetically blocking NH_3 assimilation. Such strains are auxotrophic (require glutamine or glutamate for growth) and excrete the ammonia produced by fixation of N_2 into the culture medium. The successful development of *nif*-derepressed strains raises the possibility of developing biological processes for commercial nitrogen fixation. This work examines the potential of *nif*-derepressed strains of *A. vinelandii* and *K. pneumoniae* for commercial nitrogen fixation.

EXPERIMENTAL

Materials and Methods

Microorganisms

Klebsiella pneumoniae SK-25 was obtained from Dr. Raymond C. Valentine. This strain was chosen due to its high ammonia production characteristics compared to other *nif*-derepressed strains of *Klebsiella* [17]. This strain is also among the most stable strains obtained by Valentine's group with a reversion frequency of less than 1 per 10^{10} cells. *Azotobacter vinelandii* UW590 was obtain from Dr. Winston J. Brill. This organism was chosen from among the *Azotobacter* mutants [15] since it exhibits nitrogenase levels in the presence of ammonia that are characteristic of the parent strain growing on N_2 (fully derepressed) [20].

Culture Conditions

All experiments employed 5 liter culture in a 7.5-liter New Brunswick bench-top fermentor (model 19) equipped with a pH controller and an O_2 monitor. *Azotobacter* experiments employed modified Burke's media [21] at pH 7.2 and 30°C except that glucose (autoclaved separately) was substituted for sucrose. These cultures were inoculated with approximately 5 ml log phase shake flask cultures at densities around $A_{420} = 0.7$. The stirring rate was maintained at 350 rpm, foam was controlled with Dow Corning Antifoam A spray, and aeration with air varied as described in the test. The *Klebsiella* experiment employed the minimal media of Yoch and Pengra [22] as modified by Streicher et al. [23] except that the phosphate concentration was doubled. Glutamine was added to approximate the conditions of Andersen and Shanmugam [18]. Ten $\mu g/ml$ were added initially and an additional 1 $\mu g/ml$ was added at 24 hr intervals, whereas Andersen and Shanmugam employed 4–8 $\mu g/ml$ initial and a continuous feed of 0.5 $\mu g/ml$ per day. The *Klebsiella* culture was maintained at pH 7.2 and 25°C with a stirring rate of 400 rpm and a sparging rate (Matheson ultrapure N_2) of 150–180 ml/min. Inoculum (1%) was prepared by growing a culture on Luria Broth and inoculating (1%) a flask of minimal media supplemented with 100 $\mu g/ml$ glutamine. This inoculum was used in the early stationary phase after the glutamine was utilized. This procedure was required to minimize the carry-over of Luria Broth nitrogen into the 5-liter vessel.

Oxygen Measurement

Oxygen concentration was measured by a recording New Brunswick dissolved oxygen analyzer. The instrument was zeroed prior to inoculation by sparging the culture with N_2 (Matheson ultrapure). The media was then saturated with air and the instrument spanned to 100%.

Growth Measurement

Where necessary, culture samples were diluted with distilled water to give absorbance values less than 0.6. The absorbance was measured on a Hitachi–Perkin Elmer–Coleman 139 spectrophotometer at 420 nm. Dry weight measurements were obtained by filtering 10 to 30 ml culture through a previously dried and weighted 0.45 μm Metricel millipore filter. The filter was then dried at 105°C and weighed.

Glucose Analysis

Cells were removed from culture samples by centrifugation at 12,000 $\times$ *g* for 15 min, and the supernatants were stored frozen until analyzed by the Sigma 510 glucose assay (glucose oxidase and peroxidose enzyme assay).

Ammonia Analysis

The ammonia from 1 ml culture supernatant was collected by microdiffusion and analyzed by the nesslers procedure as described by Burris [24].

Hydrogen Analysis

A 125-ml culture flask (actual volume = 135 ml) was stoppered with a butyl rubber stopper and evacuated. A hypodermic needle was inserted into the fermentor exhaust tubing and connected to a second needle with a latex tube. Sampling was accomplished by inserting the second needle into the stopper of the evacuated flask. Samples were then withdrawn from the flask and analyzed on a Hewlett Packard 5830A gas chromatograph equipped with a thermal conductivity detector and a molecular sieve column. The integral of the H_2 peak was then compared to standard mixtures of H_2 prepared by injecting pure hydrogen into stoppered culture flasks filled with air. No significant leakage of H_2 from tightly stoppered standard mixtures was observed over a 10-day period.

Reversion Analysis

Reversion was monitored by spreading 0.5 ml culture on agar plates made with the minimal media described above supplemented with 1 g/liter NH_3.

Results and Discussion

K. pneumoniae SK-25

Andersen and Shanmugam have studied a number of derepressed mutants of *Klebsiella pneumoniae* [18]. In these studies the bacteria were maintained in a dialysis bag (25 ml) suspended in approximately 10 volumes of media to provide for dissipation of waste products and better feeding of the cell suspension. The dialysis system gave significant increases in ammonia production efficiency over previous experiments [16]. The dialysis experiments yielded a value of about 4 mol glucose utilized/mol NH_3 produced during the peak efficiency period and 7–9 mol glucose/mol ammonia for the overall fermentation. The moles of ammonia produced per mole of glutamine consumed calculated for *K. pneumoniae* SK-25 were approximately 40–60 (0.28–0.41 mol NH_3/g glutamine).

While the dialysis apparatus of Andersen and Shanmugam was advantageous for their experimental goals, it is not amenable to scale up for ammonia production. To approximate their conditions on a larger scale, a conventional fermentation was run without the inclusion of the dialysis bag as described in materials and methods. The results are presented in Figure 1.

Because of the low glutamine concentration of the medium, the cells entered a stationary phase at a low cell density (A_{420} = 0.057) and remained in this stationary phase until revertants capable of growth on ammonia overran the culture at around 120 hr. In order to increase the ammonia levels of the medium, fermentations at higher optical densities (OD) were run (data not presented). The maximum efficiency of glucose utilization of such cultures was found to be as low as one-third that obtained in Figure 1. We suspect that this decrease in efficiency at higher OD is due to inhibitory fermentation products that accumulate in the medium.

The maximum rate of ammonia production occurs in the early stationary phase. A maximum efficiency of 4.4 mol glucose/mol ammonia formed occurs between 80 and 90 hr. The overall efficiency of glucose utilization from 0–115 hr is 6.6 mol glucose/mol NH_3 produced. During this period the

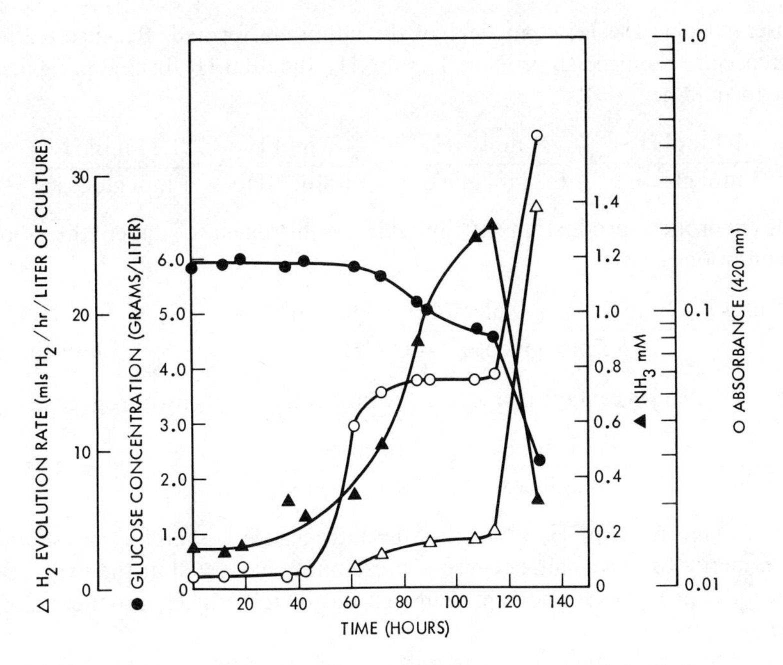

Fig. 1. Production of ammonia and hydrogen by *K. pneumoniae* SK-25.

culture consumed 13 μg glutamine/ml to yield 13.3 mol ammonia/mol glutamine used. The glucose efficiency values from Figure 1 show excellent agreement with the data of Andersen and Shanmugam and indicate that the dialysis apparatus is not necessary for attaining a high efficiency system.

Even though SK-25 is quite stable, less than one revertant per 10^{10} organisms [25], 5-liter cultures routinely revert, thus prohibiting large fermentations with this organism unless very inexpensive methods of growth control can be devised. While the slightly altered glutamine addition schedule (see Methods) may have adversely affected the efficiency of glutamine usage, the low value compared to that of Andersen and Shanmugam is likely due to reversion of the culture.

The rate of hydrogen production per liter of culture in Figure 1 parallels culture density as expected. The overall hydrogen production from 0–131 hr is 1.1 mol H_2/mol glucose consumed. Since nitrogenase is only responsible for about 0.65 mol H_2/mol NH_3 produced [18] and the production of 1 mol NH_3 consumes 6.6 mol glucose, the majority of the H_2 formed (91%) comes from the conversion of pyruvate to acetate + H_2 + CO_2 [26]. One possible application of *nif*-derepressed heterotrophs is to enhance the production of H_2 from a carbon substrate. An estimate of the maximum obtainable increase in H_2 production due to nitrogenase can be calculated from the above data by assuming that a similar culture is grown in the absence of N_2 to divert all energy and reductant to the production of H_2 and that 1 mol NH_3 is equivalent to 1.5 mol H_2. The total H_2 produced would then be that observed plus the H_2 equivalent of the ammonia formed. Because 6.2 mol glucose are required to produce 1 mol NH_3, the total H_2 formed per glucose used would be

$$\frac{1.1 \text{ mol } H_2}{1 \text{ mol glucose}} + \frac{1 \text{ mol } NH_3}{6.2 \text{ mol glucose}} \times \frac{1.5 \text{ mol } H_2}{1 \text{ mol } NH_3} = \frac{1.34 \text{ mol } H_2}{1 \text{ mol glucose}}$$

The hydrogen production attributable to nitrogenase under the above assumptions is

$$\frac{1.5 \text{ mol } H_2/1 \text{ mol } NH_3 + 0.65 \text{ mol } H_2 \text{ (side rxn)}/1 \text{ mol } NH_3}{6.2 \text{ mol glucose}/1 \text{ mol } NH_3} = \frac{0.35 \text{ mol } H_2}{1 \text{ mol glucose}}$$

The expected percentage increase in the production of H_2 by nitrogenase is

$$\frac{0.35 \text{ mol } H_2/1 \text{ mol glucose}}{1.34 \text{ mol } H_2/1 \text{ mol glucose} - 0.35 \text{ mol } H_2/1 \text{ mol glucose}} \times 100 = 35\%$$

The enhancement of H_2 production by nitrogenase is therefore substantial. The energy to power nitrogenase is presumably harvested at the expense of cell mass production and represents a valid increase in H_2 production efficiency.

A theoretical value for ammonia (or H_2) production can be made by assuming 2 ATP's used per $2e^-$ transferred and ignoring requirements for cell maintenance energy. *Klebsiella pneumoniae* SK-25 can obtain 2 mol of ATP and 2 mol of NADH by the fermentation of glucose to pyruvate and one additional ATP per mole of glucose form the clastic cleavage of pyruvate to H_2, CO_2, and acetate [26]. Assuming $3e^-$ transferred per molecule NH_3 produced and 2 ATP molecules hydrolyzed per $2e^-$ transferred, 1 mol of glucose would yield sufficient ATP and more than enough reductant to produce 1 mol of NH_3 (or 1.5 mol of H_2). Although this value cannot be achieved in practice, it does represent an upper limit to the efficiency of the *Klebsiella* system.

A. vinelandii UW590

While *Azotobacter vinelandii* UW590 is fully derepressed for nitrogenase synthesis [20], all of the nitrogen fixed is used for cell growth. Measurements of free ammonia in growing and stationary cultures routinely yielded values below 0.1 m*M* ammonia even at high cell densities. The only product of nitrogen fixation considered for this organism, therefore, is cell mass. Because the cultures are grown with N_2 as the sole nitrogen source, cell growth is an indirect measure of nitrogen fixation. Figure 2 shows the time course of cell mass production at various air flow rates indicating that the production of high cell densities requires high aeration. The maximum doubling time for the organism on N_2 appears to be around 3 hr based on the optical density data. Detailed data on the 8.5 liters of air per minute experiment is presented in Figure 3.

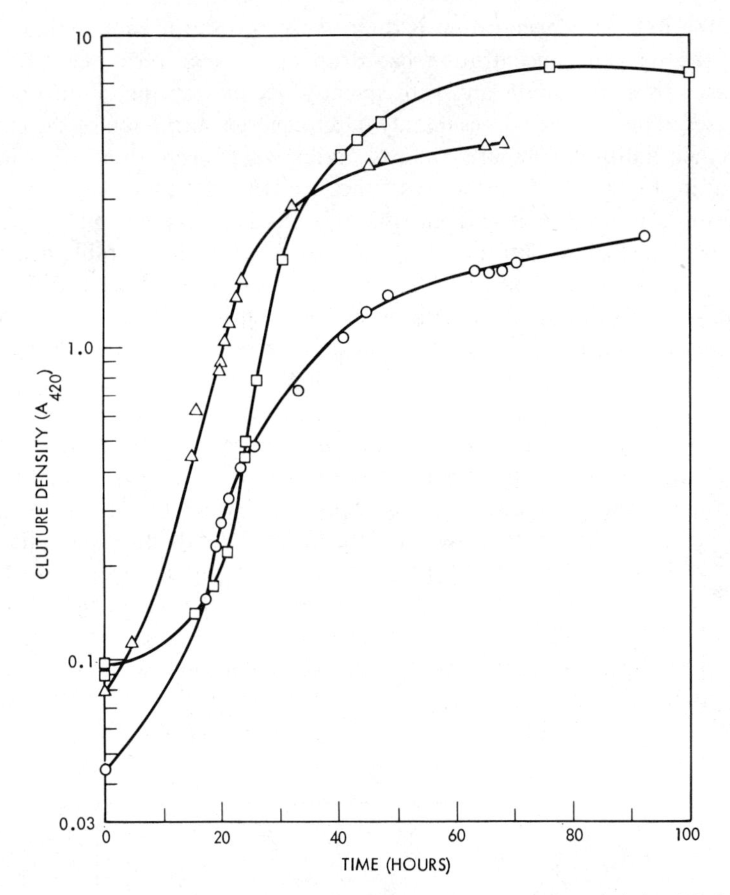

Fig. 2. Time course of growth of *A. vinelandii* UW590. (○) 0.5 liter air/min; (△) 5.0 liter air/min; (□) 8.5 liter air/min.

Oxygen concentration is known to markedly affect the activity of nitrogenase in the *Azotobacter* genus. These organisms are obligate aerobes and fix nitrogen only in the presence of oxygen. High oxygen concentration on the other hand inhibits the nitrogenase enzyme, hence a distinct optimum O_2 concentration is observed. This optimum is variable and depends on the previous oxygen levels to which the culture was adapted [27].

The efficiency of growth in Figure 3 is calculated as the grams of cells produced per gram of glucose used. This curve was generated using data obtained by extrapolating the dry weight curve on the basis of optical density. (A_{420} = 1.0 corresponds to a culture dry weight of 0.38 g/liter.) Above optical densities of 1.0, cultures show a linear relationship between optical density and dry weight. Large deviations from linearity were observed, however, with some cultures below optical density 1.0. This phenomenon appears to be associated with rapid growth rates and/or growth at high O_2 concentrations. During the initial phase of the culture, while the oxygen concentration is high, the growth efficiency remains low. Once the oxygen concentration has dropped to near zero, the efficiency increases to a maximum and subsequently declines as the culture density increases. This decline is presumably due to energy starvation of the culture by oxygen limitation. Figure 4 shows similar data for all three experiments shown in Figure 1. In each case the growth efficiency decreases with increasing culture density. No maximum is observed in the two low aeration runs because O_2 concentration had gone to zero and measurable decreases in the media glucose occurred at or past the period of peak efficiency. Figure 4 indicates that the operation of high-density cultures at reasonable efficiencies requires considerable aeration of the medium. The shift of the efficiency curve to higher culture density with increasing aeration indicates that the efficiency decrease with culture density is due to O_2 limitation.

If one assumes that the cell mass produced in these experiments contains 10% nitrogen content, the maximum efficiency of fixation is around 50 mg of nitrogen fixed per gram of glucose consumed, which is at the high end of reported values for *Azotobacter* [28, 29]. Leahy recently observed efficiency values with *A. vinelandii* UW590 approaching 100 mg of nitrogen per gram of glucose consumed which is roughly twice as efficient as the best reported values to date [30]. Since Leahy's values were based on cultures with very small consumption of carbohydrate, additional work appears appropriate to substantiate this unusually efficient value. If this work is confirmed, Leahy's conditions are the best reported values for *Azotobacter* to date.

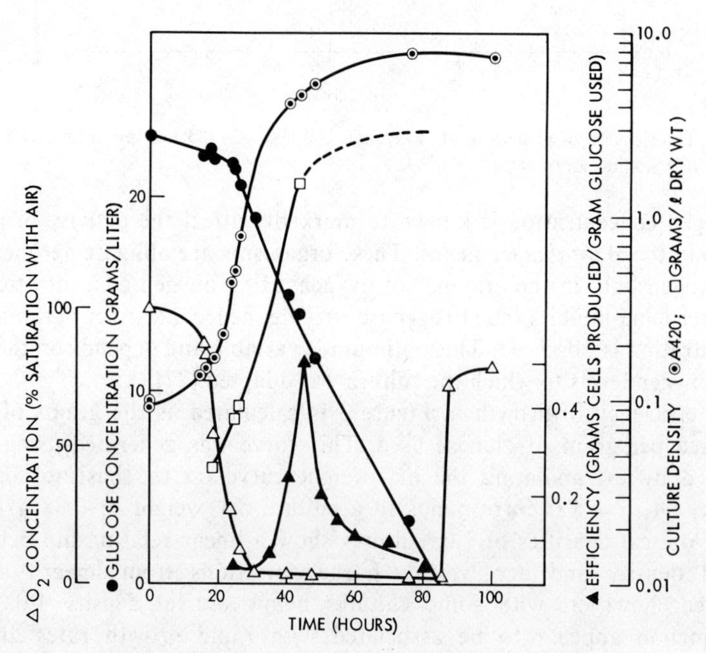

Fig. 3. Efficiency and growth of *A. vinelandii* UW590.

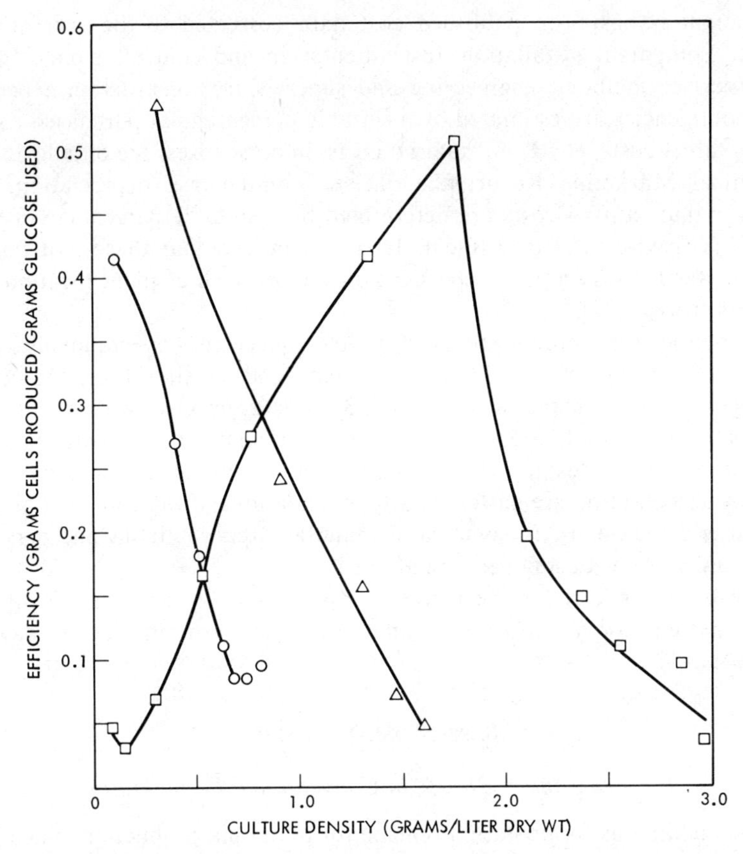

Fig. 4. Effect of culture density on the growth efficiency of *A. vinelandii* UW590. Symbols same as in Fig. 2.

Mulder [31] has proposed a theoretical efficiency limit of 280 mg of nitrogen fixed per gram of glucose consumed (0.28 mol glucose/mol ammonia formed) for aerobic systems by assuming that two molecules of ATP are hydrolyzed per $2e^-$ transferred to nitrogenase, that $6e^-$ are required for the reduction of one N_2 molecule and that all of the energy from the oxidation of glucose (38 ATP equivalent) is available for nitrogen fixation.

ECONOMIC EVALUATION

Economic Model

Estimation of fixed-capital investment and total product cost are based on procedures discussed by Peters and Timmerhaus [32]. These approximate cost figures are basically predesign cost estimates suitable for determining feasibility of proposed investment. Estimated cost of processing equipment is based on published cost data corrected to the current cost index. Equipment installation, instrumentation and control, piping, buildings, service facilities, engineering and supervision, construction expenses, and contigencies are estimated by a suitable percentage of purchased equipment, direct costs, etc. Raw material costs, in most cases, are obtained from Chemical Marketing Reporter. Utilities, maintenance, depreciation, and other manufacturing costs are determined by a suitable percentage of such items as fixed-capital investment. It is emphasized that these cost figures are only estimates and are dependent on factors such as plant location and type of process [32].

Economics for ammonia recovery at low concentrations employing selective ion exchange are based on estimated costs at the Upper-Occoquan sewage treatment plant in Virginia [33]. For ammonia recovery at high concentration, cost figures are based on equipment requirements from the coking of coal industry [4]. Estimated capital and operational costs for lagoon fermentation are patterned after calculations of Benemann et al. [34] for large algae ponds. In some cases manufacturers' bulletins and personal communication with suppliers are utilized.

The total capacity for the various fermentation systems was 10 tons of NH_4^+ per day for *K. pneumoniae* and 10 tons of fixed nitrogen per day for *A. vinelandii.*

Results and Discussion

K. pneumoniae, SK-25, in Conventional Fermentation

The continuous fermentation scheme for ammonia production is depicted in Figure 5. For the purpose of economic assessment several assumptions were desirable:

(1) Glutamine costs are not included even though SK-25 requires glutamine or mixed amino acids for growth. Glutamine or glutamate requirements of available *Klebsiella* mutants exceed the value of the fixed nitrogen products; therefore, a cheap amino acid source is needed. This requirement might be met economically with hydrolyzed fish scrap ($162.50/ton at 60% protein) or possibly by hydrolyzed cell mass from the fermentation, but these costs are not estimated because the use of these materials is speculative.

(2) Sterilization costs are not included as steam sterilization of millions of gallons of media is prohibitively expensive for a cheap product such as ammonia.

(3) Only carbohydrate, water, nitrogen gas, and recovery chemicals are included in raw material costs.

(4) Settling ponds are suitable for cell separation and recovery. Definitive data are not available to indicate the feasibility of this; however,

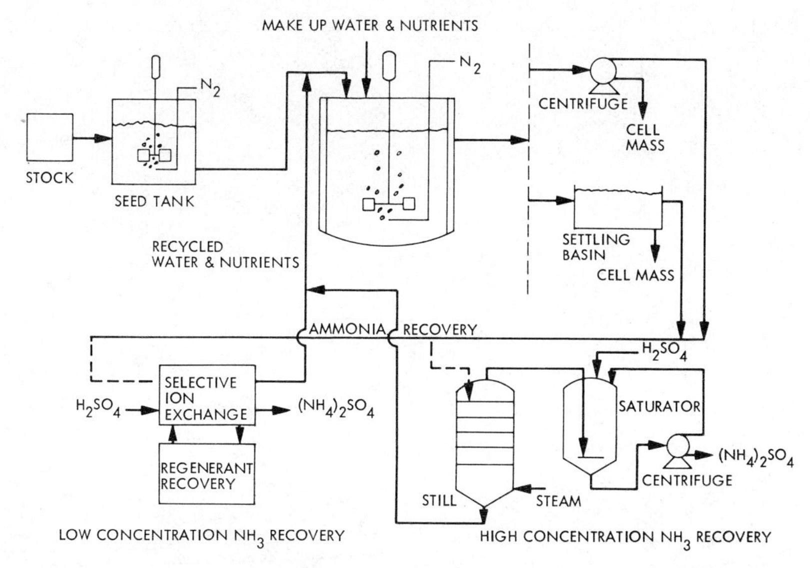

Fig. 5. Continuous fermentation by *K. pneumoniae* for ammonia production and recovery.

preliminary settling studies indicate that settling is possible with proper pH adjustment. Centrifugation for cell recovery is prohibitively expensive [35].

(5) Ammonia recovery is by selective ion exchange for ammonia concentrations of 500 mg/liter and less, and steam stripping with sulfate precipitation is the method of choice for a concentration of 1800 mg/liter.

(6) Eighty percent recycle of water and unconsumed nutrients is assumed. Figure 6 represents the effect of ammonia concentration on the product cost for glucose, molasses, and a free carbon source for a capacity of 10 tons of NH_4^+ per day. The experimental data from this study at high cell density correspond to an NH_4^+ concentration of about 97 mg/liter. For this NH_4^+ concentration, $(NH_4)_2SO_4$ cost is in excess of \$3000/ton if the carbohydrate is free. The current market price of $(NH_4)_2SO_4$ is \$65/ton. The lowest product cost in Figure 6 is represented by free carbohydrate and an NH_4^+ concentration of 1800 mg/liter corresponding to the maximum level before growth inhibition was observed in test tube studies (data not presented). Even assuming that this concentration could be attained in a fermentation system, that the carbohydrate is free, and that the assumptions listed above are reasonable, the economics are still grossly unfavorable.

Figure 7 depicts the effect of energy consumption on product cost for glucose and free carbohydrate. The energy consumption of 13.98 mol of glucose per mol of NH_4^+ corresponds to experimental data of this study at high cell density while 3.8 mol glucose/mol NH_4^+ corresponds to data from the literature [18] at low cell density. Calculations for the highest

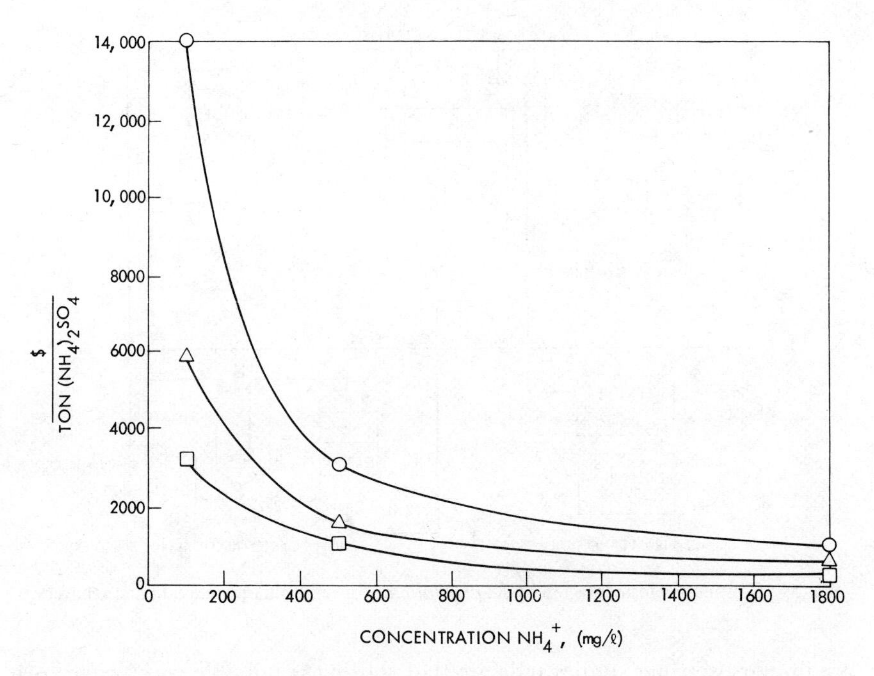

Fig. 6. Cost of ammonia production by *K. pneumoniae* as a function of ammonia concentration in conventional fermentation. (○) Glucose; (△) molasses; (□) free carbon source.

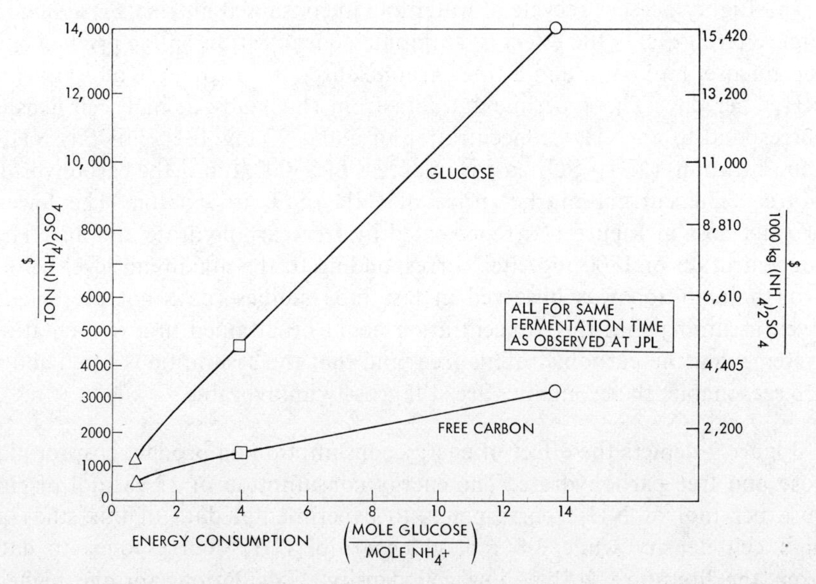

Fig. 7. Cost of ammonia production by *K. pneumoniae* as a function of reciprocal efficiency in conventional fermentation.

energy consumption noted in Figure 7 are based on a hypothetical NH_4^+ concentration of 1800 mg/liter with fermentation time and glucose consumption assumed to be the same as observed in this study; this corresponds to an efficiency of 1.37 mol NH_4^+/mol glucose, which is in excess of the theoretical limit of 1.0 for anaerobic systems.

From Table I, representing fixed-capital investment (FCI) and manufacturing cost for various NH_4^+ concentrations, it is apparent that FCI is excessive for a conventional fermentation scheme. Large fermentation equipment is needed when producing a relatively cheap material such as ammonia in dilute solution. Manufacturing costs are extremely dependent on raw material requirement.

K. pneumoniae, SK-25, in Lagoons

The ammonia production scheme for *K. pneumoniae* growing in a lagoon is similar to that depicted in Figure 5 for conventional fermentation except for replacement of the stirred fermentor with a shallow plastic-lined lagoon. Some concrete support for the lagoon was considered in the economic assessment as well as a plastic cover for the anaerobic system. Settling ponds for cell separation as well as ammonia recovery techniques are the same as for the conventional fermentation. The basic assumptions are the same for the economic assessment of ammonia production by *K. pneumoniae* in lagoons as in the conventional fermentation.

The effect of NH_4^+ concentration in the lagoon on product cost is presented in Figure 8. Product costs based on data of this study at high cell density are represented at an NH_4^+ concentration of 97 mg/liter. An NH_4^+ concentration of 1800 mg/liter corresponds to an upper limit before growth inhibition. At the upper limit of NH_4^+ concentration, with a free carbon, and with the same basic assumptions stated above, the product cost of $(NH_4)_2SO_4$ is \$91/ton. For lower concentrations in the lagoon, production costs increase rapidly.

TABLE I

Capital Investment and Manufacturing Costs for *K. pneumoniae* in Conventional Fermentation

NH_4^+ (mg/ℓ)	F.C.I. \$ MILLION	MANUFACTURING COST \$ MILLION		
		GLUCOSE	MOLASSES	FREE CARBON
97	144.2	199.8	81.4	44.9
500	34.4	43.9	21.2	14.4
1800	15.7	15.0	8.6	6.7

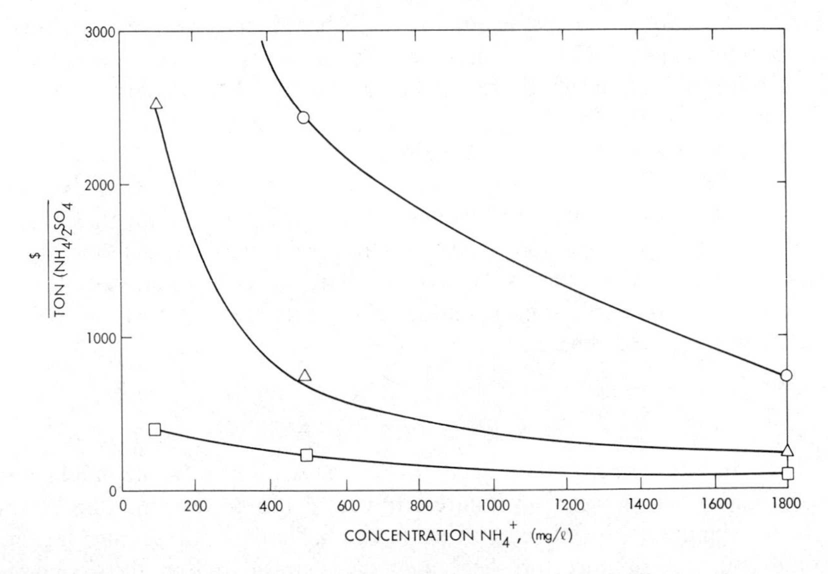

Fig. 8. Cost of ammonia production by *K. pneumoniae* as a function of ammonia concentration in a lagoon system. Symbols same as in Fig. 6.

The effect of energy consumption on product cost is shown in Figure 9. Once again, the data of this study at high cell density correspond to an energy consumption of 13.9 mol glucose/mol NH_4^+. The lowest energy consumption considered, that is, 0.73 mol glucose/mol NH_4^+, corresponds to 1800 mg NH_4^+/liter with the assumption that the same amount of carbon would be consumed as was observed in the fermentations of this study.

Because capital costs are much less for a lagoon than for agitated steel vessels, the economics for the lagoon system become much more dependent on raw material costs. As a result of this, the difference in product cost for glucose and a free carbon source is even more significant than for a conventional fermentation system. Table II represents the results of NH_4^+ concentration on fixed capital investment and manufacturing costs for the lagoon system.

A. vinelandii, UW590, in Conventional Fermentation

The economic assessment of *A. vinelandii* in both the lagoon and conventional fermentation required several assumptions:

(1) Sterilization costs are not included. This may not be a severe restriction or limitation on the actual economic feasibility since nitrogen is not included in the media and, therefore, only nitrogen fixing organisms could contaminate.

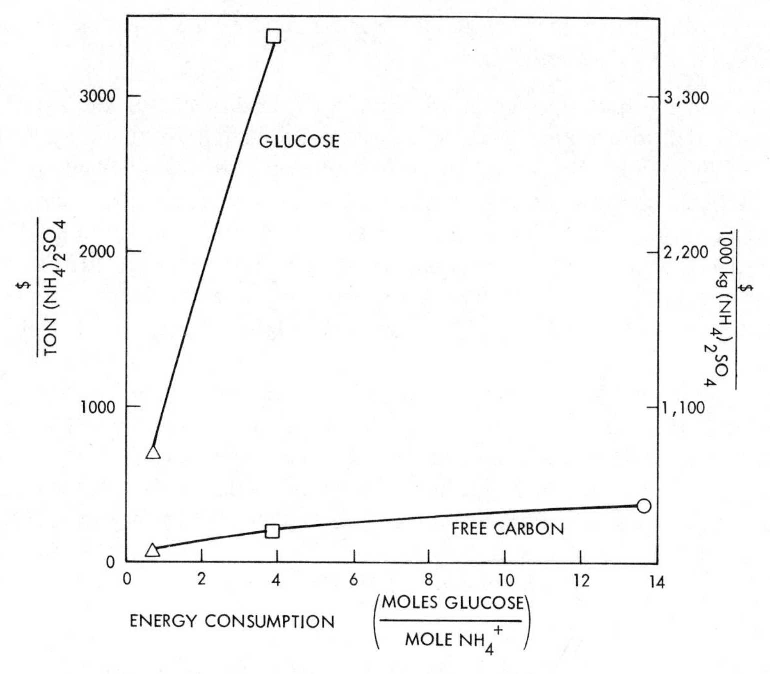

Fig. 9. Cost of ammonia production by *K. pneumoniae* as a function of reciprocal efficiency for a lagoon system.

(2) Settling ponds are suitable for cell separation and recovery.

(3) Eighty percent of the water and unconsumed nutrients are recycled.

(4) Surface aeration is adequate for cell growth in the lagoon. Experimental data show that final cell density increases with aeration; therefore, surface aeration in a lagoon may be totally inadequate for high yield of cellular mass.

TABLE II

Capital Investment and Manufacturing Costs for *K. pneumoniae* in a Lagoon System

NH_4^+ (mg/ℓ)	F.C.I. $ MILLION	MANUFACTURING COST $ MILLION		
		GLUCOSE	MOLASSES	FREE CARBON
97	8.7	152	38.6	5.2
500	2.85	32.6	9.9	3.07
1800	3.05	9.41	3.08	1.22

(5) Growth rates in the lagoon are assumed to be one-half the rate observed in a stirred fermentor.

For conventional fermentation, Figure 10 demonstrates the effect of nitrogen fixation efficiency on the cost of producing fixed nitrogen. The data of this study correspond to an overall fermentation efficiency of 18.7 mg of fixed nitrogen per gram of carbohydrate. Higher efficiencies were noted at various times during the course of the batch fermentations. For a free carbohydrate, the cost estimates indicate approximately \$15/lb fixed nitrogen for the efficiency of 18.7. Fixed nitrogen for fertilizer purposes is priced at about \$.10/lb and is considerably higher if intended for animal feed.

According to Mulder [31], the theoretical maximum efficiency for free-living nitrogen fixation in an aerobic system is 280 mg nitrogen/g carbohydrates. At this efficiency and for a free carbohydrate, the cost of fixed nitrogen by conventional fermentation is \$1.14/lb fixed nitrogen. This assumes similar fermentation times as observed at an efficiency of 18.7.

A. vinelandii, UW590, in Lagoons

From the graph of fixed nitrogen cost as a function of nitrogen fixation efficiency (Fig. 11), it is obvious that a free or extremely cheap car-

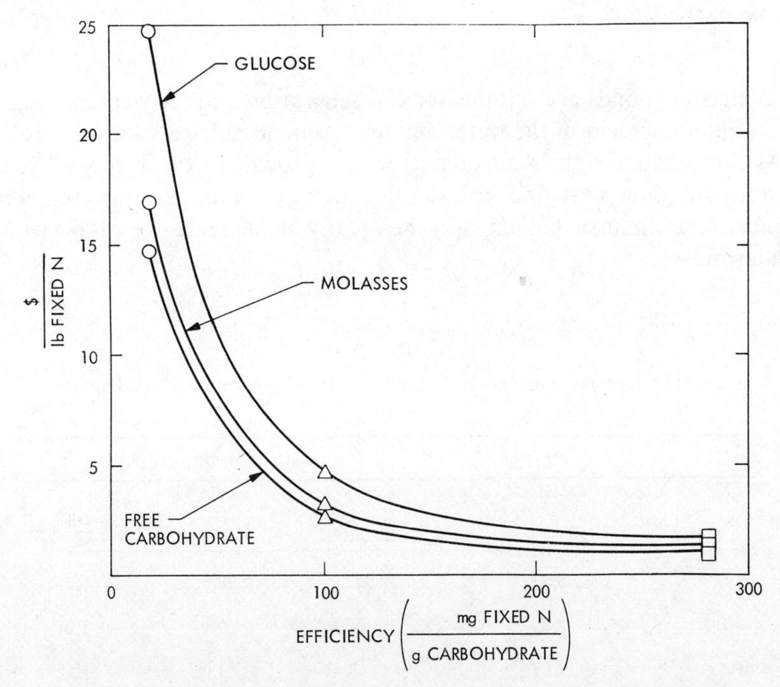

Fig. 10. Cost of fixed nitrogen as a function of efficiency for *A. vinelandii* in conventional fermentation. (○) Observed at JPL; (□) theoretical efficiency.

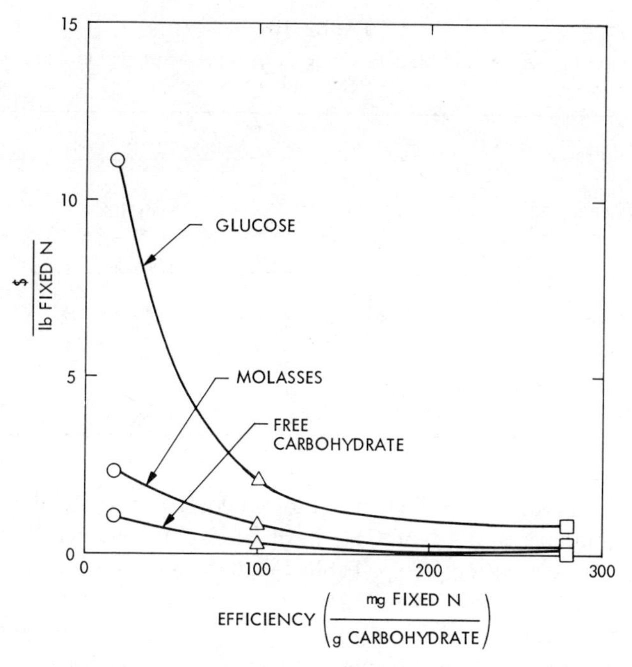

Fig. 11. Cost of fixed nitrogen as a function of efficiency for *A. vinelandii* in a lagoon system. (O) Based on JPL data; (□) theoretical efficiency.

bohydrate is mandatory for any chance of favorable economics. From the data of this study which correspond to an efficiency of 18.7 g fixed nitrogen per gram of carbohydrates, the cost of fixed nitrogen is in excess of $1/lb even for free carbohydrate. If the theoretical efficiency of 280 mg fixed nitrogen per gram of carbohydrate could be attained and a free carbohydrate used, then the cost of fixed nitrogen becomes about $.10/lb or comparable in price to present fertilizer nitrogen. However, the severe assumptions imposed on the economic assessment must be realized; they may correspond to an unrealistic and impractical situation as data are not available on *A. vinelandii* adaptation in lagoons, settling properties of the organism, growth rates, and nitrogen fixation efficiencies in lagoon conditions, etc.

Capital investment and manufacturing costs for *A. vinelandii* growth in lagoons are tabulated in Table III for a capacity of 10 tons of fixed nitrogen per day.

Economics of Nitrogen Fixation versus Anaerobic Digestion

Table IV lists the value of the product, either $(NH_4)_2SO_4$, cell mass, or methane, per 100 lb of glucose digestion (theoretical value assuming digester gases are 50% CH_4 and 50% CO_2). The theoretical methane yield

TABLE III

Capital Investment and Manufacturing Costs for *A. vinelandii* in a Lagoon System

EFFICIENCY	F.C.I.	MANUFACTURING COST		
(mg N) / (g CARBOHYDRATE)	$ MILLION		$ MILLION	
		GLUCOSE	MOLASSES	FREE CARBON
18.66	1.67×10^6	81.1	17.1	8.1
100	410,000	15.3	4.8	1.7
280	168,000	5.6	1.85	0.75

for glucose is about 6 ft^3 of CH_4 per lb of glucose. For waste materials such as poultry manure, 4.0 ft^3 of CH_4 have been obtained per lb of volatile solids [36]. It is apparent from Table IV that anaerobic digestion of carbonaceous material yields methane of greater value than the nitrogenous products of nitrogen fixation by the presently available derepressed mutants. By increasing the nitrogen content of digestor effuent and residue, it may be possible to enhance the value of these materials as fertilizers without inhibiting methanogenesis. Ammonia concentrations for swine [37] and poultry manure [36] digestion reach levels of 1500–3000 mg/liter. However, digester operation has been observed to cease for swine manure digestion at an ammonia concentration of 2000 mg/liter [37]. Ammonia loss from the digester effluent when exposed to the air would also decrease its value as a fertilizer. From these facts it would appear that increasing

TABLE IV

Comparison of Product Value for Nitrogen Fixation Processes and Anaerobic Digestion

	VALUE OF PRODUCT/100 LB. GLUCOSE CONSUMED
KLEBSIELLA PNEUMONIAE	$(NH_4)_2SO_4$
DATA FROM JPL (HIGH CELL DENSITY)	$0.09
DATA FROM JPL (LOW CELL DENSITY)	0.18
THEORETICAL	1.19
AZOTOBACTER VINELANDII	(CELL MASS)
DATA FROM JPL	$0.19
THEORETICAL	2.80
METHANE FROM ANAEROBIC DIGESTION	$1.20

nitrogen content in the digester would have no benefit and may even be detrimental for wastes of high nitrogen content. However, for wastes of low nitrogen concentration, enhancement of nitrogen fixation may prove beneficial.

Hydrogen Production by K. pneumoniae

The most promising application of *nif*-derepressed mutants (of *Klebsiella*) appears to be the enhancement of hydrogen production since the value of hydrogen produced in the fermentation (1.1 mol H_2 per mol of glucose consumed) at $2/lb is roughly 20-fold greater than that of the ammonia produced, and the H_2 equivalent of ammonia (assuming 1.5 mols H_2 per mol of ammonia) is 3.5 times that of ammonia. In addition, the activity of nitrogenase in present strains provides a significant increase over wild-type H_2 production. Hydrogen production would also be simplified since contaminating microbes would be unable to use the product in anaerobic environment. Using a nonauxotroph, culture growth could be limited by addition of ammonia, and problems of reversion, amino acid supplementation, and sterilization could be largely eliminated. While wild-type cultures could presumably be employed, *nif*-derepressed control mutants similar to *A. vinelandii* UW590 would be preferred since nitrogenase would not be repressed by fixed nitrogen contamination of the carbon material utilized. A by-product of this process would be acetate, which could be used as a fermentation substrate or presumably harvested as methane. Hydrogen production in conjunction with anaerobic digestion should be explored, possibly in a two-stage fermentation system, as a potential application for suitable *Klebsiella* mutants.

CONCLUSIONS

K. pneumoniae

Ammonia production

Several factors render the present strains of *K. pneumoniae* unsuitable for application. First, the amino acid requirement places a significant burden on the economics, assuming that a satisfactory, low-cost amino acid substitute for glutamate/glutamine could be obtained. Otherwise, the burden is prohibitive. Secondly, the genetic stability of the present mutants is insufficient for even 5-liter scale ammonia production. Presumably this could be corrected by obtaining the appropriate deletion mutants, but a third consideration, namely, sterilization, would still render ammonia production uneconomical. Any contaminating microorganisms would quickly overrun the fermentation since the high efficiency production of ammonia relies on stationary cultures at low optical densities. If the rate of ammonia production were increased markedly and high density cultures could be used, batch processes relying on high initial inoculum and short duration runs may avoid sterilization requirements for the final batch media. However, the economic projections presented, which assumed such improvements, do not indicate that such development of the organism for ammonia production alone is worthwhile. Development of a very inexpensive method of inhibiting microbial growth, which would not affect nitrogenase activity, could be employed after the culture entered stationary phase, but again the economic projections presented do not justify such activity.

Other Products

The production of ammonia by *Klebsiella* does not use all of the energy available in the carbon substrate. Products other than ammonia include cell mass, H_2, acetate, and other organic materials. The organic remainder could possibly find application as media for an aerobic fermentation or in the production of methane. Separation of these organics from the solution is likely uneconomical due to the low concentration involved.

As discussed previously, the value of the H_2 produced in the medium greatly exceeds the value of the fixed nitrogen and even the H_2 equivalent of NH_3 is worth more than the ammonia. The methane value of the carbon energy source also exceeds the value of the fixed nitrogen produced. These considerations suggest that ammonia should be considered the by-product and attention should focus on the production of H_2 and other materials.

A. vinelandii

Large-scale production of fixed nitrogen by existing strains does not appear economical. Realistic efficiencies for *A. vinelandii* are in the range of 10–50 mg of fixed nitrogen per gram of carbohydrate consumed. If efficiencies approaching theoretical limits could be achieved in inexpensive lagoons, with minimum aeration and a free carbon source, economics of the process may become competitive. At theoretical efficiency the value of the fixed nitrogen as fertilizer exceeds the value of the carbon substrate for methane production.

The use of a derepressed mutant in such a system is beneficial only if the waste material contains a significant amount of fixed nitrogen which would repress nitrogenase synthesis. Sterilization is not required because the only product is cell mass and contaminating strains are expected to have only a marginal negative impact when the material employed contains fixed nitrogen.

General

Capital investment is prohibitive for conventional fermentation employing stainless-steel vessels. This is primarily due to the vast capacity required in dealing with a low-value product in dilute solutions. Any application of free-living microorganisms for commercial nitrogen fixation will require considerable ingenuity in devising low-cost, "low-technology" methods. Fortunately, microorganisms are amenable to such low-technology methods and may find application in foreign settings where expensive equipment is unobtainable.

While the economics of nitrogen fixation by *Azotobacter* might become economical by improving the efficiency of the organism and by price increases of commercial nitrogen, the occurrence of truly free carbon materials are unlikely in a commercial setting. Free carbon on the other hand may be available on a small-scale such as to the individual farmer who needs fixed nitrogen. This would be particularly true in the foreign setting, and efforts to increase the efficiency of *Azotobacter* or other aerobic nitrogen fixers are needed.

One possible application of *nif*-derepressed microbes that was not considered was the *in situ* production of ammonia. Experiments to quantitate the survival of control-type *nif*-derepressed mutants in the field are required in order to evaluate the benefit of soil inoculation with these strains. Owing to the high carbohydrate requirements of heterotrophs, *nif*-derepressed strains of blue-green algae would appear to be the organisms of choice for *in situ* fertilization. Phototrophs, however, are limited to surface environments so *nif*-derepressed heterotrophs may still be useful in fixing nitrogen in dark environments.

The authors thank Wayne Schubert and Allen Hatter for accomplishing the majority of the experimental work presented and NSF-RANN for financial support. We are also indebted to Drs. Alva A. App, Jack Newton, Donald Isenberg, Raymond C. Valentine, K. T. Shanmugam, Kjell Andersen, and Winston J. Brill, for advice and assistance with various aspects of the work.

References

[1] B. F. Greek, *Chem. Eng. News*, *53*, 10 (1975).
[2] R. W. F. Hardy and U. D. Havelka, *Science*, *188*, 633 (1975).
[3] R. W. F. Hardy, "Increasing crop productivity: Agronomic and economic considerations on the role of biological nitrogen fixation," presented at 1977 Annual ACS Meeting, Chicago, Illinois, Aug. 28–Sept. 22, 1977.
[4] R. N. Shreve, *Chemical Process Industries*, 3rd ed. (McGraw-Hill, New York, 1967).
[5] D. E. Nichols and G. M. Blouin, "Economic considerations of chemicals nitrogen fixation," presented at 1977 Annual ACS Meeting, Chicago, Illinois, Aug. 28–Sept. 2, 1977.
[6] M. Rosenzweig and S. Ushio, *Chem. Eng.*, *81*, 62 (1974).
[7] C. P. Lattin, *Chem. Eng. News*, *53*, 10 (1975).
[8] K. J. Skinner, *Chem. Eng. News*, *54*, 22 (1976).
[9] G. Daesch and L. E. Mortenson, *J. Bacteriol.*, *110*, 103 (1972).
[10] R. W. Detroy and P. W. Wilson, *Bacteriol. Proc.*, *113*, (1967).
[11] G. W. Strandberg and P. W. Wilson, *Can. J. Microbiol.*, *14*, 25 (1968).
[12] J. Oppenheim, R. J. Fisher, P. W. Wilson, and L. Marcus, *J. Bacteriol.*, *102*, 292 (1970).
[13] P. W. Wilson, J. F. Hull, and R. H. Burris, *Proc. Natl. Acad. Sci. USA*, *29*, 289 (1943).
[14] G. J. Sorger, *J. Bacteriol.*, *98*, 56 (1969).

[15] J. K. Gordon and W. J. Brill, *Proc. Natl. Acad. Sci. USA*, *69*, 3501 (1972).
[16] K. T. Shanmugam and R. C. Valentine, *Proc. Nat. Acad. USA*, *72*, 136 (1975).
[17] K. T. Shanmugam, I. Chan, and C. Morandi, *Biochem. Biophys. Acta*, *408*, 101 (1975).
[18] K. Andersen and K. T. Shanmugam, *J. Gen. Microbiol.*, *103*, 107 (1977).
[19] D. Gauthier and C. Elmerich, *FEMS Microbiol. Lett.*, *2*, 101 (1977).
[20] W. J. Brill (personal communication).
[21] G. W. Strandberg and P. W. Wilson, *Proc. Nat. Acad. Sci. USA*, *58*, 1404 (1967).
[22] D. C. Yoch and R. M. Pengra, *J. Bacteriol.*, *92* (1966).
[23] S. Streicher, E. Gurney, R. C. Valentine, *Proc. Nat. Acad. Sci. USA*, *68*, 1174 (1971).
[24] R. H. Burris, *Methods in Enzymol.*, *24*, 415 (1972).
[25] K. T. Shanmugam (personal communication).
[26] K. Andersen, K. T. Shanmugam, and R. C. Valentine, in *Genetic Engineering for Nitrogen Fixation*, A. Hollaender, et al., Eds. (Plenum, New York, 1977).
[27] J. Postgate, in *The Chemistry and Biology of Nitrogen Fixation*, J. R. Postgate, Ed. (Plenum, New York, 1976).
[28] H. Dalton and J. R. Postgate, *J. Gen. Microbiol.*, *54*, 463 (1969).
[29] H. Dalton and J. R. Postgate, *J. Gen. Microbiol.*, *56*, 307 (1969).
[30] T. M. Leahy III, "Continuous culture of *Azotobacter vinelandii*," Masters thesis, University of Virginia, 1977.
[31] E. G. Mulder, in *Nitrogen Fixation by Free-Living Microorganisms*, W. D. P. Stewart, Ed. (Cambridge Univ. Press, New York, 1975).
[32] M. S. Peters and K. D. Timmerhaus, *Plant Design and Economics for Chemical Engineers* (McGraw-Hill, New York, 1968).
[33] U.S. Environmental Protection Agency, *Process Design Manual for Nitrogen Control*, October 1975.
[34] J. R. Benemann et al., "A systems analysis of bioconversion with microalgae," presented at Clean Fuels from Biomass and Wastes Conference, Orlando, Florida, January 25–28, 1977.
[35] D. I. C. Wang, *Chem. Eng.*, *75*, 99, August 26, 1968.
[36] A. E. Hassen, et al., "Energy recovery and feed production from poultry wastes," in *Energy, Agriculture and Waste Management*, W. J. Jewell, Ed. (Ann Arbor Science, Ann Arbor, Mich., 1975).
[37] J. R. Fishcher et al., in *Energy, Agriculture and Waste Management*, W. J. Jewell, Ed. (Ann Arbor Science, Ann Arbor, Mich., 1975).

Minimizing Residential Energy Growth*

ERIC HIRST

Energy Division, Oak Ridge National Laboratory, Oak Ridge, Tennessee 37830

1. INTRODUCTION

This paper examines alternative "futures" for residential energy use between now and the year 2000. In particular, we develop a future that minimizes energy growth during this period, subject to certain technological and economic constraints. The purpose is to evaluate future residential energy paths assuming continued economic growth, no major lifestyle changes, continuation of ongoing demographic trends, and no premature scrappage of capital stocks (structures, appliances, equipment). The basic principle used to construct the minimum energy future is minimization of energy-related lifecycle costs for operation of U.S. households.

The scenarios developed suggest that it is technically and economically feasible to reduce U.S. residential energy growth from an average rate of about 1.8% to as low as 0.4% per year. This translates into an energy saving of almost 7 Qbtu in 2000.† The economics of reducing energy growth are also likely to be quite favorable. The present worth (at a real interest rate of 8%) of the reduction in fuel bills exceeds the extra capital cost of improved structures and equipment by 35 billion dollars. This is equivalent to an average saving of almost 400 dollars per household.‡

These analyses were conducted with a detailed engineering–economic model of residential energy use developed at the Oak Ridge National Laboratory (ORNL) [1]. This model simulates household energy use at the national level for four fuels (electricity, gas, oil, other); eight end uses (space heating, water heating, refrigeration, freezing, cooking, air conditioning, lighting, other); and three housing types (single-family, multifamily, mobile homes). Each of these 96 fuel-use components is calculated for each year of

* Editors note: This paper was presented as an after-dinner address during the Symposium on Biotechnology in Energy Production and Conservation. Although it does not directly address biotechnology, it does present a very interesting viewpoint on energy conservation and was of significant interest to the Symposium attendees.

† Quantities are given in British units. 1 Qbtu = 1 quad = 10^{15} btu. 1 btu = 1055 J. Electricity use figures are in terms of primary energy (11,500 btu/kWhr); that is, they include losses in generation, transmission, and distribution. Figures for gas and oil do not include losses associated with refining and transportation.

‡ All monetary figures are given in terms of 1975 dollars. Use of "constant" dollars corrects for the effects of inflation.

Biotechnology and Bioengineering Symp. No. 8, 175–189 (1978)

a simulation from 1970 through 2000 as functions of stocks of occupied housing units and new construction, average housing size, equipment ownership by fuel and end use, thermal performance of new and existing housing units, unit energy requirements for each type of equipment, and usage factors that reflect household behavior. The model also calculates annual fuel expenditures, equipment costs, and capital costs for improving thermal performance of structures at the same level of detail.

The next section discusses several technologies that can be used to reduce residential energy use in a cost-effective manner. Section 3 presents our baseline projection of residential energy use. Then we examine the energy and economic effects of implementing the residential conservation programs proposed in the April 1977 *National Energy Plan* (NEP). The effects of introducing advanced residential technologies are evaluated in Section 5. Section 6 examines the consequences of a future that includes the regulatory/incentive programs of NEP, new technologies due to federal and private research, development, and demonstration (RD&D), and the assumption that manufacturers and consumers make optimal decisions concerning investments in structures and equipment beginning in 1980. Optimal decisions are those that minimize the lifecycle cost of owning and operating systems. The final section summarizes our results and discusses the feasibility and likelihood of minimizing energy growth in the residential sector.

2. IMPROVED RESIDENTIAL TECHNOLOGIES

Two kinds of options exist to reduce energy use. The first involves changes in the way households operate existing systems (behavioral changes such as setting thermostats at a higher temperature in the summer). The second involves purchase of more efficient systems (technological changes such as increased insulation in new homes).* Traditionally, designs for residential buildings and equipment were developed with little regard for energy use. This was reasonable given the long history of low and declining fuel prices. Because of the energy inefficiencies built into most systems, the opportunities to save energy by redesign are enormous.

Because space heating accounts for more than half of residential energy use, we begin with an examination of residential space-heating equipment. Table I shows the distribution of energy through a typical gas furnace [2]. Only about 60% of the energy actually appears within the living space to provide heat. Energy is lost because of steady-state inefficiencies, pilot losses, cycling losses, infiltration of outside air, and energy and mass losses through the warm air ducts (if the ducts are located outside the conditioned space).

* A third possibility is changes in the composition of household functions over time. For example, household preferences might shift away from summer air conditioning to the use of attic fans. Such lifestyle changes are not included in the present analysis.

TABLE I

Approximate Distribution of Energy through a Typical Gas Furnace

Steady state stack loss	23%
Cycling loss	5
Pilot light	4
Infiltration	5
Duct loss	3
Net useful heat to living space	60
Total	100

Each of these losses can be reduced or eliminated through better design. In the past, many heating systems were oversized by as much as 150% [2]. Compared to a double oversized furnace, sizing the furnace to the design load of the house could cut cycling and pilot losses and cut fuel use by about 9%, which would also reduce capital cost by almost 12%. Most furnaces are installed in the conditioned space, thus adding to the infiltration loss. A sealed combustion furnace brings outside air directly to the furnace for fuel combustion. The outside air can also be preheated with the combustion products. A sealed combustion furnace with preheated combustion air could cut energy use 12% with a 13% increase in capital cost. Increasing steady-state efficiency by redesigning the heat exchanger and adding induced draft could cut energy use 12% while increasing capital cost 7%. Replacing the standing pilot with electric ignition would cut energy use 6% and increase furnace cost 7% [3].

A combination of design changes could cut gas furnace energy use by more than one-fourth, as shown in Figure 1. The cost to the homeowner for the furnace would increase about 22% ($280). The reduction in annual heating bills (at the 1976 gas price) is $45 for a single-family home in Philadelphia. Thus, the investment is repaid in six years. Because the furnace is likely to last at least 15 years and the price of gas is almost certain to increase in the future, this is a very attractive investment.

Many of the same improvements can be made to an oil furnace. Because an oil furnace does not have a pilot light, reduction in energy use is somewhat less (about 20% rather than 25% for a gas furnace). Investing in a more efficient oil furnace is likely to be repaid within six years [3].

Because of natural-gas-availability problems, many new homes use electricity for space heating. Use of an electric heat pump instead of an electric furnace (with resistance heating elements) can cut energy use 40%. At today's electricity prices, this investment is repaid within three years [3].

Table II summarizes recent analyses of improvements to residential equipment and appliances [2–5]. The table shows typical maximum potential savings with today's technologies and payback periods to recover investment in more efficient equipment. These paybacks are calculated with 1976 fuel prices; as fuel prices continue to increase in the future, payback

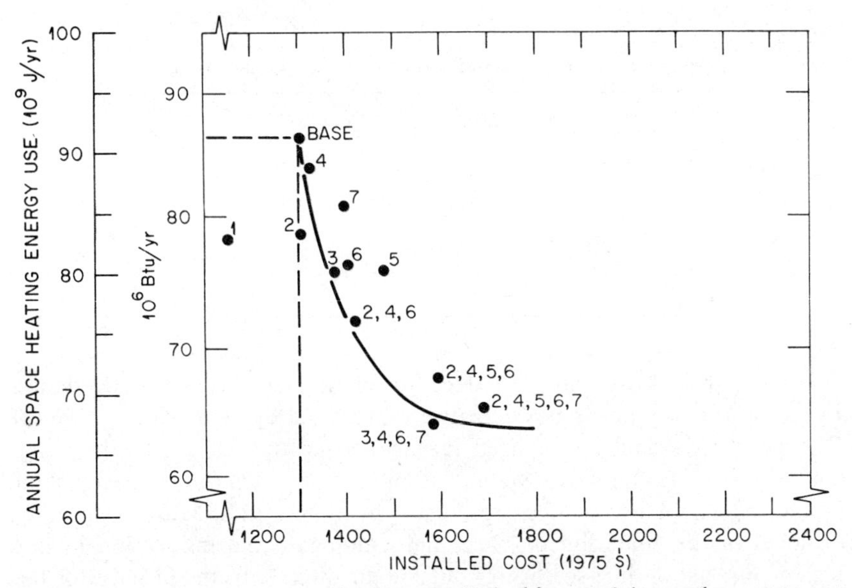

Fig. 1. Cost of gas furnace energy savings. 1, properly sized furnace, 2, bonnet thermostat set at 5°F above room temperature; 3, automatic flue damper; 4, increased duct insulation; 5, sealed combustion, 6, increased steady-state efficiency (to 84%); 7, electric ignition.

periods will become shorter. Lesser improvements in equipment efficiency will also yield shorter payback periods.

Changing the "thermal performance" of homes can also reduce fuel use. Thermal performance refers to the structure's ability to minimize heat losses in the winter and gains in the summer. Adding insulation to the attic, walls, and floor; using storm windows and storm doors; properly orienting the home on the building site; and careful caulking and weatherstripping around all doors, windows, and other joints can cut winter heat losses by 50% or more.

As an example, consider a new gas-heated mobile home constructed in accordance with the latest federal standards (promulgated by the Department of Housing and Urban Development in 1976); see Figure 2. Investing 500 dollars more on weatherization than required by the HUD standards could cut space heating energy use by an additional 40%. Depending on where in the country the mobile home is located, this investment will be repaid in two (e.g., Minnesota) to eight (e.g., Georgia) years. Payback periods will be shorter for electrically heated mobile homes [6].

An examination of the major residential energy using systems (space heating, air conditioning, water heating, refrigeration) suggests that opportunities exist to save substantial amounts of energy through better design of these systems. These design changes rely only on known technologies; they do not involve the use of emerging technologies such as advanced heat pumps or solar systems. Typically, unit energy use can be cut

10–50%; investment in these design changes is typically repaid within a few years.

3. BASELINE PROJECTION

Inputs to the ORNL energy use model required to develop a projection include: population, fuel prices, per capita income, and specifications for government conservation programs (e.g., appliance efficiency standards, tax incentives for retrofitting homes, fuel price increases). Each of these inputs must be provided for the 1970–2000 period.

We assume that population grows according to the Bureau of the Census Series II projection [7]. Per capita income is derived from a Data Resources, Inc. projection of Gross National Product [8] and the Series II population projection.

Projections of household formation, stocks of occupied housing units, and average housing unit size are from our housing model. In developing our estimates of housing stocks, we assume that trends in housing choices (among single-family, multifamily, and mobile homes) between 1960 and 1970 will continue through the end of the century.

Table III shows the values of population, households, housing choices, and incomes from 1970 through 2000 used in all projections discussed here. Between 1976 and 2000, population grows at an average annual rate of 0.8%, while the number of households grows a 1.6%. Higher growth in household formation is due to income growth (2.4% per year) and the changing age composition of the population.

Residential fuel prices are from the Department of Energy and the Brookhaven National Laboratory [8]. As Figure 3 shows,* these projections

TABLE II

Potential Energy and Economic Effects of Equipment Design Changes with Present-Day Technologies

	Reduction in annual energy use (%)	Payback period (years)
Space heating		
Gas furnace	25	6
Oil furnace	20	6
Heat pump	40	2–9
Water heating		
Electric	15	2
Gas	25	3
Refrigerator	50	2
Room air conditioner	35	6

* The higher gas and oil prices (NEP) are discussed in Sec. 4.

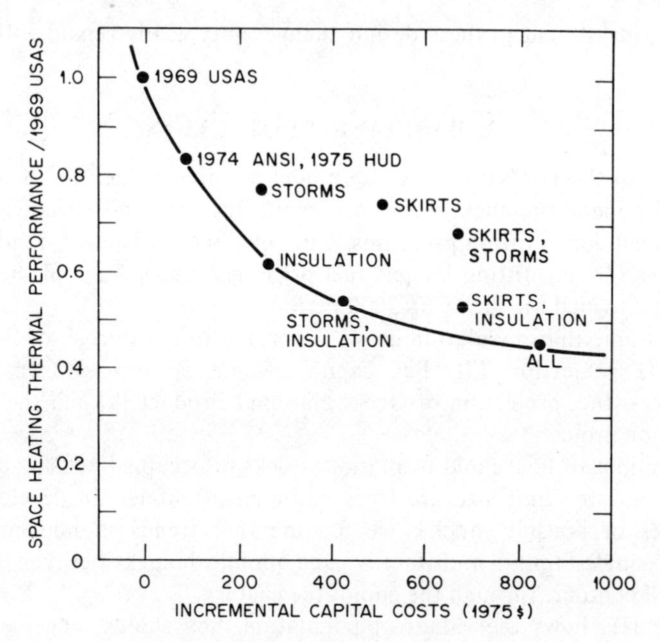

Fig. 2. Costs of gas-heated mobile home.

indicate a substantial increase in real gas prices (average annual growth of 3.1% between 1976 and 2000) and moderate increases in electricity (1.0% per year) and oil (1.7% per year) prices.

Finally, we assume that there are no government programs that require or encourage households to reduce energy use. In other words, we ignore programs mandated by the 1975 *Energy Policy and Conservation Act* [9], the 1976 *Energy Conservation and Production Act* [10], and those proposed

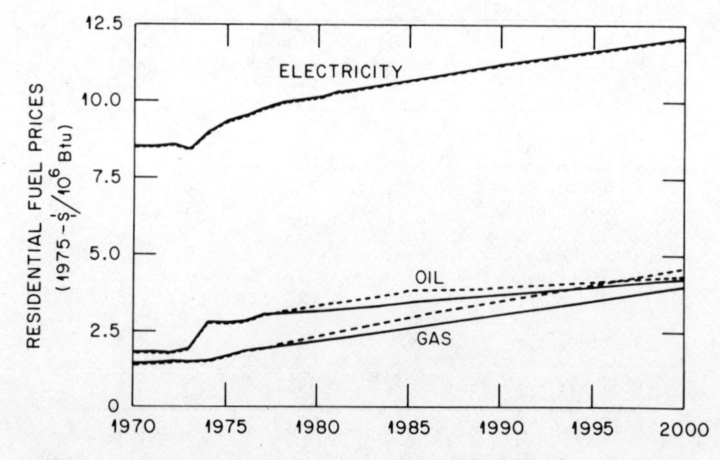

Fig. 3. Projections of residential fuel prices. (——) Baseline; (---) NEP.

TABLE III

Inputs Assumed for All Projections of Residential Energy Use

	Population (10^6)	Households (10^6)	Distribution of occupied housing units (%)			Per capita personal income (1975-$)
			single-family	multi-family	mobile home	
1970	205	63	69	27	3	5,420
1975	214	71	67	29	4	5,850
1976	215	73	67	29	4	6,050
1980	222	80	65	31	5	7,150
1985	233	88	63	32	5	7,970
1990	244	95	62	32	6	8,890
2000	260	106	61	33	6	10,570

in the 1977 *National Energy Plan* [11]; these programs are discussed in Section 4. We also assume that no new technologies are developed and offered to the marketplace; efficiency improvements are limited in the baseline to those possible with today's technologies (Table II). This constraint is relaxed in Section 5. Thus the baseline represents a free-market case, one in which only voluntary responses with current technologies are assumed.

Figure 4 shows the projection of energy use to the year 2000 produced by our model using the inputs discussed above. The model estimates that fuel use will grow from 16.3 Qbtu in 1976 to 17.7 Qbtu in 1980, 21.6 Qbtu in 1990, and 24.9 Qbtu in 2000. Average annual growth rate in energy use from 1976 to 2000 is 1.8%, compared with 3.6% per year from 1950 to 1976 [12]. Energy use per household grows at an average annual rate of 0.2% between 1976 and 2000, compared with 1.6% per year between 1950 and 1976 [12].

The contribution of different fuels to the total changes during the projection period. Because of the sharp increase in petroleum prices during the early 1970s and consumer preference for gas and electricity, the fraction of household energy use accounted for by oil declines from 14% in 1976 to 10% in 2000. Electricity's share increases from 47% to 68% for the reasons given above and because of growing ownership of electric air conditioners and electric food freezers. The contribution of gas to the total declines from 34% to 21% during this period; other fuels also contribute a declining portion of the total, down from 5% to 1%.

Energy use grows more slowly in the baseline than historically for several reasons. First, fuel prices are assumed to rise in the future while, historically, fuel prices declined. Second, the effects of the fuel price increases in the early 1970s are felt slowly over time and dampen energy growth in the future as households replace equipment and structures with systems that are more efficient. A third reason relates to "saturation." Between 1950 and 1975, household ownership of air conditioners, refrigerators, freezers, heat-

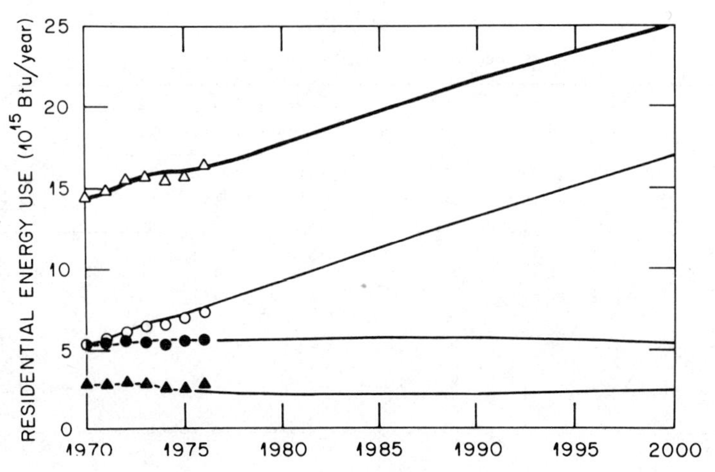

Fig. 4. Projection of energy use to year 2000. (△) Total; (○) electricity; (●) gas; (▲) oil.

ing systems, and water heating systems increased dramatically. By 1975, almost all households had heating and water heating equipment; more than half of all households had air conditioning systems. Thus, the potential for increasing ownership of known energy-using systems is slight.* Fourth, growth in population and households is expected to be slower between now and 2000 than it was between 1950 and 1976. Finally, incomes are assumed to grow more slowly in the future than they have in the past.

Even though fuel prices increase, rising incomes are sufficient to increase household ownership of freezers, air conditioners, lighting fixtures, and "other" uses. Freezer ownership increases from 28% in 1970 to 51% in 2000. Air conditioner ownership increases from 35% to 82%. The number of lights per household increases 13% between 1970 and 2000. Ownership of "other" equipment increases 41%.

4. NATIONAL ENERGY PLAN

Here we evaluate the energy and economic effects of the residential energy conservation programs authorized by the 94th Congress [9, 10] and expanded upon by the present administration [11]: appliance efficiency targets to be implemented in 1980, thermal performance standards for construction of new residences to be implemented in 1978 with stronger standards in 1980, and several programs to encourage weatherization (retrofit) of existing housing units. This case also includes higher gas (about

* Our energy model does not explicitly allow for introduction of new end uses (e.g., sidewalk deicing, swimming pool heating). However, the model does include an "other" end use, and this is allowed to grow as incomes rise (depending on growth of fuel prices). In the baseline, energy use for other purposes increases from 1.6 to 2.9 Qbtu between 1976 and 2000.

15%) and oil (about 5%) prices because of the proposed crude oil equalization tax and changes in the regulated price of natural gas in NEP [8]: see Figure 3.

Figure 5* shows the effects on residential energy use of implementing the three NEP programs listed above [13]. Energy use is reduced, relative to the baseline, by 1.1 Qbtu in 1980, 2.0 Qbtu in 1990, and 2.0 Qbtu in 2000. These NEP programs are estimated to cut residential energy growth from 1.8% to 1.4% per year. The cumulative energy saving of 42 Qbtu (see Table IV) represents 8% of the baseline.

Table IV also shows the cumulative direct economic effects of these programs on the nation's households. The NEP programs reduce the present worth of household fuel bills by $31 billion. This is partly offset by higher capital costs for improved equipment and structures of $29 billion. Thus, the net economic benefit is 2 billion dollars.† Were it not for the higher gas and oil prices due to NEP, the net economic benefit would be increased from 2 billion to 27 billion dollars.

The lifestyle changes implied by the NEP programs are minor and all positive. The reduced operating costs due to the programs increase the intensity with which households operate equipment and slightly increase ownership of energy-using equipment. For example, 82% of the households in 2000 own air conditioning systems in the baseline compared to 84% with NEP.

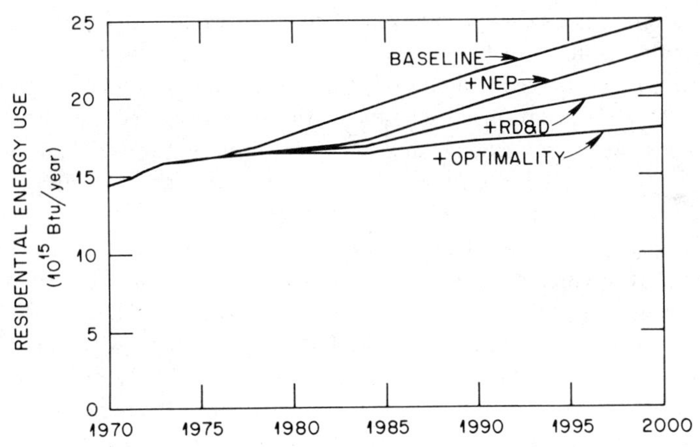

Fig. 5. Effects on residential energy use of implementing NEP, RD&D, and optimality programs.

* The two bottom curves in Figure 5 are discussed later.

† Future expenditures on fuels, equipment, and structures are discounted to 1977 using a real (corrected for inflation) discount rate of 8%. This is equivalent to a nominal rate of 14% if inflation is 6% per year. A higher discount rate would reduce the net benefits of conservation programs.

TABLE IV

Cumulative (1977–2000) Benefits of NEP Programs

Energy	(QBtu)	Economic	(billion 1975-$)
Electricity	13	Fuels	31
Gas	22	Equipment	- 8
Oil	7	Structures	-21
Total	42	Net	2

5. RESEARCH, DEVELOPMENT, AND DEMONSTRATION

Both private industry and the federal government (Department of Energy) are conducting RD&D programs to bring to the market new systems for satisfying household end uses.* These new systems are likely to be much more energy efficient than are existing ones. For example, gas-fired heat pumps are expected to provide annual space heating requirements in a typical house with only about half the natural gas consumption of a conventional furnace. Electric heat pump water heaters are likely to require only half as much energy for water heating as do today's electric water heaters. However, these high-efficiency systems are also likely to cost more than conventional systems.

We used our energy model to evaluate the extent to which households might adopt these new residential energy systems and the energy and direct economic consequences of doing so [14]. Figures 5 and 6 show how the energy savings due to development of these new technologies increase over time. The bottom curve in Figure 6 is the energy savings due to implementation of the NEP programs discussed in the preceding section. The middle curve shows the energy savings due to the combination of NEP and RD&D. Therefore, the area between these two curves represents the energy savings due to RD&D, given implementation of NEP.

The dynamics of RD&D benefits differ substantially from those due to NEP. Energy savings due to NEP increase rapidly through the early 1980s and then remain essentially constant from 1985 to 2000. RD&D benefits, on the other hand, grow slowly at first, but then (in the 1990s) grow much more rapidly than do the NEP benefits. In the year 2000, the energy saved because of RD&D (2.4 Qbtu) is more than the NEP energy savings (2.0 Qbtu); in 1985 the RD&D energy saving is only a fifth of the NEP saving. Together, the two programs reduce energy use in the year 2000 by almost 20%. The cumulative savings from 1977 through 2000 of NEP (42 Qbtu) is more than double the cumulative RD&D savings (21 Qbtu); see Tables IV and V.

* We assume that the sole effect of government RD&D programs is to accelerate the commercial availability of new systems; that is, private industry would develop such systems, but not as quickly as with government involvement.

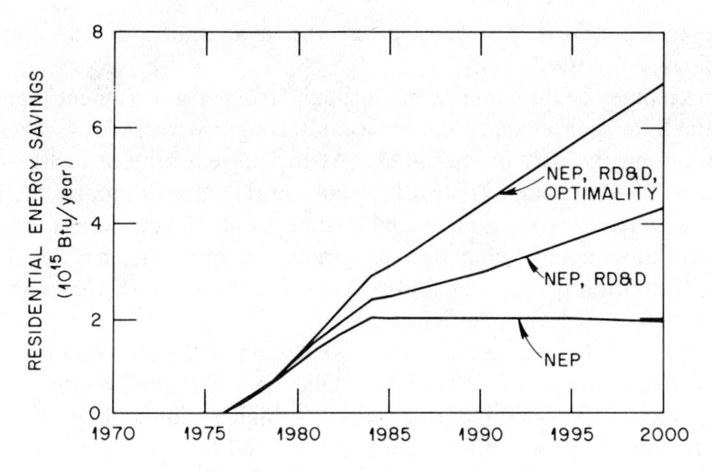

Fig. 6. Energy savings over time.

Table V summarizes the direct economic effects on the nation's households of these RD&D programs, exclusive of government costs. The present worth of fuel bill reductions is $23 billion dollars. This is partially offset by increases in capital costs for more efficient equipment and structures of 2 billion dollars.* The net economic benefit is about 21 billion dollars.

6. NEP, RD&D, AND NO MARKET IMPERFECTIONS

In our final case we allow the model to select the mix of equipment and structures that minimizes lifecycle costs to consumers beginning in 1980. The federal programs discussed in Sec. 4 (NEP) reduce lifecycle costs for consumers but they do not minimize these costs. As time goes on and fuel prices rise (Fig. 3), the NEP programs become less effective. This case examines the energy and direct economic implications of programs that

TABLE V

Cumulative (1977–2000) Benefits of New Technologies

Energy	(QBtu)	Economic	(billion 1975-$)
Electricity	17	Fuels	23
Gas	2	Equipment	- 2
Oil	3	Structures	0
Total	22	Net	21

* The extra capital cost associated with these new technologies is quite small. This suggests that a major effect of new systems is to reduce the cost of meeting the NEP standards, discussed in the preceding section. Without NEP, the increased capital cost due to the availability of new systems is 6 billion dollars, rather than the 2 billion dollars shown in Table V.

"optimize" purchase decisions given the availability of the improved technologies discussed in Sec. 5.

Such changes could come about through stronger government regulatory programs. Or such changes might occur through increased awareness and motivation on the part of consumers. At the present time, it is difficult for consumers to collect and process information they need to make "rational" decisions concerning equipment and structures efficiency vs cost tradeoffs. However, government education programs, energy efficiency labels, and other information programs could provide that data and thereby encourage consumers to choose more efficient systems.

Relative to the preceding case (NEP and RD&D), energy savings increase from 0.6 Qbtu in 1985 to 1.4 Qbtu in 1990 and 2.6 Qbtu in 2000. Energy use in 2000 (18.0 Qbtu) is only 11% higher than in 1976; this is our minimum energy growth future; see Figures 5 and 6.

The cumulative energy saving relative to the preceding case is 28 Qbtu and the cumulative net economic benefit is 13 billion dollars. Table VI shows the cumulative effects of the combined NEP, RD&D, and optimal program relative to the original baseline. Again, these economic benefits do not include the costs of government regulatory, information, and RD&D programs. However, government costs are likely to be only a small fraction of the net benefit to consumers.

This combination of programs yields a cumulative energy saving of 92 Qbtu between 1977 and 2000, almost ⅕ of the baseline cumulative energy use. Almost 60% of the energy saving is electricity; gas accounts for 30% and oil for the remainder.

The net economic benefit of these programs is 35 billion dollars. The overall benefit/cost ratio to society is 1.7 (that is the fuel bill reductions exceed the extra capital costs by 70%). Thus the programs are very cost effective. Were it not for the higher NEP gas and oil prices, the economic benefits would be much higher (almost 60 billion dollars rather than 35 billion dollars), yielding a benefit/cost ratio of 2.3.

7. SUMMARY

We used the ORNL engineering–economic model of residential energy use to evaluate the energy and direct economic effects of four different futures. These include a baseline in which no government conservation programs are implemented, an NEP case in which the regulatory/incentive programs authorized by the 94th Congress and proposed by President Carter are implemented, an RD&D case in which new technologies are developed and offered, and a final case which includes the NEP programs, new technologies, and additional programs to ensure that consumers minimize lifecycle costs when they purchase structures and household equipment.

TABLE VI

Cumulative (1977–2000) Benefits of NEP, RD&D, and Elimination of Market Imperfections

Energy	(QBtu)	Economic	(billion 1975-$)
Electricity	52	Fuels	83
Gas	28	Equipment	-22
Oil	12	Structures	-26
Total	92	Net	35

These results suggest that a minimum energy growth future (0.4% per year) is feasible both technically and economically. The cumulative energy saving that would occur because of these programs is more than 90 Qbtu between now and the year 2000. This is equivalent to six years of present-day residential energy use. The economic benefit of the programs is 35 billion dollars between now and the end of the century, almost 400 dollars household.

However, these results do not suggest that this is the *likely* future for residential energy use. A future of minimum residential energy growth requires strong public support; a set of dynamic, cost-effective, timely regulations for efficiencies of new structures and equipment; continued private and government RD&D programs to develop improved residential technologies; accurate, detailed education programs to inform consumers about their energy-related choices; active cooperation from manufacturers of residential appliances and equipment and from organizations and individuals involved in the design, construction, and financing of residential structures (architects, builders, contractors, suppliers, banks); and policies to undo the damage of historical fuel pricing regulations (i.e., natural gas and oil price controls). All these social, political, and institutional requirements suggest that we may not minimize energy growth during the rest of this century. Nevertheless, considerable progress will surely be made during the next two decades in developing and implementing effective conservation programs; the benefits of progress are likely to be substantial.

Our results are presented as if the future could be predicted with certainty. This is not the case. Our results are based on a number of assumptions and our engineering–economic model contains many simplifying assumptions [1]. Therefore, our results should be considered "suggestive" rather than "definitive." For example, lack of both theory and data prevents us from constructing an adequate model of consumer behavior with respect to the purchase and use of household equipment and structures. This makes it difficult to estimate the relative contributions to our minimum energy future of voluntary (free-market) behavior and government conservation programs. Also, the relationships between equipment and structure energy efficiency and capital cost are based on limited engineering analysis and preliminary cost estimates. These performance/cost relationships strongly influence estimates of the economic attractiveness of different futures.

Despite these uncertainties in our analysis and results, it seems clear that energy conservation regulatory, incentive, and RD&D programs can substantially reduce energy use between now and the year 2000. Such programs would also save money for households, reduce the adverse environmental effects of energy production and use, and provide more time to develop new energy sources.

Research sponsored jointly by the Office of Conservation and Solar Applications and the Energy Information Administration, U.S. Department of Energy under contract W-7405-eng-26 with the Union Carbide Corporation.

References

[1] E. Hirst and J. Carney, *The ORNL Engineering-Economic Model of Residential Energy Use*, Oak Ridge National Laboratory, ORNL/CON-24, July, 1978; also E. Hirst et al., *An Improved Engineering-Economic Model of Residential Energy Use*, Oak Ridge National Laboratory, ORNL/CON-8, April 1977.

[2] E. Hise and A. Holman, *Heat Balance and Efficiency Measurements of Central, Forced-Air, Residential Gas Furnaces*, Oak Ridge National Laboratory, ORNL-NSF-EP-88, October 1975; also U. Bonne, J. Janssen, and R. Tonberg, "Efficiency and Relative Operating Cost of Central Combustion Heating Systems: IV. Oil-Fired Residential Systems," *1977 Transactions of the American Society of Heating, Refrigerating and Air Conditioning Engineers, 83*(1), 1977.

[3] D. O'Neal, *Energy and Cost Analysis of Residential Space Heating Systems*, Oak Ridge National Laboratory, ORNL/CON-25, July, 1978.

[4] R. Hoskins and E. Hirst, *Energy and Cost Analysis of Residential Water Heaters*, Oak Ridge National Laboratory, ORNL/CON-10, June 1977; also A. D. Little, Inc., *Study of Energy-Saving Options for Refrigerators and Water Heaters, Volume 2: Water Heaters, May 1977.*

[5] R. Hoskins and E. Hirst, *Energy and Cost Analysis of Residential Refrigerators*, Oak Ridge National Laboratory, ORNL/CON-6, January 1977; also A. D. Little, Inc., *Study of Energy Saving Options for Refrigerators and Water Heaters, Volume 1: Refrigerators*, May 1977.

[6] P. Hutchins and E. Hirst, *Engineering-Economic Analysis of Mobile Home Thermal Performance*, Oak Ridge National Laboratory, ORNL/CON-28, October, 1978.

[7] Bureau of the Census, "Projections of the Population of the United States: 1975–2050," *Current Population Reports*, Series P-25, No. 704, U.S. Department of Commerce, July 1977.

[8] S. Carhart, Brookhaven National Laboratory, personal communication, January 17, 1978; also R. Sastry, U.S. Department of Energy, personal communication, January 27, 1978.

[9] 94th Congress, *Energy Policy and Conservation Act*, PL 94-163, December 22, 1975.

[10] 94th Congress, *Energy Conservation and Production Act*, PL 94-385, August 14, 1976.

[11] The White House, *The President's Energy Program*, April 20, 1977; also Executive Office of the President, *The National Energy Plan*, April 29, 1977.

[12] Bureau of Mines, "Annual U.S. Energy Use Up in 1976," press release, U.S. Department of the Interior, March 14, 1977; also E. Hirst and J. Jackson, "Historical Patterns of Residential and Commercial Energy Uses," *Energy*, *2*(2), June 1977.

[13] E. Hirst and J. Carney, *Residential Energy Use to the Year 2000: Conservation and Economics*, Oak Ridge National Laboratory, ORNL/CON-13, September 1977; also E. Hirst and J. Carney, "Effects of Federal Residential Energy Conservation Programs," *Science*, *199*(4331), February 24, 1978.
[14] E. Hirst, *Energy and Economic Benefits of Residential Energy Conservation RD&D*, Oak Ridge National Laboratory, ORNL/CON-22, February 1978.

Biological Removal and Recovery of Trace Heavy Metals

RANDOLPH T. HATCH and ARUN MENAWAT*
Chemical and Nuclear Engineering Department, University of Maryland, College Park, Maryland 20742

INTRODUCTION

The occurrence of microorganisms in nature with the ability of oxidizing metal salts has been well established in the literature. Of the many microorganisms with the ability to insolubilize metal oxides, *Sphaerotilus* (a filamentous bacterium) and *Leptomitus* (a true fungus) have been studied extensively due to their appearance in waste sludges and polluted waters [1–4]. These microorganisms have been shown to oxidize iron to form insoluble precipitates. Both microorganisms have sheaths composed of a protein–polysaccharide–lipid complex and tend to accumulate internal sulfur granules when grown in the presence of hydrogen sulfide [5]. Ferric and manganic salts tend to impregnate the sheath of *Leptomitus*. Although mineral salts are also accumulated by *Sphaerotilus*, the deposits are found in a mucilage layer outside the sheath. After several cell divisions, the *Sphaerotilus natans* bacteria are liberated from the end of the sheath, and the individual cells remain motile by subpolar flagella until they strike a solid object. The bacterial cell will then attach itself and form a new sheath. This cycle is somewhat slower because of the long doubling times (6–10 hr [6]) and may require 36 hr or more.

The fact that the nutritional requirements of *S. natans* are well established [5], has led to the growth of this bacteria in pure culture on a defined medium. A wide variety of carbohydrates may be used as the carbon source. Inorganic nitrogen may be utilized; however, the growth is stimulated by a number of amino acids. The basal mineral salt medium must include calcium as well as magnesium, potassium, and iron. The optimum growth temperature range is 25–30°C. The optimum pH for growth is approximately neutral. Although it is an aerobe, *S. natans* has been shown to grow at dissolved oxygen concentrations as low at 0.1 mg/liter [7].

Sphaerotilus natans is of particular interest for the recovery or removal of various mineral salts due to its ability to deposit the minerals outside the

* Present Address: Owens Corning Fiberglas, Materials Technology Laboratory, Granville, Ohio 43023.

Biotechnology and Bioengineering Symp. No. 8, 191–203 (1978)

0572-6565/78/0008-0191$01.00

cell. This ability permits the potential recovery of minerals from dilute aqueous streams by the use of this bacteria. Some of the questions to be resolved before an economical process can be designed concern the following: (1) the range of mineral salts which can be insolubilized, (2) the rates of mineral salt deposition, (3) the quantitites of cell mass produced, and (4) the nature of the attachment of mineral salt to the exterior cell surface.

The purpose of this research is the investigation of *S. natans* as a potentially useful element of a microbial process for the recovery or removal of trace metals from aqueous waste streams.

EXPERIMENTAL PROCEDURE

The bacteria were purchased from American Type Culture Collection and grown in shake flasks in order to study batch kinetics. The growth medium used was 0.05 g beef extract, 0.05 or 0.01 g metal salt, and 1.0 liter tap water.

The metal salts that were used for the study were sulfates and chlorides of Fe, Mg, Cu, Co, Cd, Ni, and Cr.

The growth medium was autoclaved before use to ensure sterilization before inoculation with the bacteria. Triplicate flasks were used for each metal concentration. One of the three flasks was used as a control flask, and the other two were inoculated with *S. natans* from an agar slant. The shake flasks were incubated at 30°C in an environmental shaker maintained at the speed of 150 rpm.

Batch kinetics was studied by taking 2-ml samples of the growth medium every 60 hr from each flask. These samples were filtered through 0.2-μm

Fig. 1. Clump of *S. natans* filaments; Mg, 50 ppm; 120×.

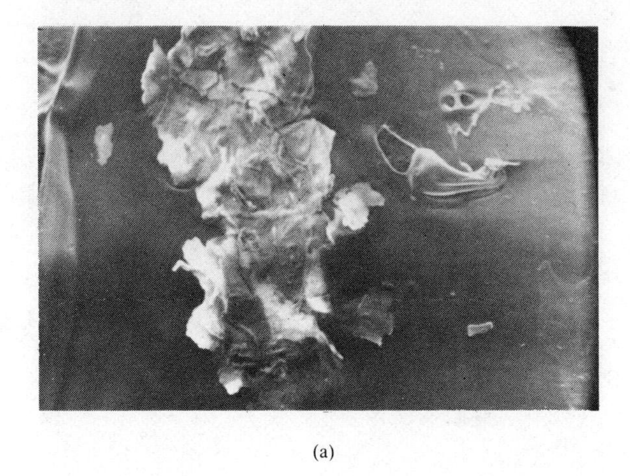

(a)

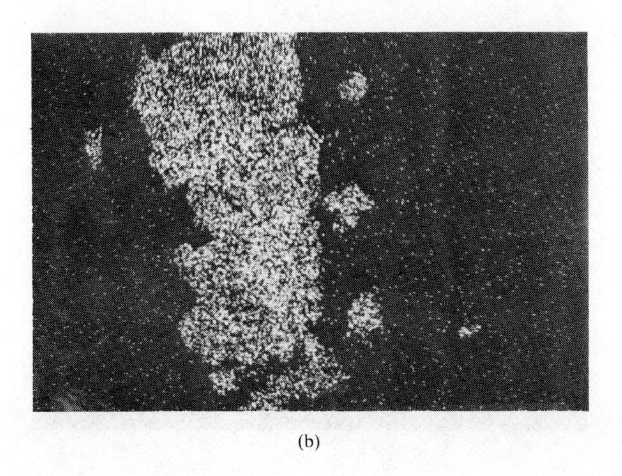

(b)

Fig. 2. Mass of *S. natans* after growth on iron. (a) 50 ppm Fe; 34×. (b) 50 ppm Fe; Fe x-ray.

Millipore filters to remove the majority of the solids. The concentration of the metal in the samples was measured using an atomic absorption spectrometer.

Scanning electron microscopic studies of the bacterial colonies were undertaken (see Figs. 1–5) to locate the insoluble metal. The bacterial colonies were dried on carbon disks and were coated with carbon. Pictures of the whole colonies as well as just single filaments of the bacteria were taken at different magnifications. X-ray studies were also performed to locate the zones of metal particles.

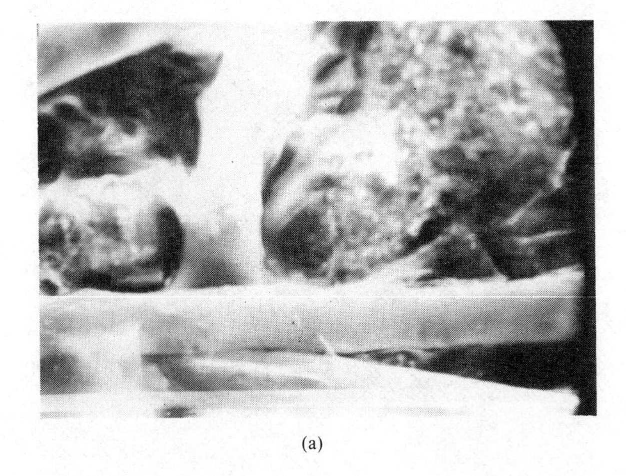

(a)

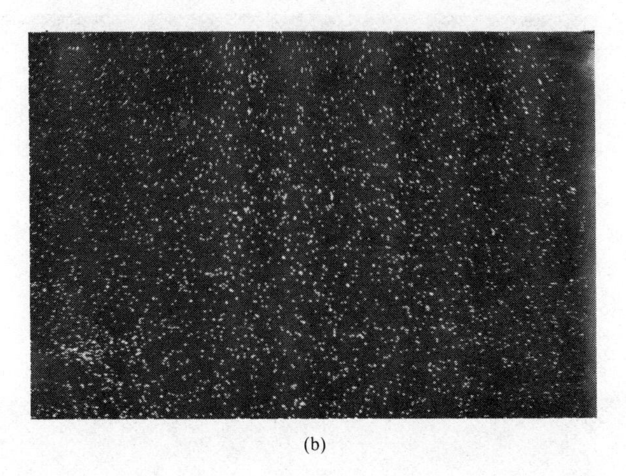

(b)

Fig. 3. *Sphaerotilus natans* filaments after growth on iron. (a) 50 ppm Fe, 1100×; (b) 50 ppm Fe, Fe x-ray.

RESULTS

The filamentous structure of *S. natans* is shown in the scanning electron micrograph (SEM) of Figure 1. The bacteria tends to grow in the form of large "cottony" clumps of sufficient size to kill fish by plugging their gills. To demonstrate the location of the metal precipitate, x-ray analysis was also used. In Figure 2(a), a large mass of bacteria is shown at a magnification of 34 after growth with ferric sulfate at 50 mg/liter. The corresponding x-ray micrograph (Fig. 2(b)) shows the uniform deposition of iron precipitate over the cell mass. A SEM of individual *S. natans* filaments is shown in Figure 3(a) at a magnification of 1100. The corresponding x-ray micrograph (Fig. 3(b)) shows the even distribution of iron particles. Because no iron concentration appears at the cell surface, the iron preciptate is clearly extracellular in the form of a fine precipitate. Similar x-ray analyses were performed after growth at a copper concentration of 50 mg/liter. Again, the copper deposit showed up as a uniform precipitate in the cell mass in Figures 4(a) and 4(b) at a magnification of 45. Under closer exami-

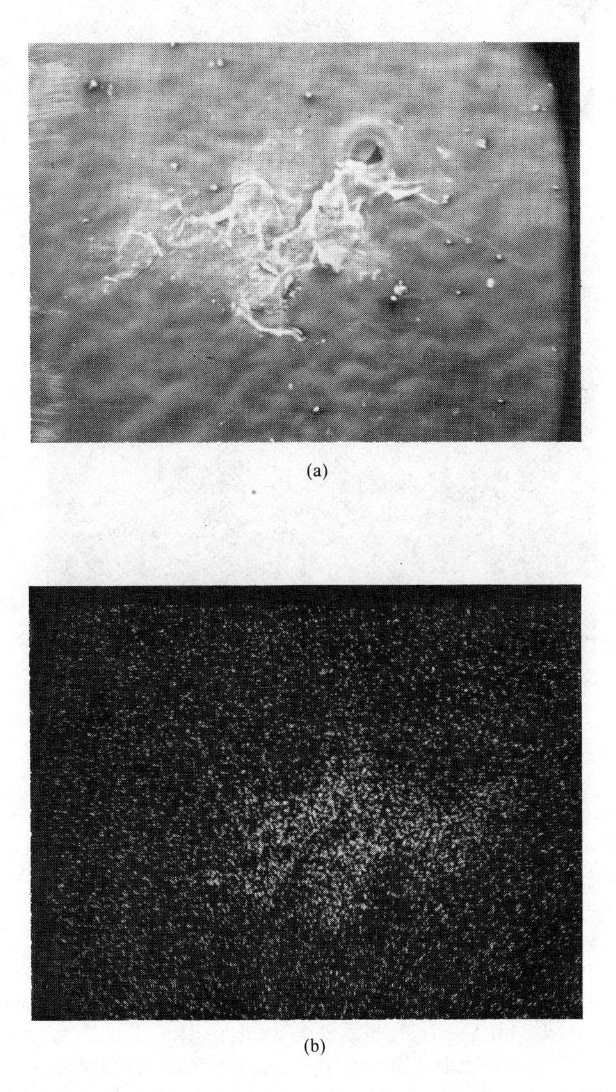

Fig. 4. Mass of *S. natans* after growth on copper. (a) 50 ppm Cu, 45×; (b) 50 ppm Cu; Cu x-ray.

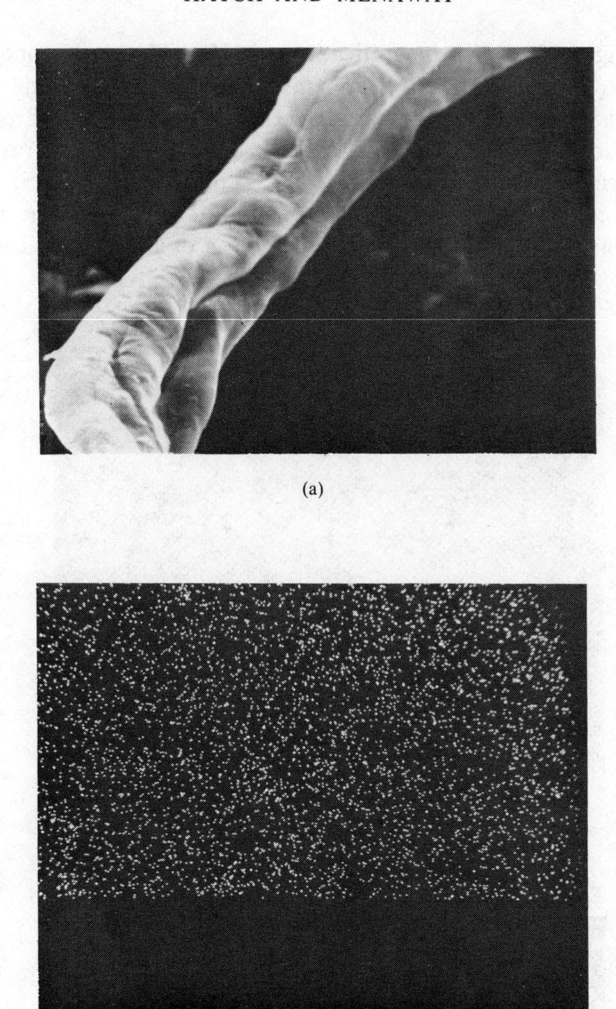

(a)

(b)

Fig. 5. *Sphaerotilus natans* filaments after growth on copper. (a) 50 ppm Cu; 1300×. (b) 50 ppm Cu, Cu x-ray.

nation at a magnification of 1300 in Figures 4(a) and 4(b), the copper appears as a very fine precipitate, uniformly distributed around the *S. natans* filament. From these studies it is evident that the *S. natans* does not accumulate the metal salts. Since this is in agreement with other results reported in the literature, it is possible that this microbial precipitation of metal salts could be utilized to recover certain trace metals.

Of the seven metals tested, growth of *S. natans* was only found with the sulfates of iron, magnesium, copper, cobalt, and cadmium. The bacteria did not exhibit growth in the presence of sulfates of nickel and chromium nor with the chlorides of iron, magnesium, copper, and cobalt.

The concentration of the soluble metal salt was followed after the inoculation of replica shake flasks over a period of 11 days. In Figures 6–10, the metal concentration is plotted as a function of time on semilogarithmic graphs. For the cases of iron sulfate (Fig. 6) and magnesium sulfate (Fig. 7), the first-order kinetics are indicated down to a concentration of approximately 2 ppm. The growth of *S. natans* lowered the iron, magnesium, copper, cobalt, and cadmium concentrations down to approximately 0.6, 0.4, 2.0, 1.0, and 0.3 ppm, respectively, from 10 ppm during 11 days of batch growth. The rate constants calculated from the data presented in Figures 6–10 are in Table I. For copper, cobalt, and cadmium, the rate constants were found to be independent of initial mineral concentration. The yield as shown in Table I was determined for the starting concentration of 50 ppm using 500 ml broth volume. This was necessary due to the exceptionally low amount of dry cell mass produced over the 11-day growth period. The yield of cell mass produced per mass of metal removed ranged from 0.043 to 0.051. The notable exception was iron which

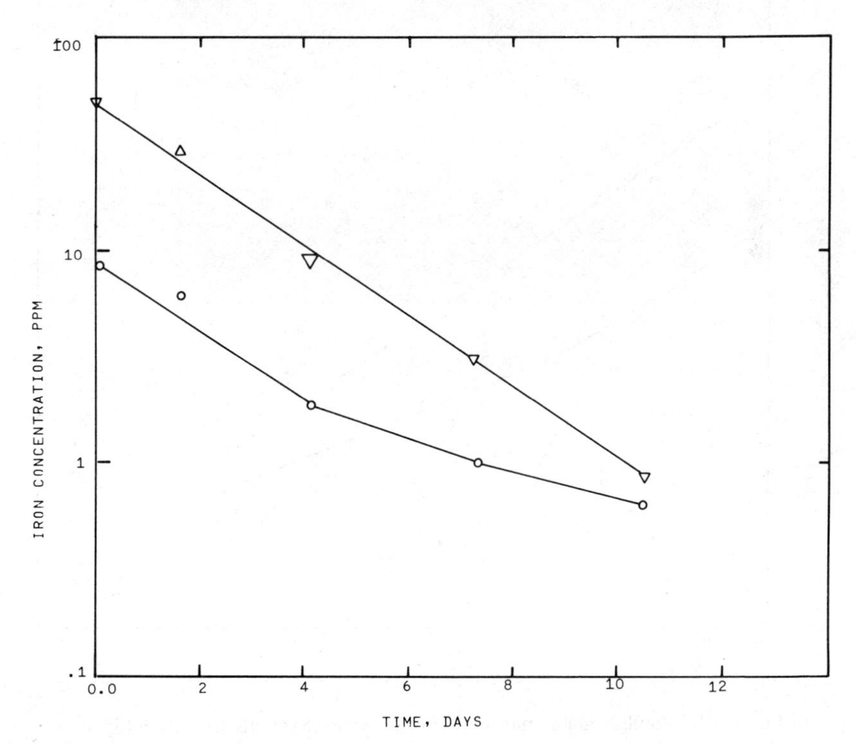

Fig. 6. Insolubilization of iron by *S. natans*. Initial metal concentration (ppm): (▽) 50; (○) 10.

TABLE I
Kinetics of Insolubilization

METAL	INITIAL CONC. PPM	RATE CONSTANT $k*10^3$ (min^{-1})	YIELD $\frac{\text{gm Cell Mass}}{\text{gm Metal}}$	YIELD $\frac{\text{gm Cell Mass}}{\text{gm Mole Meta}}$
Fe	50	.27	2.06×10^{-2}	1.15
	10	.25-.20		
Mg	50	.26	5.13×10^{-2}	1.25
	10	.27-.15		
Cu	50	.11	4.67×10^{-2}	2.97
	10	.11		
Co	50	.15	4.28×10^{-2}	2.52
	10	.15		
Cd	50	.25	4.30×10^{-2}	4.83
	10	.25		

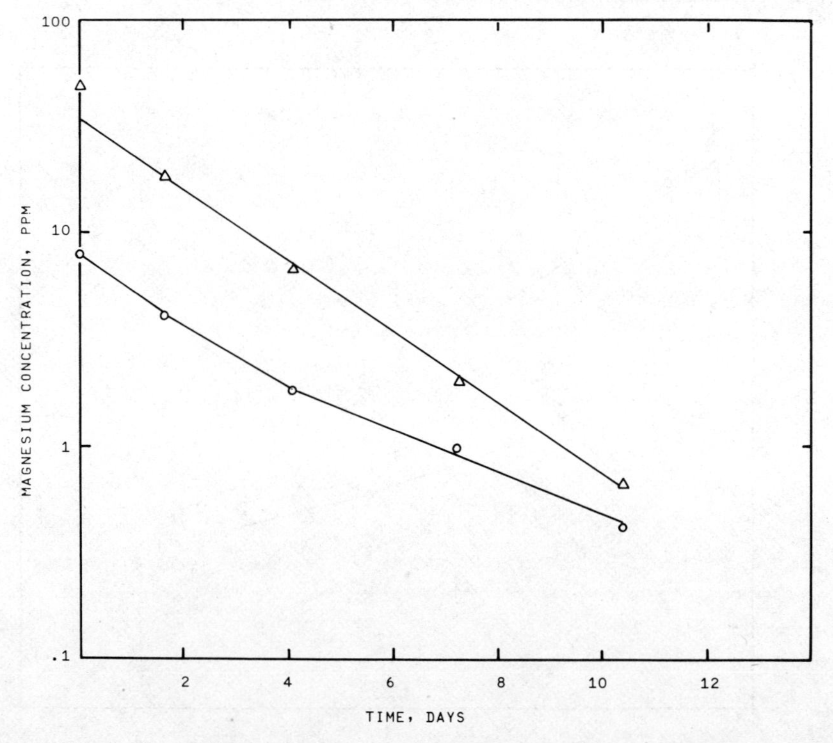

Fig. 7. Insolubilization of magnesium by *S. natans*. Symbols same as in Fig. 6.

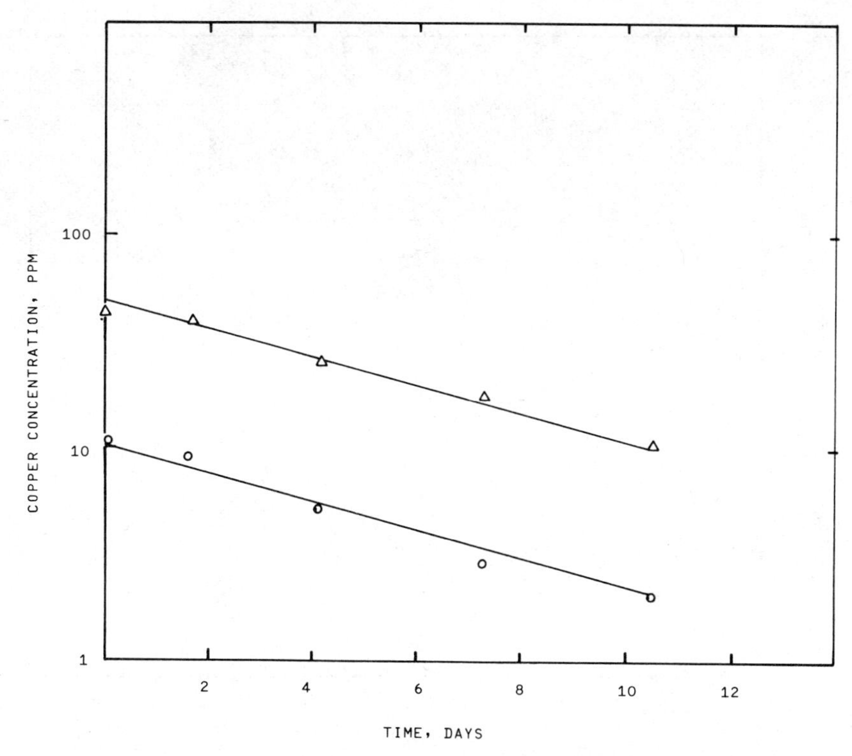

Fig. 8. Insolubilization of copper by *S. natans*. Symbols same as in Fig. 6.

was only 0.021. Although there is no good explanation for this, it may be due to the metabolic requirement for iron, which results in an enzyme system subject to feedback inhibition by the products of iron deposition. When the cell yield is based upon the moles of metal insolubilized, the yield appears to increase with molecular weight.

In order to model the kinetics of metal deposition or insolubilization, simple first-order insolubilization kinetics appears to apply:

$$-\frac{dC_m}{dt} = kC_m \tag{1}$$

This may be due to either a kinetic limitation or a mass transfer limitation. For the case of growth limitation, Monod kinetics may be assumed:

$$\frac{dx}{dt} = \mu x = \frac{\mu_{\max} C_m x}{K_s + C_m} \simeq \frac{\mu_{\max} C_m x}{K_s} \tag{2}$$

The rate of deposition of metal would, in turn, be related to the rate of cell

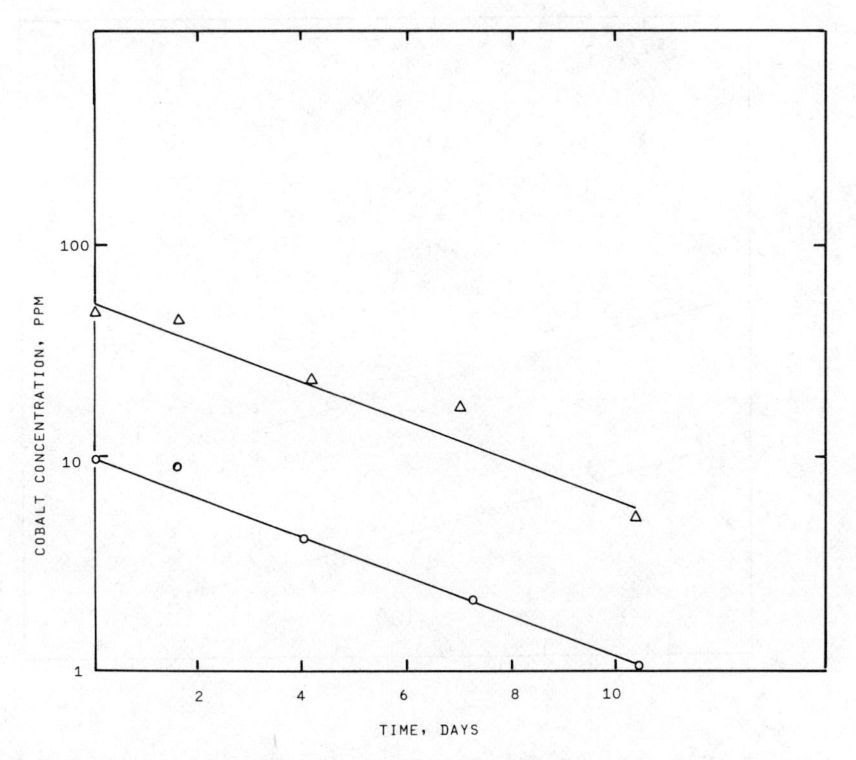

Fig. 9. Insolubilization of cobalt by *S. natans*. Symbols same as in Fig. 6.

mass production by a yield constant:

$$-\frac{dC_m}{dt} = \frac{1}{Y}\frac{dx}{dt} = \left(\frac{\mu_{\max}x}{K_s Y}\right) C_m \tag{3}$$

The first-order rate constant (k) would increase as the viable cell mass increased. Since the data shown in Figures 6–10 indicate a constant or decreasing rate constant, and it is physiologically unlikely that these metals would be growth limiting in these concentration ranges, the data do not appear to support the case of growth limitation.

First-order kinetics would also be manifested under mass transfer limitation. The steady-state mass transfer flux to the surface of the *S. natans* pellet may be appoximated by

$$N_A = [-D(C_{m_s} - C_m)]/r_s \tag{4}$$

for mass transfer from an infinite medium to a spherical surface. It would then follow that

$$-\frac{dC_m}{dt} = \frac{4\pi r_s^2}{V_L} N_A = \frac{4\pi r_s D_m}{V_L}(C_m - C_{m_s}) \tag{5}$$

Upon integration, the time dependency of the metal concentration becomes

$$-\ln\left(\frac{C_m - C_{m_s}}{C_{m_i} - C_{m_s}}\right) = k_d t \tag{6}$$

where

$$k_d = 4\pi r_s D_m / V_L$$

For this case of mass transfer limitation, the data can be reanalyzed by regression analysis to determine C_{m_s} and the diffusion rate constant (k_d). As shown in Table II, the diffusion rate constant is found to vary from 0.007 hr^{-1} for copper deposition to 0.020 hr^{-1} for iron deposition. The approximate effective molecular diffusivity can then be calculated for the liquid volume of 30 ml and the *S. natans* clump diameter of 2 mm. The calculated diffusivities are found to vary from 6.6×10^{-5} cm^2/sec for iron to 2.3×10^{-5} cm^2/sec for copper. Since these values are within a factor of 2 of the actual molecular diffusivities for these metal irons [8], it is apparent that the insolubilization process is mass transfer limited. The metal concentration at

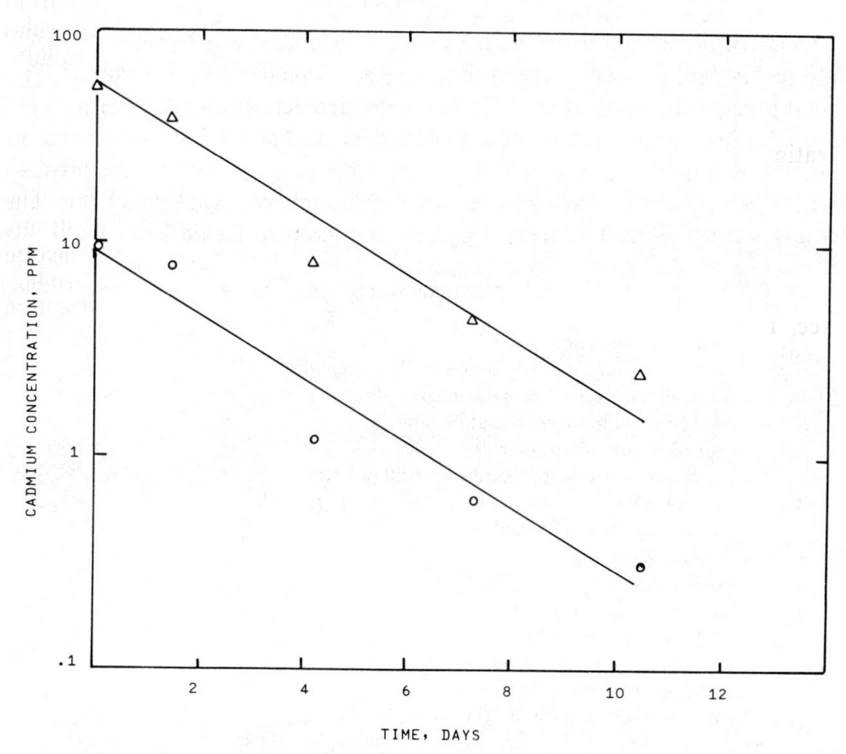

Fig. 10. Insolubilization of cadmium by *S. natans*. Symbols same as in Fig. 6.

TABLE II

Analysis of Mass Transfer Limitation

Metal	Diffusion Rate Constant hours^{-1}	C_{m_s} ppm	Regression Co-efficient
Iron	0.020	0.57	0.988
Magnesium	0.019	0.37	0.946
Copper	0.007	0.00	0.961
Cobalt	0.010	0.00	0.956
Cadmium	0.014	0.10	0.879

the surface of the *S. natans* clump (C_{m_s}) is shown in Table II to vary from approximately zero for copper and cobalt to 0.57 ppm for ion. This provides an estimate of the lower limit of the metal concentration for this insolubilization process.

SUMMARY

Sphaerotilus natans have been shown to insolubilize a number of metal sulfates including iron, magnesium, copper, cobalt, and cadmium. The metal precipitates in the form of a finely divided deposit outside the cell surface. The insolubilization process continues according to first-order reaction kinetics at mineral concentrations below 1 ppm and results in the production of less than 0.1 g cell mass/g metal insolubilized. Analysis of the data indicates that the insolubilization process is mass transfer limited.

Nomenclature

C_m	metal concentration (g-liter)
C_{m_s}	metal concentration at *S. natans* surface (mg/liter)
D_m	molecular diffusivity of metal ion (cm^2/hr)
k	rate constant for insolubilization (min^{-1})
k_d	diffusion rate constant (hr^{-1})
k_s	Michaelis constant for limiting nutrient (g/liter)
N_A	diffusive flux of metal ion to *S. natans* clump (g/cm^2 hr)
r_s	radius of *S. natans* clump (cm)
V_L	liquid volume (ml)
Y	yield constant on metal (g/cell mass/g metal)
x	cell density (g/liter)
x_i	initial cell density (g/liter)
Greek	
μ	cell growth rate (hr^{-1})
μ_{max}	maximum cell growth rate (hr^{-1})

The authors wish to express their appreciation to the National Science Foundation for its partial support of this research through grant number ENG76-18480.

References

[1] G. J. Farquhar and W. C. Boyle, *J. Water Pollut. Control Fed.*, *43*(5), 779 (1971).
[2] J. A. Servizi, D. W. Martens, and R. W. Gorden, *J. Water Pollut. Control Fed.*, *43*(2), 278 (1971).
[3] E. J. C. Curtis, *Water Res.*, *3*, 289 (1969).
[4] W. S. Mueller and W. Litsky, *Water Res.*, *2*, 289 (1968).
[5] J. D. Phaup, *Water Res.*, *2*, 597 (1968).
[6] J. D. Phaup and J. Gannon, *Water Res.*, *1*, 523, (1967).
[7] C. C. Ruchhoft and J. F. Kachmar, *J. Sewage Works*, *13*, 3 (1941).
[8] R. A. Robinson and R. N. Stokes, *Electrolyte Solutions*, 2nd ed. (Academic, New York, 1959).

Bioconversion of Agricultural Wastes into Animal Feed and Fuel Gas

M. MOO-YOUNG, A. R. MOREIRA, A. J. DAUGULIS, and C. W. ROBINSON

Biochemical and Food Engineering Group, Department of Chemical Engineering, University of Waterloo, Waterloo, Ontario, Canada, N2L 3G1

INTRODUCTION

Vast quantities of agricultural and forestry wastes accumulate each year, resulting in a deterioration of our environment and a loss of potentially valuable resources. At the same time mankind is facing two other crucial issues: food shortages and a decline in renewable energy reserves.

A report by the "Workshop on Alternative Energy Strategies" done at the Massachusetts Institute of Technology [1] indicates that world oil demand will exceed supply by an amount that will increase at a rate of 1.8×10^6 bbl day^{-1} year^{-1} beginning in 1987. This means that the next decade is of critical importance if alternative solutions to the energy supply are to be established on the basis of sound technological and economic studies. In a similar pattern, world food reserves are now insufficient to meet a one-month supply, and even the United States and Canada do not have enough food stocks to respond to world emergencies [2]. This situation is the result of a squandering of precious reserves of fossil fuels and cereal grains in man's desire for a high standard of living and a plentiful supply of status foods.

The biological conversion of agricultural and forestry residues into feed and fuel products can be a partial solution to these problems. One estimate [3] indicates that the annual net yield of photosynthesis is 1.8×10^{12} tons of biodegradable substances which are about 40% cellulose. In addition, in the major feedlot operations of the United States, 150–240 $\times 10^6$ tons of manure (on a dry basis) are produced per year, which must be collected and stored for disposal [4]. If only a fraction of these materials were to be converted into fuel gas and feed protein, with a concurrent reduction in the volume of material to be disposed of finally, a significant contribution could be made to the overall problem of resources recycle and conservation.

At the University of Waterloo research is being conducted to develop technically and economically feasible fermentation processes for the simultaneous production of feed protein and fuel gas from the agricultural wastes which are readily available in Canada. The intent is to develop

Biotechnology and Bioengineering Symp. No. 8, 205–218 (1978)

0572-6565/78/0008-0205$01.00

processing schemes with relatively low-level technology whose simplicity would allow their incorporation into normal farming operations. It is expected that the capital investment and the running costs for these processes will be low, making the process economics and operations attractive for the average-sized farm in Ontario (the equivalent of between 100 and 1000 head of cattle). Installation and operation of these processes would also generate new jobs without requiring highly qualified manpower and would thus improve the employment situation in Canada.

WASTE CHARACTERIZATION

Wide varieties of agricultural and forestry wastes are available as potential candidates for biological conversion into feed and fuel products. In a recent report [5], Detroy and Hesseltine identified the most important residues from the processing and harvesting of agricultural commodities. Table I, adapted from their report, shows a list of such residues.

TABLE I

Low-Value Agricultural Residues which are Potentially Useful In the Proposed Bioconversion Process[a]

Commodity	Residues
Wheat, rice, barley, oat, grass seed	Straw
Corn	Stalks, husks, cobs
Processed Grains	
Corn	Wastewater
Wheat	Bran
Rice	Bran
Soybean	Soy Waste
Other Commodities	
Cattle, Swine	Animal Waste
Sugarcane	Bagasse
Wood pulp	Sulfite liquor, clarifier sludges
Fruits and vegetables	Seeds, peels, husks, cones, stones, rejected whole fruit, juice
Potatoes	Peels, starch water
Oils and oilseeds: groundnuts, cottonseed, soybean, palm coconut, etc.	Shells, husks, lint, fiber, sludge, presscake, wastewater
Beverage Processing	
Molasses	Spent molasses liquor stillage
Coffee	Pulp, wash water

[a] Adapted from Detroy and Hesseltine [5].

TABLE II

Solid Waste Production in the United States[a]

Waste Type	Amount (10^6 Ton/year)
Agricultural and food wastes	400
Manure	200
Urban refuse	150
Logging and other wood wastes	60
Industrial wastes	45
Municipal Sewage solids	15
Miscellaneous organic wastes	70
TOTAL	940

[a] From Humphrey [6].

Table II shows the estimated annual solid waste production in the United States [6]. It can be seen that wastes of agricultural origin make the biggest contribution to the total solid waste output. The major crops generating residues are corn, soybeans, and wheat [7], while cattle and swine are the main producers of manure [5]. One of the major problems associated with utilizing these residues is the economics of residue collection and storage. With a few exceptions, such as sugarcane bagasse and animal wastes from the large feedlots in the central and southern states of the United States west of the Mississippi River, most of the agricultural residues are left scattered in the fields where they are grown. To avoid this disadvantage, we are looking into wastes which are either relatively concentrated at the source or are abundant in Canada and appear to be relatively easy to collect.

Table III shows a list of the residues that are being investigated in our laboratories as carbon and noncarbon nutrient sources. Although all of these residues are composed of cellulose, hemicelluloses, and lignin, the relative proportions of these basic constituents may change significantly from residue to residue. Crop materials contain 35–60% cellulose, 10–30% hemicelluloses, 4–18% lignin, and 5–20% ash [8]. The composition of animal wastes is strongly influenced by the type of manure, the diet of the animals, and the method used for manure collection. Table IV shows the composition of feedlot wastes from cattle which have been fed on a high-energy corn ration [9]. Pig wastes contain more starches and have a higher nitrogen content than cattle wastes (from 2.8–7.5% nitrogen on a dry weight basis) [10]. Both types of manure are good sources of nitrogen, phosphorous, and potassium (Table V). Forestry residues also have variable

TABLE III
Residues Selected for the Proposed Bioconversion Processes

Main Carbon Source(s)	Non-Carbon Nutrient Source(s)
Hardwood sawdust	Cattle manure
Clarifier sludges from pulp & paper industries	Pig manure
Cornstover	Poultry manure
Wheat straw	Soil fertilizers
Barley straw	
Oat straw	
Bagasse	

composition, depending on their origin. Published data for North American woods [11] indicates that they are 41.0–53.3% α-cellulose, 16.3–32.5% lignin, 2.9–4.8% uronic anhydride, 1.1–4.4% acetyl, and 0.2–0.4% ash.

This paper deals mainly with the upgrading of agricultural wastes into valuable feed and fuel products.

PROCESS DESCRIPTION

The basic design for the proposed process is shown in Figure 1. It consists essentially of three interconnected subsystems: an anaerobic digestor utiliz-

TABLE IV
Composition of Feedlot Wastes (Cattle Manure)[a]

Component	Fiber (%DM)	Soluble & fines (%DM)	Total (%DM)
Cellulose	22-27	2-4	10
Carbohydrate	45-53	7-11	25
Lignin	6-10	6-10	6-10
Ash	11-12	29-42	19
Protein N	1.8-2.0	4.5-6.0	3.6
Dry Weight (%)	48-50	50-52	100

[a] From Bellamy [9].

TABLE V
N, P, and K Content of Pig and Cow Manures (% DM)[a]

Type of Manure	N %	N Range	P %	P Range	K %	K Range
Dairy Cattle	3.3	1.9-5.5	0.35	0.1-0.4	2.0	1.0-3.0
Beef Cattle	2.0	1.5-4.0	0.65	0.3-0.7	1.6	1.0-3.0
Pig	4.0	2.8-7.5	1.00	0.2-1.5	1.2	0.2-1.6

[a] Adapted from Stewart and Chaney [10].

ing manure and crop residues, a chemical hydrolyzer where the crop wastes are treated with a dilute acid solution to remove the hemicelluloses, and an aerobic fermentation stage where yeast cells are cultivated on a mixture of the sugar solution from the chemical hydrolyzer and the nutrient-rich liquor from the anaerobic digestor.

Both the anaerobic digestion unit and the manure-holding tank are envisioned as underground, concrete-lined vessels. The holding tank usually already exists in most farms in Canada and will not be an additional cost to the process. The digestor gas, which is composed primarily of methane, is stored in floating-top, semiunderground tanks, and the fuel gas is used to supply heat to the digestor and other process operations that require energy

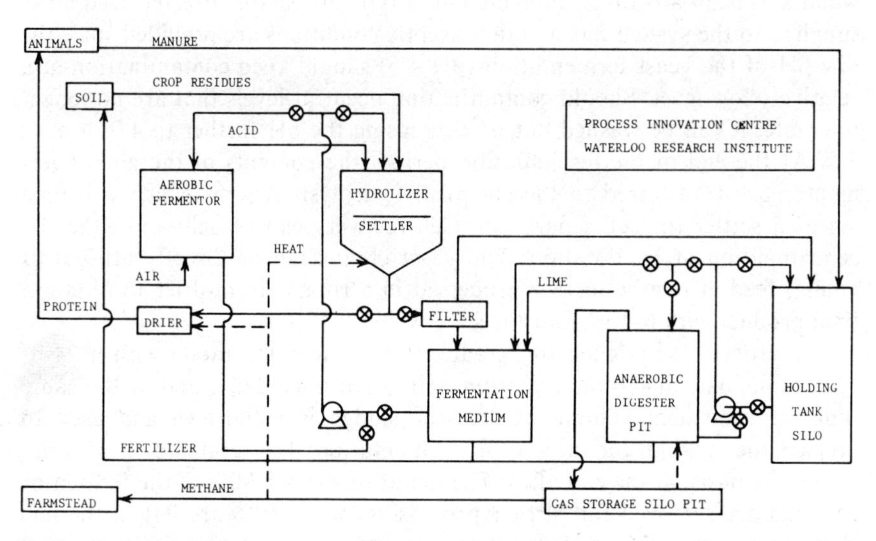

Fig. 1. Process for the concurrent production of fuel gas and protein-rich animal feed from agricultural wastes. Watfood B process for the concurrent production of fuel gas and protein-rich animal feed from agricultural wastes. Once-a-day fill-and-dump cyclic-batch operation.

and/or to the farmstead in general. The digestor sludge, withdrawn from the bottom of the digestor, is used as a nutrient source for the aerobic yeast fermentation. The sludge is suitable for this purpose largely because of its high ammonia nitrogen, and volatile fatty acids content. Consequently, the operating conditions for the digestor are such that a balanced trade-off is reached between the gas production rate and the concentrations of nitrogen and volatile fatty acids in the liquor. These considerations make the performance of the anaerobic digestion unit quite different from conventional anaerobic digestors which are usually designed to maximize the gas production rate.

The chemical hydrolysis of the crop wastes is performed in a conical-bottom reactor vessel made either of fiberglass or of clad or stainless steel. The reactor is charged with chopped lignocellulosic material, and $0.25N$ sulfuric acid solution is added until the desired solids to liquid ratio is reached. The reactor contents are heated to 100°C by steam injection or electric heating and are kept at this temperature for 1 hr. The slurry is withdrawn from the bottom of the reactor, filtered, and the liquor is either recycled to the hydrolyzer in order to treat a second batch of material or stored in the fermentation-medium tank where it is mixed with the anaerobic liquor. Lime is used to adjust the pH of the fermentation medium to 4.5. The acid-treated wastes retained in the filter are washed with water to recover some of the entrapped sugars and to decrease the acidity and are then used either to increase the gas production rate in the anaerobic digestor or to formulate ruminant feed rations through the use of an alkali treatment step.

The aerobic fermentor for the cultivation of yeast biomass is made of wood and is an air-lift or bubble-column type of reactor. Presterilized air is supplied to the system but no other aseptic conditions are provided since the low pH of the yeast fermentation (pH 4.5) should keep contamination at a relatively low level. Should contamination occur at levels that are no longer tolerable, it can be washed out by decreasing the pH further to 4.0 or even 3.5. At the end of the fermentation period, the contents of the air-lift fermentor are transferred to the chemical hydrolysis reactor which will then act as a settler for cell separation. Cell recovery can be achieved either by centrifugation or by flotation. The yeast cream can be directly utilized as animal feed or can be further processed in a rotary drum-drier to obtain a final product with 6–10% moisture.

The process is designed to operate in a cyclic-batch mode with a 24-hr cycle time. The anaerobic digestion unit is fed once daily, and at the same time an equivalent volume of digestor sludge is withdrawn and used to prepare the medium for the aerobic fermentation. The fermentor is also run on a daily basis; at the end of the fermentation period, 90% of the fermentor contents are removed for further processing, while 10% are left in the fermentor as inoculum for the next batch. The sugar solution from the acid hydrolysis of crop residues can also be prepared on a daily basis or, to save energy, a one-week hydrolysate supply can be prepared in one day and then used during the following week. The fermentation medium, however, should be prepared on a daily basis to minimize the chances for contamination.

Owing to its modular design, the process can easily be converted into alternative configurations in case there is an equipment breakdown or a different feed ration is desired. Table VI lists some of these alternative schemes. These options can utilize either fertilizer mixtures or straight manure as a nutrient source in the event of a malfunction with the anaerobic digestion unit. With a suitable design of the chemical hydrolyzer, it is possible to use the acid treatment to remove not only the hemicelluloses but also some or almost all of the cellulose and consequently increase the concentration of sugars in the fermentation medium. Fungi which utilize cellulose, such as *Chaetomium cellulolyticum* [12–16], can be alternatively used as the protein source. In this case, a conical-bottom fermentor should be used to eliminate the settling of the solids.

A pilot plant has been designed and built to evaluate the technical feasibility of the process. The design is flexible, and each piece of equipment can perform several functions, as shown in Table VII.

TABLE VI

Scenarios for Various Fermentation Processes to Upgrade Farm Wastes

	SCP FEED					
	Organism	C-Substrate(s)	Other Nutrients	Carbohydrate Feed	Animal Feed Type	By-Product Credit
1	Yeast	Hemicelluloses	Fertilizer	NaOH-cellulose	ruminants	-
2	Yeast	Hemicelluloses & some cellulose	Fertilizer	NaOH-cellulose	ruminants	-
3	Yeast	Hemicelluloses & cellulose	Fertilizer	-	all	-
4	Yeast	Hemicelluloses	Manure	NaOH-cellulose	ruminants	-
5	Yeast	Hemicelluloses & some cellulose	Manure	NaOH-cellulose	ruminants	-
6	Yeast	Hemicelluloses & cellulose	Manure	-	all	-
7	Yeast	Hemicelluloses	Anaerobic Liquor	NaOH-cellulose	ruminants	CH_4
8	Yeast	Hemicelluloses & some cellulose	Anaerobic Liquor	NaOH-cellulose	ruminants	CH_4
9	Yeast	Hemicelluloses & cellulose	Anaerobic Liquor	-	all	CH_4
10	Fungi	Hemicelluloses & some cellulose	Manure	NaOH-cellulose	ruminants	-
11	Fungi	Hemicelluloses & some cellulose	Anaerobic Liquor	NaOH-cellulose	ruminants	CH_4
12	Mixed Culture	(Combinations of 1,2,4,5,7,8,10,11)				

TABLE VII
Characteristics of the Pilot Plant Facilities

Equipment	Purposes
Stainless Steel Reactor Capacity: 130 liters Working volume: 110 liters	Anaerobic digestor
Stainless Steel Reactor Capacity: 220 liters Heating: Steam and one electrical heater Operating pressure: atmospheric	Hydrolyzer Settler Fermentor (fungi)
Acrylic Bubble Column Capacity: 90 liters Working volume: 60 liters	Fermentor (yeast)
Stainless Steel Rietz Thermascrew Blancher Capacity: 200 liters	Hydrolyzer Fermentor (solid-state)

EXPERIMENTAL RESULTS AND DISCUSSION

Anaerobic Digestion Unit

The operating conditions of the anaerobic digestion unit when using cattle manure are inlet feed concentration, 7.5% DM; inlet feed pH 6.4; temperature, 39°C; pH 7.1; retention time, 10–14 days. The performance of the digestor during a two-week period is shown in Figure 2.

The digestor was fed daily with a 7.5% dry matter cow manure slurry, and the appropriate volume of digested sludge withdrawn to keep the reactor volume constant. After this two-week period the gas production rate stabilized at 5 liter/hr, with 29% solids reduction. This gas production is equivalent to 0.23 m^3/kg added solids, which is slightly lower than the values for the solids-conversion efficiency in anaerobic digestors at 35–40°C working with municipal refuse [17]. Although the gas production could be increased by increasing the digestor temperature, it was decided to work in the mesophilic range to take advantage of lower energy requirements and better stability of the reactor. The average gas composition was 60% methane, 30% carbon dioxide, and lower percentages (3–4%) of nitrogen and hydrogen.

The anaerobic liquor contained 1.2 to 1.8 g/liter total Kjeldahl nitrogen and 1.0 to 1.5 g/liter volatile fatty acids (measured as acetic acid equivalent).

Chemical Hydrolyzer

The hydrolysis reactor now is being operated at a solids to liquid ratio of 1:20 by charging it with 2 kg of chopped barley straw and 40 liters of 0.25*N*

H_2SO_4. The reactor contents are heated to 100°C using low-pressure steam and kept at this temperature for 1 hr. After filtering the slurry and washing the acid-treated straw with tap water, about 42 liters of a sugar solution containing 10 g reducing sugars/liter are recovered. This amount corresponds to a sugar yield of 0.21 g of sugar/g straw, which duplicates sugar yield data obtained at the laboratory scale (typically between 0.17 and 0.20 g reducing sugars/g straw). In this sense, the chemical hydrolyzer performs satisfactorily and reproduces the results obtained at the smaller scale.

In an attempt to increase the concentration of sugars in the straw hydrolysate, the liquor was recycled to the hydrolyzer and a second batch of straw was processed in a similar manner as described before. However, the final reducing sugars concentration did not double but only reached 15 g/liter. This phenomenon seems to reflect the degradation of sugars, probably xylose from long exposure to high temperatures, as other investigators have observed [18]. At the moment studies to develop ways of increasing the sugar concentration while minimizing the secondary degradation effects by using a semisolid system are being conducted.

Aerobic Yeast Fermentation

Candida utilis (ATCC 9226) and *Ch. cellulolyticum* ATCC 32319) are used in this process as the single-cell protein sources. The yeast was chosen because of its capability to utilize hexoses and pentoses as the carbon and energy sources and also because of its known acceptability as a food and feed additive. The fungus was chosen for its good growth characteristics on lignocellulosic materials [12–14]. Furthermore, feeding trials with *Ch. cellulolyticum* used as a protein supplement have not shown any adverse effects.

Preliminary laboratory and pilot-plant experiments were performed in shake flasks, in a 2-liter fermentor (New Brunswick Scientific Co., New

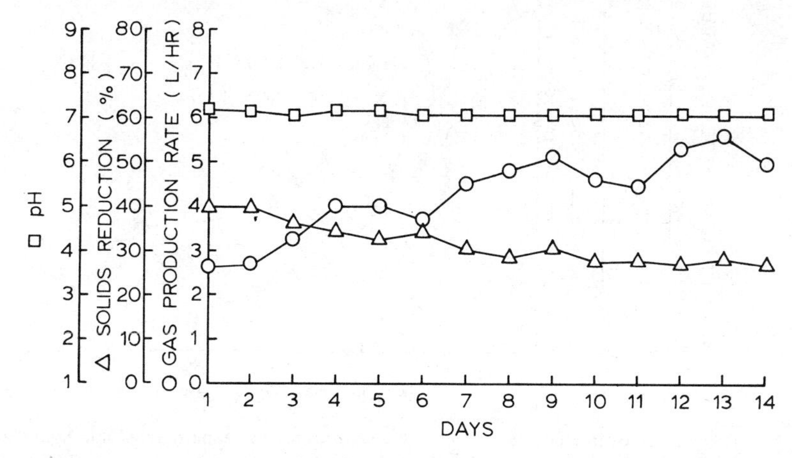

Fig. 2. Performance of the anaerobic digestor over a two-week period. Operating conditions given in text.

Jersey), and in the bubble-column fermentor. The experiments were run at 30°C and pH 4.5. The air flow rate was 1 v/v/m; foam was controlled by manual addition of a silicone antifoam agent (G. E. antifoam 71 nonionic silicone emulsion). A 10% preadapted inoculum was used in all the experiments to simulate the cyclic-batch operating mode that was envisioned at the plant level. The C/N ratio in the fermentation medium was 10:1.

Figure 3 shows some preliminary laboratory data obtained in the 2-liter fermentor (1.3-liter working volume) for growth of *C. utilis* on a mixture of cattle manure, anaerobic liquor, and a sugar solution containing 16 g of reducing sugars/liter (reducing sugars were determined by the dinitrosalicylic acid assay [19]). A fraction of these sugars (6 g/liter) was from barley straw hydrolysate, while 10 g/liter was from added glucose. The purpose of this particular experiment was to determine whether the anaerobic liquor contained enough nutrients to support high cell mass concentrations without further supplementation and whether any by-product of the yeast metabolism would accumulate at such levels that it would become inhibitory. In this work protein is taken as 6.25 × Kjeldahl nitrogen, and, for the calculation of cell mass data, the cells are assumed to be 50% protein.

After a 3-hr lag phase, the cells began growing exponentially at a specific growth rate of 0.28 hr^{-1}. After 13 hr of fermentation, the dissolved oxygen concentration began to rise, indicating a decline in the metabolic activity of the microorganisms. The culture then was supplemented with yeast extract,

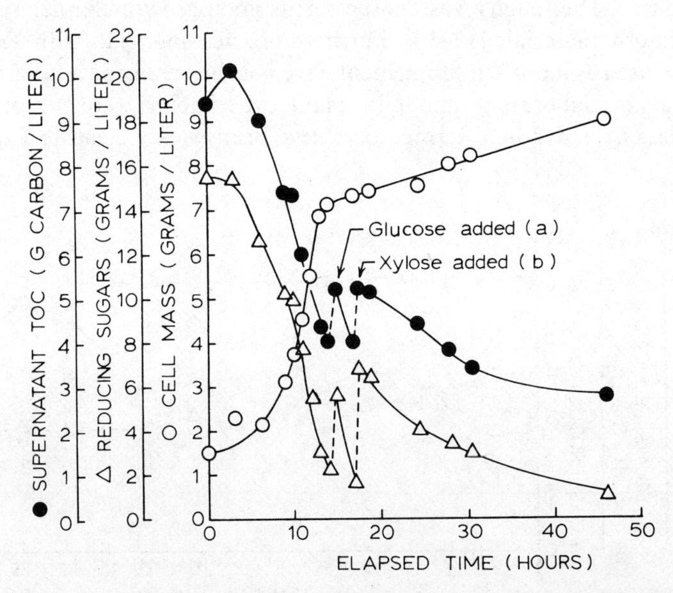

Fig. 3. Fermentation time courses for *C. utilis* grown on cow manure anaerobic liquor and barley straw hydrolysate supplemented with glucose: (a) 4 g glucose added, (b) 4 g of xylose added.

KH_2PO_4, and $(NH_4)_2SO_4$ added in sequence during the following hour; no response to these additions was observed, indicating that the decrease in metabolic activity was not due to limitation by any of these nutrients. After 14.5 hr, 4 g of glucose were added to the fermentor. An instantaneous response was observed, with a very sharp decrease in the dissolved oxygen level. This decrease indicates that, between 13 and 14.5 hr, the cells had little or no glucose available and were apparently utilizing xylose, for which the growth rate is much less (0.033 hr^{-1}). This induced period of high cell metabolic activity ended at 16 hr of elapsed fermentation time. When the dissolved oxygen began to rise, 4 g of xylose were added to the culture vessel. However, no significant response was observed, indicating that the yeast cells were utilizing xylose very slowly. In fact, the cell mass data in Figure 3 correspond to a growth rate of 0.033 hr^{-1} during the period of xylose uptake (16.5–46 hr).

Figure 3 also presents the data for total organic carbon (TOC) concentration in the culture supernatant and as determined using a Beckman Model 915 total organic carbon analyzer; the TOC results follow the same trends as the reducing sugar data. Assuming a 40% carbon content for the reducing compounds, it can be calculated there are 3.04 g of carbon per liter of nonreducing materials at the beginning of the run. A similar calculation for the endpoint of the fermentation indicates the presence of 2.70 g carbon/liter. We conclude, therefore, that only reducing sugars were utilized for the growth of the organism. In fact, during the entire experiment 23.3 g of reducing sugars/liter were utilized, which corresponds to 96.8% of the total reducing sugars added. The yield of cells, determined on the basis of the reducing sugars utilized, was 0.32 g of cells/g reducing sugars.

A sample of the data collected in the bubble-column fermentor is shown in Figure 4. In this run, *C. utilis* was grown on a mixture of cattle manure anaerobic liquor, and barley straw hydrolysate (obtained by treating two batches of straw with the same 0.25*N* sulfuric acid solution at a liquid–solids ratio of 20:1), and supplemented with 4 g glucose, 2 g xylose, 0.2 g KH_2PO_4, and 0.5 ml NH_4OH [28%(w/w)NH_3] per liter. The yeast exhibited the same diauxic growth characteristics as described before; however, the specific growth rate during the initial glucose uptake period (from 0 to 9 hr) was only 0.17 hr^{-1} and decreased to 0.013 hr^{-1} during the period of xylose metabolism (9 to 36 hr). The lower values for the specific growth rate observed in the bubble-column fermentor, as compared to the 2-liter stirred tank fermentor may be due to lower mass transfer rates in the bubble-column fermentor and/or inhibition by some compound which reaches inhibitory levels when the same acid solution is used to treat two batches of straw.

Of the 15.9 g of reducing sugars/liter initially present, approximately 3 g/liter remained in the culture broth at the end of the fermentation. This fact seems to indicate that, when two batches of straw are processed with the same sulfuric acid solution, some sugar degradation occurs. Although

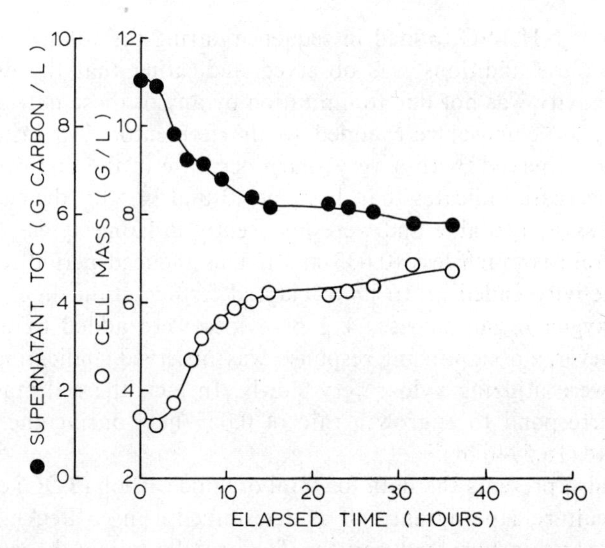

Fig. 4. Fermentation time courses for *C. utilis* grown on cattle manure anaerobic liquor and barley straw hydrolysate (bubble-column fermentor run).

the new compounds formed still have reducing characteristics they cannot be metabolized by the yeast. Intensive research studies are being performed to resolve this problem, since these degradation effects represent a significant loss of sugars and consequently will be detrimental to the process economics.

PROCESS ECONOMICS

The ultimate practical success of any process depends on its economic capability to compete with alternative processes which are already well established for production of the same or similar goods. Consequently, detailed economic studies are being performed in parallel with the experimental work in order to screen for the processing schemes which are most attractive economically.

In our economic analysis the production cost (in dollars per pound of protein) is based on the expenses for raw materials, labor, maintenance and repairs, operating supplies, utilities, financing, and depreciation. Equipment costs are obtained from published cost data and updated to 1977 using the Marshall and Stevens cost indexes, and chemicals prices are obtained from current chemicals selling price lists. A credit is given to the process for the production of fuel gas (compared on a heating value basis to LPG, commonly used on farms) and for the acid-alkali treated straw (as compared to hay forage carbohydrate).

Table VIII shows a sample of the data which have been generated for the production cost of yeast SCP using wheat straw and cow manure. These prices refer to a process where a 3:1 liquid to solids ratio is used for the

TABLE VIII

Estimated Production Cost for Yeast SCP[a] (1977 US $/lb protein)

# Head of cattle	Basic cost	Cost crediting forage carbohydrate	Cost crediting digestor gas	Cost crediting forage carb. & digestor gas
100	0.70	0.43	0.52	0.26
500	0.55	0.29	0.38	0.12
1,000	0.52	0.25	0.35	0.08
5,000	0.48	0.21	0.31	0.04

[a] Based on a 10% final moisture for the yeast cells.

acid hydrolysis of straw followed by a 4% (w/w) NaOH treatment. Straw was charged at 1 ¢/lb. The gas production rate in the anaerobic digestor was assumed to be 50 ft^3 per head of cattle per day [20]. Although some of these conditions are slightly different from the ones used in the experiments described before, we believe that they will be achieved as the process advances to more refined stages.

The data clearly show the effect of scale on the estimated production cost. Because soymeal protein is currently selling for 25 ¢/lb [21], the process would be unlikely to be economically competitive without the forage-carbohydrate and digestor-gas credits. However, when these credits are taken into consideration, the estimated protein cost for the case of a 100-head farm size is 26 ¢/lb and becomes significantly lower than 25 ¢/lb at farm sizes greater than 100 head of cattle. This means that the process should be economically very attractive for the average farm size in Ontario (equivalent to 500 head of cattle).

CONCLUSIONS AND FUTURE WORK

The work performed to date has shown the technical feasibility of the proposed process to produce both protein and carbohydrate constituents for animal feed and fuel gas from agricultural wastes. The anaerobic liquor has proven to be a good source of nutrients, supporting a cell biomass of at least 7 g/liter. The aerobic fermentor has been utilized successfully to grow yeast cells on anaerobic liquor and straw hydrolysate without special sterilization precautions other than "keeping the house clean" being taken, and no major contamination problems have been noted.

Some problems remain to be resolved. When two batches of straw are treated with the same acid solution, a decrease in the sugar yield is observed, and a significant amount of reducing materials is not utilized during the yeast fermentation. To solve this problem, research is being conducted on the kinetics of the hydrolytic process. Furthermore, the cell growth rate is lower than the normal values (between 0.4 and 0.5 hr^{-1}) for yeast grown on synthetic medium, which may indicate the presence of some inhibitory compound in the fermentation medium.

Further studies also are being performed on the optimization of the solids to liquid ratio in the hydrolysis stage, the influence of addition of treated straw to the anaerobic digestor on the gas production rate and on the nitrogen content of the anaerobic liquor, the development of inexpensive techniques for anaerobic sludge solids and yeast biomass separation, the design of process equipment, the assessment of the potential of semi-solid-state fermentation, and the process economics in order to derive the most attractive strategies.

This work was supported by a grant from the National Research Council of Canada. Our thanks are also due to Professor J. M. Scharer for his valuable discussions and to D. Vlach, I. Vlach, M. Bonga, and P. Hryb for their capable assistance with the laboratory analyses.

References

[1] W. Lepkowski, *Chem. Eng. News*, *55*(22), 10–14 (May 30, 1977).
[2] C. J. Rogers, *Resour. Recovery Conserv.*, *1*, 271 (1976).
[3] T. K. Ghose, *Adv. Biochem. Eng.*, *6*, 39 (1977).
[4] G. R. Stephens and G. H. Heichel, *Biotechnol. Bioeng. Symp*, *5*, 27 (1975).
[5] R. W. Detroy and C. W. Hesseltine, "Availability and utilization of agricultural and agro-industrial wastes," paper presented at the AIChE Annual Meeting, New York City, New York, November 13, 1977.
[6] A. E. Humphrey, *Chem. Eng.*, *81*(26), 98 (1974).
[7] J. H. Sloneker, *Biotechnol. Bioeng. Symp.*, *6*, 235 (1976).
[8] C. E. Dunlap, in *Single Cell Protein II*, S. R. Tannenbaum and D. I. C. Wang, Eds. (MIT Press, Cambridge, Mass., 1975), pp. 244–262.
[9] W. D. Bellamy, in *Single Cell Protein II*, S. R. Tannenbaum and D. I. C. Wang, Eds. (MIT Press, Cambridge, Mass., 1975)., pp. 263–272.
[10] R. A. Stewart and R. L. Chaney, *Proc. Soil Conserv. Am.*, *30*, 160 (1976).
[11] H. F. J. Wenzl, *The Chemical Technology of Wood*, (Academic Press, New York, 1970).
[12] M. Moo-Young, D. S. Chahal, and D. Vlach, *Biotechnol. Bioeng.*, *20*, 107 (1978).
[13] N. Pamment, M. Moo-Young, C. W. Robinson, and J. Hilton, "Solid state cultivation of *Chaetomium cellulolyticum* on alkali-pretreated sawdust," *Biotechnol. Bioeng. 20*, 1735 (1978).
[14] N. Pamment, M. Moo-Young, F.-H. Hsieh, and C. W. Robinson, "Growth of *Chaetomium cellulolyticum* on alkali-pretreated hardwood sawdust solids and pretreatment liquor," *Appl. Environ. Microbiol.*, *36*, 284 (1978).
[15] N. Pamment, C. W. Robinson, and M. Moo-Young, "Pulp and paper mill solid wastes as substrates for SCP production," *Biotechnol. Bioeng.* (in press).
[16] N. Moo-Young, C. W. Robinson, N. Pamment, J. Hilton, and F.-H. Hsieh, "Conversion of pulp mill wastes into protein-enriched animal feed supplements," CPAR Project No. 47O, Progress Report to March 31, 1977, University of Waterloo Research Institute, Waterloo, Ontario.
[17] J. T. Pfeffer and J. C. Liebman, *Resour. Recovery Conserv.*, *1*, 295 (1976).
[18] G. A. Grant, Y. W. Han, A. W. Anderson, and K. L. Frey, *Develop. Ind. Microb.*, *18*, 599 (1977).
[19] G. L. Miller, *Analytical Chem.*, *31*, 426 (1959).
[20] P. L. Silveston, *AIChE Symposium Series*, *72*, No. 158, 33 (1976).
[21] D. Mowat, University of Guelph, personal communication, February 20, 1978.

Thermochemical Pretreatment of Nitrogenous Materials to Increase Methane Yield

DAVID C. STUCKEY and PERRY L. McCARTY

Environmental Engineering and Science, Department of Civil Engineering, Stanford University, Stanford, California 94305

INTRODUCTION

Owing to the projected shortage of fossil fuels, research is being conducted on the production of methane from organic material by anaerobic digestion. Current estimates [1] indicate that the digestion of refuse and agricultural residues could provide as much as 20% of the 1970 demand for natural gas. However, an economic analysis [2] revealed that the cost of methane from anaerobic treatment is a sensitive function of the biodegradability of the organic materials treated. This research on thermochemical treatment of organics is being conducted in an effort to increase biodegradability and hence reduce costs of methane production.

The Stanford studies are concerned primarily with lignocellulosic and nitrogenous organics since they are among the most prevalent organic residuals available. This paper summarizes current results on the effect of thermochemical pretreatment on the anaerobic biodegradability of nitrogen-containing materials.

Waste activated sludge (WAS) from biological treatment of municipal wastes consists largely of bacterial cells and was selected as a representative residual organic with a high nitrogen content. This material is known generally to be only 30–50% biodegradable by anaerobic treatment. Previous work [3] has shown that thermochemical pretreatment of WAS can increase its biodegradability considerably; however, it was found that toxic materials were produced.

The objectives of this work were then

(1) To obtain an understanding of why nitrogenous components of bacterial cells are relatively refractory (poorly biodegradable);

(2) To evaluate the effect of thermochemical pretreatment variables on the degradability of bacterial cells and their nitrogenous constituents; and

(3) To evaluate the possible production of toxic materials during thermochemical treatment.

Biotechnology and Bioengineering Symp. No. 8, 219–233 (1978)

0572-6565/78/0008-0219$01.00

BACKGROUND

Refractory Organics

For this evaluation, refractory organic materials are defined as those which are not degraded anaerobically by an active anaerobic culture within 25 days. Alexander [4] suggested that an organic compound may persist without degradation if (1) the environment is not conducive to microbial life or to the degradation of the compound, or (2) the substance itself is resistant, either totally or partially, to biodegradation under all circumstances. With respect to bacterial cell organics, four hypotheses are being evaluated to explain their resistance to biodegradation under anaerobic conditions:

(1) Lack of an essential nutrient for degradation, e.g. oxygen.
(2) Need for different organisms—organisms may not be present which can degrade the organics.
(3) Inaccessibility of organics to enzymes.
(4) Inherent refractory nature of organics.

The first two hypotheses relate to environmental factors, and the last two to inherent characteristics of the organic materials. Some information regarding these hypotheses was provided by this study.

Composition of Bacterial Cells and Effect of Heat Treatment

A typical composition of bacteria (dry weight basis) is [5] carbon, 50 ± 5%; nitrogen, 8–15%; hydrogen, 10%; oxygen, 20%; ash, 5%; lipids, 10–15%; protein, 50%; carbohydrates, 10–30%; RNA, 10%; DNA, 3–4%. These percentages can vary considerably, however, depending on the species and the stage of growth of the cell, that is, whether it is in the log growth or stationary phase. As given above, the nitrogenous materials (protein, DNA, RNA) comprise about two-thirds of the bacterial cell.

The bacterial cell wall accounts for 20–30% of the dry weight of the cell. The rigid structural framework consists of parallel polysaccharide chains covalently cross-linked by peptide chains. This peptidoglycan structure is resistant to the action of peptide hydrolyzing enzymes, which do not attack peptides containing D-amino acids [6].

In addition, gram-positive cell walls are low in lipids (<2%) and high in polysaccharide (40–60%), while gram-negative walls are high in lipids (20%) and lower in polysaccharides (12–30%). Gram-positive walls contain 10–30% amino sugars, while gram-negative cells have only 2–7% amino sugars [5].

Under thermochemical treatment WAS may generally react in the following ways.

(a) Lipids: hydrolyzed easily under acid or alkaline conditions to glycerol and fatty acids.

(b) Carbohydrates: bacterial polysaccharides would be expected to hydrolyze to simpler polysaccharides, or sugars.

(c) Proteins: hydrolyzed by acid solutions to amino acid monomers. Some peptide bonds, those of valine, isoleucine, and leucine, for example, are more stable than others and require longer hydrolysis times or stronger acids. Peptide bond cleavage is noticeably faster in hydrochloric acid than in sulfuric [7]; amino acids can be further degraded to ammonia and organic acids. Under alkaline conditions proteins can also be hydrolyzed; however, the rate and extent is generally less than with acid.

(d) Nucleic acids: RNA and DNA would be expected to hydrolyze to produce constituent bases, sugars, and orthophosphate.

Various intermolecular reactions may also be expected, for example, the "Browning" reaction [8], which involves the polymerization of carboxyl groups with amino groups to form brown nitrogenous polymers and copolymers, termed melanoidins. High temperature and extremes in pH increase the rate of this polymerization. A disadvantage of melanoidin formation is that due to their structure and complexity, they are likely to be difficult to degrade.

EXPERIMENTAL PROCEDURES

Source of Material

The thermochemical treatment of air-thickened WAS from the San Jose–Santa Clara Water Pollution Control Plant was evaluated. By weight the sludge was 4.3% total solids, of which 72% were volatile. Sufficient WAS was collected at one time and stored at 3°C to prevent deterioration during the study.

Pretreatment

The heat-treatment reactor (Parr Instrument Co., No. 4561) had a Teflon liner and nominal capacity of 300 ml and was constructed of Monel, with exposed parts in Hastelloy-C to reduce potential heavy-metal toxicity. The reactor was stirred at 400 rpm. The reactor required from 45 to 60 min to reach the desired pretreatment temperature, was held there for 1 hr, and then was cooled rapidly to room temperature by cooling coils. Nitrogen was used to flush the reactor before the treatment to prevent oxidation with air.

In order to investigate the effect of mechanical cell lysis, WAS was ultrasonically lysed with a Bronwill "Biosonik III." Ten-milliliter samples were subjected to treatment for 15 min.

Biodegradability and Toxicity

The biodegradability of complex organic materials such as bacterial cells is difficult to measure directly. When a biodegradable organic is consumed by bacteria, a portion is converted to end-products, such as methane, and another portion is transformed into bacterial cells. Hence, methane production alone can only be used as an indicator of the extent of biodegradation as it gives no measure of that portion converted to bacterial cells. Under anaerobic conditions the portion converted to bacterial cells may vary from 2–25% but under aerobic conditions it may be as high as 30–60%. Similar problems result when other measures such as decrease in COD, protein, or RNA content are used. Because of this problem, measures such as "convertibility to methane" or "decrease in COD" are used only as indicators of biodegradability. "Biodegradability" itself is used as a direct measure only when the absolute consumption of a material, corrected for conversion to cellular material, is considered.

The convertibility of organics to methane was measured by the Biochemical Methane Potential (BMP) analysis [9]. Here, 250-ml serum bottles were initially gassed with a mixture of 70% N_2 and 30% CO_2, and 7.5 ml of thermochemically pretreated WAS was added. A combined nutrient media and anaerobic seed (143 ml) was then transferred anaerobically to the serum bottles, serum caps were inserted, and the bottles were incubated at 35°C. Gas production and composition were monitored regularly during a test period of 25 days.

The percent convertibility of a sample to methane can be estimated from the BMP results if the initial COD is known [9], as the theoretical yield for 100% conversion to methane is 0.35 m^3 CH_4 (STP) per kilogram of COD converted.

Aerobic biodegradabilities were assessed similarly, except the bottles were incubated with a cotton plug. Anaerobic seed was used to provide the same initial population. Ten days were used to assess aerobic biodegradability.

Toxicity was evaluated with the Anaerobic Toxicity Assay (ATA) [9]. Pretreated WAS was added together with nutrient media, deoxygenated bicarbonate buffer, anaerobic seed, and 2 ml of an acetate–propionate substrate in 125-ml serum bottles. The bottles were incubated at 35°C and gas production was monitored regularly. Inhibition was determined from the ratio of the maximum rate of gas production (ml/day) in the samples to that of controls (maximum rate ratio, MRR).

Analytical Procedures

Soluble constituents were determined on centrifuged samples after filtration through a 0.45 μm filter; COD was determined according to *Standard Methods* [10]. Organic and ammonia nitrogen were determined by the Kjeldahl digestion and distillation methods, respectively [10]. Total and soluble protein were determined by the Lowry method [11]. Insoluble proteins were released by hydrolyzing a diluted sample ($1/10$ to $1/100$) in 0.1*N* NaOH for 30 min at 100°C.

Both total and soluble RNA were determined by the cupric-ion-catalyzed orcinol reaction [12]. This measures only ribose and purine-bound ribose, and depends on the conversion of the pentose in the presence of hot acid to furfural. Total RNA was determined on the hydrolyzed sample prepared for total protein analysis since RNA is also solubilized in dilute hot NaOH.

The modified ninhydrin colorimetric analysis was used for free amino acids [13]. Ninhydrin reacts strongly with ammonia [6] and the method of Hattingh et al. [14] was used to reduce ammonia in the sample to low concentrations before analysis.

RESULTS

Effect of Thermochemical Treatment on WAS

The results are summarized in Table I and Figure 1. From 5 to 15% of the TKN and COD were not recovered after heat treatment, perhaps due to volatilization or incomplete recovery of solids. The general effect of heat treatment was an initial solubilization of organic nitrogen (SON) or release of NH_3. As the temperature increased, SON remained relatively constant up to 225°C, while NH_3 increased indicating hydrolysis of total organic nitrogen (TON) to SON, which concomitant destruction of SON to NH_3.

Protein was also solubilized, but the soluble fraction was hydrolyzed significantly to amino acids and NH_3 at 225°C and above. The RNA concentration decreased rapidly with temperature, especially when HCl was used.

Soluble COD increased slowly with temperature from 150°C to 225°C. At 250°C it decreased in concentration, perhaps indicating the precipitation

TABLE I

Effect of Thermochemical Pretreatment on WAS Composition

Heat Treatment Conditions		ORGANIC N			PROTEIN			RNA		COD		
Temp. (°C)	Chemical added (300 meq/l)	TKN mg N/l	SON mg N/l	NH_3 mg N/l	Total mg/l	Soluble mg/l	Amino Acids mg/l	Total mg/l	Soluble mg/l	Total mg/l	Soluble mg/l	% Solubilized
Feed WAS		3,220	120	670	11,700	665	155	2,260	375	46,300	-	-
U.S. lysed WAS		3,200	185	765	11,600	1,000	225	2,290	290	46,400	9,200	20
150	None	3,100	1,600	755	9,530	7,500	1,370	1,740	2,120	39,700	20,900	45
175	None	3,060	1,490	870	9,820	8,130	1,800	1,530	1,420	43,200	22,100	48
	$Ca(OH)_2$	3,090	1,290	945	8,530	5,400	1,890	1,300	1,070	38,700	18,700	40
	NaOH	2,780	1,610	945	9,820	8,570	3,030	1,530	1,270	42,400	25,400	55
	HCl	2,920	1,790	760	8,580	5,940	5,790	140	370	45,500	24,900	54
200	None	3,300	1,650	940	9,230	7,190	2,140	260	200	46,800	23,300	50
	NaOH	3,080	1,510	1,200	9,230	7,500	3,900	1,060	735	42,300	25,100	54
	HCl	3,150	1,650	1,110	6,750	4,470	6,260	20	60	44,200	24,000	52
225	None	3,080	1,560	1,180	8,450	5,760	2,390	90	90	45,200	23,500	51
250	None	3,110	1,210	1,350	7,790	4,730	1,470	75	50	40,700	21,500	47

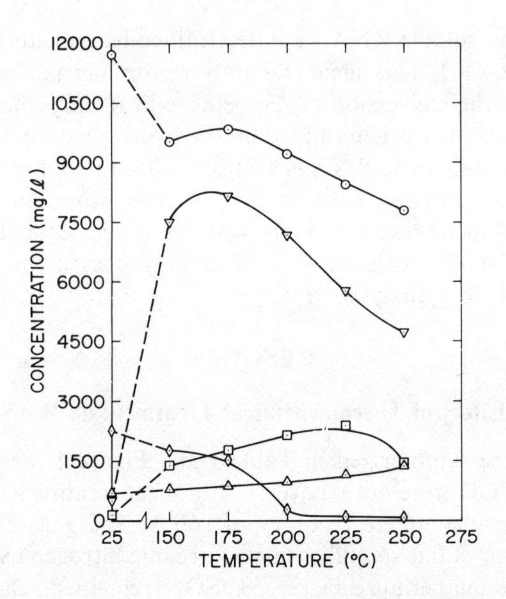

Fig. 1 Effect of pretreatment temperature on nitrogen forms (no chemical addition).(○) Total protein; (▽) soluble protein; (□) amino acids; (△) ammonia; (◇) RNA.

of an insoluble compound. SON also decreased with temperatures above 200°C.

The different chemical additions had varying effects, the most pronounced being the effect of HCl in increasing hydrolysis of protein and RNA.

Anaerobic Biodegradability

Ten pretreated WAS samples plus ultrasonically lysed and normal WAS were analyzed by the BMP procedure to obtain an indication of increased biodegradability through heat treatment. Table II and Figure 2 summarize the calculated convertibility of total COD to methane for the various treatments.

Heat treatment increased the anaerobic biodegradability of WAS considerably. During anaerobic biological decomposition, about 10% of the metabolized organics are normally converted to bacterial cells. Hence, the maximum short-term COD reduction or convertibility to methane from complete substrate utilization would be about 90%. Thus results (Table II) indicate that by thermochemical treatment a maximum of about 85% of the organics (78/0.9) were biodegradable compared with about 53% for nontreated controls.

Without chemical addition, biodegradability peaked at 175°C and decreased steadily to 250°C. With either NaOH or HCl, the degradability

TABLE II
Convertibility of Total COD to Methane

Temperature (°C)	Convertibility to Methane (percent) No Chemical	$Ca(OH)_2$ 300 meq/l	NaOH 300 meq/l	HCl 300 meq/l
25	48	–	–	–
150	65	–	–	–
175	68	60	78	75
200	60	–	75	76
225	51	–	–	–
250	43	–	–	–
U.S. lysis	53			

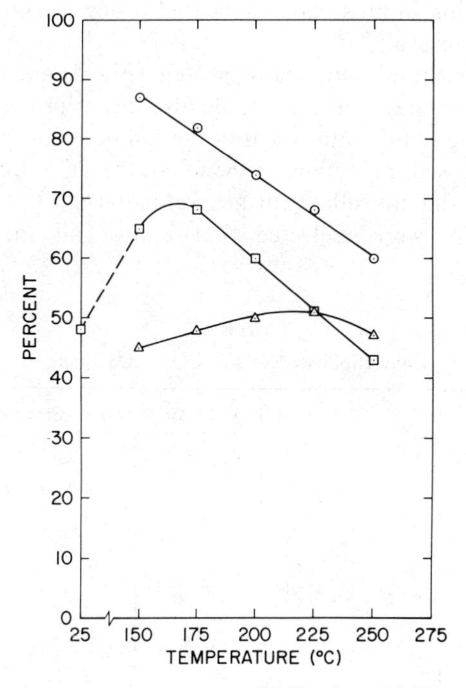

Fig. 2. Effect of pretreatment temperature on solubilization of COD and biodegradability as indicated by percent convertibility of total and soluble COD to methane (no chemical addition). (⊙) Soluble; (□) Total; (△) Soluble as percent of total.

was greater than with no chemical addition. On the other hand, pretreatment without chemicals at 175°C was better than with $Ca(OH)_2$.

The biodegradability of soluble COD was indicated by the reduction in its concentration during the BMP analysis. The results (Table III, Figure 2) indicate biodegradability decreased linearly with increasing temperature. Addition of chemicals at 175°C had little effect but at 200°C increased the biodegradability significantly.

Effect of Thermochemical Pretreatment on Anaerobic Biodegradability of Nitrogen Fraction

The biodegradability of remaining nitrogenous organics in general and protein in particular after heat treatment is indicated by the reduction in the concentrations of these components during the BMP test. A substantial portion of the total and soluble organic nitrogen (Fig. 3) appears to be biodegradable after heat treatment. However, as the temperature of pretreatment rose above 175°C, the degradable portion decreased significantly. One exception was higher total organic nitrogen destruction at 250°C, which, because of consistent trends in other data, appears to be due to an experimental error.

Chemical addition in most cases appeared to lower TON removal, while increasing SON removal.

Again, remaining total and soluble protein (Fig. 4) and amino acids (not shown), were substantially biodegradable after lower pretreatment temperatures, but as the temperature rose, the amount of reduction in the components decreased, reflecting, in the main, the TON trend. The effect of chemical addition did not follow any general pattern.

Data for RNA were collected, but almost all the fractions were

TABLE III

Convertibility of Soluble COD to Methane

Temperature (°C)	Convertibility to Methane (percent)			
	No Chemical	$Ca(OH)_2$ 300 meq/1	NaOH 300 meq/1	HCl 300 meq/1
25	–	–	–	–
150	87	–	–	–
175	82	78	81	84
200	74	–	80	84
225	68	–	–	–
250	60	–	–	–
U.S. lysis	89	–	–	–

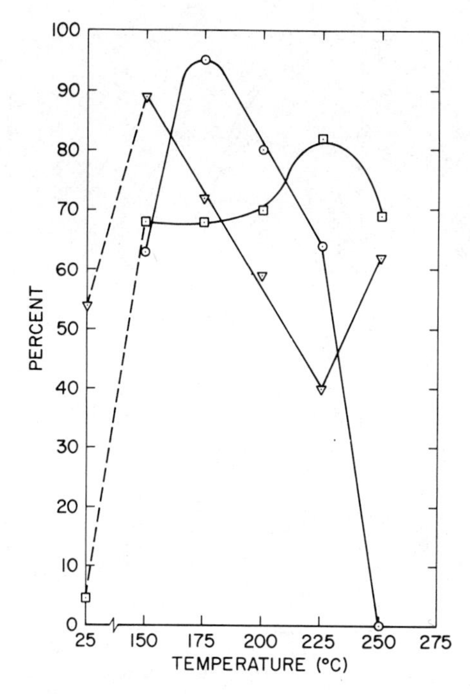

Fig. 3. Solubilization and anaerobic biodegradability of total and soluble nitrogenous organics as indicated by reduction in organic nitrogen concentration during the BMP analysis (no chemical addition). (□) Soluble as percent of total; (○) soluble; (△) total.

substantially destroyed (Table I), indicating at minimum that the pyrimidine (purine) ribose bond was cleaved by mild thermochemical treatment.

Effect of Thermochemical Pretreatment on Toxicity of WAS

The results of toxicity evaluation are presented in Table IV. As temperature increased above 150°C, the substrate became more toxic. The addition of NaOH seemed to increase the toxicity only marginally, while HCl treatment made the WAS extremely toxic. Acclimation to the toxicity occurred in some cases after seven days, and in others as long as 130 days were needed.

Gas chromatographic analysis of inhibited samples after 45 days revealed the presence of substantial quantities of C2 to C6 volatile acids. The concentration of individual acids were as high as 20m*M*, indicating that the inhibitory material affected the ability of the methanogenic microorganisms to ferment these materials.

Effect of Thermochemical Pretreatment on Aerobic Biodegradation

Aerobic and anaerobic reduction of COD was evaluated with identical pretreated samples to determine if differences in biodegradation were sig-

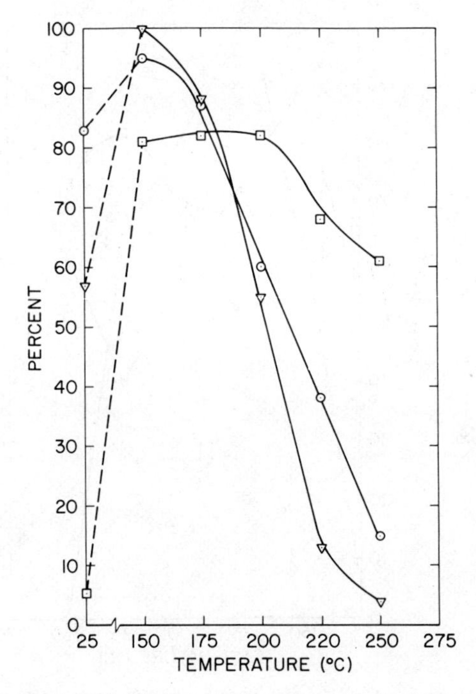

Fig. 4. Solubilization and anaerobic biodegradability of total and soluble protein as indicated by reduction in protein concentration during the BMP analysis (no chemical addition). (▽) Total; (○) soluble; (□) soluble as percent of total.

nificant. Nearly identical nutrient media and anaerobic seed were used to ensure degradation conditions were similar. It is believed that sufficient aerobic or facultative bacteria were present in this seed to make the comparison valid.

During aerobic biodegradation cellular growth can be significant so that COD reduction alone would be a poor indicator of biodegradability. For this reason, the actual percentage decrease in COD was divided by 0.7 to obtain the percent biodegradability of total COD listed in Table V. This procedure was to correct for an assumed yield factor of 30%. Such a correction is not needed for the soluble portion because soluble COD reduction measures both oxidation and synthesis to particulate cellular matter.

Soluble COD biodegradation was high at low temperature but decreased linearly with increase in pretreatment temperature. Total COD biodegradability reached a maximum at 175°C and then decreased with increasing temperature. The addition of chemicals at 175°C appears to have decreased biodegradability, while at 200°C the effect was mixed.

The effect of aerobic biodegradability as determined by reduction in various nitrogen-containing organics is indicated in Table VI. Again, as in anaerobic degradation, there is a trend to decreasing biodegradability with increasing temperature.

TABLE IV
Maximum Rate Ratio (MRR) for Heat-Treated WAS

Heat Treatment		Dilution (sample/total)*		
Temp. (°C)	Chemical added (300 meq/1)	2/50	10/50	25/50
Feed WAS		-	-	1.35
U.S. lysed WAS		-	-	1.26
150	None	1.19	1.20	1.24
175	None	1.18	1.02	0.73 A14
	$Ca(OH)_2$	1.06	1.28	0.87 A7
	NaOH	0.98	0.13 A20	0.17
	HCl	0.54 A16	0.14	0.21
200	None	1.13	0.16 A24	0.19 A69
	NaOH	1.06	0.13 A20	0.17
	HCl	0.56 A20	0.14 A131	0.17
225	None	0.88 A8	0.11 A57	0.16
250	None	0.63 A7	0.10	0.13

* A14 denotes acclimation after 14 days, etc.

TABLE V
Aerobic Biodegradability of WAS

Temperature (°C)	Biodegradability (percent)			
	No Chemical	$Ca(OH)_2$ 300 meq/1	NaOH 300 meq/1	HCl 300 meq/1
Total COD*				
25	51	-	-	-
150	53	-	-	-
175	67	56	58	57
200	59	-	67	51
225	54	-	-	-
250	56	-	-	-
Soluble COD				
25	-	-	-	-
150	84	-	-	-
175	81	74	75	79
200	70	-	76	80
225	69	-	-	-
250	60	-	-	-

* Calculated using a 30-percent yield factor for total biodegradability.

TABLE VI

Aerobic Reduction in Nitrogenous Materials (Percent)

Heat Treatment		Organic N		Protein		RNA	
Temp. (°C)	Chemical added (300 meq/l)	Total	Soluble	Total	Soluble	Total	Soluble
Feed WAS		53	-	56	68	17	74
150	None	35	100	71	92	3	83
175	None	52	87	72	83	11	80
	$Ca(OH)_2$	44	71	65	60	21	80
	Na(OH)	43	79	53	77	-	79
	HCl	47	97	47	76	-	76
200	None	40	82	50	71	-	40
	NaOH	55	99	57	74	-	60
	HCl	39	90	34	70	-	10
225	None	54	78	45	63	-	22
250	None	47	70	44	53	-	-

DISCUSSION

The effect of lower temperature (150°C) pretreatment of WAS appears to be lysis of the cell and partial hydrolysis of organics. Increased temperatures of treatment increased the solubilization of organics slightly up to a maximum of 51% at 225°C. At higher temperatures the amount of solubilized organics decreased, suggesting the formation of larger molecules through polymerization.

The addition of chemicals increased the solubilization marginally, if at all.

The effect of pretreatment on nitrogenous organics was as expected. Even after 150°C pretreatment, most of the organic nitrogen forms were solubilized (protein, amino acids, RNA). As the temperature rose, the SON hydrolyzed first to amino acids, and then presumably the amino nitrogen, together with RNA, was hydrolyzed or otherwise converted to ammonia. HCl increased the extent of hydrolysis significantly, as expected, while with NaOH the increase was slight. The amino acid concentration increased with increasing temperature, but at 250°C the concentration dropped sharply, possibly as a result of the Browning reaction.

Heat treatment of WAS increased biodegradability substantially up to a maximum at 175°C treatment. This increase came from the insoluble as well as the soluble fraction. Hence, heat treatment altered the structure of the insoluble fraction to make it more amenable to biodegradation also.

The decreased biodegradability of WAS at temperatures above 175°C could be due either to the formation of refractory compounds during heat treatment or to the inhibition of the microorganisms by the treated WAS. While the data do not allow firm conclusions, the inhibition data in Table IV reveal that when samples treated at 200°C or above were diluted approximately the same as in the BMP test, only the samples treated at 225°C and 250°C were inhibitory. Hence it appears that the decreased biodegradation at 200°C was due primarily to the formation of refractory compounds, both soluble and insoluble, rather than inhibition. Acclimation occurred at both 225°C and 250°C after only eight days, which would also suggest that refractory materials were the main cause for lower biodegradability at higher temperatures.

Chemical addition, except with $Ca(OH)_2$, resulted in substantial increases in biodegradability. These increases cannot be explained by solubilization alone and hence could be due to alteration of the insoluble organics making them more amenable to attack by extracellular enzymes. The addition of $Ca(OH)_2$ lowered the solubilization and overall biodegradability, the cause for which was not determined.

Unfortunately, heat treatment also produced strongly inhibitory compounds. This is obviously an important consideration as it limits the concentration of the heat-treated WAS that can be digested. After 175°C pretreatment and no chemicals, a 2.2% concentration of WAS can probably be digested without significant inhibition. Higher concentrations might be digested after organism acclimation to the inhibitory materials.

Comparison between aerobic and anaerobic biodegradability of the total and soluble fractions of thermochemically pretreated WAS (adjusted for cell yield) (Fig. 5) reveal few large differences. Aerobic conditions did not lead to higher biodegradabilities, in fact they were lower in all cases except

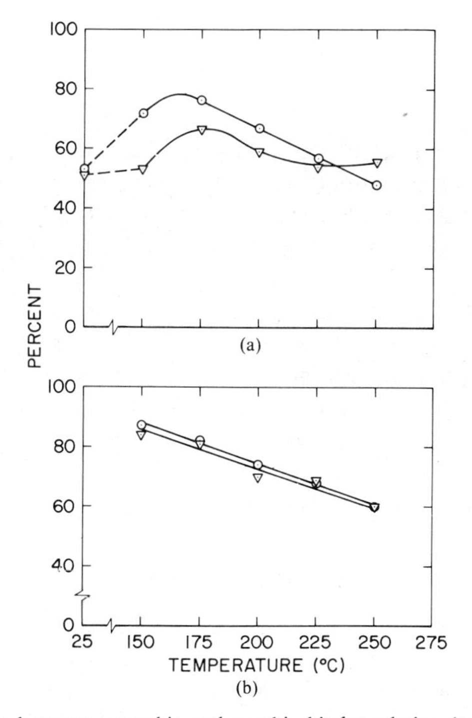

Fig. 5. Comparison between anaerobic and aerobic biodegradation (no chemical addition). (⊙) Anaerobic; (▽) aerobic. (a) Total COD biodegraded; (b) soluble COD biodegraded.

TABLE VII

Comparison Between Thermal and Ultrasonic Pretreatment

Pretreatment		Convertibility to Methane, Percent
Untreated WAS		48
Ultrasonic Lysis		53
Heat Treatment:	150°C	65
	175°C	68

at 250°C. This result is important, as it illustrates that oxygen is not critical for the biodegradation of the organic material present in WAS.

With regard to the question of why bacterial cells are relatively refractory, four hypotheses were advanced. From the preceding discussion it is apparent that oxygen is not a critical factor controlling degradability. From the literature [15–17], it appears that anaerobic organisms present at different temperatures (mesophilic vs thermophilic) do not affect biodegradability significantly. While this is perhaps not sufficient to dispel the second hypothesis, the last two hypotheses would seem to be the most plausible explanations for the low biodegradability of WAS.

The third hypothesis states that the refractory nature of the cells is due to the inaccessibility of organics to enzymes. Simple cleavage of the cell wall should make the internal components of cells more accessible and hence more biodegradable. Ultrasonic lysis of the cells did increase biodegradability slightly (Table VII). However, pretreatment at 150 and 175°C led to significantly higher biodegradation indicating other factors besides simple lysis may be more important for increasing biodegradability.

The ultrasonic lysis data, however, are not yet felt to be conclusive. Additional evaluations are necessary to be sure the procedures used resulted in efficient rupture of cell walls. However, if the data are accurate, it would appear that the refractory nature of bacterial nitrogen compounds is not due to protection by the cell walls alone, but also to their inherent structure. Thermochemical pretreatment caused substantial hydrolysis of the large nitrogenous macromolecules, resulting in a significant increase in biodegradability. However, this increase due to hydrolysis can be offset if subsequent reactions cause repolymerization of small molecules into either refractory or inhibitory materials.

CONCLUSIONS

Nonoxidative thermochemical pretreatment can lead to substantial increases in the mesophilic anaerobic biodegradability of WAS. This increase occurs in both the soluble and insoluble portions and is dependent on the temperature of treatment, and on the type of chemical used.

The effect of thermochemical pretreatment on WAS is to promote hydrolysis and to split complex nitrogen polymers into simpler constituent molecules. However, as treatment conditions become more severe, nitrogen-containing organic material is formed which is refractory under the conditions studied.

Thermochemical pretreatment causes the formation of toxic compounds, and this effect becomes more pronounced as treatment conditions become more severe. However, given sufficient time, methanogenic consortiums can acclimate to some extent to this inhibition.

Oxygen appears not to be a controlling factor in the degradation of WAS.

The refractory nature of nitrogenous materials in bacterial cells appears to be due in part to the protection afforded by the cell wall, but mostly to the inability of exocellular enzymes to hydrolyze complex nitrogenous macromolecules into simple biodegradable components.

This research was supported by the Department of Energy, Grant No. DOE-EY-76-5-63-0326-PA-44.

References

[1] G. L. Christopher and N. L. Krascella, Report M 911599-2, NSF Grant No. GI-34991 (July 1973).
[2] R. G. Kispert, S. E. Sadek, L. C. Anderson, and D. L. Wise, Dynatech Report No. 1258, NSF/RANN/SE/C-827/PR/74/5 (January 1975).
[3] R. T. Haug, D. C. Stuckey, J. M. Gossett, and P. L. McCarty, *J. Water Pollut. Control Fed.*, *50*(1), 73 (1978).
[4] M. Alexander, *Adv. Appl Microbiol.*, *7*, 35 (1965).
[5] S. E. Luria, in *The Bacteria*, Vol. I, H. Gunsalus and R. Y. Stanier, Eds. (Academic, New York, 1960), p. 13.
[6] A. L. Lehninger, *Biochemistry*, 2d ed. (Worth, New York, 1975).
[7] D. Roach and C. W. Gehrke, NASA Contract Report 110881, Univ. of Missouri (1969).
[8] J. E. Hodge, *Agri. Food Chem.*, *1*(15), 928 (1953).
[9] W. F. Owen, D. C. Stuckey, J. B. Healy, Jr., L. Y. Young, and P. L. McCarty (unpublished).
[10] American Public Health Association, *Standard Methods for the Examination of Water and Wastewater*, 13th ed. (USGPO, Washington, D.C., 1971).
[11] O. H. Lowry, N. J. Rosebrough, A. L. Farr, and R. J. Randall, *J. Biol. Chem.*, *193*, 265 (1951).
[12] R. I. San Lin and O. A. Schjeld, *Anal. Biochem.*, *27*, 473 (1969).
[13] H. Rosen, *Arch. Biochem. Biophys.*, *67*, 10 (1957).
[14] W. H. Hattingh, P. G. Thiel, and M. L. Siebert, *Water Res.*, *1*, 185 (1967).
[15] J. Maly and H. Farus, *J. Water Pollut. Control Fed.*, *43*(4), 641 (1971).
[16] J. F. Malina, in *Proceedings of the 16th Conference on Industrial Wastes* (Purdue University, Lafayette, Ind., 1961), p. 232.
[17] C. G. Golueke, *Sewage Ind. Wastes*, *30*, 1225 (1958).

Solids Recovery from Methane Fermentation Processes*

PAUL H. BOENING and JOHN T. PFEFFER
Department of Civil Engineering, University of Illinois, Urbana, Illinois, 61801

1. INTRODUCTION

Several systems for recovering energy from organic residues or from biomass crops are in various stages of development. The production of methane by an anaerobic microbial fermentation process is one of these systems. The organic material is slurried with water for addition to the fermentation tanks. After discharge from the fermentors, this slurry must be dewatered, recycled, or returned to the land. This paper will discuss the results of studies of the characteristics of residual solids from methane fermentation of beef manure and corn stover.

This work was conducted in small pilot scale fermentation systems [1]. The residues used in this study were passed through a feed mill for size reduction before being slurried. This unit was equipped with four different-sized screens: 3.2, 6.3, 19, and 39 mm. The maximum particle size could be reduced to any of the above sizes. The final sizes used were 3.2 and 6.3 mm. The relatively dry feed along with the desired amount of water or recycle liquor was added to the slurry tank (ST). Figure 1 shows a schematic of one pair of reactors. A second pair operates in parallel. After complete wetting of the organic material it was discharged into mixed holding tanks (HT). The slurry was then fed to the reactors by progressing cavity pumps (P1) that were controlled by time switches (TS).

The fermentation reactors consisted of four completely mixed stainless-steel tanks with a total volume of 0.91 m^3 each. The liquid volume of each reactor was 0.775 m^3. Heating was accomplished by circulating hot water from a hot water heater through external jackets on each tank. A temperature controller (TC) activated a solenoid valve (SV) to maintain the desired temperature. A level controller (LC) activates the progressing cavity effluent pumps (P2), which discharged the fermented slurry into holding tank (T1) for any additional processing. Gas from the reactors was passed through wet test meters (M) for flow measurements.

* Work supported by U.S. Department of Energy Contract EY-76-S-02-2917.

Biotechnology and Bioengineering Symp. No. 8, 235–255 (1978)

0572-6565/78/0008-0235$01.00

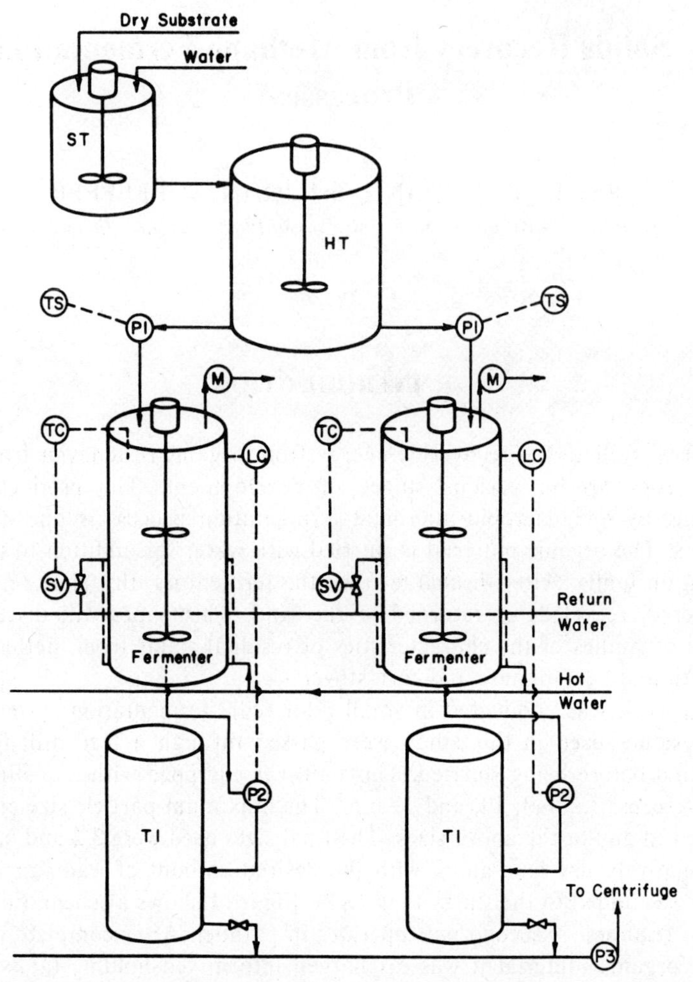

Fig. 1. Schematic of one set of reactors: slurry tank = ST, holding tank = HT, temperature control = TC, solenoid valve = SV, level controller = LC, feed pump = P1, effluent pump = P2, centrifuge feed pump = P3, gas meter = M, time switch = TS, and effluent tank = T1.

2. PARTICLE SIZE DISTRIBUTION

The particle size distribution of the fermented solids plays a major role in determining the solids recovery potential. Also, the composition of these different-size fractions will be important in evaluation of by-product recovery systems. The particle size determination was made using a wet sieving technique described in detail elsewhere [1]. A known volume of effluent slurry was passed through a series of sieves. The solids retained on each sieve were washed with a fixed volume of distilled water with the wash water being passed through each successive sieve. The sieves and the retained solids were oven dried, and the dry weight of the retained solids was determined. The solids passing through all of the sieves were determined by sampling the water collected from the sieves. Standard dry sieving techniques were utilized to determine the particle size distribution for the dry feed material after passage through either the 3.2- or 6.3-mm screen.

Typical size distribution for the feed and reactor effluent slurries are given in Table I. The reactor processing the manure was operated at 58°C and a 10-day residence time with a total solids content of 5.34%. The reactor processing corn stover was operated at 58°C and a 12-day residence time. The total solids content in the reactor was 3.44%. The raw manure slurry contained solids passed through a 6.3-mm screen. The total solids content of the manure slurry was 4.49%. The raw corn slurry solids were

TABLE I

Particle Size Distribution

Sieve Size	Percent of Solids Retained			
	Total Solids	Volatile Total Solids	Suspended* Solids	Volatile* Sus. Solids
	Reactor #1 Effluent (Manure Feed)			
10	12.1	14.7	25.4	29.5
20	21.9	26.6	46.0	53.2
50	35.0	41.4	74.5	82.6
100	40.6	45.4	85.3	90.6
200	47.6	50.1	100.0	100.0
Pan	100.0	100.0	--	--
	Reactor #3 Effluent (Corn Feed)			
10	61.4	74.0	78.0	83.0
20	65.8	78.0	83.0	87.0
50	72.4	82.0	92.0	94.0
100	76.3	85.0	97.0	98.0
200	79.1	90.0	100.0	100.0
Pan	100.0	100.0	--	--
	Raw Manure Slurry Mixing Tank #2			
10	34.0	35.5	59.6	61.2
20	38.7	41.0	67.9	85.2
50	49.6	52.3	87.0	86.1
100	53.8	57.0	94.4	93.5
200	57.0	60.3	100.0	100.0
Pan	100.0	100.0	--	--
	Raw Corn Slurry Mixing Tank #1			
10	41.7	52.0	75.9	77.9
20	46.2	57.8	84.1	86.6
50	51.2	58.0	93.1	93.8
100	53.4	60.4	97.2	97.4
200	54.9	62.1	100.0	100.0
Pan	100.0	100.0	--	--

* Suspended solids are considered to be those solids retained on the 200 mesh sieve

passed through a 3.2-mm screen and the slurry solids concentration was 6.73%.

These data show that a significant portion of the total solids and total volatile solids in the feed as well as the fermented slurry passed the 200-mesh screen. These solids represent small suspended and colloidal solids plus dissolved solids. There were some significant differences in the size distribution of these slurries. The proportion of raw manure that passed through the large screens was greater than for the raw corn stover, even though the corn was passed through the smaller (3.2-mm) screen. The proportion of large solids in the manure slurry decreased even more upon digestion. The opposite was true for the corn stover after digestion. The solids appeared to increase in size. This may have been due to the absorption of additional water during the relatively long residence time in the reactors.

The distribution of suspended solids as defined for this study is of more importance. The insoluble solids passing the 200-mesh screen (74 μm) will generally require chemical coagulation before they can be removed, which will greatly increase the cost of recovering these solids. The majority of the "suspended" solids can be removed as a relatively large particle. In all of the manure slurries tested, 85–90% of the solids retained on the 200-mesh screen were retained on the 100-mesh screen. With the exception of the reactors operated at long retention times (10 days at 58°C and 15 days at 40°C), the 50-mesh screen retained 90% of the "suspended" solids.

An even greater percentage of the fermented corn stover was retained on the large sieves. The 10-mesh sieve captures 80–90% of the "suspended" solids in the effluent from the reactors receiving corn stover. There were two distinct size ranges in the corn stover slurry. Approximately 42% of the total solids were retained by the 10-mesh screen. An additional 13% of the total solids were retained on the 200-mesh screen while 45% passed through the 200-mesh screen.

This relationship is shown graphically in Figure 2 for the manure slurries where the percent of "suspended" solids passing a given size screen is plotted against screen size. The dry manure, after passage through the mill with a 6.3-mm screen, was subjected to a dry sieve analysis resulting in the size distribution shown in Figure 2. When the manure was slurried with water, the size distribution changed substantially as shown by the curve for the feed slurry. This curve is for only those solids that are equal to or greater than the 200-mesh size. This substantial change in size distribution can be attributed in part to an increase in particle size due to absorption of water.

Figure 3 shows the particle size distribution of corn milled on both the 3.2- and 6.3-mm screen. The manure milled on the 6.3-mm screen is included as a comparison. The dry corn milled in the 6.3-mm screen is retained to a greater degree on the three largest mesh sizes. The majority of the wetted feed and digested effluent solids are either retained on the 10-mesh screen or pass the 200-mesh screen. Approximately 75% of the volatile

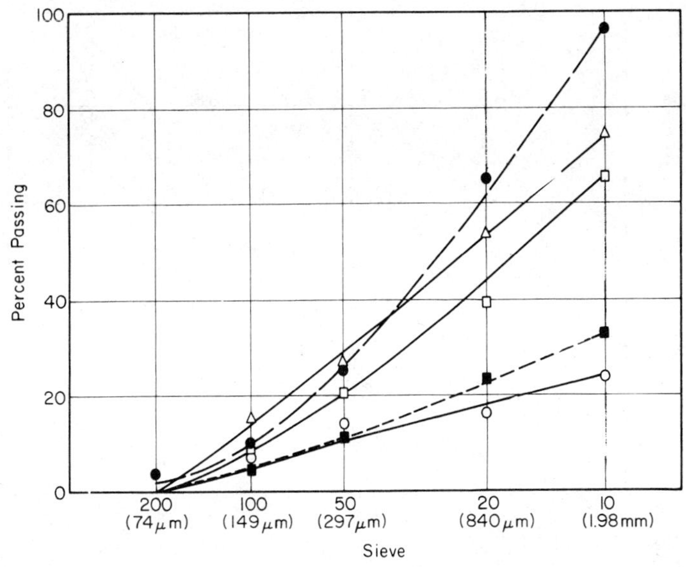

Fig. 2. Manure-particle size distribution at various process stages: (●) dry feed manure; (■) feed slurry; (△) θ = 10 day, 58°C; (□) θ = 10 day, 40°C; (O) θ = 2.5 day, 58°C.

total solids and of the "suspended" solids can be removed by 2.0-mm screen.

Figure 2 also illustrates the variation in size distribution resulting from increased stabilization. More organics were fermented to gas at the longer retention times resulting in smaller particles. Of the two curves marked at 10-day retention time, the upper curve was for a 58°C fermentation temperature, while the lower curve was for a 40°C temperature. A highly loaded thermophilic system as evidenced by the curve marked as a 2.5-day retention time shows a particle size distribution only slightly different from the feed slurry.

3. COMPOSITION OF DIFFERENT-SIZE FRACTIONS

The potential for by-product recovery depends on the composition of the residue from the fermentors. In particular, the composition of the various particle sizes will be of significance in establishment of the type of solids recovery system required. Samples for fiber analysis [2] and nitrogen analysis [3] were taken from the various size sieves after the particle size distribution had been completed. Additional analyses for calcium carbonate and iron [4] were conducted on the filtrate obtained from the cellulose hydrolysis step employed in the fiber analysis.

Table II shows the average of eight sets of analyses conducted on the effluents from reactors processing manure. The data in Table III are the

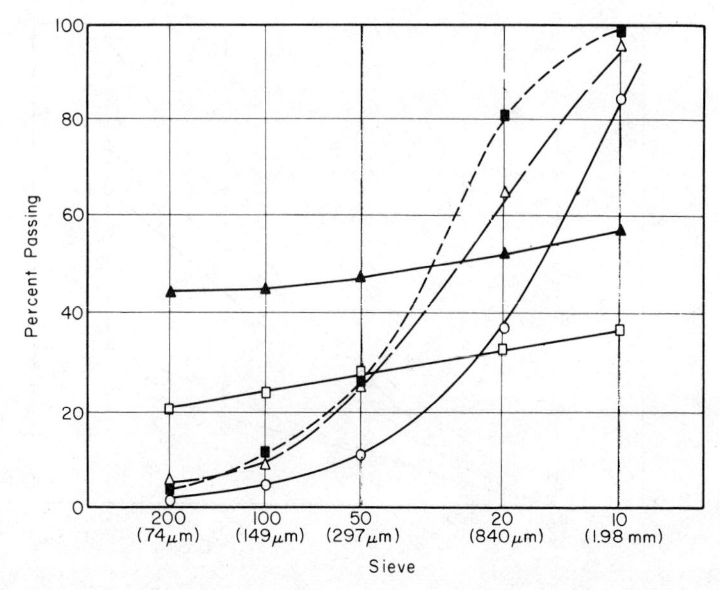

Fig. 3. Corn–particle size distribution at various process stages: (▲) corn feed slurry; (■) 3.2 mm dry corn feed; (△) 6.3 mm dry manure feed; (○) 6.3 mm dry cornfeed; (□) θ = 15 day, 58°C.

average values of two sets of analyses conducted on the manure feed slurry. The extractable material as measured by an alcohol–benzene extraction was approximately 1% in the reactor effluents. The composition of the remaining solids is generally considered to be holocellulose, lignin, and ash. In any sample containing nitrogeneous compounds, a correction for the protein content is applied. The protein correction is required because the acid hydrolysis of cellulose will cause the protein molecules to precipitate and be measured as lignin.

TABLE II

Manure Fiber Analysis—Manure Reactor Effluent

Sieve Size	Tot. Solids Retained on Sieve - %	Percent of Solids Retained on Sieve				
		Ext.	Cellulose	Lignin	Protein	Fixed Solids
10	27.2	1.1	54.2	14.6	1.4	30.0
20	8.6	1.4	53.8	9.2	1.8	34.7
50	8.8	1.3	47.8	9.7	2.6	46.1
100	4.7	1.1	43.6	17.3	2.1	55.4
200	4.5	1.0	43.2	12.4	1.5	51.1
Pan	46.4	---	----	----	---	----

TABLE III

Manure Fiber Analysis—Manure Feed Slurry

Sieve Size	Tot. Solids Retained Sieve - %	Percent of Solids Retained on Sieve				
		Ext.	Cellulose	Lignin	Protein	Fixed Solids
10	39.2	1.6	53.3	14.3	1.6	29.3
20	5.6	1.6	53.3	22.4	1.4	21.4
50	9.5	2.4	55.9	16.1	2.4	23.2
100	4.3	3.6	43.0	26.3	3.3	23.8
200	3.0	4.0	47.1	23.1	2.1	23.7
Pan	38.5	---	----	----	----	----

The fixed solids (ash) content determination was made on samples taken from the sieves. The fixed solids percentage was used to calculate the lignin by deducting this percentage from the solids remaining after cellulose hydrolysis. As can be seen from Table II, the sum of the percentage of individual components in the fiber frequently exceeded 100%.

An attempt to explain this problem that occurred in the reactor effluent slurry resulted in the following. The manure contained a quantity of crushed limestone that was added to various low areas in the feed lot. When this manure was milled, the limestone was reduced in size. As the manure was processed through the system, the large particles of limestone were removed; however, the finer particles passed through the system. This grit also caused wear on the various metal surfaces of the process equipment resulting in a measurable quantity of iron in the sludge. When the limestone and iron were subjected to the acid conditions used in the cellulose hydrolysis, the limestone and iron dissolved also. Analysis of the hardness and iron content of the liquid remaining after cellulose hydrolysis showed a significant calcium and iron content. These data are shown in Table IV.

TABLE IV

Manure Hardness and Iron Analysis—Percent Solids Retained on Sieve

Sieve Size	Feed Slurry		Reactor Effluent	
	$CaCO_3$	Fe	$CaCO_3$	Fe
10	0.9	0.3	1.1	0.4
20	1.4	0.3	2.5	0.6
50	2.4	0.3	2.7	0.8
100	2.8	0.4	4.6	0.9
200	7.2	0.4	5.2	1.0

TABLE V

Manure Corrected Fiber Analysis—Reactor Effluent

Sieve Size	Tot. Solids Retained on Sieve - %	Percent of Solids Retained on Sieve				
		Ext.	Cellulose	Lignin	Protein	Fixed Solids
10	27.2	1.1	52.7	14.8	1.4	30.0
20	8.6	1.4	50.7	11.7	1.8	34.7
50	8.8	1.3	44.3	5.7	2.6	46.1
100	4.7	1.1	38.1	3.3	2.1	55.4
200	4.5	1.0	37.0	9.4	1.5	51.1
Pan	46.4	---	----	---	---	----

The calcium and iron content represent a significant contribution to the cellulose content in the manure fiber analysis. The data in Table V have been corrected by deducting the calcium and iron from the cellulose and recalculating the lignin content. The data in this table show that most of the cellulose and lignin occur in the larger-size particles, generally in excess of 0.3-mm size. These data also show that the protein content of these solids is very low, less than 3.0%. Washing of these solids may have removed the microbial mass from the solids and caused these protein rich solids to pass through the 200-mesh screen.

Sufficient fiber analysis on the corn slurries have not been completed. Preliminary data show that the solids retained on the 10-mesh sieve are about 2% protein. Because most of the total solids are captured on the 10-mesh sieve or pass the 200-mesh sieve, the majority of the protein is captured on the 10-mesh sieve or is in the filtrate passing the 200-mesh sieve. Cellulose accounts for 45–55% of the solids on most sieve sizes, especially, those 0.3-mm and larger. This is the same range as for manure in Table V.

These data clearly show that for both slurries essentially all of the recoverable solids exist as particles greater than 0.3 mm. The ash content of these solids varied from 30 to 46.1%. The cellulose and lignin content decreased as the ash content increased with the smaller particles. In general, the protein content was less than 3%. It was, however, relatively uniform with particle size distribution. Based on these analyses, the most practical system for solids recovery appears to be a mechanical screen sized to capture 0.25-mm particles. The screened solids can then be pressed to remove additional water if desired.

4. CENTRIFUGE DEWATERING STUDY

The objective of the centrifuge dewatering study is to evaluate the efficiency of the centrifuge in solids capture and solids dewatering. A Sharpless

Mark III solid bowl–basket centrifuge with a variable speed drive was used to conduct the dewatering tests. This unit does not provide for continuous cake discharge; therefore, a predetermined amount of slurry which will ensure that the bowl is not overloaded with solids was fed to the centrifuge. The centrate retained in the bowl at the end of the run was withdrawn by the use of a skimmer pipe.

The results of the centrifuge tests on the fermented beef manure slurry are shown in Figures 4 and 5. Cake solids (Fig. 4) are a function of the centrifuge speed. Both the percent of total solids and volatile total solids are

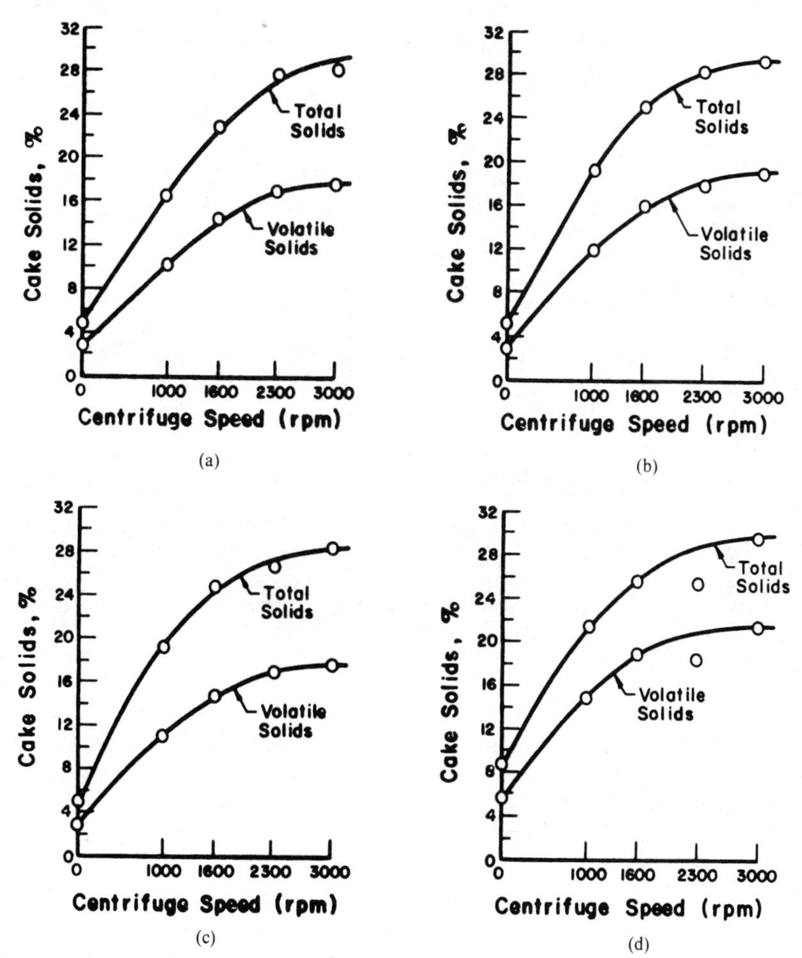

Fig. 4. Cake solids produced during centrifuge tests (manure slurries). (a) Reactor No. 1 effluent (6/15/77), volume centrifuged = 40 liter; (b) reactor No. 2 (6/14/77), volume centrifuged = 40 liter; (c) reactor No. 1 effluent (7/20/77), volume centrifuged = 40 liter; (d) reactor No. 4 effluent (7/27/77), volume centrifuged = 40 liter.

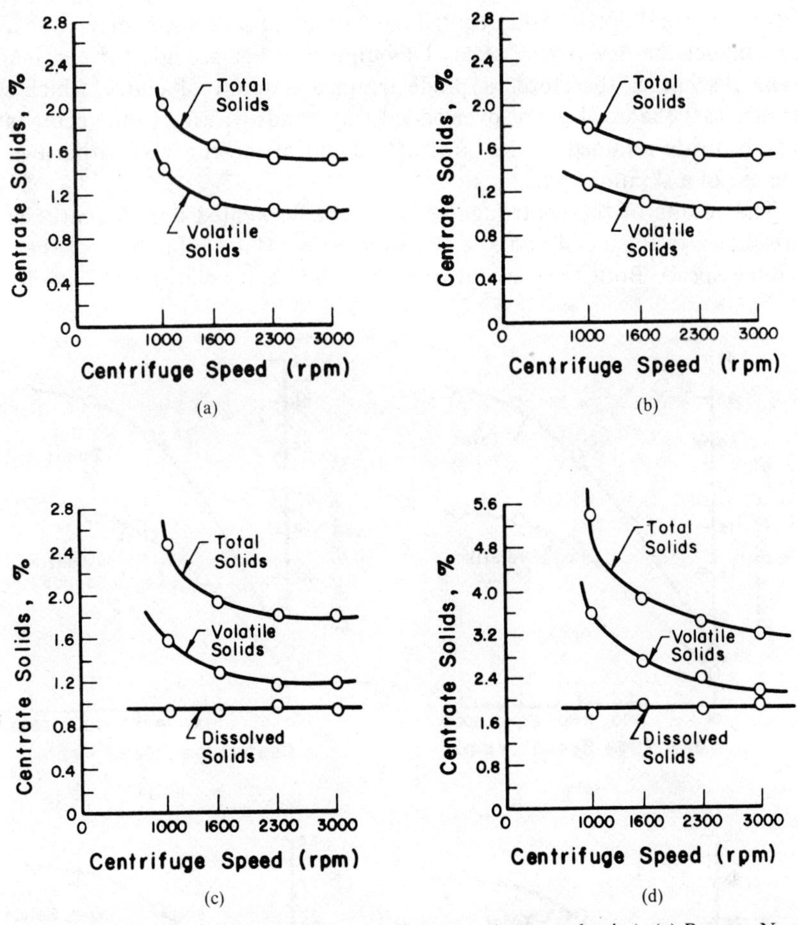

Fig. 5. Centrate solids produced during centrifuge tests (manure slurries). (a) Reactor No. 1 effluent (6/15/77); (b) reactor No. 2 effluent (6/14/77); (c) reactor No. 1 effluent (7/20/77); (d) reactor No. 4 effluent (7/27/77).

shown in this figure. At low speeds and *g* forces, the cake contained a considerable amount of water. At 1000 rpm (200 × *g*), the cake solids were found to be only 16–19% when the feed slurry solids were approximately 5.0%. With a speed of 2300 rpm (790 × *g*), the cake solids increased to between 22 to 27%.

The maximum cake solids were obtained at 3000 rpm (1700 × *g*) with a high value of 30%. The curves show that very little additional water was removed after the speed exceeded approximately 2300 rpm. At the higher speeds the cake was very dry and was not easily removed from the bowl. The fermentor retention time apparently was not a factor in the dewatering tests.

The results for a reactor receiving a feed slurry of approximately 9.0% solids are shown for Reactor 4 (Fig. 4). As one might expect, the solids capture increased with increasing speed. From inspection of this curve, it would appear that the optimum speed was again 3000 rpm or less, as with the 5.0% feed slurry.

The centrate percent solids are plotted against rotational speed for the same runs and are shown in Figure 5. Again, both total and volatile solids are plotted. With a 5.0% feed slurry, the lowest centrate solids percentage of 1.5 to 1.8% was obtained at the highest speeds. There was virtually no improvement in centrate solids capture after 2300 rpm. As with the cake solids, the fermentor retention time did not seem to influence centrate solids capture.

However, the concentration of the feed slurry has a pronounced effect on the centrate solids, as illustrated by Reactor 4 in Figure 5. Doubling the feed concentration has the effect of more than doubling the solids concentration at the lowest speeds, and increasing the values at the higher speeds by at least one-half. Inspection of Figure 5 would indicate that the optimum speed was not obtained. The maximum speed of the Mark III unit is 3400 rpm. Therefore, tests at higher speeds were not possible.

Similar centrifugation studies were conducted on the fermented corn stover slurry with the results shown in Figure 6. Reactor 1 was receiving a feed of corn stover that had been thermochemically pretreated with 10% by weight (dry solids) of lime at 115°C for 4 hr. The feed slurry for Reactor 3 was untreated. Both reactors were operating at 58°C and a 15-day residence time. The maximum cake solids were obtained at 2400 and 2600 rpm, the maximum speeds used on the samples from Reactors 1 and 3, respectively. It would appear that higher speeds would produce a drier cake. The thermochemical pretreatment significantly improves the dewatering characteristics of this slurry. Also, the slurry from the fermentation of the corn stover dewaters better in the centrifuge than does the beef manure.

There was a marked difference in the centrate solids from the two reactors. This may be due, in part, to the high solids content of Reactor 1 (6.06%) compared to Reactor 3 (3.01%). Also, the lime added for the pretreatment will increase the dissolved solids. The flat curves show that there is little possibility of improved solids capture with increasing centrifuge speed.

5. VACUUM FILTRATION STUDIES

A common technology for slurry dewatering involves filtration mechanisms ranging from simple screens to filter belt presses. One such system that has evolved is the vacuum filtration process. This process has been widely used to dewater slurries from many sources. A laboratory filter leaf test procedure was used for evaluation of the filtration characteristics of

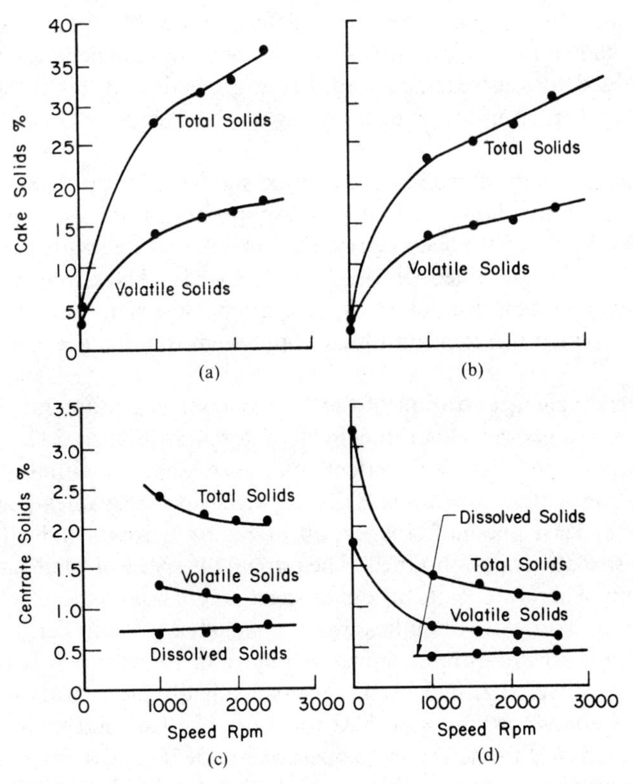

Fig. 6. Cake solids and centrate solids produced during centrifuge tests (corn slurries). (a) Reactor No. 1 effluent (4/4/78); (b) reactor No. 3 effluent (4/19/78); (c) reactor No. 1 effluent (4/4/78); (d) reactor No. 3 effluent (4/1978).

these slurries [5]. This procedure employs a section of filter cloth that is commercially available for installation on a vacuum filter. The test cloth is subjected to the same sequence of operations encountered in an operating vacuum filter. The data obtained from these tests can be used for predicting the response of the full-scale system.

The dewatering characteristics of the slurry can be changed by chemical conditioning. Various chemicals can be used to improve the filtration characteristics. Iron salts, especially ferric chloride, are effective in this chemical conditioning. The addition of organic polymers has also been found to be of assistance.

A series of tests using the filter leaf test procedure was undertaken using three different media as follows. These media were obtained from the Eimco BSP Division of Envirotech Corp.

PO-808: polyethelene monofilament yarn, 1/1 plain weave, 40 × 23 thread count at 10.5 oz/yd^2.

NY-415: nylon monofilament yarn, 1/1 plain weave, 40 × 40 thread count at 5.7 oz/yd^2.

POPR-859: polypropylene monofilament yarn, 2/2 twill weave, 68 × 30 thread count at 8.5 oz/yd^2.

The NY-415 was the most porous of the three media used, while POPR-859 had the tightest weave. As would be expected, the media plays a key role in determining the response of the slurry to this dewatering technique. The data shown in Table VI provide some interesting information on the filterability of the corn and manure slurries.

The slurry solids concentration varied somewhat, but in general was in the 40–60 g/liter range. For the media tested, POPR-859 produced a satisfactory filtrate quality, about 1.5 g/liter of suspended solids for the fermented manure slurries. Unfortunately, this required a massive chemical dose, and the resulting filter cake had a very high moisture level. The chemical flocculation of the fine particles results in a large quantity of water that is retained by the cake when removed by this medium. The same chemical treatment resulted in a cake of 21.1% solids for the PO-808 medium. However, the filtrate suspended solids concentration was significantly higher. The cake solids content for the PO-808 medium was 29.2% with lower chemical addition, but again, the filtrate suspended solids concentration was 10 g/liter.

The data show that higher cake solids can be obtained with limited chemical addition when using a coarse weave media. The data also show that the higher cake solids are produced at the expense of the solids capture. Solids capture is a mass balance expressed as the percentage of the total

TABLE VI

Filter Leaf Test Results[1]

Filter Media	Iron[2] g/l	Polyelec.[3] mg/l	Cake Solids %	Filtrate Sus. Solids - %
Manure				
POPR-859	7.0	10(A22)	14.6	0.15
PO-808	6.0	10(A22)	21.1	0.69
PO-808	2.1	0	29.2	1.01
PO-808	2.1	5(A23)	21.2	0.88
PO-808	2.1	10(A23)	18.0	0.44
NY-415	1.7	0	24.0	0.92
Corn Stover				
NY-415	0.0	0	17.4	0.75
NY-415	0.5	0	18.8	0.95
NY-415	11.7	0	22.1	0.94

[1] Form time = 30 sec, drying time = 120 sec.

[2] Iron Dosage as $FeCl_3 \cdot 6H_2O$.

[3] Polyelectrolyte—Dow Purifloc A22 and A23.

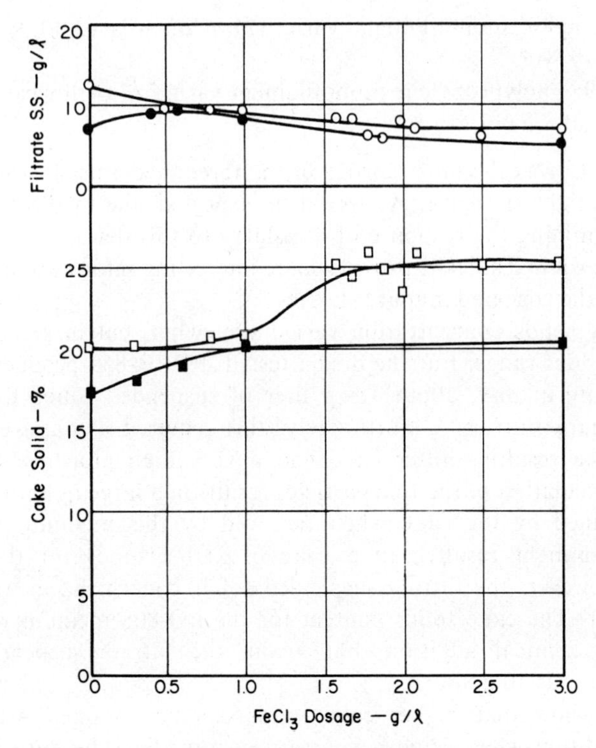

Fig. 7. Cake and filtrate solids obtained with filter leaf test: form time = FT, drying time = DT. Media NY-415; FT = 30 sec; DT = 120 sec; (■,●) corn; (□,O) manure.

solids retained on the filter out of the total solids applied to the filter. The filtrate-suspended solids are significantly higher when a drier cake is produced. Attempts to use fine media such as the POPR-859 without chemical conditioning were total failure. The medium clogged immediately, and it was not possible to pull any liquid through the test leaf.

As a result of several tests, it became apparent that the most effective filter media for solids dewatering would be a coarse medium like NY-415. The effect of ferric chloride on slurry filterability is shown in Figure 7. A manure cake solids of 20% was obtained without any chemical addition. A dosage of 1.7 g/liter of $FeCl_3 \cdot 6H_2O$ (68 lb per ton of dry solids) produced the maximum manure filter cake solids of 25%. The filtrate-suspended solids were reduced from 12.5 g/liter at 0 g/liter iron to 8 g/liter at a iron dosage of 1.7 g/liter.

The corn stover slurry filterability was similar to the manure slurry. Figure 7 shows the results of both slurries. The manure slurry produced higher cake solids for a given iron dosage. Because of the high concentration of alkalinity in the manure slurries (5–9 g/liter as $CaCO_3$), the pH shift resulting from the iron addition was minimal. The final pH for the conditioned manure slurry was in the 6.5–7.0 range. However, the corn stover slurry alkalinity was low. A significant pH shift occurred with iron addition. These data are shown in Table VII. The impact of pH on filterability of the corn stover slurry is being investigated. Adjustment of pH on the manure slurry was not investigated because of the high buffer capacity of this slurry. Chemical requirements would be excessive.

Solids capture is one criterion for evaluation of filter performance. The data in Figure 8 show the solids capture obtained with two media with various chemical dosage, form times and drying times. In this figure, the open symbols are for 30-sec form times while the closed symbols are for 15-sec form times. The thicker manure cakes produced with the long form time resulted in higher solids capture due to the deeper media filtration effect of the thicker cake. Longer drying times reduced the solids capture. This may have resulted from the extraction of more water and fine solids from the cake with the longer drying times and cake compression. High solids capture was possible but at a cost. Large chemical dosages and finer filter media yielded up to 97% solids capture.

The corn stover cake solids are about the same as the manure cake solids. The filtration rate is higher for corn, but the solids capture is lower. This is because no chemicals were used to capture the fines on the relatively coarse NY-415 media. The corn curve for 30-sec form time in Figure 8 shows poorer solids capture than the curve for 15-sec form time. This may be the result of cake compression after 15-sec of pulling fines through the medium, reducing solids capture.

The expected vacuum filter performance based on the filter leaf test analysis is shown in Table VIII. The finer medium, PO-808 provided a higher solids capture when a massive dosage of iron was added. Filtration rates were low and the cake moisture was low. The coarse medium NY-415 did not yield as high a solids capture, but the resulting cake was somewhat drier. The filtration rate was also better for NY-415 than for the finer medium.

TABLE VII

Effect of Iron Dosage on pH

Manure			Corn Stover		
Iron	pH		Iron	pH	
g/l	Initial	Final	g/l	Initial	Final
1.7	8.1	7.4	1.0	7.6	6.7
2.1	8.3	7.4	4.7	7.7	4.2
7.0	7.8	6.4	6.7	7.7	2.8

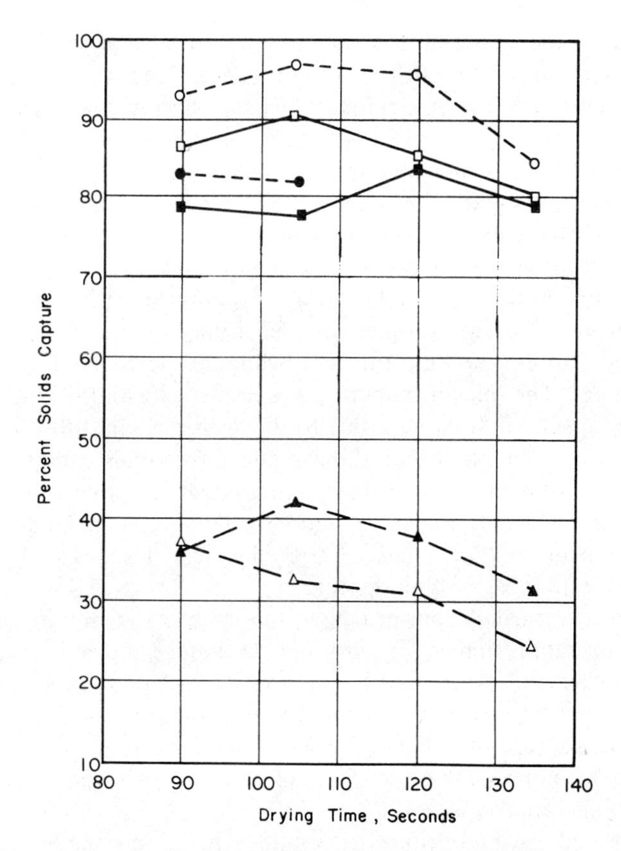

Fig. 8. Solids capture using filter test leaf procedure. (△,▲) NY-415, corn slurry; (○,●) $FeCl_3$ = 70 g/liter, A-23 = 10 mg/liter, PO-808; (□,■) $FeCl_3$ = 1.7 g/liter, NY-415, manure slurry.

Based on the results of this study, vacuum filtration does not appear to be a viable technique for processing these slurries. If dry cake and reasonable solids capture are the primary objectives, centrifugation will be more effective. If solids capture is not important, the larger solids can be removed from the slurries with screens. Presses can then be used to produce a dry cake. A vacuum filter with a coarse medium may also be effective if cake solids in excess of 20% are not required.

6. EVALUATION OF PROCESS RESIDUALS

The fermentation process removes only carbon, hydrogen, and oxygen from the substrate being processed. All forms of plant nutrients (nitrogen, phosphorus, potassium, and minerals) are conserved, and the nitrogen is partially upgraded to protein nitrogen in the form of bacterial cells (single cell protein). In order to evaluate the recovery potential of the residue from the system, one must know if these materials are soluble or insoluble and the associated concentrations. A series of tests was conducted to determine the nitrogen, phosphorus, and potassium content of the feed and effluent slurries and of the centrifuge cake. The ammonia and organic nitrogen in these fractions were also determined. The analytical techniques employed are presented in detail elsewhere [1].

A. Nitrogen Composition

A summary of the results of the nitrogen analysis is shown in Table IX. These numbers are the average of the analysis of four separate samples for the manure slurry and two separate samples for the corn stover. Fermenta-

TABLE VIII

Expected Vacuum Filter Performance

Form Time Sec.	Dry Time Sec.	Cake Solids %	Filter Loading $kg/m^2/hr$	Solids Capture %
NY-415, $FeCl_3$ - 1.7 mg/l Manure Slurry				
15	90	17.6	21.5	78.6
	105	17.0	17.6	77.4
	120	22.2	28.3	83.5
30	90	22.0	32.7	86.1
	105	21.9	28.8	90.2
	120	24.0	24.4	85.5
PO-808, $FeCl_3$ - 7 g/l A 23-10 mg/l Manure Slurry				
15	90	19.5	24.4	83.0
	105	21.3	18.1	82.0
30	90	19.4	25.9	93.5
	105	21.9	22.5	97.0
	120	22.3	18.6	96.0
	135	21.0	15.6	85.0
NY-415 Corn Slurry				
15	90	14.6	39.7	36.0
	105	15.9	55.6	42.6
	120	16.0	39.3	38.0
	135	16.2	22.3	31.6
30	90	15.0	47.2	37.3
	105	16.5	24.5	32.2
	120	17.4	28.8	31.2
	135	20.2	27.7	24.4

TABLE IX

Nitrogen Distribution, mg N per gram of Total Solids

Temp °C	θ Days	Organic Nitrogen Feed	Organic Nitrogen Eff.	Organic Nitrogen Cake	Ammonia Nitrogen Feed	Ammonia Nitrogen Eff.	Ammonia Nitrogen Cake
	Manure Slurry						
40	4.8	15.1	20.7	40.1	0.2	2	0.3
	9.5	18.1	16.9	39.7	1.5	3	0.2
	13.7	28.4	32.7	39.2	1.3	2.3	0.4
58	3.9	29.2	21.8	19.1	2.3	2.4	0.02
	4.7*	54.8	53.8	48.6	3.9	4.0	2.2
	4.9	26.8	21.3	21.3	1.6	2.2	0.01
	5.7*	44.4	53.9	-	4.5	5.5	4.0
	6.4	26.8	19.4	19.3	1.6	2.0	0.02
	9.5	29.2	28.8	22.1	2.3	2.9	0.03
	Corn Stover						
58	15.0	15.6	16.6	-	5.1	3.9	-
(6% lime)	15.0	6.3	13.8	-	3.5	4.1	-
(10% lime)	15.0	6.3	11.5	-	1.2	3.7	-

tion temperature and retention time do not appear to be significant in determining the nitrogen distribution for the manure. The impact of the manure quality greatly overshadows the effect of the above variables. The retention times identified by the asterisks in Table X show this effect. The feed manure during this period was relatively fresh, having a high nitrogen content and a large fraction of unstable organics.

Since supplemental nitrogen is added for the fermentation of corn stover, the nitrogen in the feed is controlled by the amount added as NH_4Cl or recycled along with the recycled liquid from the slurry dewatering step. It appears that the organic nitrogen content of the fermentor effluent is greater than the influent. This would be expected since bacterial protoplasm syn-

TABLE X

Phosphorus Distribution, mg P per gram of Total Solids

Temp °C	Total Phosphorus Feed	Total Phosphorus Eff.	Total Phosphorus Cake	Soluble Phosphorus Feed	Soluble Phosphorus Eff.	Soluble Phosphorus Cake
40	8.4	9.4	6.1	1.8	0.8	1.8
58	8.3	8.1	5.8	5.2	1.1	2.0

TABLE XI

Potassium Distribution—mg K per gram of Total Solids

Temp °C	Total Phosphorus			Soluble Phosphorus		
	Feed	Eff.	Cake	Feed	Eff.	Cake
40	8.4	9.4	6.1	1.8	0.8	1.8
58	8.3	8.1	5.8	5.2	1.1	2.0

thesis is occurring in the fermentor. However, the amount of organic nitrogen captured with the centrifuge cake is very low. The total Kjeldahl nitrogen in the cake was found to be less than 1.0 mg N per gram of total solids.

The organic nitrogen content of the centrifuge cake from the fermented manure slurry ranged between 1.91 and 4.86% of the total solids. The crude protein (6.25 N) was between 12 and 30%. This may have value for refeed. Because of the excessively low conversion of corn stover to methane, the resulting cell growth is minimal. The protein content of this centrifuge cake is less than 1%. Recovery as an animal food is impractical.

The effluent from the manure fermentation contained between 0.08 to 0.6% total nitrogen, depending on the raw manure quality and the feed slurry solids concentrations. The ammonia nitrogen content of this effluent ranged between 200 and 1000 mg/liter. The nitrogen fertilizer value of the entire effluent slurry is high. Solids recovery for refeed leaves an ammonia-rich water for disposal.

B. Phosphorus Distribution

A summary of the phosphorus analysis for the manure slurry is shown in Table X. The phosphorus distribution was independent of the fermentation time. Therefore, only the averages for the mesophilic and thermophilic fermentations are presented. The only apparent trend in the phosphorus data is the decrease in soluble phosphorus in the effluent. This suggests that the phosphorus was becoming insoluble in the fermentation step, but the particles were not of sufficient size to be removed by the centrifuge. In fact, the total phosphorus content of the solids in the centrifuge cake was significantly lower than the solids in the reactor effluents.

Phosphorus recovery from the manure fermentation will be greatest if the total effluent stream is utilized. Considering an 80% recovery of total solids in the centrifuge cake, the phosphorus recovery in this cake will be approximately 56.6%. The balance will be retained with the centrate.

Phosphorus is added to the corn stover feed slurry at the rate of 3.9 mg P per gram of total solids. This phosphorus is added in a soluble form, and the centrifuge studies show that 80–90% of the phosphorus is found in the centrate. The centrifuge cake contains less than 1 mg P per gram of total solids. Therefore, recycle of the centrate to reduce the quantity of phosphorus added would be the most practical application of this liquid residue.

C. Potassium Distribution

A summary of the potassium analysis for the manure slurries is given in Table XI. The distribution was independent of the fermenter retention time, so only the averages for mesophilic and thermophilic fermentation are presented. Most of the potassium in these samples was in the soluble form. Therefore, the potassium content of the centrifuge cake was very low. Effective utilization of the potassium from manure fermentation will require utilization of the liquid fraction of the slurry.

Only limited data have been collected on the potassium content of the corn stover slurry. As with the manure, most of the potassium is in the soluble form. Potassium is added along with the phosphorus (K_2HPO_4) to the corn stover slurry.

That the corn stover contains very little potassium is evidenced by the potassium content of the centrifuge cake (less the 1.0 mg per gram of total solids). Also, the potassium content of the various particle sizes greater than the 200-mesh sieve was generally less than 1 mg per gram of total solids.

7. SUMMARY

Conventional slurry dewatering processes can be applied to these slurries. High solids capture with vacuum filtration requires significant dosage of coagulating chemicals. However, the larger solids can easily be removed with a coarse filter media. Cake solids in excess of 20% can be obtained without chemical addition. Centrifugation produces a drier cake and achieves a reasonable solids capture without chemical addition. For relatively high solids capture and dry cake, a centrifuge appears to be the process of choice. If only limited solids capture is required, a simple screen with openings of approximately 0.25 mm can be used. The screen discharge can then be pressed to a very dry cake.

By-product recovery from the fermentation residue has limited potential. From manure fermentation, a cake with a crude protein content varying between 12 and 30% can be produced. The value of this cake as an animal feed is questionable since this protein is associated with indigestible fibers. The liquid (centrate) from this system has a high concentration of nitrogen (ammonia), phosphorus, and potassium. Land disposal can result in a fertilizer credit for either the centrate or the entire stream from the manure fermentation.

There appears to be no by-product recovery potential from the corn stover slurry. Solids recovery is necessary to recycle the liquid fraction. This reduces the cost of chemicals for nutrients and pH control and greatly reduces the quantity of residue for disposal. The protein content of the solids is very low. Therefore, the only use for these solids is return to the land. Since the plant nutrients in the initial raw material is very low, the fertilizer value is minimal, consisting primarily of chemicals added for the microbial nutrition.

References

[1] J. T. Pfeffer, *Biological Conversion of Biomass to Methane—Beef Lot Manure Studies*, U.S. Department of Energy Report No. COO-2917-9, Dept. of Civil Engineering, University of Illinois, Urbana, June 1978.

[2] American Society for Testing and Materials, *Annual Book of ASTM Standards*, Part 22, ASTM, Philadelphia, 1974.

[3] *Methods for Chemical Analysis of Water and Wastes* (U.S. Environmental Protection Agency, Washington, D.C., 1974).

[4] *Standard Methods for the Examination of Water and Wastewater*, 14th Ed., American Public Health Association, Washington, D.C., 1975.

[5] R. N. Hill, and M. M. Kaiser, *Eimco Filtration Manual* (Eimco BSP, Division of Envirotech, Salt Lake City, Utah, 1966).

Enhancement of Methane Production in the Anaerobic Digestion of Sewage Sludges

R. R. SPENCER

Battelle Pacific Northwest Laboratory, Battelle Boulevard, Richland, Washington 99352

INTRODUCTION

Recent renewed interest in the anaerobic digestion process as a source of methane has been a consequence of shortages and increased costs of natural gas. The technology and economics of resource recovery from sewage sludge digestion, as well as digestion of agricultural and municipal solid waste organics, have been the subjects of a number of research programs.

Although anaerobic digestion has a long history of use, it has a reputation for poor process stability and has never been universally accepted as an effective wastewater sludge handling technique. In addition, the rate of anaerobic digestion is relatively slow. For digesters operating in the mesophilic temperature range (30–37°C), a solids residence time (SRT) of 10–30 days is typically required at municipal facilities.

In an effort to improve the digestion process, Battelle Pacific Northwest Laboratory is currently conducting a research program for the U.S. Department of Energy assessing the effect of powdered activated carbon on the anaerobic digestion of sewage sludge. The primary objective is to develop a technique for increasing methane production in anaerobic systems. The impact of carbon on volatile solids destruction, sludge dewatering characteristics, and process stability is also being examined. As warranted, design data will be developed for application of the carbon addition process to full-scale systems. The program is being accomplished through a series of laboratory experiments, pilot studies, and field demonstrations.

This paper presents the results of two experiments involving carbon addition to bench-scale digesters. The effect of various doses of powdered activated carbon in stressed systems is evaluated. The effectiveness of carbon, flyash, and powdered coal in unstressed digesters is also assessed. Results of ongoing laboratory and pilot plant experiments, and future field studies, will be the topics of subsequent reports.

PREVIOUS WORK

The use of powdered activated carbon as an additive to anaerobic digesters dates back over 40 years. In 1935 Rudolfs and Trubnick [1]

Biotechnology and Bioengineering Symp. No. 8, 257–268 (1978)

0572-6565/78/0008-0257$01.00

reported on two field studies and a laboratory investigation involving the addition of carbon to digesters. All units were unmixed and operated at temperatures below the mesophilic range. It was found that the activated carbon accelerated the stabilization of poorly digesting sludge. Increases in gas production and a greater reduction in volatile solids were observed. A U.S. patent issued to Statham [2] in 1936 claimed similar benefits.

Following the 1930s, virtually no mention of the application of carbon to anaerobic treatment appears until the studies conducted by Adams [3, 4]. In a laboratory study, municipal sludge was digested in a batch treatment process. It was reported that after three weeks, total methane production in a digester dosed with carbon was five times that of a control unit. The optimum carbon dose was found to be approximately 5% of sludge solids. Three field studies described by Adams used about the same carbon dose. Increases in gas production and volatile solids destruction were observed, as well as higher percentages of methane in the digester gas.

More recently, Spencer [5] reported a two- to four-fold increase in methane production in stressed digesters operating at SRTs ranging from 2.5 to 10 days. A study conducted by ICI United States, Inc., [6], concluded that carbon addition can improve the operation of stress-loaded digesters, regardless of whether the stress comes from hydraulic overload or chemical toxicity. Furthermore, enhancement in gas production depends largely on the state of the digester.

Based on a review of the literature, it is evident that carbon is responsible for improved digestion under certain operating conditions. However, these studies do not constitute a thorough experimental evaluation of this phenomenon. None of the laboratory investigations involved the operation of unstressed digesters on a continuous or semicontinuous feed basis. Also, there is no agreement as to the mechanism(s) responsible for enhanced digestion. Several theories have been proposed: (1) carbon provides sites for the anaerobic reaction to occur; (2) carbon adsorbs toxic materials that may inhibit the digestion process; (3) alkaline carbons increase the buffering capacity of the digestion system; and (4) carbon contributes trace metal nutrients that are required by the anaerobic bacteria. It is apparent that further research is required to establish the action of carbon in anaerobic digesters and to efficiently apply the carbon addition process to actual practice.

RESEARCH APPROACH

Twelve bench-scale anaerobic digesters were used in the experimental studies. Each vessel consisted of a 4-liter glass reaction flask, having a working volume of 3.5 liters. The digesters were maintained at 35 ± 1°C by means of individually controlled heating mantles. In the first test, mechanical mixing devices were used to continuously stir the substrate. In order to eliminate occasional gas leakage problems encountered at the rubber seals where the mixer shafts entered the digester vessels, a gas recirculation mixing system was installed for the second study. Mixing rates were maintained at uniform levels for all digesters in each experiment. In all cases mixing was sufficient to keep the sludge well mixed. Gas produced in each digester was collected in calibrated plexiglass columns by displacement of an acidified, saturated salt solution. Gas measurements were corrected for atmospheric pressure and temperature.

The test digesters were fed on a semicontinuous basis: once per day when operating at SRTs of 10 days or greater. Additives, such as powdered activated carbon, were incorporated with the feed sludge. Immediately prior to feeding, an amount of substrate equal to the feed volume was withdrawn from each vessel. Substrate samples were routinely analyzed for pH, alkalinity, volatile acids, total solids, and volatile solids, as described in "Standard Methods" [7]. The composition of the digester gas was determined periodically by gas chromatograph analysis.

Feed sludge, procured from the Richland, Washington, sewage treatment plant, consisted of a mixture of raw and trickling filter sludge collected in the primary clarifier. Typically, this material had a total solids content of about 4%, of which approximately 75% was volatile solids. Digested seed sludge, used during start-up of the bench-scale digesters, was obtained from the plant's primary digester. Influent sewage to the Richland treatment facility originated almost entirely from domestic sources.

Carbon Dose in Stressed Digesters

This study evaluated the effect of carbon doses, ranging from 500 to 10,000 mg/liters, on "stressed" anaerobic digesters. Stressed systems are defined here as those that are functioning at less than normally expected levels of efficiency. Examples of operating conditions that may produce stress include hydraulic overloading, toxic materials in the feed substrate, and poor temperature control.

The conversion of volatile organic acids to methane gas is generally considered to be the most sensitive step in the anaerobic process. In a stressed system this step is usually inhibited, resulting in an accumulation of volatile acids in the digester. For the purposes of this research, sewage sludge digesters were considered to be significantly stressed when total volatile acid concentrations were in excess of 1000 mg/liter.

Nine bench-scale digesters were utilized in the experiment. All units were initially charged with 1500 ml digested seed sludge and 2000 ml primary sludge. Powdered activated carbon (Hydrodarco H–ICI United States, Inc.) was added to eight of the digesters to establish carbon concentrations from 500 to 10,000 mg/liter. An additional unit received no carbon and served as a control. The digesters were then operated at a 10-day SRT. Carbon concentrations were maintained at initial levels. The solution pH was controlled between 6.7 and 7.4 by the periodic addition of sodium bicarbonate ($NaHCO_3$).

After four weeks of operation, detailed operating data were collected over a 10-day period. Such parameters as gas production, gas composition, and total volatile acids were closely monitored. Owing to the high proportion of primary sludge initially in the digesters and a relatively short detention time (10 days), all of the units were stressed during the initial four-week period. Volatile acid concentrations were slowly decreasing, however, and ranged from 2090 to 4150 mg/liter when data collection was initiated.

Carbon, Coal, and Flyash in Unstressed Digesters

This experiment examined the effect of powdered activated carbon, coal, and flyash on relatively unstressed anaerobic digesters. The carbon was the same as that used in the previous study. Coal was a subbituminous variety, crushed so that 100% passed through a 70-mesh screen. Flyash was obtained from a coal-fired power plant.

Eight bench-scale digesters were initially filled with 1500 ml digested sludge and 2000 ml primary sludge. Four of the units were operated at a 10-day SRT, while the remaining four functioned at a 20-day SRT. After about 100 days, total volatile acids concentrations in nearly all of the units were 250–750 mg/liter. One digester at each detention time was then dosed with 1500 mg/liter carbon, another with 4000 mg/liter coal, and a third with 4000 mg/liter flyash. The final unit in each set served as a control.

The initially established additive concentrations were maintained during a 37-day data collection period. During this time, total volatile acids averaged less than 750 mg/liter in all digesters, except the 20-day SRT unit dosed with coal, which averaged 1220 mg/liter. Thus, for the most part, the digesters were not significantly stressed. The addition of chemicals for pH control was not required.

Volatile Acid Material Balance

Anaerobic digestion is normally considered to be a two-stage process. In the first stage, complex organics are converted to simpler forms by the action of a group of facultative and anaerobic bacteria called "acid formers." The primary end products of this step are volatile organic acids, the most common of which are acetic and propionic acids. Stabilization of the influent substrate occurs when the acids are transformed to methane and carbon dioxide by the "methane formers," which are strict anaerobes.

Based on the results of the two laboratory studies, it was desired to determine the effect of carbon on each of the two steps of the digestion process. Gas production data were available for the tests; however, there was no direct information regarding the rates of volatile acid generation. These were estimated by performing a volatile acid material balance on each digester over the duration of the data collection period. A schematic diagram of the material balance is shown in Figure 1.

To apply the material balance to the experimental data, it was necessary to make several assumptions. First, it was assumed that all of the methane produced in the digesters was generated via the acetic acid intermediate, as follows:

$$CH_3COOH + HCO_3^- \rightarrow CH_3COO^- + H_2O + CO_2\uparrow \quad (1)$$

$$CH_3COO^- + H_2O \rightarrow CH_4\uparrow + HCO_3^- \quad (2)$$

Equation (1) is a simple chemical neutralization reaction, in which acetic acid is rapidly converted to acetate. In Eq. (2) methane is biochemically produced from acetate by the action of methane forming bacteria. This second step is the slowest and therefore rate limiting. Based on the overall reaction, 1 mol CH_4 and 1 mol CO_2 are produced from each mol CH_3COOH. At standard conditions (0°C, 760 mmHg) this is equivalent to 0.37 liter CH_4 and 0.37 liter CO_2/g CH_3COOH. For the purpose of the material balance, exactly one-half of the volumetric gas production recorded in the experimental studies was considered to be CH_4 generated as shown in Eq. (2). Higher methane compositions that were actually observed were presumably due to the reduction of CO_2 to CH_4:

$$CO_2 + 4H_2 \rightarrow CH_4 + 2H_2O \quad (3)$$

Another assumption was that volatile acids were not present in the feed sludge. Typically, acid concentrations in the sludge from the Richland sewage treatment plant range from 500 to 800 mg/liter. At the 10 and 20 day SRTs examined in this study, the daily sludge feed volumes were relatively small—10 and 5% of digester volume, respectively. At these conditions, the total quantity of volatile acids in the feed would not be significant in comparison to the other material balance parameters listed in Figure 1.

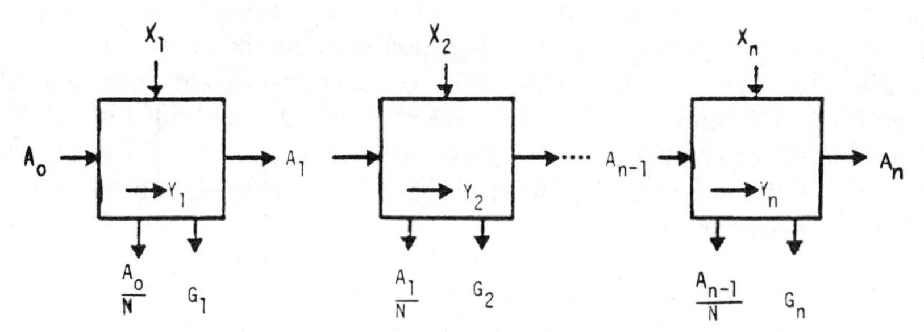

Fig. 1. Schematic diagram of volatile acid material balance. Note that X_n = volatile acids in feed sludge added during nth day, g; A_n = volatile acids in digester at the end of nth day, g; G_n = methane generated (in volatile acid equivalents) during nth day, g; N = mean solids residence time in digester, days; n = length of data collection period, days; Y_n = volatile acids generated in digester during nth day, g; A_n/N = volatile acids removed from digester with discharged sludge during nth day, g.

Thus, the exclusion of this value from the acid balance would have little effect on the results. Omitting the X_n term in Figure 1, the following expression is obtained:

$$Y_{\text{avg}} = \frac{\sum_{n=1}^{h} A_{n-1}}{(n)(N)} + \frac{\sum_{n=1}^{n} G_n}{n} + \frac{A_n - A_0}{n} \tag{4}$$

where Y_{avg} is the average daily production of volatile acids during the data collection period, g/day.

Values of Y_{avg} were calculated for the experimental data using Eq. (4). Average daily methane production (G_{avg}) was considered to be equal to 50% of the observed rate of total gas production.

Based on Y_{avg}, G_{avg}, and L_{avg} [average volatile solids (VS) loading rate, g/day] the efficiencies of the acid-forming and methane-forming steps were estimated. A volatile acid formation efficiency factor (E_a) and a methane formation efficiency factor (E_m) were defined as follows:

$$E_a = Y_{\text{avg}}/L_{\text{avg}}$$

where E_a is dimensionless, Y_{avg} is in grams of HAc per day, and L_{avg} is in grams of VS per day;

$$E_m = G_{\text{avg}}/Y_{\text{avg}}$$

where E_m is dimensionless, G_{avg} is in liters of CH_4 per day, and Y_{avg} is in equivalent liters of CH_4 per day. E_a provides an indication of the efficiency of the conversion of volatile solids to organic acids, while E_m represents the efficiency with which volatile acids are transformed to methane. At a theoretical steady-state condition in which all volatile acids are converted to methane, $E_m = 1$. If all the volatile feed substrate is converted to volatile acids, $E_a = 1$.

Clearly, discretion must be used when applying the efficiency factors to experimental data. The factors are merely indicators of process efficiency and may have little quantitative meaning in themselves. The research presented in this paper compared digestion systems that were operating in parallel at the same loading rates, solids residence times, and environmental conditions. The only variables were the type of additive and dose. In this situation, the calculated efficiency factors should provide an indication as to the step of the anaerobic digestion process being effected, and the relative extent of the enhancement.

RESULTS

Effect of Carbon in Stressed Digesters

The impact of carbon on gas production in stressed anaerobic digesters is illustrated in Figure 2. All units were operated at a 10-day SRT and an average loading rate of 2.7 kg VS/day/m^3 (0.17 lb VS/day/ft^3). A trend of

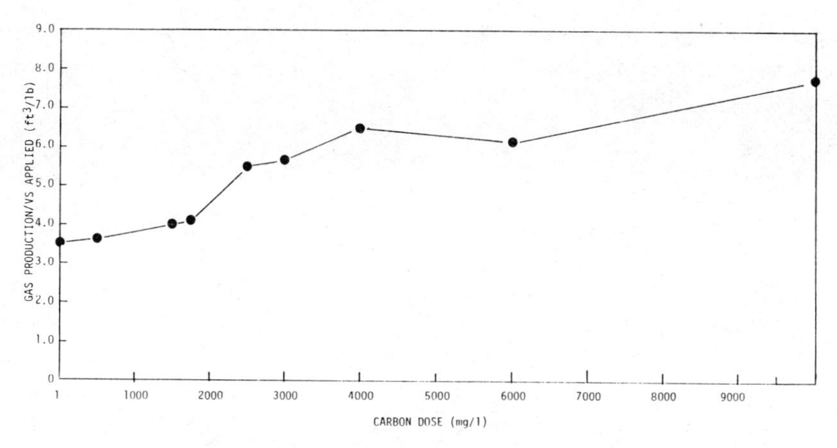

Fig. 2. Average gas production per volatile solids applied at various carbon doses in stressed anaerobic digesters.

enhanced gas production with increasing carbon was evident over the entire dose range studied. Specifically, a maximum enhancement of over 100% was achieved with the addition of 10,000 mg of carbon per liter.

In addition to increasing total gas production, carbon was responsible for a higher methane content in the digester gas. As shown in Figure 3, a maximum of about 69% methane was observed at a dose of 6000 mg/liter, as compared with less than 57% for the control digester. In general, the quality of the off-gas improved with increasing carbon dose.

The relationship between volatile acid concentration and carbon dose is displayed in Figure 4. Process stability, as evidenced by lower volatile acids, was consistently enhanced with increasing levels of carbon. Average acid concentrations ranged from 3420 mg/liter in the control unit to 1120

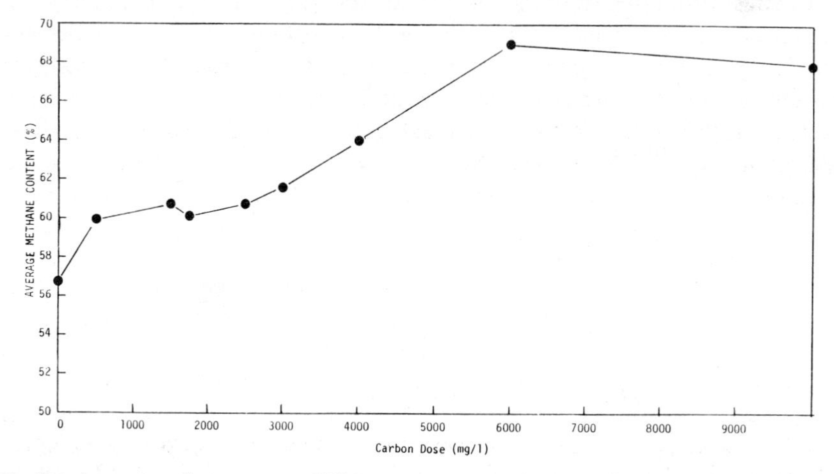

Fig. 3. Average methane content of digester gas at various carbon doses in stressed anaerobic digesters.

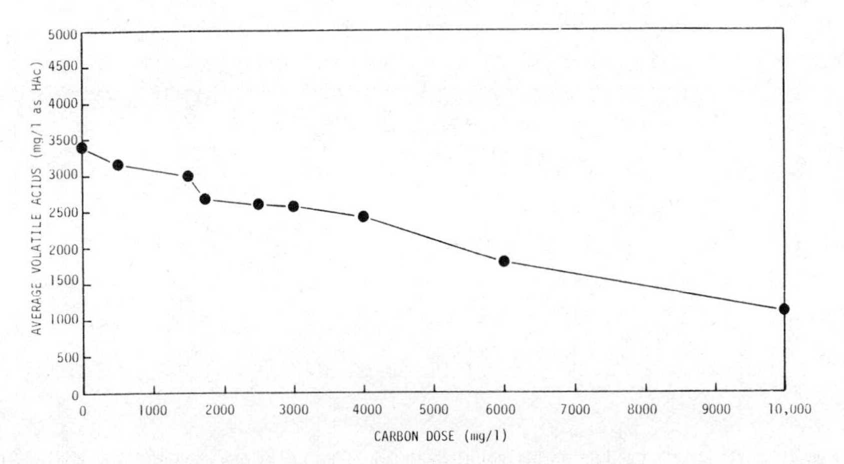

Fig. 4. Average volatile acids concentration at various carbon doses in stressed anaerobic digesters.

mg/liter in the highest dosed digester. At the end of the data collection period, volatile acids were less than 750 mg/liter in the three units containing the most carbon.

Important test parameters, including efficiency factors, are listed in Table I. From the table, the value of the methane formation efficiency factor (E_m) increases steadily with carbon dose. In fact, at higher concentrations the quantity exceeds unity, indicating that volatile acids were being consumed more rapidly than they were being produced. A rapid decrease in the volatile acid inventory was occurring in these digesters. Based on the results, it appears that carbon did enhance the methane-forming step of the digestion process.

The last column in Table I lists the volatile acid formation efficiency factors (E_a). Values range from 0.37 for the control unit to 0.61 for the digester containing 10,000 mg of carbon per liter. The data indicate that carbon was responsible for improved efficiency in the conversion of organic material to volatile acids. Thus, in this experiment both steps of the digestion process were apparently aided by the addition of carbon.

Effect of Carbon, Coal, and Flyash in Unstressed Digesters

A comparison of gas production in relatively unstressed digesters containing various additives is shown in Figure 5. All units were operated at a 10-day SRT and an average loading rate of 3.4 kg VS/day/m³ (0.21 lb VS/day/ft³). It can be seen that the carbon-dosed digester consistently produced more gas than the other digesters. Enhancement over the control unit averaged 12%. Average gas production in the coal, flyash, and control digesters varied from one another by only 2%. The digester gas in all the units consisted of 65% methane.

TABLE I

Test Summary for Stressed Anaerobic Digesters Operating at a 10-day SRT

Dig. No.	Carbon Dose (mg/l)	Initial-Final Average Volatile Acids (mg/l as HAc)	G_{avg} CH4 Prod. (liter/day)	Y_{avg} VA Prod. in CH4 eq. (liter/day)	E_m CH4 Form. Eff. Factor	Y_{avg} VA Prod. (gm/day as HAc)	L_{avg} VS Loading (gm/day)	E_a VA Form. Eff. Factor
1	None	4,150 - 2,690 (3,420)	1.08	1.34	0.81	3.58	9.66	0.37
2	500	3,760 - 2,670 (3,170)	1.08	1.35	0.80	3.63	9.66	0.38
3	1,500	3,620 - 2,380 (2,990)	1.22	1.45	0.84	3.87	9.66	0.40
4	1,750	3,420 - 2,180 (2,690)	1.23	1.42	0.87	3.79	9.66	0.39
5	2,500	2,990 - 2,130 (2,580)	1.62	1.85	0.88	4.96	9.66	0.51
6	3,000	3,220 - 1,620 (2,580)	1.69	1.81	0.93	4.85	9.66	0.50
7	4,000	3,330 - 630 (2,410)	1.92	1.88	1.02	5.05	9.66	0.52
8	6,000	2,240 - 340 (1,790)	1.85	1.84	1.01	4.93	9.66	0.51
9	10,000	2,090 - 90 (1,120)	2.31	2.20	1.05	5.88	9.66	0.61

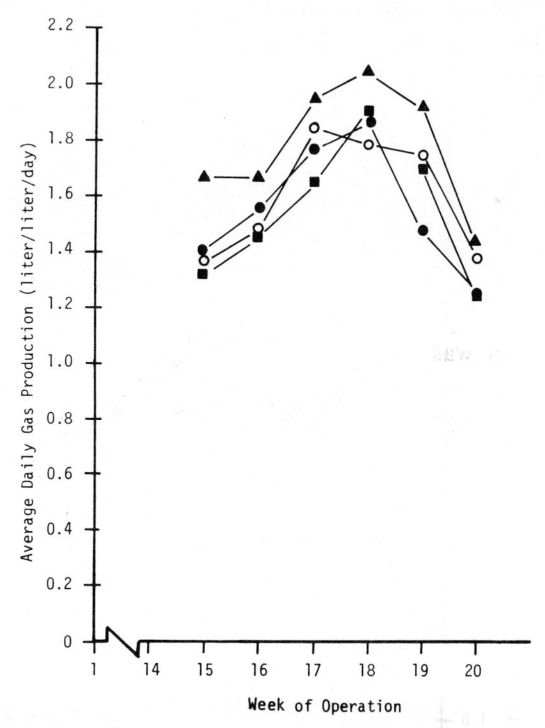

Fig. 5. Average daily gas production in unstressed digesters operating at a 10-day SRT: (○) control; (▲) 1500 mg carbon/liter; (●) 4000 mg coal/liter; (■) 4000 mg flyash/liter.

TABLE II

Test Summary for Unstressed Anaerobic Digesters Operating at a 10-day SRT

Dig. No.	Additive/ Dose	Initial-Final Average Volatile Acids (mg/l as HAc)	G_{avg} CH4 Prod. (liter/day)	Y_{avg} VA Prod. in CH4 eq. (liter/day)	E_m CH4 Form. Eff. Factor	Y_{avg} VA Prod. (gm/day as HAc)	L_{avg} VS Loading (gm/day)	E_a VA Form. Eff. Factor
1	None	740 - 120 (740)	2.85	2.92	0.97	7.82	11.7	0.67
2	4,000 mg/l Coal	733 - 143 (303)	2.80	2.82	0.99	7.55	11.7	0.64
3	4,000 mg/l Flyash	217 - 917 (458)	2.78	2.87	0.97	7.69	11.7	0.66
4	1,500 mg/l Carbon	250 - 190 (475)	3.19	3.25	0.98	8.70	11.7	0.74

A summary of the experimental data is presented in Table II. The methane formation efficiency factors (E_m) for the test digesters ranged between 0.97 and 0.99. The consistency of these values indicates that none of the additives had a significant effect on the conversion of volatile acids to methane. The volatile acid formation efficiency factors (E_a) were 0.64 to 0.67 for the coal, flyash, and control units. The value for the digester containing carbon was somewhat higher, 0.74. Thus, the increased gas production that was observed with carbon was probably attributable to enhancement of only the acid-forming step.

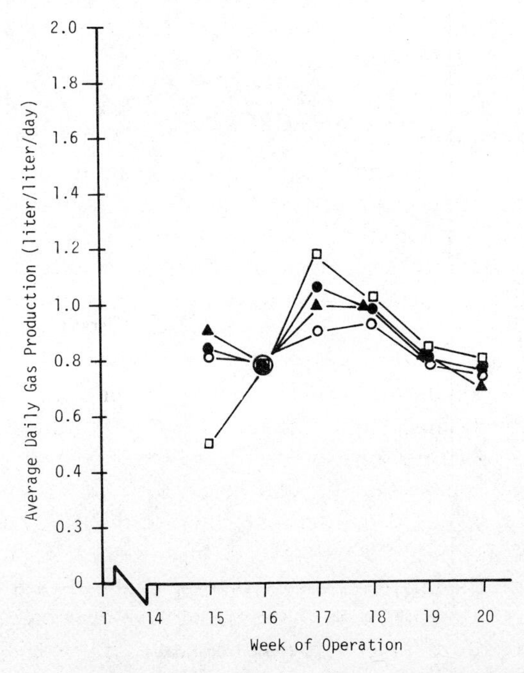

Fig. 6. Average daily gas production in unstressed digesters operating at a 20-day SRT: (○) control; (▲) 1500 mg carbon/liter; (●) 4000 mg coal/liter; (□) 4000 mg flyash/liter.

TABLE III

Test Summary for Unstressed Anaerobic Digesters Operating at a 20-day SRT

Dig. No.	Additive/ Dose	Initial-Final Average Volatile Acids (mg/l as HAc)	G_{avg} CH4 Prod. (liter/day)	Y_{avg} VA Prod. in CH4 eq. (liter/day)	E_m CH4 Form. Eff. Factor	Y_{avg} VA Prod. (gm/day as HAc)	L_{avg} VS Loading (gm/day)	E_a VA Form. Eff. Factor
5	None	594 - 570 (412)	1.46	1.52	0.96	4.06	5.83	0.70
6	4,000 mg/l Coal	411 - 309 (341)	1.54	1.58	0.97	4.23	5.83	0.73
7	4,000 mg/l Flyash	1,680 - 404 (1,220)	1.52	1.64	0.93	4.39	5.83	0.75
8	1,500 mg/l Carbon	433 - 285 (342)	1.54	1.58	0.97	4.22	5.83	0.72

Gas production data for relatively unstressed digesters operating at a 20-day SRT are displayed in Figure 6. (Note: The flyash-dosed digester experienced operational problems during the data collection period. Because average volatile acids were somewhat high, 1220 mg/liter, the experimental data for this unit are not truly representative of an unstressed digester.) All units functioned at a loading rate of 1.6 kg VS/day/m^3 (0.10 lb VS/day/ft^3). Average gas production varied by less than 6% among the four units. The fraction of methane in the off-gas was consistently 65–66%.

The values of E_m and E_a, along with other test information, are listed in Table III. The consistency of the efficiency factors among the test digesters indicates that none of the additives had an important effect on either the acid-forming or methane-forming steps.

CONCLUSIONS

A portion of the research evaluated the effect of powdered activated carbon on stressed anaerobic digesters utilizing a sewage sludge substrate. The addition of carbon resulted in increased methane production and greater process stability. The degree of enhancement appeared to be proportional to carbon concentration over the dose range studied (500–10,000 mg/liter). A maximum increase in methane production of about 150% was observed at the highest carbon dose.

The effect of 1500 mg/liter carbon, 4000 mg/liter coal, and 4000 mg/liter flyash on relatively unstressed digesters was also examined. Units using a sewage sludge substrate were operated at 10- and 20-day SRTs. A 12% increase in methane production was observed in a carbon-dosed digester functioning at a 10-day detention time. Enhancement was not evident with carbon at a 20-day SRT. No significant improvement in methane production was obtained in any of the digesters using coal or flyash as additives.

Using the experimental data, a technique was developed for estimating the efficiencies of the methane-forming and acid-forming steps in the anaerobic digestion process. The results indicated that in stressed systems both stages of the digestion process were enhanced by the addition of powdered carbon. In the relatively unstressed systems, when enhancement did occur, only the acid-forming step was affected. This information will supplement current research at Battelle Pacific Northwest Laboratory aimed at determining the mechanism(s) by which carbon enhances the digestion process.

Based on the results of this study, it appears that the benefits of carbon addition are greatest in stressed systems. Only very moderate increases in methane production would probably be attainable in well-operating digesters. Coal and flyash do not seem to be effective in enhancing gas production in unstressed systems. However, their effectiveness has not been tested in stressed situations.

This work was supported by funds provided by the U.S. Department of Energy, Contract EY-76-C-06-1830.

References

[1] W. Rudolfs and E. H. Trubnick, *Sewage Works J.*, *7*, 852(1935).
[2] N. Statham, U.S. Patent 2,059,286, November 3, 1936.
[3] A. D. Adams, *Water Sewage Works*, *122*(8), 46(1975).
[4] A. D. Adams, *Water Sewage Works*, *122*(9), 78(1975).
[5] R. R. Spencer, "The addition of powdered activated carbon to anaerobic digesters: Effects of methane production," M.S. Thesis, University of Washington, Seattle, 1976.
[6] *Improved Anaerobic Digester Performance with Powdered Activated Carbon*, Report No. 903-4, ICI United States, Inc., 1977.
[7] *Standard Methods for the Examination of Water and Wastewater*, 14th ed. (American Public Health Association, Washington, D.C., 1976).

Kinetics of Methane Fermentation

Y. R. CHEN and A. G. HASHIMOTO

U.S. Meat Animal Research Center, Science and Education Administration, U.S. Dept. of Agriculture, Clay Center, Nebraska 68933

INTRODUCTION

Recently, there has been an increasing interest in using anaerobic fermentation for stabilizing high strength residues such as municipal refuse and livestock residues. Anaerobic fermentation of these materials would produce methane and protein biomass [1, 2] and may offer a partial solution to disposal and pollution problems. Anaerobic fermentation kinetics must be understood in order to evaluate the potential of the anaerobic process for producing methane and protein biomass and to be able to analytically compare different anaerobic systems.

Anaerobic treatment is generally described as a two-step process [3]. In the first step, complex organic compounds are converted to less complex soluble organic compounds by enzymatic hydrolysis. These hydrolysis products are then fermented to simple organic compounds, predominantly volatile fatty acids, by a group of facultative and anaerobic bacteria called "acid formers." In the second step, the simple organic compounds are fermented to methane and carbon dioxide by a group of substrate specific, strictly anaerobic bacteria called "methane formers."

In this paper, the kinetic equations for methane fermentation will be derived, and the kinetic constants will be determined based on reported methane production data. Fermentation kinetics of sewage sludge, municipal refuse, and livestock residues will be compared.

FERMENTATION KINETICS

For a completely mixed continuous flow system without solids recirculation, the rates of change of the bacterial cell mass and substrate concentration are governed by the following two differential equations:

$$\frac{dM}{dt} = \mu M - \frac{M}{\Theta} \tag{1}$$

and

$$\frac{dS}{dt} = -\dot{F} + \frac{S_0 - S}{\Theta} \tag{2}$$

Biotechnology and Bioengineering Symp. No. 8, 269–282 (1978)
Published by John Wiley & Sons, Inc.

where M is the cell mass concentration (mass/volume), μ is the specific growth rate of the micro-organism (time^{-1}), $\dot{F}$ is the volumetric substrate utilization rate (mass volume^{-1} time^{-1}), S is the biodegradable effluent substrate concentration (mass/volume), S_0 is the biodegradable influent substrate concentration (mass/volume), Θ is the hydraulic retention time, and t is time. Most models used to describe biological waste treatment processes incorporate a bacterial decay term in Eq. (1) to account for disappearance of the bacterial mass through endogenous respiration and lysis. In this study, the bacterial decay term is not included in order to keep the model as simple as possible.

For a completely mixed system, the average solids retention time equals the hydraulic retention time.

The relationship between $\dot{F}$ and μ is given by

$$\mu = Yq = (Y/M)\,\dot{F} \tag{3}$$

where q is the specific substrate utilization rate and Y is the growth yield constant.

Under steady state, Eqs. (1)–(3) yield

$$\mu = 1/\Theta \tag{4}$$

$$\dot{F} = (S_0 - S)/\Theta \tag{5}$$

$$M = Y(S_0 - S) \tag{6}$$

Thus, the specific growth rate is numerically equal to the reciprocal of the average solids retention time. The volumetric substrate utilization rate can be calculated using Eq. (5).

We have derived a substrate utilization kinetic model to relate the specific growth rate to the substrate concentration. This kinetic equation is given by [4] the following:

$$\frac{\mu}{\mu_m} = \frac{S/S_0}{K + (1 - K)S/S_0} \tag{7}$$

where μ_m is maximum specific growth rate of the microorganism and K is a kinetic constant. This model shows that the μ_m occurs at washout (i.e., $\mu \to \mu_m$ as $S \to S_0$). Thus the minimum retention time is numerically equal to the reciprocal of μ_m:

$$\Theta_m = 1/\mu_m \tag{8}$$

Equation (7) also shows that μ is zero when substrate is not available.

For a completely mixed, continuous-flow system under steady state, Eqs. (4) and (7) give

$$\Theta = 1/\mu_m + (K/\mu_m)(S_0 - S)/S \tag{9}$$

Thus, μ_m and K can be graphically determined by plotting Θ vs $(S_0 - S)/S$, where the intercept is equal to $1/\mu_m$ and the slope is equal to K/μ_m.

When μ_m and K are determined for a particular substrate, S can be predicted by

$$\frac{S}{S_0} = \frac{K}{\Theta/\Theta_m - 1 + K} \tag{10}$$

Equation (10) shows that the effluent substrate concentration depends on the influent substrate concentration.

To study the kinetics of methane fermentation of complex substrates such as industrial, municipal, and animal residues, two different approaches are generally used: one approach is to find the rate-limiting substrate for the kinetic evaluation [5, 6]; another approach is using chemical oxygen demand (COD) or volatile solids concentration (VS) as an indicator of the substrate concentration [4, 7]. The later approach evaluates the overall process performance and is convenient to use. However, a portion of the COD or VS is not available to the microbes as substrate. The methods to determine the refractory portion and the kinetic constants using VS or COD as gross substrate were previously described in detail [4].

There are difficulties in using COD or VS as the gross substrate. The laboratory test for COD of high strength residues requires at least 100 times dilution which generally yields unreliable data. Also, some of the volatile acids in the effluent are volatilized during the VS determination. Because the volatile acids are precursors of methane fermentation, their volatilization during the VS determinations causes errors in the calculated amount of substrate utilized.

Methane production is directly correlated with COD reduction. Since no oxidizing agent is added, the only way COD reduction can occur is through the removal of organic material from the waste, such as through the evolution of methane and carbon dioxide. The other avenues of COD reduction through hydrogen sulfide and hydrogen gas evolution are insignificant [8].

A reduction of 1 g COD is equivalent to the production of 0.35 liters of methane at STP [8]. Knowing the COD loading to the fermentor and the volume of methane produced, the remaining COD in the digester can be calculated.

If B denotes the liters of CH_4 at STP produced/g COD added to the digester and B_0 is the liters of CH_4 at STP/g COD produced at infinite retention time, the biodegradable COD in the fermentor will be directly proportional to $B_0 - B$, and B_0 will be directly proportional to the biodegradable COD loading. From Eq. (10)

$$\frac{B_0 - B}{B_0} = \frac{K}{\Theta/\Theta_m - 1 + K} \tag{11}$$

or

$$B = B_0 \left[1 - \frac{K}{\Theta/\Theta_m - 1 + K} \right] \tag{12}$$

Equation (12) shows that, if $\Theta/\Theta_m > |1 - K|$, the plot of B vs. $1/\Theta$ should be a straight line with $B \rightarrow B_0$ as $\Theta \rightarrow \infty$. This method can be used to determine B_0 for each residue studied. Note that, when K is equal to unity, the inequality, $\Theta/\Theta_m > |1 - K|$, is always satisfied. Since B_0 is the liters of CH_4 produced per gram of COD added after an infinite retention time, it is the finally attainable methane production per gram of COD of the given substrate.

Equation (11) can be rearranged to give:

$$\Theta = \Theta_m + \Theta_m K[B/(B_0 - B)] \tag{13}$$

From Eq. (13), the plot of Θ vs. $B/(B_0 - B)$ yields a straight line with the intercept equal to Θ_m and the slope equal to $K\Theta_m$. Therefore, this equation can be used to determine the kinetic constants μ_m and K.

Since B is the methane production/g COD added, the volumetric methane production rate (γ_V) equals B times the loading rate:

$$\gamma_V = \frac{BS_{T_0}}{\Theta} = \frac{B_0 S_{T_0}}{\Theta} \left[1 - \frac{K}{\Theta/\Theta_m - 1 + K} \right] \tag{14}$$

where γ_V is in vol CH_4/vol digester/time and S_{T_0} is the total influent COD concentration. The maximum volumetric methane production rate $\gamma_{V_{max}}$ is obtained by taking the derivative of γ_V with respect to Θ and equating it to zero. So,

$$\gamma_{V_{max}} = \frac{B_0 S_{T_0}}{\Theta} \left[\frac{1 - K/(K + K^{1/2})}{1 + K^{1/2}} \right] \tag{15}$$

which occurs at

$$\Theta = \Theta_m(1 + K^{1/2}) \tag{16}$$

The COD/VS ratio for a given residue is generally constant. Since VS can be determined more easily and accurately than COD for concentrated complex substrates, it is more convenient to use quantities expressed in terms of VS. By defining the following constituents as follows:

B' = liters CH_4 at STP/g VS added

B_0' = liters CH_4 at STP/g VS added as $\Theta \rightarrow \infty$

S_{T_0}' = influent total VS concentration, g/liter

then Eqs. (11)–(16) will still be valid by substituting B', B_0', and S_{T_0}' for B, B_0, and S_{T_0}, respectively.

KINETIC CONSTANTS FOR COMPLEX SUBSTRATES

Table I gives the kinetic constants, $\gamma_{V_{max}}$, and B_0 (or B_0') for anaerobic fermentation of sewage sludge [5], municipal refuse [2], and livestock residues [7, 9, 10].

Equation (12) was used to determine B_0 (or B_0'). Figure 1 shows a plot of B vs. $1/\Theta$ for $\Theta \geq 10$ days. Since the plot of B (or B') vs. $1/\Theta$ was found to be linear for $\Theta \geq 10$ days, linear regressions were used to determine the intercept B_0 (or B_0'). The exception was the data of Bryant et al. [9] and Varel et al. [10], since the retention times they used were only 3, 6, and 9 days. The B' vs. $1/\Theta$ plot of their data did show fairly good linearity for $S_{T_0}' = 59$ and 78 g VS/liter for beef residue and $S_{T_0}' = 60$ g VS/liter for dairy residue. The average B_0' (0.280 liter CH_4/g VS) determined for $S_{T_0}' = 59$ and 78 g VS/liter for beef residue was assumed for all cases of beef residue fermented at 60°C, and B_0' (0.169 liter CH_4/g VS) determined for $S_{T_0}' = 60$ g VS/liter for dairy residue was assumed for all cases of dairy residue fermented at 60°C.

After determining B_0 (or B_0'), linear regressions of Θ vs. $B/(B_0 - B)$ were used to find Θ_m and K, according to Eq. (13). Linear regressions of the previously mentioned data of Varel [10] and Bryant [9] showed μ_m to be close for each data set. Since it is logical that μ_m is constant at the same tempera-

TABLE I

Kinetic Constants for Methane Fermentation of Complex Wastes

INFL. VS CON. (g/ℓ)	Temp (°C)	B_0 (ℓ CH_4/g COD)	B_0' (ℓ CH_4/g VS)	μ_m (day^{-1})	K	$\gamma_{V_{max}}$ (ℓ CH_4/ℓ/day)	Wastes and Data Source
18.4	35	0.227	0.350	0.33	0.26	0.93	Sewage Sludge
18.4	25	0.243	0.375	0.13	0.34	0.39	O'Rourke
18.4	20	0.263	0.406	0.114	1.03	0.21	(1968)
—	35		0.182[a]	0.33	0.30	0.75[b]	Municipal
—	40		0.299[a]	0.74	1.13	1.20[b]	Refuse
—	45		0.202[a]	0.77	1.26	1.03[b]	
—	50		0.256[a]	0.50	0.71	1.13[b]	Pfeffer
—	60		0.290[a]	0.63	0.68	1.62[b]	(1974)
34.9	32.5	0.215	0.271	0.28	.96	0.68	Dairy Waste
52.3	32.5	0.199	0.251	0.28	1.07	0.89	
69.8	32.5	0.187	0.236	0.33	1.56	1.08	Morris
87.2	32.5	0.176	0.222	0.33	1.39	1.35	(1976)
60	60		0.169	0.79	0.75	2.30	Dairy Waste
80	60		0.169	0.79	1.08	2.57	
100	60		0.169	0.79	1.62	2.58	
120	60		0.169	0.79	3.57	1.92	Bryant et al.
140	60		0.169	0.79	9.04	1.16	(1976)
59	60		0.280	0.77	0.87	3.41	Beef Waste
78	60		0.280	0.77	0.93	4.36	
97	60		0.280	0.77	1.86	3.74	Varel et al.
117	60		0.280	0.77	2.90	3.45	(1977)

[a] 61% CH_4 in the total gas assumed (based on results of Gossett et al. [13] for identical refuse)

[b] $S_{T_0}' = 18.4$ g/liter assumed.

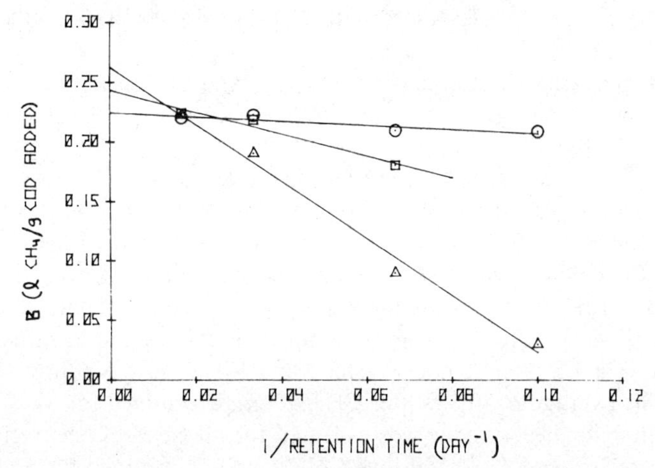

Fig. 1. Effect of retention time on methane production for sewage sludge ($\Theta \geq 10$ days). Sewage sludge O'Rourke data (1968): (⊙) 35°C; (⊡) 25°C; (△) 20°C.

ture and for the same substrate, μ_m was assumed to be constant for each data set. Values for K were then calculated using Eq. (13) and averaged for each S'_{T_0}. Table I shows that K is close to unity for $S'_{T_0} = 59$ and 78 g VS/liter for the beef residue and for $S'_{T_0} = 60$ g VS/liter for the dairy residue. Equation (12) shows that when K is close to unity, B vs. $1/\Theta$ approaches a linear relationship. This explains why these data sets were linear even for the short retention time and justifies their use for estimating B'_0.

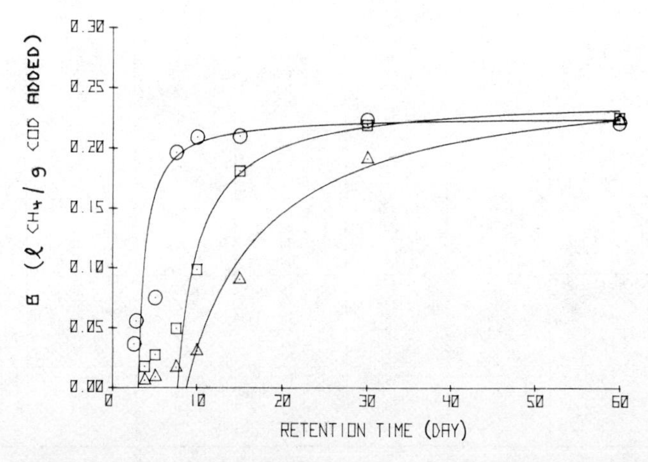

Fig. 2. Comparison of calculated methane production with experimental results for sewage sludge. Sewage sludge O'Rourke data (1968), influent concentration = 28.4 g COD/liter: (⊙) 35°C; (⊡) 25°C; (△) 20°C.

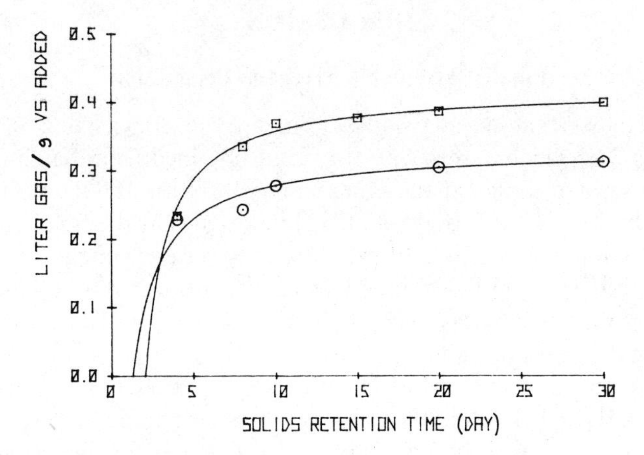

Fig. 3. Comparison of calculated gas production with experimental results for municipal refuse. Municipal refuse (Pfeffer data (1974): (⊙) 45°C; (⊡) 50°C.

Figures 2–4 graphically compare the predicted methane (or gas) production with experimental data. Correlations between the calculated curves and the laboratory data were generally good. The higher than predicted methane production at retention times shorter than the predicted minimum retention (O'Rourke's data) may be due to the presence of adhered growth on the fermentor walls [5]. This may occur when the mixing is inadequate.

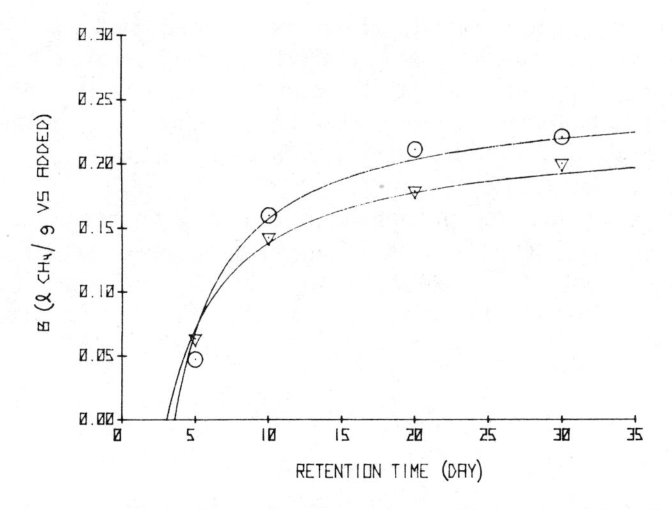

Fig. 4. Comparison of calculated methane production with experimental results for dairy residue. Dairy wastes (Morris data 1976): (O) 52.3 g/liter influent VS concentration; (▽) 87.2 g/liter influent VS concentration.

DISCUSSION

Finally Attainable Methane Production

Table I shows that sewage sludge fermented at 20°C has a B'_0 of 0.406 liter CH_4/g VS, which agrees with the result obtained from batch fermentation of the sewage sludge at the same temperature for 186 days reported by Malý and Fadrus [11]. Hatfield et al [12] reported that the finally attainable CH_4 production from batch fermentation of sewage sludge at 35.2 and 25.3°C were 0.385 and 0.375 liter CH_4/g VS, respectively, assuming that the volume fraction of CH_4 in the total gas is 68% [5]. These values agree with the results given in Table I.

Table I shows that the beef residues fermented at 60°C had a B'_0 equal to 0.280 liter CH_4/g VS, while the dairy residues fermented at 60°C had a B'_0 equal to 0.169 liter CH_4/g VS. Since the cattle used in these experiments were dairy breeds, it is probable that the difference in B'_0 was caused by the difference in ration. The beef steers were fed a high-grain finishing ration, while the dairy manure came from cattle being fed 72% roughage.

Pfeffer's data for municipal refuse showed that B'_0 varies with fermentation temperature (Table I). B'_0 is highest at 60°C (0.290 liter CH_4/g VS) and lowest at 35°C (0.182 liter CH_4/g VS). Assuming that the g COD/g VS = 1.39 for raw refuse [13], the B_0 at 35°C is 0.214 liter gas/g COD, which is close to the value of 0.190 liter gas/g COD reported by Gossett et al. [13] for municipal refuse at the same temperature.

Minimum Retention Time

Table I shows that, for municipal refuse and sewage sludge at 35°C, the minimum retention time is about three days. This agrees with the minimum retention time reported by Torpey [14] and Andrews et al. [15] for process failure at that temperature. This is also the same minimum retention time that Lawrence and McCarty [16] reported for methanogenic bacteria catabolizing volatile acids.

Figure 5 shows the effect of temperature on the maximum specific growth rate for the data from Table I. A postulated curve was drawn through the data and conforms to the generally accepted optimum temperatures of 40–42°C and 60–65°C for mesophilic and thermophilic microorganisms, respectively.

Kinetic Constant K

Results from dairy and beef residues showed that K varied with influent VS concentration (S'_{T_0}). Figure 6 shows the effect of the influent VS concentration on K for dairy and beef residues fermented at 60°C. The good agreement between dairy and beef residues as shown in Figure 6 may be due to the similar mixing employed in their experiments. Both used plat-

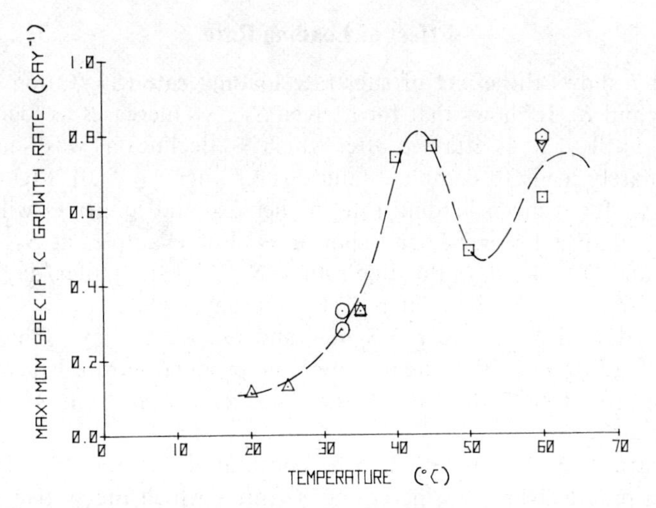

Fig. 5. Effect of temperature on maximum specific growth rate. (⊡) Pfeffer (1974); (△) O'Rourke (1968); (⊙) Morris (1976); (▽) Varel et al. (1977); (⬡) Bryant et al. (1976). Dashed curve is postulated curve.

form shakers to mix the fermentor contents [9, 10]. Figure 6 also shows that K increases gradually at low S'_{T_0}. However, for high S'_{T_0}, K increases rapidly as S'_{T_0} increases.

Figure 6 also shows that the dairy residue fermented at 32.5°C had a slightly higher K than the daily and beef residues fermented at 60°C. This difference may be due to the lower temperature or the minimal mixing (manual mixing, once a day) employed by Morris [7].

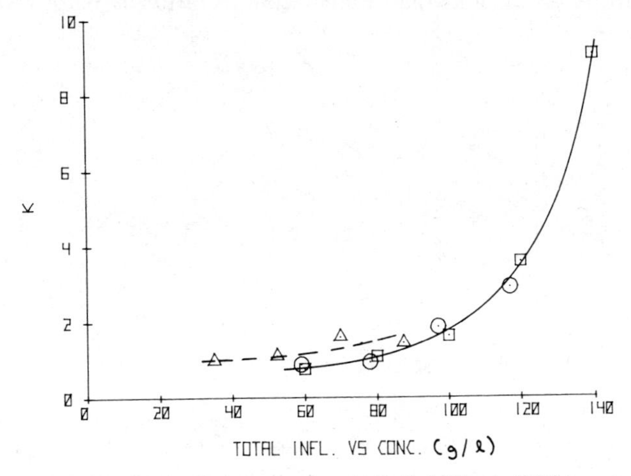

Fig. 6. Effect of total influent VS concentration on K. Solid line is 60°C curve; dashed line is 32.5°C curve. (⊡) Dairy wastes (Bryant et al., 1976); (⊙) beef wastes (Varel et al., 1977); (△) dairy wastes (Morris, 1976).

Effect of Loading Rate

Figure 7 shows the effect of substrate loading rate (S_{T_0}/Θ) on γ_V for a given μ_m and K. It shows that for a given S_{T_0}, γ_V increases as loading rate increases until $\gamma_{V_{max}}$ is attained after which γ_V declines as wash-out begins and ultimately leads to complete failure (i.e., when $\gamma_V = 0$). Figure 7 also shows that, for a given loading rate, higher S_{T_0} and longer Θ will yield a higher γ_V than a lower S_{T_0} and shorter Θ. For example, at $S_{T_0} = 70$ g VS/liter and $\Theta = 4.7$ days (loading rate $= S_{T_0}/\Theta = 15$ g/liter/day), Figure 7 shows that $\gamma_V = 3.14$ liter CH_4/liter/day, while $\gamma_V = 2.33$ liter CH_4/liter/day at $S_{T_0} = 40$ g VS/liter and $\Theta = 2.67$ days. Thus, at the identical loading rate, 35% greater methane production rate is achieved by increasing the total influent substrate concentration from 40 to 70 g VS/liter.

In Figure 7, K was assumed to be constant for various S'_{T_0}. However, there is a practical limit to increasing S'_{T_0} after which the system becomes overloaded. For example, Table I shows that, for dairy residue at 60°C, $\gamma_{V_{max}}$ increases from 2.30 to 2.58 liter CH_4/day as S'_{T_0} increases from 60 to 100 g VS/liter. However, as S'_{T_0} is increased further from 100 to 140 g VS/liter, $\gamma_{V_{max}}$ drops from 2.58 to 1.16 liter CH_4/liter/day. This suggests that there is a limit to $\gamma_{V_{max}}$. This limit may be due to reduced mass transfer at the higher concentration, or presence of substances at inhibitory concentrations.

Predicting Methane Production of a Pilot-Scale Fermentor

A pilot-scale thermophilic, anaerobic fermentor is in operation at the U.S. Meat Animal Research Center [1]. The working volume is 5.1 m³, temperature is 55°C, and the influent concentration is 65 g VS/liter. The

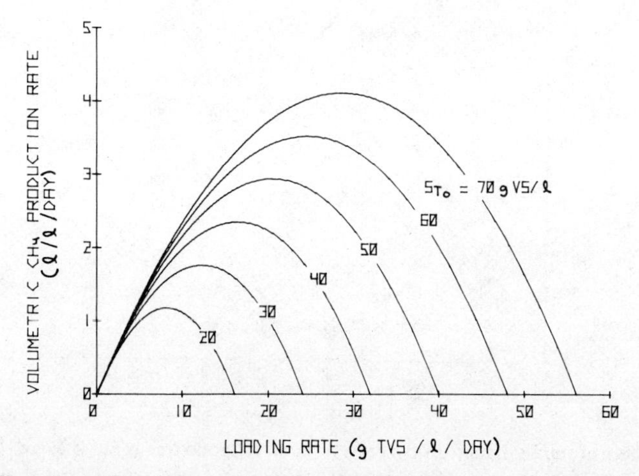

Fig. 7. Effect of loading rate on methane production rate. Data for beef cattle: $B_0 = 0.280$ liter CH_4/g VS; $K = 0.9$; $\mu_m = 0.8$ day^{-1}.

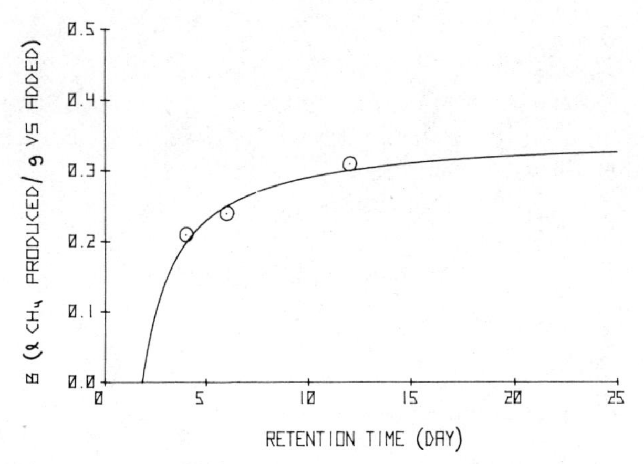

Fig. 8. Comparison of predicted methane production with result from a pilot-scale fermentor. Beef wastes: pilot scale digester (volume is 5.1 m^3); temperature = 55°C, B_0 = 0.35 liter CH_4/g VS. Solid line is predicted curve with μ_m = 0.55 day^{-1} (at 55°C) and K = 0.9.

maximum specific growth rate is estimated to be 0.55 day^{-1} from Figure 5. From Figure 6, K is estimated to be 0.9 at S'_{T_0} = 65 g VS/liter. B'_0 was experimentally determined to be 0.35 liter CH_4/g VS.

The methane production from the fermentor can then be predicted by using these values in Eq. (12). Figure 8 shows the actual methane production from the fermentor along with the predicted production from Eq. (12). The agreement between the predicted and actual values is excellent. Thus, the relationships shown in Figures 5 and 6 are useful for predicting methane production of other fermentors operating under similar conditions (i.e., temperature, influent concentration, mixing, types of substrates, etc.).

Predicting Effluent COD Concentration

Using the kinetic constants obtained from methane production data, the effluent COD concentration (S_T) can be calculated [4]:

$$S_T = S_{T_0}\left[R + \frac{(1-R)K}{\Theta/\Theta_m - 1 + K}\right] \qquad (17)$$

where R is the refractory coefficient. Figure 9 graphically compares the predicted effluent COD concentration with experimental effluent COD concentration for sewage sludge fermented at 35, 25, and 20°C. The refractory coefficient of 0.36 given by O'Rourke [5] was used in this calculation. Figure 9 shows good agreement between the predicted and measured COD concentrations. Thus, the kinetic constants not only predict the rate of methane production, but also the rate at which the substrate is utilized. This is important when substrate stabilization is of concern as in the digestion of sewage sludges.

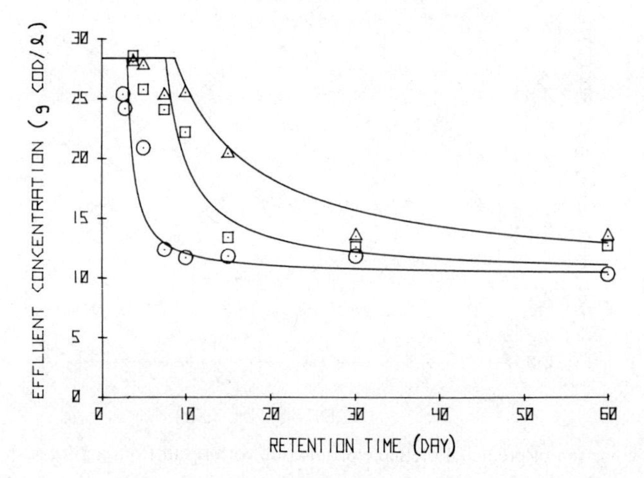

Fig. 9. Comparison of predicted effluent COD concentration with measured effluent COD concentration. Sewage sludge O'Rourke data (1968), influent concentration is 28.4 g COD/liter, refractory coefficient is 0.36: (⊙) 35°C; (⊡) 25°C; (△) 20°C.

CONCLUSIONS

The following conclusions can be drawn from this study:

Methane production data under steady state can be used to study the kinetics of the methane fermentation. The constants determined were maximum specific growth rate (μ_m), kinetic constant (K), and finally attainable methane production (B_0 or B_0').

B_0 (or B_0') varies with temperature and type of substrate. B_0' for municipal refuse fermented at 60°C was 60% higher than that at 35°C. Residue from the animals fed high roughage diets had lower B_0' than residue from animals fed higher energy rations. The B_0' obtained for sewage sludge at 35, 25, and 20°C agreed with those previously reported for batch fermentation of the sewage sludge at similar temperatures.

The maximum specific growth rate varies with temperature. Optimum values for μ_m were postulated at 40–42°C for mesophilic bacteria and 60–62°C for thermophilic bacteria.

When S'_{T_0} is low, K is practically constant. However, when S'_{T_0} is high, K increases rapidly as S'_{T_0} increases.

Increasing substrates loading rate increases methane production rate until the maximum utilization rate is achieved. For a given loading rate, higher influent concentration and longer retention time yields higher methane production rate than lower influent substrate concentration and shorter retention times. However, there is a practical limit to the loading rate and influent substrate concentration where the system becomes overloaded.

The kinetic constants obtained from laboratory studies were successfully used to predict the results of a pilot-scale fermentor.

Nomenclature

B	liter CH_4 (at STP) leaving the fermentor/g COD added
B'	liter CH_4 (at STP) leaving the fermentor/g VS added
B_0	liter CH_4 (at STP) produced/g COD as $\Theta \to \infty$
B_0'	liter CH_4 (at STP) produced/g VS as $\Theta \to \infty$
COD	chemical oxygen demand (g/liter)
$\dot{F}$	substrate utilization rate (mass/vol/time)
K	kinetic constant
M	cell mass concentration (mass/vol)
R	refractory coefficient (dimensionless)
S	biodegradable digester or effluent substrate concentration (mass/vol)
S_0	biodegradable influent substrate concentration (mass/vol)
S_T	effluent total COD concentration (mass/vol)
S_{T_0}	influent total COD concentration (mass/vol)
S_{T_0}'	influent total VS concentration (mass/vol)
VS	volatile solids concentration (mass/vol)
Y	growth yield constant (g cell mass/g substrate)
q	specific substrate utilization rate (mass/vol/time/cell mass concentration)
t	time
Θ	hydraulic or average solids retention time (time)
Θ_m	minimum hydraulic or average solids retention time (time)
μ	specific growth rate of microorganism ($time^{-1}$)
μ_m	maximum specific growth rate of microorganism (at $S = S_0$) ($time^{-1}$)
γ_V	volumetric methane production rate (vol/vol/time)
$\gamma_{V_{max}}$	maximum volumetric methane production rate

References

[1] A. G. Hashimoto, Y. R. Chen, and R. L. Prior, *Proceedings, Energy from Biomass and Wastes* (IGT, Chicago, 1978), p. 379.

[2] J. T. Pfeffer, *Biotechnol. Bioeng.*, *16*, 771 (1974).

[3] P. L. McCarty, *Public Works* (September–December), 107 (1964).

[4] Y. R. Chen and A. G. Hashimoto, "A model for substrate utilization kinetics," *Trans. Am. Soc. Agr. Eng.*, to be published.

[5] J. T. O'Rourke, "Kinetics of anaerobic waste treatment at reduced temperatures," Ph.D. dissertation, Stanford University, 1968.

[6] P. L. McCarty, *Principles and Applications in Aquatic Microbiology*, H. Henkelekian and N. C. Donodero, Eds. (Wiley, New York, 1964), p. 314.

[7] G. R. Morris, "Anaerobic fermentation of animal wastes: A kinetic and empirical design evaluation," M.S. thesis, Cornell University, 1976.

[8] P. L. McCarty, "Anaerobic Process," presented at the Birmingham Short Course on Design Aspects of Biological Treatment, Int'l Assoc. of Water Pollution Research, Birmingham, England, 1974.

[9] M. P. Bryant, V. H. Varel, R. A. Frobish, and H. R. Isaacson, *Seminar on Microbial Energy Conversion*, H. G. Schlegel, Ed. (E. Goltz KG, Göttingen, Germany, 1976).

[10] V. H. Varel, H. R. Isaacson, and M. P. Bryant, *Appl. Environ. Microbiol.*, *33*, 298 (1977).

[11] J. Malý and H. Fadrus, *J. Water Pollut. Control Fed.*, *43*(4), 641 (1971).

[12] W. D. Hatfield et al., *Ind. Eng. Chem.*, *20*, 174 (1928).

[13] J. M. Gossett, J. B. Healy, Jr., W. F. Owen, D. C. Stuckey, L. Y. Young, and P. L. McCarty, "Heat treatment of refuse for increase anaerobic biodegradability,"

NSF/RANN/SE/AER-74-17940-A01/PR/76/2, Technical Report No. 212, Civil Engineering Department, Stanford University, 1976.

[14] W. N. Torpey, *Sewage Ind. Waste*, *27*(2), 121, (1955).

[15] J. F. Andrew, R. D. Cole and E. A. Pearson, "Kinetics and characteristics of multistage methane fermentations," SERL Rept. 64-11, University of California, Berkeley, 1964.

[16] A. W. Lawrence and P. L. McCarty, "Kinetics of methane fermentation in anaerobic waste treatment," Tech. Rept. No. 75, Stanford University, 1967.

Solar Energy Conversion with Hydrogen-Producing Cultures of the Blue-Green Alga, *Anabaena cylindrica*

PATRICK C. HALLENBECK,* LEON V. KOCHIAN, JOSEPH C. WEISSMAN,* and JOHN R. BENEMANN
Sanitary Engineering Research Laboratory, College of Engineering, University of California, Berkeley, California 94804

INTRODUCTION

The production of hydrogen from water by solar energy can be accomplished through a variety of multistep chemical or physical processes (e.g., photovoltaic cells followed by electrolysis). A catalyst capable of carrying out the decomposition of water in a single-step process would be desirable. Two key problems in developing such catalysts are the relatively low energy content of the solar spectrum and the rapid recombination of "nascent" hydrogen and oxygen molecules. The reativity of oxygen with activated hydrogen (in a diffusion-controlled reaction) does not allow the operation of a water-splitting–oxygen-evolving reaction in direct proximity of the hydrogen-producing reaction. Photosynthesis—biological solar energy conversion—utilizes a two-photon process in series to accomplish water splitting to yield oxygen and reduced ferredoxin, which is capable of producing hydrogen in the presence of a hydrogenase enzyme. A study of such a chloroplast–ferredoxin–hydrogenase reaction [1] failed to demonstrate simultaneous evolution of hydrogen and oxygen and indicated a strong inhibition of hydrogen formation by the oxygen evolved during the process. Only by the additional use of oxygen-consuming reactions was the rate of hydrogen evolution comparable to normal photosynthetic reaction rates [1, 2].

Heterocystous blue-green algae consist of at least two types of cells—the heterocysts and vegetative cells—typically arranged in long filaments. Numerous studies [3–8] have demonstrated that heterocysts contain the nitrogenase enzyme system responsible for nitrogen fixation in these organisms. The ability of nitrogenase to catalyze hydrogen evolution in the absence of molecular nitrogen, or in the presence of carbon monoxide, is well known [9]. Oxygen and hydrogen evolution are separated at the microscopic level in heterocystous blue-green algae; *Anabaena cylindrica* was used for the demonstration of simultaneous production of hydrogen and oxygen [10].

* Graduate Group in Biophysics, University of California, Berkeley, California.

Biotechnology and Bioengineering Symp. No. 8, 283–297 (1978)

0572-6565/78/0008-0283$01.00

The lack of oxygen inhibition of the nitrogenase-catalyzed hydrogen evolution is thought to be because of the inability of oxygen to diffuse rapidly through the heterocyst cell wall and its uptake due to the relatively high respiratory rate in the heterocysts. The presence of a small amount of hydrogen production by a reversible hydrogenase activity and a substantial rate of hydrogen uptake by an uptake hydrogenase [11] are additional factors complicating this system. A model of hydrogen production and metabolism by *A. cylindrica* is shown in Figure 1.

Recently, the original experiment [10] was scaled up 1000-fold (from 2 ml to 2 liters) and sustained hydrogen production (for three weeks) was obtained by using *A. cylindrica* cultures continuously sparged with argon-carbon dioxide [12]. The cultures turned yellow due to nitrogen starvation-induced degradation of the blue-green pigment phycocyanin, the major light-harvesting pigment in the algae. This change decreased rates of photosynthesis (as measured by oxygen evolution or carbon dioxide fixation) by about 90%. Heterocyst frequency increased about three-fold (up to 15% of the total cell number), and rates of hydrogen evolution (per milligram of dry weight or per microgram of chlorophyll) increased proportionately. The cultures could be maintained in a condition of active production of hydrogen and approximately stoichiometric amounts of oxygen for more than three weeks through repeated additions of small amounts of ammonium (assumed to be required for cell maintenance). Limitations on the rates of hydrogen evolution and durabilities of the cultures were ascribed to reductant limitations (due to lowered photosynthesis in the vegetative cells) and to filament breakage at the heterocysts known to result in loss of nitrogenase activity [6]. We have now overcome some of these limitations through use of 1% N_2 in the argon gas phase and have also demonstrated the continuous, catalytic production of hydrogen from water and sunlight by *A. cylindrica* in an outdoor biophotolysis converter.

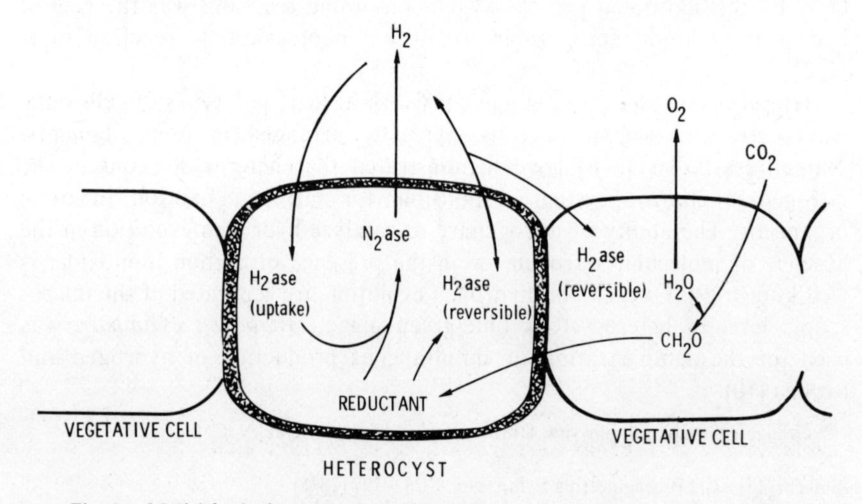

Fig. 1. Model for hydrogen evolution by nitrogen-starved cultures of *A. cylindrica*.

MATERIALS AND METHODS

Organism and Culture Methods

Anabaena cylindrica was grown as previously described [12]. In indoor laboratory experiments, when the cultures had reached a density of 0.105–0.210 mg/ml, hydrogen evolution and increased heterocyst differentiation and nitrogenase activity were initiated by sparging with argon, nitrogen and carbon dioxide (0.3% carbon dioxide, 1%, 2%, or 3% nitrogen and the balance argon). In outdoor experiments, anaerobic sparging was commenced at a density of about 0.72 mg/ml. Gases were mixed with Matheson Rotameters and the percent composition calculated from Rotameter calibration charts provided by the company. Cultures were sparged via manifolds fitted with valves to regulate flow rates. Flow rates were measured using a soap-bubble flow meter. Gas flow rates through the cultures were in the range 1–9 liter per hour.

Illumination

Indoor laboratory experiments were conducted in a two-sided light box with four duro-lite Naturescent fluorescent bulbs on each side. One-liter capacity Roux bottles filled with water lined each side and served as heat filters. Culture temperatures were usually within the range 26–29°C. The light intensity, measured with a Lambda Li-Cor model LI-185 light meter, at all heights on the sides of the 2-liter capacity cylindrical culture vessel (outer diameter 12.5 cm) was 2.0×10^4 ergs/cm^{2}sec. Light intensity values were not corrected for reflection, transmission or incident angle losses. Outdoor culture vessels were 4.5 cm in diameter, held approximately 0.8 liter of blue-green algal culture, and were placed upright at a 65° angle facing southwest. Daily insolation data were collected with an Eppley model 8-48A pyranometer.

Gas Analysis

Hydrogen production and acetylene reduction activity were assayed as previously described [12]. Oxygen production by the cultures was measured with a Varian A-90-P3 gas chromatograph equipped with a gas sample loop. Operational parameters were the same as those used for hydrogen.

RESULTS AND DISCUSSION

Laboratory Cultures

In Table I are shown the relative rates of hydrogen production and acetylene reduction by a nitrogen-starved algal culture. Rates of hydrogen evolution by nitrogenase (as judged either by acetylene reduction or as hydrogen produced under 7% carbon monoxide) are sufficient to account for nearly all the hydrogen evolution under argon. The rate of hydrogen

TABLE I

Hydrogen Production and Acetylene Reduction of Nitrogen-Starved Cultures[a]

GAS PHASE	H_2 µl/mg dry wt./hr	C_2H_4 µl/mg dry wt./hr
ARGON	27.7	--
ARGON + CO (7%)	31.1	--
ARGON + C_2H_2 (14%)	0.9	35.6
*Argon	1.0	--
*Argon + C_2H_2 (14%)	1.0	0.16

[a] Klett = 70, 23 hr after start of nitrogen starvation. *A. cylindrica* 629 was grown as described previously [12] in modified Allen and Arnon medium. Samples were withdrawn from cultures and sparged in the dark for 10 min with argon (99.7%) and carbon dioxide (0.3%) to remove any hydrogen dissolved in the liquid phase. Two-ml aliquots were injected into 5.7-ml micro-Fernbach flasks which had been previously fitted with serum stoppers, flushed with argon, and made up to the appropriate gas mixture. Assays ran for 20 min and were terminated with the injection of 0.2 ml of 25% TCA. Hydrogen or ethylene produced was quantified using gas chromatographs equipped with, respectively, a thermal conductivity detector or a flame ionization detector [12].

* Tungsten-grown cultures were cultured in the absence of Mo and in the presence of 8 ppm W (as Na_2WO_4).

evolution from reversible hydrogenase [13] was measured using cultures grown under the same conditions [12] except with tungsten substituted for molybdenum and was found to be much less than the rate of hydrogen evolution catalyzed by nitrogenase. This experiment suggests that an active reversible hydrogenase is present in the heterocysts and confirms some aspects of the model of heterocyst function presented in Figure 1.

Figure 2 demonstrates the time course of hydrogen production in the laboratory by 2-liter cultures of *A. cylindrica* sparged with argon-N_2-CO_2 (98.7%-1%-0.3%). The average rates of hydrogen evolution (3.75 ml per liter of culture per hour) were about twice as high as previously obtained with cultures sparged with argon-CO_2 and supplemented with ammonia [12]. Apparently, under our conditions, 1% N_2 allows sufficient biosynthesis and cell maintenance to achieve and sustain higher rates of hydrogen production with only limited cell growth and no significant inhibition of the hydrogenase activity of nitrogenase. It is interesting in this regard to note that observed *in vivo* K_{N_2} values lie in the range 0.02–0.07 atm [14]. Thus, by being below the half-saturation pN_2, the efficiency of nitrogen reduction by nitrogenase remains at a low level, a level that presumably allows cell maintenance, but very little cell growth. The concentrations of nitrogen used are also well below the K_i of nitrogen inhibition of hydrogen evolution (0.12

atm) [15] and, therefore, use of 1% N_2 in the gas phase should not significantly reduce the rate of hydrogen production. Of course, nitrogen limitation is affected by the rate of transfer of N_2 into the liquid phase, as well as the composition of the sparging gas phase. Under the conditions of our experiments (low gas flow rates, low liquid agitation, and large gas bubbles), the transfer of nitrogen into the liquid probably limited nitrogenase activity as much as the low gas phase concentrations of nitrogen. In a culture vessel characterized by a larger transfer factor (such that transfer is no longer limiting) even lower concentrations of nitrogen would be required.

Relatively high photosynthetic efficiencies were achieved with cultures sparged with argon, carbon dioxide, and 1% nitrogen. The efficiency of photosynthetic conversion of light to hydrogen is defined as the free energy of the total amount of hydrogen produced divided by the total energy of the light incident on the culture vessel, times 100. From the data obtained with culture L13, the calculated photosynthetic efficiency (averaged over the length of the experiment) is 1.5%. (This is probably a low estimate as the light intensity received by the sides of the culture vessel was assumed to be full light intensity.) This is well above values reported previously for nitrogen-starved cultures of *A. cylindrica* [12]. With these cultures (maintained on 1% nitrogen), H_2–O_2 ratios were in the range 0.75:1 to 0.95:1. While the ratios obtained are below the desired ratio of 2:1, at least some of the reductant not used for hydrogen production may be required for culture maintenance (similar to the requirement for N additions).

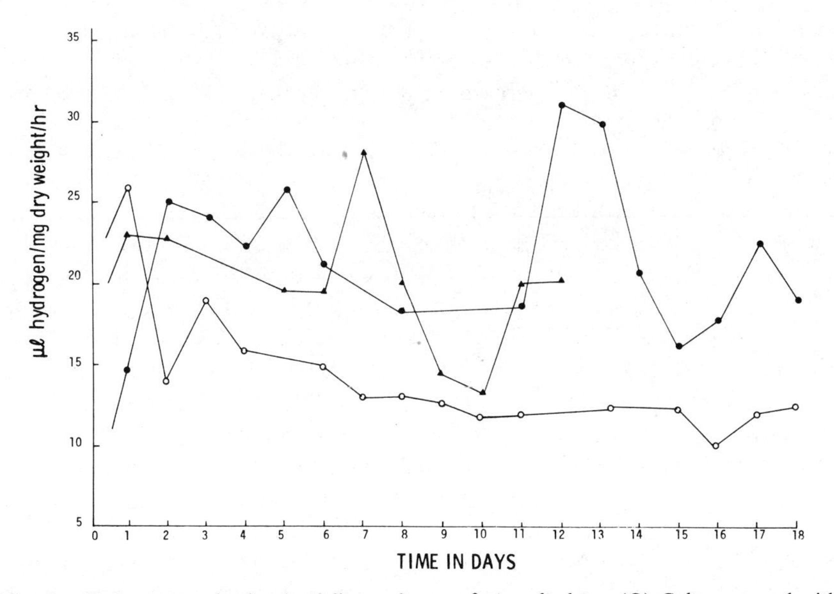

Fig. 2. Hydrogen production by 2-liter cultures of *A. cylindrica*. (○) Culture gassed with argon and carbon dioxide only (from [12]); (▲) and (●) culture L15 and L13 gassed with argon, carbon dioxide, and 1% nitrogen.

Filament Breakage

The basis for the effect of the added nitrogen in the gas phase is not entirely known. Possible factors limiting hydrogen production in cultures sparged with argon and carbon dioxide were filament breakage and decline in the reductant supply. We have found that low gas phase concentrations of nitrogen effect a significant reduction in filament breakage. As Table II shows, hydrogen-producing cultures of *A. cylindrica* bubbled with nitrogen at low concentrations exhibit a greatly reduced amount of filament breakage as compared to cultures sparged with argon and carbon dioxide, filament lengths after one or two days were only 10% of the day 0 filaments. Filament breakage in cultures that received a gas phase enriched with 1% or 2% nitrogen was much less severe. After two days, filament lengths were generally at least 50% of the day 0 filaments. After longer periods of nitrogen starvation, filament length usually decreased further, but on the average was still about 25% of the unstarved culture. (At this time the extent of filament breakage varied from culture to culture probably due to other factors such as differences in mechanical mixing and gas flow rates.) The cultures bubbled with 2% nitrogen generally showed less filament breakage than those receiving 1%. In fact, one culture (L20) showed an increase in filament length of 133% after five days. This is consistent with the moderate growth observed with cultures receiving higher concentrations of nitrogen (2–3%). (At times, some cultures sparged with 1% N_2 also exhibited slow growth.) Inclusion of nitrogen in the gas phase had no significant effect on heterocyst frequency, which was usually 13–16%. Thus, inclusion of 1% nitrogen in the gas phase is effective in reducing filament breakage while growth is still severely limited. Higher concentrations have been found to lead to increased rates of growth (ultimately reducing

TABLE II

Filament Lengths of Nitrogen Starved Algal Cultures

GASSING MIXTURE	CULTURE	ABSOLUTE FILAMENT LENGTH (MICRONS)	KLETT	HYDROGEN PRODUCTION μℓ/mg/hr	PROPORTIONAL FILAMENT LENGTH (% of day = 0 length)
Ar/CO_2		Day 0 = 800		0	
		Day 2 = 80		14.2	Day 2 = 10% Average of several cultures(12)
$Ar/1\%N_2/CO_2$	L-19	Day 1 = 66	24	8.2	
		Day 5 = 185	46	10.0	Day 5 = 28%
		Day 9 = 55	148	0	Day 9 = 8%
	L-22	Day 0 = 233	24	0	
		Day 3 = 121	50	7.6	Day 3 = 52%
		Day 16= 35	69	9.1	Day 16= 15%
	L-23	Day 0 = 280	22	0	
		Day 6 = 140	41	13.2	Day 6 = 50%
		Day 7 = 70	46	9.2	Day 7 = 25%
$Ar/2\%N_2/CO_2$	L-20	Day 0 = 168	37	0	Day 1 = 100%
		Day 1 = 168	54	15.7	Day 5 = 133%
		Day 5 = 224	116	13.6	
	L-21	Day 0 = 322	32	0	
		Day 1 = 361	50	9.3	Day 12 = 31%
		Day12 = 99	220	2.14	

hydrogen production). It is not clear whether the primary effect is to allow broken filaments to grow and regenerate to some extent or to permit the biosynthesis of structural components which prevent filament breakage and the production of inactive free heterocysts.

Light Intensity Effects

Although these cultures were induced for hydrogen production, and increased heterocyst differentiation and nitrogenase activity under light intensities (2.0×10^4 ergs/cm^2/sec) which were saturating (i.e., supported exponential growth on air, CO_2 of these cultures), after this process higher light intensities are required to saturate nitrogenase activity (Fig. 3). This effect is also shown in a shift-up experiment (Fig. 4) in which a blue-green algal culture was induced at medium light intensities (2×10^4 ergs/cm^2/sec) and then incubated under a high light intensity (6×10^4 ergs/cm^2/sec) for 72 hr. After this time the culture was again incubated at the lower light intensity. After switching to the high light intensity, rates of hydrogen evolution were almost double the initial low light intensity rate. The high rate of hydrogen production was maintained throughout the high light intensity regime and returned to the level of the initial rate soon after switching the culture back to the original light intensity.

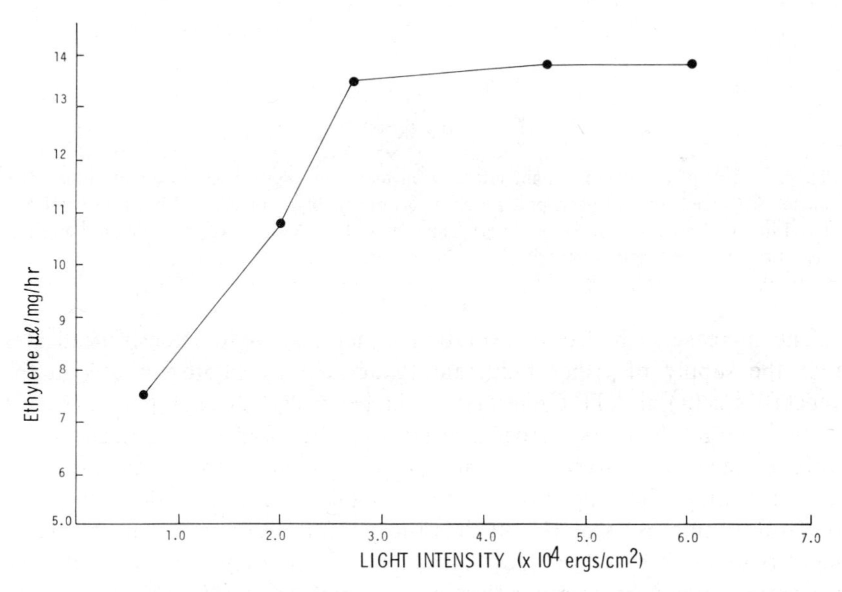

Fig. 3. Nitrogenase activity at different light intensities; Two-ml aliquots of a nitrogen-starved, 2-liter culture were incubated in the presence of 15% acetylene at the light intensities shown for 20 min. Various light intensities were obtained by placing neutral density filters (Rohm and Haas Solar Control Plexiglass) directly under the microfernbach flasks. The culture density was 0.245 mg dry wt/ml.

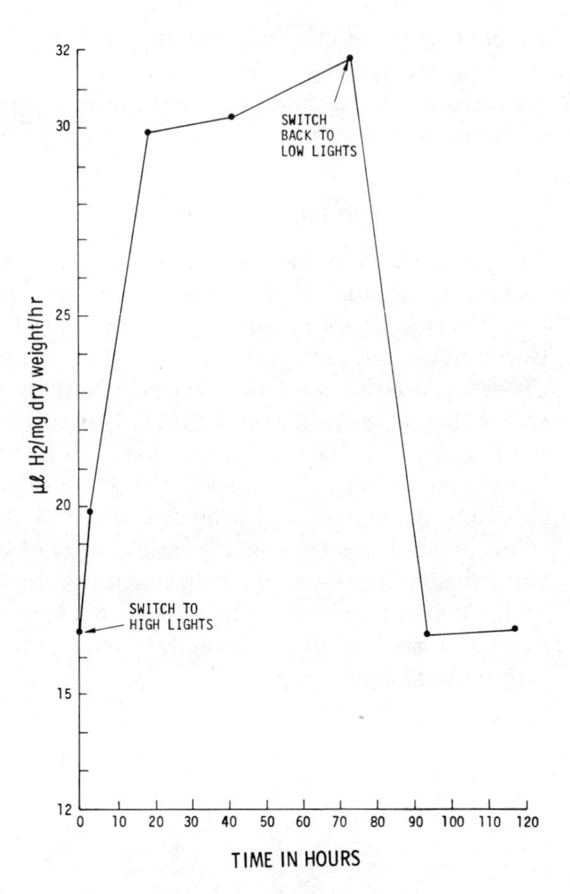

Fig. 4. Effect of a shift-up in light intensity on hydrogen production. Two-liter culture that had been induced for hydrogen production at a light intensity of 2.0×10^4 erg/cm²/sec was placed in a light intensity of 6.0×10^4 erg/cm²/sec at 0 hr. At 72 hr the lighting on the culture was returned to the original intensity.

The increase in hydrogen evolution under high light intensity indicates that the supply of either reductant (generated by photosynthesis in the vegetative cells) or ATP (generated in the heterocyst through photosystem I activity) or both limits nitrogenase activity. Because as a consequence of nitrogen starvation most of the accessory pigments of photosystem II are absent, it might be supposed that higher light intensities would be required to saturate photosystem II and thus the generation of reductant. Evidence has already been presented [12] that, during the first few days after nitrogen starvation, pools of stored reductant are high, and thus the increase in hydrogen production upon increasing the light intensity can be attributed to stimulation of photosystem I activity in the heterocyst. After longer periods of nitrogen starvation, reductant is in shorter supply, and nitrogenase activity is more immediately dependent on photosystem II activity. This dependency is demonstrated in Figure 5, which shows the results of assaying a culture, that had been producing hydrogen for two weeks, under low and high light intensities in the presence and the absence of DCMU. By this time the supply of stored reductant has been depleted and DCMU has an immediate effect on nitrogenase activity. It is likely that under these conditions at least part of the stimulatory effect of increased light intensity is through increased generation of reductant.

Outdoor Studies

For outdoor studies, a simple array of 1-liter-capacity glass tubes was placed upright at a 65° angle facing southwest. Aeration with the desired gas phase (and mixing)was via a gas stream introduced at the bottom of the tubes. No effort was made to keep the cultures sterile; previous laboratory experience had demonstrated that there was no significant difference between axenic (sterile) or bacterially contaminated cultures. The data shown in Figure 6 are representative of the results obtained. The volume of

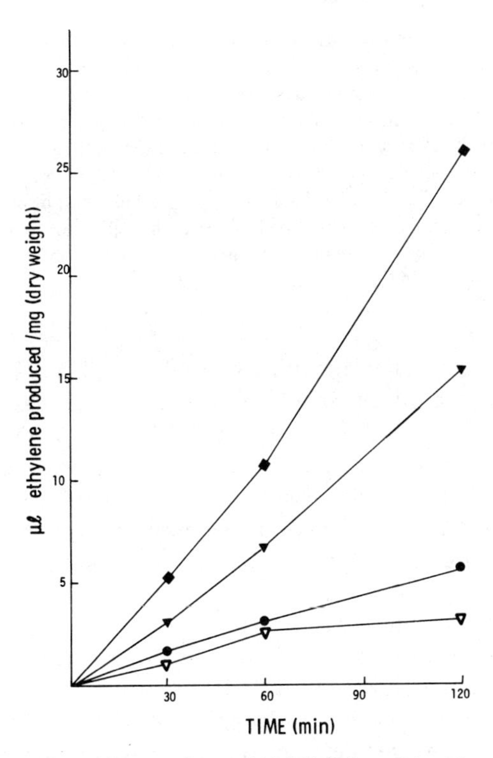

Fig. 5. Effects of DCMU additions at low and high light intensities; Two-ml samples of a culture that had been producing hydrogen for two weeks were incubated in the presence of 15% acetylene at the light intensities indicated, with and without $2 \times 10^{-5} M$ DCMU [3-(3,4-dichlorophenyl)-1,1-dimethyl urea]. (♦) High light; (▼ low light; (●) high light + DCMU; (Δ) low light + DCMU.

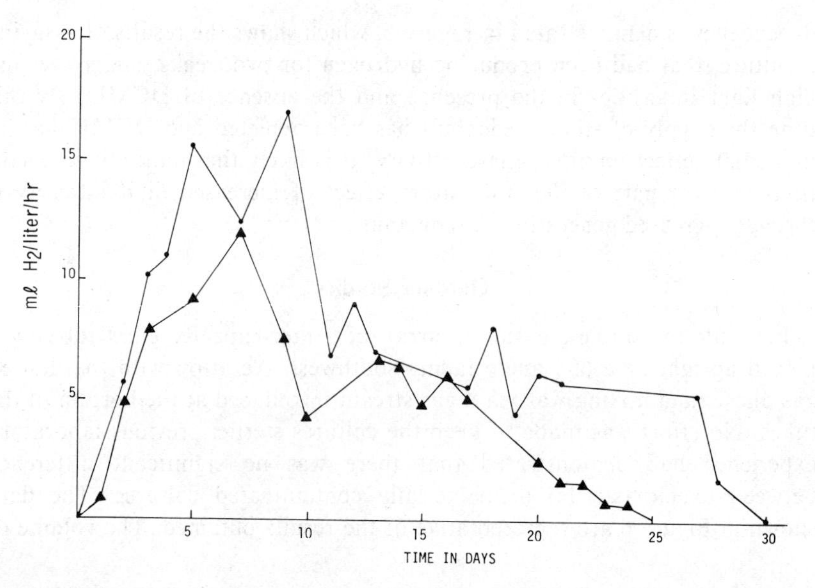

Fig. 6. Hydrogen production by outdoor cultures sparged with argon (98.7%), N_2 (1%), and CO_2 (0.3%). Cultures (*A. cylindrica* 629) were grown on modified Allen and Arnon media in the array of 1-liter-capacity glass tubes described in the text. After reaching densities of 0.72 mg dry weight/ml, the cultures were sparged with argon (98.7%), N_2 (1%), and CO_2 (0.3%). During the course of the experiment, the cell densities of the cultures increased to about 1.52 mg dry weight/ml. During daylight hours, the outdoor cultures received an average light intensity of 4.29×10^5 ergs/cm²/sec (535 Langleys/day). (Daily total insolation was calculated by integration of the continuous output of an Eppley model 8-48A pyranometer and then averaged over the days of the experiment to obtain an average light intensity.) Temperature of the cultures was usually within the range 22° to 26°C (daytime only).

hydrogen produced per liter of culture (6.8 ml H_2/1/hr) is higher in the outdoor experiments than in the indoor experiments. The reason is that much higher densities were used for outdoor cultures since the incident light intensity was greater. However, as algal density was not optimized, specific activities (per milligram of dry weight or chlorophyll) were lower. (Specific activities, averaged over the length of the experiment, were for indoor cultures, 22.8 μl H_2/mg dry weight/hr and for outdoor cultures, 5.96 μl H_2/mg dry weight/hr.) Some of the variability in the data and the eventual cessation of activity result from several factors, such as changes in gassing rates and clumping of algae, but some of the fluctuations could not be correlated with changes in temperature, density, pH, or insolation. The cause of these changes is at present unknown. Calculation of photosynthetic efficiency is difficult because of the geometry of the tubes; it is between 0.1 and 0.2% (total solar). Laboratory cultures maintained under artificial light have a photosynthetic efficiency of 1.5%. With proper selection of a converter geometry and optimal algal density, this level of efficiency should also be attainable with cultures maintained in sunlight.

Diurnal Variations in Hydrogen Production

The measurement of the diurnal variations in output from any solar collector can help determine important operational parameters such as response to fluctuations in solar flux, the time and duration of periods of peak conversion, and the solar flux that saturates the converter. Four outdoor cultures of *A. cylindrica* were assayed for H_2 at 2-hr intervals for the 24-hr period beginning at 8:00 AM on 10 June. The cultures were grown outdoors to densities varying from 0.96 to 1.2 mg dry weight/ml and then sparged with an atmosphere of argon, CO_2, and N_2. The cultures were producing H_2 for two days at the time of the assay and thus were young cultures.

As the graph of the assay (Fig. 7) demonstrates, rates of hydrogen evolution rose steadily from 8:00 AM to 2:00 PM and peaked between 1:00 PM and 3:00 PM. As the sun dropped in the sky and the incident radiation decreased, the cultures showed a concomitant decline in hydrogen production. In fact, the variation in hydrogen evolution over the day shows almost exactly the same dependence on time before and after solar noon as the

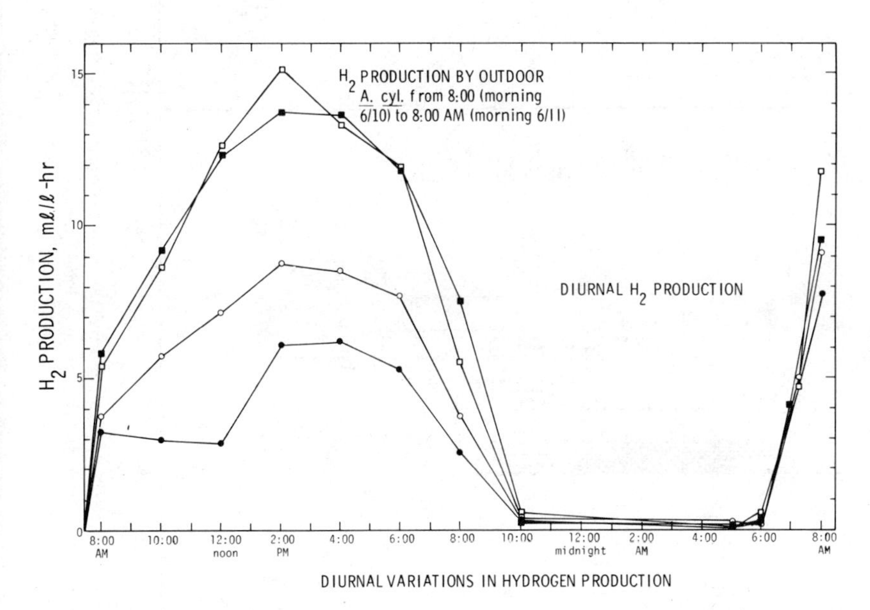

Fig. 7. Diurnal variations in hydrogen production. Four outdoor cultures of *A. cylindrica* were assayed for hydrogen production at 2-hr intervals for the 24-hr period beginning 0800 on 10 June. The cultures had been grown to densities ranging from 0.96 to 1.2 mg dry weight/ml and then sparged with an atmosphere of argon, CO_2 and N_2 at gas flow rates in the range of 7.2 to 9.0 liter/hour. At the time of the experiment, the culture densities were (in mg dry weight/ml): RT 10, 1.07; RT 11, 1.12; RT 12, 1.02; RT 13, 0.95. The cultures had been producing hydrogen for two days at the time of the assay and thus were young cultures. (●) RT 13; (○) RT 12; (□) RT 11; (■) RT 10.

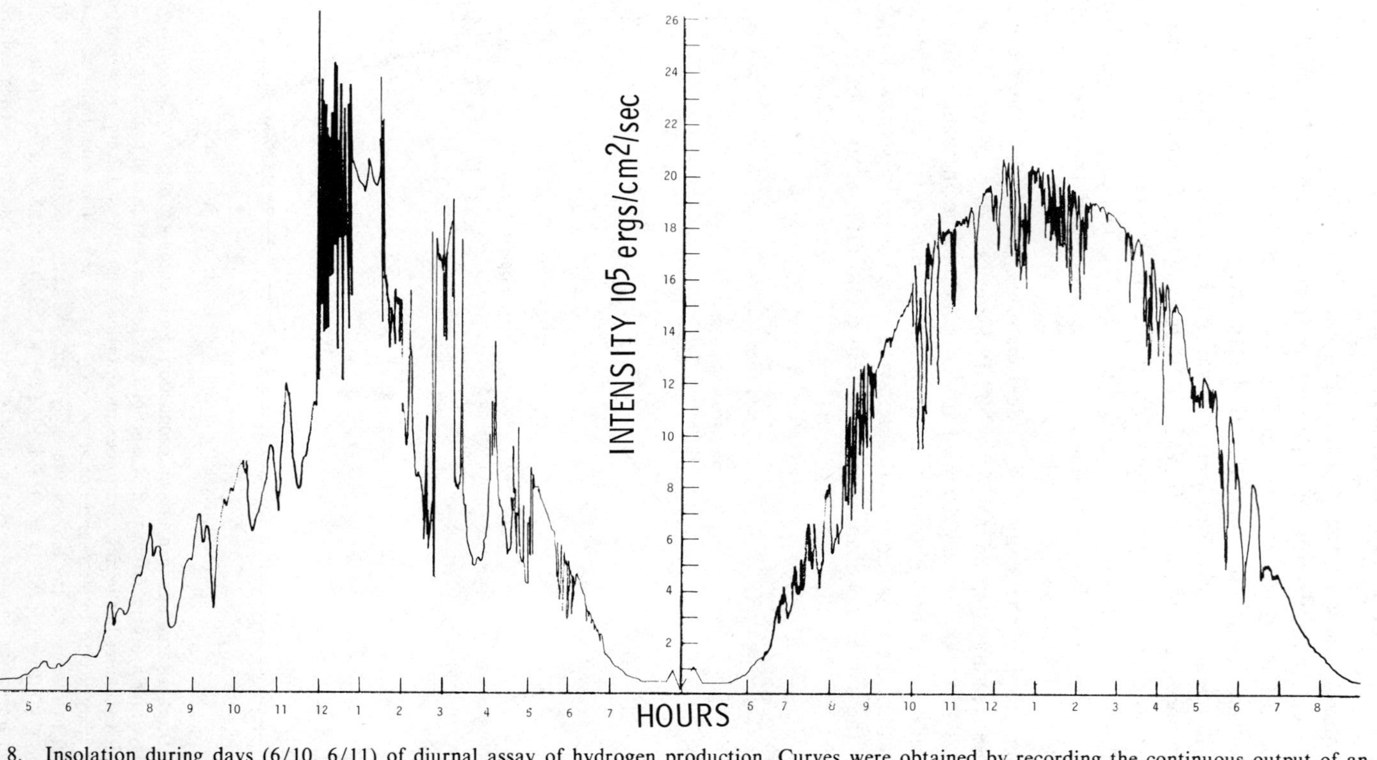

Fig. 8. Insolation during days (6/10, 6/11) of diurnal assay of hydrogen production. Curves were obtained by recording the continuous output of an Eppley model 8-48A pyranometer. Time of day is given as local time.

incident light intensity. Figure 8 shows the incident solar flux for the days of 10 June and 11 June as recorded from the output of an Eppley 8-48A pyranometer. Both hydrogen evolution and solar flux (most clearly seen for 11 June, Fig. 8) are approximately cosine functions of the zenith distance of the sun. Several points are apparent from a comparison of Figures 7 and 8; Hydrogen production is not saturated at the lower solar fluxes and reaches maximal values during the hours of maximum solar flux. (The rates of hydrogen production might have been higher if the day had been sunnier.) Secondly, the day during which most of the data were taken (10 June) was partly cloudy with significant fluctuations in solar flux occurring over periods of up to 15 min. The lack of short-term fluctuations in hydrogen output demonstrates that the response time of the system is sufficiently slow to average over expected fluctuations in solar flux. Of course, the response time is a combination of several factors including among others, physiological response time (relatively fast), gas space within algal culture vessel, and gas flow rates.

Hydrogen evolution dropped to zero soon after the sun set and was zero before sunrise the following morning. The morning of 10 June was heavily overcast, but the morning of the 11th was quite clear and sunny, accounting (at least in part) for the increase in the rate of hydrogen evolution from the morning of the 10th to the 11th.

CONCLUSION

We have demonstrated that a catalytic, sustained production of hydrogen from water can be carried out under outdoor conditions using a simple glass converter and a stationary blue-green algal culture. This process meets the basic technical requirements of biophotolysis.

Improvement in rates of hydrogen production by this system could be achieved by selecting wild-type, blue-green algae better suited to hydrogen production and genetically improving the organism for this task. Specific requirements for the algae are tolerance of expected temperature variations, maximum nitrogenase (or, preferably, reversible hydrogenase) activity, increases in vegetative cell photosynthesis through decreased phycocyanin degradation during nitrogen starvation, and increased filament strength. One of the most significant requirements is that the cells continue exhibiting high levels of hydrogen production even when not stirred or mixed. This latter requirement derives from the very limited energy inputs allowable for operation of practicable systems. Thermophilic, mat-forming, blue-green algae might provide suitable strains. The actual culture vessels and hydrogen collectors should be arranged horizontally with a gas space, so the carrier gas need not be pushed through a liquid head.

The key constraint on a biological solar energy converter is the very low capital and operational costs allowable. A simple calculation shows that, if 3% of incident energy in the Southwest United States were converted to hydrogen, only 0.25×10^9 joules/m^2/yr (21.9×10^3 BTU/ft^2/yr) would be produced, worth about \$0.60, assuming \$2.40/10^9 joules ($\simeq$\$2.50/MBTU). Even if a higher price were to be allowed for hydrogen (assuming, for example, local industrial use or availability of inexpensive fuel cells), total capital investment into a converter system could not be much larger than \$5.40/m^2 (50¢/ft^2). A proposed design for such a low-cost culture vessel–hydrogen-collector system is an array of horizontal, thin glass tubes [16]. The glass tubes would be standard units presently manufactured for the fluorescent lighting industry (235 cm $\times$ 3.81 cm with 30–35 mil walls) and available for \$0.125 (U.S.) per unit in lots over 1000. At estimated mass production costs of connecting plastic pipes, assembly, and installation (on rooftops, during construction), a 1.22 m $\times$ 2.44 m converter panel, using 32 tubes, could cost about \$15. The necessary accessory equipment (liquid and gas recirculation pumps, hydrogen/oxygen gas separators or combustion chambers, algal growth chambers, monitoring equipment, etc.) would be subject to considerable economy of scale which, together with operational requirements (provision of algal cultures, monitoring, mixing, etc.) would necessitate the use of systems of several thousand panels.

In conclusion, biophotolysis using heterocystous blue-green algae has been demonstrated. The practical application of this system is dependent upon the development of low-cost converters and effective algal strains.

Nomenclature

pN_2 partial pressure of nitrogen

K_{N_2} partial pressure of nitrogen at which the rate of nitrogen fixation is half maximal

K_i concentration of inhibitor that causes 50% inhibition of the maximal, uninhibited reaction rate

We wish to thank Dr. William J. Oswald for encouragement and support and Thomas Tiburzi for technical assistance. This research was supported in part by the Energy Research Development Administration Grant No. E(04-3)-34 #239.

References

[1] J. R. Benemann, J. A. Berenson, N. O. Kaplan, and M. D. Kamen, *Proc. Natl. Acad. Sci. U.S.A.*, *70*, 2317 (1973).

[2] K. K. Rao, L. Rosa, and D. O. Hall, *Biochem. Biophys. Res. Commun.*, *68*, 21 (1976).

[3] P. Fay, W. D. P. Stewart, A. E. Walsby, and G. E. Fogg, *Nature*, *220*, 810 (1968).

[4] C. P. Wolk and E. Wojciuch, *Planta* (*Berl.*), *97*, 126 (1971).

[5] E. Tel-Or and W. D. P. Stewart, *Biochim. Biophys. Acta*, *423*, 189 (1976).

[6] N. M. Weare and J. R. Benemann, *Arch. Mikrobiol.*, *90*, 323 (1973).

[7] W. D. P. Stewart, A. Haystead, and H. W. Pearson, *Nature*, *224*, 226 (1969).

[8] J. Thomas, J. C. Meeks, C. P. Wolk, P. W. Shaffer, S. M. Austin, and W. S. Chien, *J. Bacteriol.*, *129*, 1545 (1977).

[9] W. A. Bulen, R. C. Burns, and J. R. LeCounte, *Proc. Nat. Acad. Sci.*, *53*, 532 (1965).

[10] J. R. Benemann and N. M. Weare, *Science*, *184*, 174 (1974).

[11] J. R. Benemann and N. M. Weare, *Arch. Microbial. 101*, 401 (1974).

[12] J. C. Weissman and J. R. Benemann, *Appl. Environ. Microbiol.*, *33*, 123 (1977).

[13] Y. Fujita, H. Ohama, and A. Hattori, *Plant Cell Physiol.*, *5*, 305 (1964).
[14] G. W. Strandberg and P. W. Wilson, *Proc. Nat. Acad. Sci.*, *58*, 1404 (1967).
[15] Calculated from Figure 4, J. C. Hwang and R. H. Burris, *Biochim. Biophys. Acta*, *283*, 339 (1972).
[16] J. R. Benemann and J. C. Weissman, in *Microbial Energy Conversion*, H. G. Schlegel and J. Barnea, Eds. (Erich Goltze, K. G., Gottingen, Germany, 1976), pp. 413–426.

Cultivation of Nitrogen-Fixing Blue-Green Algae on Ammonia-Depleted Effluents from Sewage Oxidation Ponds

JOSEPH C. WEISSMAN, DON M. EISENBERG, and JOHN R. BENEMANN

Sanitary Engineering Research Laboratory, College of Engineering, University of California, Berkeley, California 94804

INTRODUCTION

Oxidation ponds are presently used in many parts of the United States both for primary (solids removal) and secondary (oxidation) treatment of wastewaters [1]. In these ponds microalgae grow and produce the oxygen required for bacterial breakdown of the biodegradable, organic wastes. Oxidation ponds are cheaper than conventional wastewater treatment processes, such as the activated sludge process, in capital, operational, and energy costs [2]. One drawback of oxidation ponds is the often high concentration of microalgae in the effluents, which results in their failure to meet present EPA standards of a maximum of 30 mg/liter suspended solids in waste treatment plant effluents. (These standards have recently been lifted for pond systems treating below 2 million gallons of sewage per day.) Low-cost processes for microalgae removal from oxidation pond effluents are required. Both microstraining [3] and in-pond settling [2] have recently been studied. Microstrainers are rotating drums wrapped with small mesh screens and equipped with a backwash for removing the algae that is retained. Microstraining oxidation pond algae depends on the selective cultivation of filamentous or large colonial algae; in-pond settling appears to depend on inducing nutrient limitations.

The oxidation pond effluents produced after microalgae removal still contain sufficient nutrients to allow the growth of additional crops of algae, specifically the nitrogen-fixing blue-green algae, because phosphate is usually the relatively most abundant of algal nutrients in wastewaters. Filamentous, heterocystous blue-greens are the most common nitrogen-fixing algae. These have the unique ability of simultaneously generating oxygen photosynthetically and reducing N_2 to ammonia. Nitrogenase, the enzyme which catalyzes this reduction, is extremely oxygen sensitive, as is the reaction. Heterocysts are differentiated cells that are specialized to protect the site of N_2 fixation from O_2. As these blue-green algae are also the most

Biotechnology and Bioengineering Symp. No. 8, 299–316 (1978)

0572-6565/78/0008-0299$01.00

common nuisance algae in eutrophic waters, phosphate removal from waste treatment plant effluents, whether conventional or ponds, is desirable. Conceptually it should be possible to grow nitrogen-fixing blue-green algae in ponds on nitrogen-depleted wastewaters under conditions of intensive, controlled cultivation. Because these algae are filamentous, it is relatively easy to harvest them with microstrainers, resulting in an effluent that would no longer support algal growth in receiving bodies of water. Such "advanced" wastewater treatment can currently only be achieved with expensive physical-chemical processes.

The outdoor cultivation of nitrogen-fixing blue-green algae has not been studied under N_2-fixing conditions. Previous experience with outdoor cultures is limited to the preparation of blue-green algae for inoculating rice paddies; these cultures were grown on fixed nitrogen [4–6]. Thus, an objective of our research is to determine the limits of productivity and the conditions for optimizing it. Small-scale ponds diluted with synthetic media were used for this purpose.

In this report we demonstrate, on a small scale, the basic elements of a multistage microalgae, advanced wastewater treatment system. In the first stage, green algae are grown up to or approaching the limit of readily available nitrogen (ammonia). In the second stage the remaining ammonia is utilized and the algae are then settled out in a batch growth or isolation process [2]. The effluent is used to cultivate a crop of nitrogen-fixing blue-green algae. This concept is shown in Figure 1. The microalgae biomass produced in such a system is a suitable substrate for methane production by anaerobic digestion [7]. The digester effluent, containing the nutrients taken up and fixed by the algae, could be used as an organic fertilizer [8]. Therefore the system shown in Figure 1 can have the multiple function of advanced wastewater treatment–water reclamation, net fuel production, and fertilizer recycling and production.

The dual objectives of maximal algae biomass production and nutrient removal can be best achieved in high-rate ponds, which are shallow (about 25–50 cm) and mechanically mixed [9]. This pond design allows higher algae concentrations, more efficient light utilization, and better operational control than the deep (100–200 cm), unmixed facultative oxidation ponds that are the most commonly used oxidation ponds. The objective of maximal light utilization requires light to be the growth-limiting nutrient. This requirement, however, conflicts with the requirements of nitrogen and phosphorus utilization, as these must also become growth-limiting for efficient removal.

An optimization is required which depends on the relative needs and economics of wastewater treatment and of fuel–fertilizer production from the algal biomass. As carbon would quickly become a limiting nutrient for algal growth on sewage, because carbon dioxide transfer from air (0.03% CO_2) is slow, additional CO_2 must be introduced to allow maximal algae production

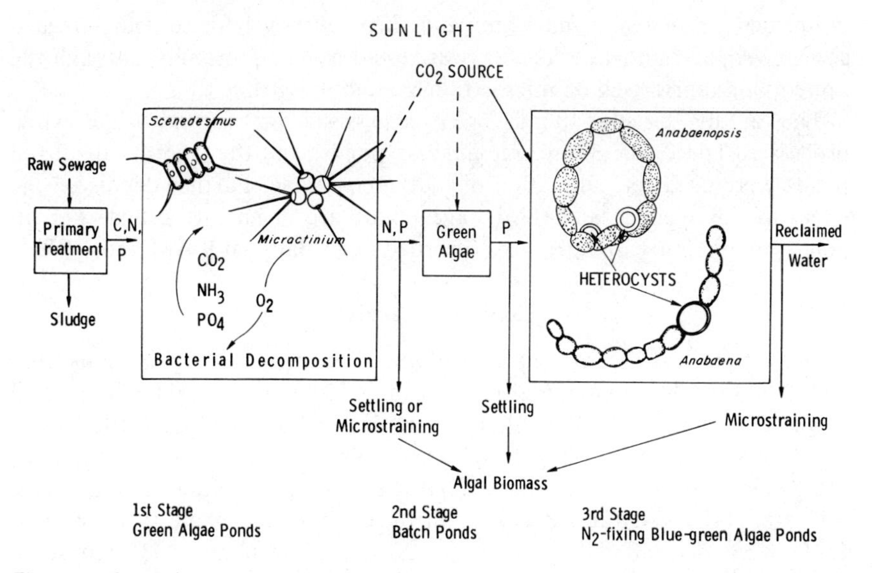

Fig. 1. Advanced wastewater treatment using a multistage ponding system. Relative pond sizes are not accurately represented.

in the nitrogen-fixing pond and, for best results, even in the first green algae pond. One important requirement is that the green algae ponds effectively remove metabolizable nitrogen to below the concentration that would allow either nonfixing algae to grow or that would inhibit the nitrogen-fixing reaction of the blue-green algae.

MATERIALS AND METHODS

Experimental Pond System

The experimental ponds used consisted of (a) 12-m² rectangular ponds mixed by paddle wheels, (b) 3.0-m² circular and rectangular ponds mixed by paddle wheels, and (c) 0.55-m² rectangular ponds mixed by air pumps. The 12-m² and circular 3.0-m² ponds were used as first stage, high-rate ponds and operated at depths from 20–30 cm. They were diluted with sewage from Richmond, California, which was settled on a 1.5-hr detention time and passed over a cascade screen with 2.8-mm slots. The 12-m² ponds were harvested over a 4–8-hr period with the inflow of sewage controlled by a float switch. Samples for laboratory analysis were either composited from the pond effluent line over the first half of the dilution period or taken as grabs halfway through the dilution. Effluents were then pumped into microstrainers (0.2-m² straining area). Nylon screens (Tetco, New Jersey) with 26-μm openings were used. Samples of microstrained effluents were also taken for analysis. The 3.0-m² ponds were diluted daily by pumping out the prescribed volume of pond water and then refilling with settled, screened sewage. Algae removal efficiency was tested by hand straining through the same nylon fabric used on microstrainers, and by settling.

The 3.0-m^2 circular ponds were also used for second stage batch processes. These, the 3.0-m^2 rectangular ponds, and the 0.55-m^2 air-mixed ponds were used for cultivation of blue-green algae. Further details on the operation of second- and third-stage ponds are given in the text. For all ponds, temperature and pH were recorded twice daily, at 9 AM and 4 PM.

Laboratory Analysis

Dry weight was determined after drying filtered samples (Whatman GfC) at 103°C for 1 hr and again after igniting at 550°C for 15 min. Chlorophyll *a* was extracted by boiling filtered samples in 90% methanol/10% water (v/v) for 45 sec and reading centrifuged (5000 × *g* for 15 mins) samples at 750 and 665 nm. MacKinney's [10] extinction coefficient was used. NH_4^+-N, NO_3^--N, total Kjeldahl nitrogen, total phosphate-P, COD, and BOD were determined according to *Standard Methods* [11] with some modifications.

Nitrogen-fixation capacity was measured using the acetylene reduction assay by incubating 2-ml pond samples under an atmosphere of 85% air/15% C_2H_2 (v/v) for 30 min in a shaker bath at 27°C under saturating light intensity. Further details are given in Ref. 12.

Phycobiliproteins were measured by the method of Bennett and Bogarad [13] with the following modification of extinction coefficients (A. Glazer, University of California, Berkeley, personal communication): $E_{1\ cm}^{1\%} = 70$ for phycocyanin at 620 nm, $E_{1\ cm}^{1\%} = 127$ for phycoerythrin at 565 nm.

Algae were counted manually using a hemocytometer. Herbivores were also enumerated after concentration using a Sedwik-Rafter chamber.

An Eppley 8-48A Pyranometer driving a strip chart recorder was used to measure insolation. The recorder tracings were integrated with a hand planimeter.

The nitrogen-fixing blue-green algae used in the experiments were isolated from bioassays performed on pond water from isolation ponds in Woodland, California, and on pond water from low nitrogen ponds at Richmond, California.

OPERATION OF 12-m^2 GREEN ALGAE PONDS FOR NITROGEN REMOVAL

The objective of these experiments was to produce an effluent suitable for cultivation of blue-green algae. The requirement was that the effluent would not be suitable for growth of green algae because of nitrogen limitation. This requirement would be sufficient to allow cultivation of nitrogen-fixing blue-green algae without interference from green algae or inhibition of their nitrogen-fixing capacity. Because only negligible nitrate concentrations were measured and effluents depleted of ammonium nitrogen would not support further growth of green algae, we used NH_4^+-N (rather than total -N) as the criteria for determining the effectiveness of the pond operations. This criterion was reasonable as ammonia is also the direct repressor of heterocyst development and nitrogen fixation in blue-green algae. Because the ammonia concentration that effectively represses nitrogen fixation was not known, an NH_4^+-N level in the pond of about 0.5 mg/liter was aimed at. However, as discussed below, a higher level of ammonia is permissible in the influent to an actively growing culture of nitrogen-fixing blue-green algae.

Two processes are responsible for the removal of ammonia in sewage oxidation ponds: uptake by algae (and other pond organisms) and outgassing. (Nitrification–denitrification known to occur in facultative ponds was not a factor in the operation of these high-rate ponds, as no significant level of nitrate was detected.) Ammonia outgassing is promoted by the relatively high pH observed in the experimental, high-rate oxidation ponds, particularly in the afternoons when the pH rose to 10 and above. The pH rise indicates that carbon can be a limiting factor for algal growth during part of the day; apparently carbon is not released fast enough from sewage by bacterial decomposition. Efficiency of ammonia removal by the algae depends on the size of the standing crop. The optimal pond detention time is longer in winter than in summer.

Table I summarizes the results obtained with the 12-m^2 ponds during ten experiments. The data in Table I demonstrate that the best NH_4^+-N removal was achieved by use of relatively long detention times and high standing algae crops. This effect can be ascribed both to increased nitrogen uptake by the algae and increased outgassing during the long detention period. For comparison, influent PO_4-P concentrations varied from 10–15 mg/liter and effluents from 6–10 mg/liter, with a removal range of 10–40%.

A number of factors affected pond operations. The drought in California resulted in significantly increased sewage strengths during the second half of 1977. This affected the detention time and standing algae crop required for effective NH_4^+-N removal in a single-stage pond system. We also observed that microalgae cultivation in small-scale intensive cultures is not a stable process. Algae species compositions can change abruptly, over the course of a few days, and, more importantly, the algae cultures are often subject to precipitous decreases in density accompanied by zooplankton grazers, fungal diseases, and other biological agents. The causes for such population changes and crashes are obscure; control measures need to be worked out as they affect the operation of the ponds and, in particular, ammonia removal.

Before the effluents from these ponds can be considered for cultivation of nitrogen-fixing blue-green algae, the green algae must be removed (e.g., harvested) and the NH_4^+-N concentration further reduced. As shown in Table I, experiments 1 through 4, microstraining can be quite effective in removing algae from the effluents of these ponds. Over 80% removal was

TABLE I

NH_4^+-N Concentrations and Algal Removal Efficiencies in 12.0-m² First-Stage Ponds[a]

Exp.) Date	Avg. Θ, days	Cell Density, mgℓ⁻¹	Productivity gm/m²/day	Chlor.a %VSS	%Removal Chlor.a by microstraining	NH_4^+-N, mgℓ⁻¹ infl. sewage	pond effl.
1) 1/17-2/21/77	15.0	167	1.7	2.6	94	27(24-32)	3.4
2) 1/17-2/21/77	6.7	134	3.1	2.1	83	27(24-32)	11.9
3) 7/4-7/15/77*	6.0	310	15.7	1.7	96	46	7.7
4) 7/4-8/3/77*	3.3	270	23	1.7	93	47	14.0
5) 11/6-11/19/77	4.0	130	8.2	2.5	12	45(27-61)	16.8
6) 11/27-12/17/77+	3.8	59	3.9	1.8	8	48(30-70)	33.7
7) 11/6-11/19/77	8.0	167	5.2	3.5	14	45(27-61)	7.4
8) 11/27-12/17/77+	7.4	39	1.4	2.2	12	48(30-70)	18.5
9) 11/6-11/19/77	16.0	193	3.1	3.9	22	45(27-61)	1.5
10) 11/20-11/29/77+	13.2	36	0.8	1.9	11	45(30-61)	8.3

[a] All ponds were mixed at a paddle wheel rotation of 1.3 rpm (5 cm/sec linear velocity at the average immersion radius at the paddle wheel) except those marked with an asterisk which were mixed at 2.4 rpm. A plus sign (+) denotes that the low pond density was accompanied by a fungal infection of pond algae. Ponds were diluted over a 4-hr period in the winter and an 8-hr period in the summer with inflow of settled sewage controlled by a float switch. Depths were 25–30 cm.

obtained during eight months of the year [14]. During this period, ponds were harvestable by microstraining only when the most common algal types, *Scenedesmus* and *Micractinium*, were present in large colonies. A long detention time was found to correlate with the formation of large colonies. Long detention times also result in a relative loss of productivity (e.g., yields and photosynthetic efficiency). In the fall and early winter (Table I), microstraining was not effective regardless of detention times. Thus, we developed a strategy for removal of both residual algae and ammonia that involved a batch pond operation following the initial continuous experimental high-rate pond.

SECOND-STAGE BATCH POND OPERATIONS

The effluents from the first-stage continuous growth ponds (Table I) were pumped into 3.0-m² circular ponds and operated in a batch process. Two different types of operations resulted. When algae removal from the first stage was high (e.g., experiments 1–4, Table I), then a substantial increase in algae density was observed, and the ponds operated as batch growth ponds. If microstrainer harvest efficiency was low (e.g., experiments 5–10, Table I), then little further algal growth was observed, and the ponds operated essentially in a "pond isolation" process. In pond isolation, algae removal results through in-pond settling occasioned by flocculation of the algae under conditions of nutrient limitation [2].

Over two dozen batch ponds were operated. Some representative results are shown in Tables II and III. Ponds were filled on day 0 from the indicated source. They were "seeded" with microstrainer concentrate to reduce lag times when microstraining efficiently removed algae from the effluents of the first-stage ponds. In some cases CO_2 was bubbled into the ponds with diffusing stones, but the relatively slow rate (below 2 liters of 0.3% CO_2 per minute) resulted in carbon remaining as a limiting nutrient in all ponds. This was evidenced by the increase in pH, which rose to about 10 in the batch growth ponds a few days after starting the experiments. Tables II and III also show the other variables in pond operation, including depth, mixing, backwash water dilution, and source of the effluent. Backwash water dilution had no effect on the growth or settling of the algae, but it did reduce the final yield somewhat.

TABLE II

Representative Data from Second-Stage Batch Ponds[a]

Day	pH am/pm	VSS, mg/ℓ	Chlor.a, mg/ℓ	NH_3^- N, mg/ℓ
0	—	40	0.55	17
1	7.9/9.2	—	—	9
3	10.2/10.8	136	2.01	—
6	9.6/ -	186/5.3	1.18/.03	0.04

Day	pH am/pm	VSS, mg/ℓ	Chlor.a, mg/ℓ	NH_3^- N, mg/ℓ
0	8.2/9.0	40	0.41	9
1	8.1/10.3	—	—	0.8
2	9.2/9.0	101	0.89	—
5	8.0/8.7	102	—	0.1
7	8.1/ -	109/3.8	0.80/0.01	—

[a] On the final day, the mixers were shut off and the ponds allowed to settle for 1–2 hr. The VSS and chlorophyll values before and after settling are given, separated by a slash. Tap water dilution (from microstrainer bachwash) was 10%. [A] Upper table: Batch No. 1. C-2; d-10 in.; filled with microstrained effluent from a first-stage pond (θ = 3.3 days) + 8-liter microstrainer concentrate (7/26–8/1/77). Lower table: Batch No. 2. C-2; d = 7 in., filled with microstrained effluent from a first-stage pond (θ = 4.6 days) + 12-liter microstrainer concentrate (8/3–8/10/77).

TABLE III

An Example of the Batch Isolation Process[a]

Day	pH am/pm	VSS, mg/ℓ	Chlor.a, mg/ℓ	NH_3^- N, mg/ℓ
0	- /10.4	144	4.10	3
3	10.2/10.6	234	5.45	0.14
7	10.01/ -	289/198	3.62/2.13	0.06

Day	pH am/pm	VSS, mg/ℓ	Chlor.a, mg/ℓ	NH_3^- N, mg/ℓ
0	-/10.9	175	1.95	0.06
3	9.1/9.5	174	1.45	0.1
7	7.6/ -	103/8.2	0.43/0.01	0.8

[a] The final supernatant from No. 23 was brown and contained rotifers. These were removed by straining after the pond was settled. [B] Upper table. Batch No. 7. C-3; d = 4 in.; filled with micro-strained effluent from a first-stage pond (θ = 8 days); no CO_2. (10/3–10/10/77). Lower table. Batch No. 23. A-2; d = 9 in.; filled with #7 and #8 supernatant, no CO_2 (10/10–10/17/77).

The pattern of growth was similar in all of the batch growth ponds (Fig. 2, Table II). Both volatile suspended solids (ash-free dry weight) and chlorophyll *a* increased, with the latter leveling off before the former, and decreasing toward the end of the growth period. The decrease in chlorophyll to volatile solids ratio and the yellowing of the cultures are typical of

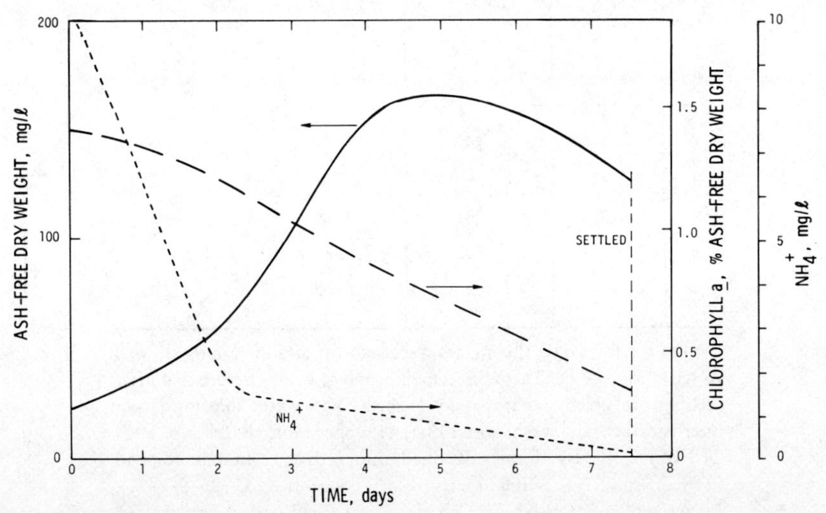

Fig. 2. Sketch of typical changes in density, chlorophyll *a* content, and NH_4^+-N concentration during batch growth in second-stage ponds. (——) VSS; (— — —) % chlorophyll *a*; (------) NH_4^+-N concentration.

nitrogen limitations, which is evident from the very low ammonia concentrations remaining in the ponds at the end of the growth period. The yellowing of the algae and disappearance of ammonia was accompanied by clumping and flocculation of the algae. At this stage the algal cultures settled very efficiently, usually within a few hours after turning off the mixing. The pond supernatants were then decanted and used in cultivation of nitrogen-fixing blue-green algae (see next section).

The above process worked well in the summer, but toward fall microstrainer harvestability of the algae in the primary ponds decreased, and settling in the secondary ponds became less reliable (Table III, batch No. 7). When the algae did not settle well, supernatants from two (or more) ponds were combined and subjected to a period of isolation (Table III, batch No. 23). Chlorophyll *a* concentrations decreased, and the algae settled out. Often rotifer blooms appeared which consumed much of the algae, yielding a clear, brown-colored supernatant after settling. When such a bloom occurred, the final ammonia concentrations were higher than normal in the settled effluents.

Table IV summarizes the results of eleven batch-pond operations. The maximum yields (maximum algal density minus starting density) ranged between 70 and 200 mg/liter. In most cases, these yields were 50–100% of

TABLE IV

Yields, Removals, and Production from Batch Growth Ponds[a]

Date/ Batch	ΔVSS Max mg/ℓ	NH_3- N %Removal	VSS % Removal	Chlor.a %Removal	Production $gm/m^2/day$
1) 7/26-8/1/77	146	99.8	97.3	97.4	7.8/7.8
2) 8/3-10	70	>99	96.3	98.8	2.3/2.3
3) 8/7-15	190	98.5	97.9	100	14.2/4.8
4) 8/10-19	170	86	80	75	8.7/3.8
5) 8/15-23	210	90	—	—	17/ -
6) 9/26-10/2	—	90	—	—	—
7) 10/3-10	145	80	31	41	2.7/2.7
8) 10/6-17	130	>99	23	47	4.3/4.3
9) 10/8-17	150	>99	< 40	< 60	- /4.2
10) 10/10-17	150	>99	40	50	6.7/3.9
11) 10/10-17	80	>99	37	46	15.4/6.6

[a] Two production values are listed. The first corresponds to the productivity attained at the onset of the stationary phase; the second is the productivity at a time when the algae could be harvested. ΔVSS maximum equals VSS at the onset of stationary phase minus the initial (inoculum) VSS.

the yields obtained in the first-stage ponds, giving a total of 300–500 mg/liter of green algal biomass grown on the influent sewage. Some algae settled during the continuous and batch growth phases; these were not measured or included in the final productivity determinations. In all ponds ammonia was removed very efficiently. Algal removal by settling became less effective in the fall and required a period of isolation, as discussed above. Phosphate was only measured occasionally; it was about 5 mg/liter, which is sufficient for blue-green algal cultivation. The productivities (g/m^2/day) were lower in the batch growth ponds than in the continuous, primary-stage ponds, as may be expected. Productivity from the continuous ponds depends on detention time. Total production at $\theta = 3$ days, during the summer was about 25 g/m^2/day in these ponds. Maximum algal production, during the summer, from batch ponds occurred after about three days and was approximately 15 g/m^2/day. Both continuous and batch ponds would have been more productive with carbon dioxide additions, but higher pH of the batch ponds indicated more severe carbon dioxide limitation. During the fall, when detention times of the continuous ponds were increased to eight days, ponds run as batch for 7–9 days produced very nearly as well as the continuous ponds.

A different approach for algae and NH_4^+-N removal from the first-stage ponds is to induce the formation of flocs, composed of algal and nonalgal solids, which can be separated by screens or rapid settling. Previous experiments [14] had demonstrated that a relatively fast mixing regime (paddlewheel rotation of 4.0 rpm in the 12.0-m^2 ponds), resulted in the formation of such flocs. In Table V is shown the results of operating a 3.0-m^2 pond with mixing sufficient to keep nearly all of the solids suspended (3.5 rpm of the paddle wheel). Algal removal by microstraining averaged above 85% and, more importantly, almost complete ammonia removal was achieved. The flocs also settled rapidly, with 75% of the volatile solids settling through 25 cm in less than 1 hr. This process is currently being investigated as the preferred method for solids removal. The reliability of mixing-induced flocculation, and the subsequent fast settling in deep clarifiers must be demonstrated. In this experiment (Table V), high algal removal efficiencies were combined with relatively high productivities. Volatile suspended solids production averaged 15 g/m^2/day during the latter part of the experiment. However, due to the presence of nonalgal solids (estimated at about 33% from the chlorophyll concentration in this and slowly-mixed ponds), actual algal productivities and photosynthetic efficiencies were lower than shown.

It is thus possible under certain conditions to produce effluents suitable for the growth of nitrogen-fixing algae using only one, semicontinuously operated pond. If the influent sewage is not too strong, then fast mixing of the first-stage, high-rate pond can be used to increase algal removal efficiency by microstraining or settling through flocculation of algal and nonalgal ponds. The fast mixing also increases NH_4^+-N removal by stimu-

TABLE V

Summary Data from a 3.0-m^2, Fast-mixed First-Stage Pond[a]

DATE:	2/5-3/4/78	3/5 - 4/1/78	4/2 - 4/27/78
Pond VSS, $mg\ell^{-1}$	184	206	263
% Chlorophyll *a*	2.1	2.2	1.9
% Removal by Microstrainers	92	87	82
Harvest, ℓ day^{-1}	95	185	185
$\theta \propto 1/(\text{avg. harv.})$, days	8.4	4.3	4.3
Productivity, Total VSS, gm m^{-2} day^{-1}	5.7	12.8	16.2
Langleys day^{-1}	240	370	470
Photo. Eff. % Total	1.3	1.9	1.9
NH_4^+ -N, $mg\ell^{-1}$	0.7	< 0.1	< 0.1
NH_4^+ -N, % removal	97	> 99	> 99
COD, $mg\ell^{-1}$	77	90	—
COD, % removal	61	67	—

[a] This pond was operated semicontinuously by pumping out the prescribed volume daily and refilling with settled sewage. Grab samples for laboratory analyses were taken shortly before dilution. Depth was 20 cm; paddle wheel rotation was 3.5 rpm.

lating outgassing and increasing pond algal densities. Recycling a fraction of the clarified effluent can be used to dilute the incoming sewage by whatever amount is necessary for complete NH_4^+-N removal in the pond.

CULTIVATION OF NITROGEN-FIXING BLUE-GREEN ALGAE ON LOW-N SEWAGE EFFLUENTS

The outdoor cultivation of heterocystous blue-green algae was initiated in the spring of 1977. Although all the early inoculations failed to exhibit any significant growth and, in most cases, disappeared within a few days, experience showed that, by keeping light intensities low (through prolonged shading of the ponds) and initial densities high, successful inoculation could be carried out. Cultivation was carried out both on sewage pond effluents low in ammonia—obtained as described above (typically the NH_4^+-N was well below 250 ppb)—and on artificial blue-green algal media (one-quarter strength Allen and Arnon media [15] supplemented with 0.2 g/liter $NaHCO_3$).

Three types of ponds were used in the experiments: circular 3.0-m^2 ponds that were 25 cm deep (C-ponds); rectangular 3.0-m^2 ponds that were 45 cm deep (A-ponds); and 0.55-m^2 ponds that were 35 cm deep (W-ponds). The C- and A-ponds were mixed by paddle wheels and the W-ponds by compressed air/carbon dioxide. The ponds were operated by semicontinuous dilution; each day the prescribed percentage of the pond volumes was withdrawn (at about 12:00 PM to 2:00 PM) and diluted with an equal volume of medium. Samples for laboratory analyses were withdrawn before dilution.

All ponds were sparged during the daytime with mixtures of air/carbon dioxide for pH control and as a carbon source. In general, this method of carbon dioxide addition proved inadequate in maintaining the pH within prescribed limits (particularly in the late afternoons).

In general, the algae grown in mass culture were first cultured in the laboratory on one-quarter strength Allen and Arnon media [15] in successively larger batch cultures up to 75 liter. These were used for outdoor pond inoculations. The cultures used included an *Anabaenopsis* sp., two *Nostoc* species, and a mixed *Nostoc-Anabaena* culture. Acetylene reduction assays were carried out in the laboratory on small samples freshly collected from the ponds (see Materials and Methods). *In situ* assays in the ponds showed a good correlation between laboratory-determined rates of nitrogen fixation and those observed below the pond surface. A summary of some of the results is given in Table VI.

The *Nostoc-Anabaena* cultures grew with tightly tangled filaments, forming macroscopic clumps or flakes. These clumps settled quickly, requiring fast mixing to keep them in suspension. Long-term cultivation on sewage effluents was not successful with these algae. Densities decreased slowly over a period of one to three months. A typical experiment is shown in Figure 3. The pond detention time was lengthened as pond density dropped. Initially, the culture contained an *Anabaena* sp. (30%) which grew freely suspended as well as at the edges of *Nostoc* clumps (comprising 60% of the biomass). The *Nostoc* clumps were very densely packed and contained both green and yellowed cells (this was also characteristic in laboratory cultures).

TABLE VI

Summary Data of Outdoor Blue-Green Algal Cultivation[a]

POND	Dates 1977-1978	AREA	ALGAE	MEDIA	Average Detention Time Days	Average Density mg/ℓ	Average Productivity $gm/m^2/day$	Visible Light Conversion Efficiency	Average N_2 Fixation Capacity n mol C_2H_4/min/mg
A-1	8/31-9/6	$3m^2$	Nostoc	low N effl.	7	120	2.9	0.3	0.7
A-2	8/31-9/16	$3m^2$	Mixed	"	7.6	130	3.1	0.4	1.0
C-3(A-2)	9/7 -9/16	$3m^2$	Anabaenopsis	"	8	90	2.0	0.3	0.6
W-1	9/29-10/17	$0.55m^2$	Anabaenopsis	Chemicals	8	150	3.8	0.6	2.2
W-1	11/7-12/12	$0.55m^2$	Anabaenopsis	Chemicals	10	145	3.2	0.9	3.8
W-2	11/22-12/12	$0.55m^2$	Anabaenopsis	Chemicals	8	58	1.4	0.4	4.0
W-1	4/23-5/6	$0.55 m^2$	Anabaenopsis	Chemicals	8	275	7.0	0.8	3.2

[a] Pond W-1 was heated except from September 29 to October 17.

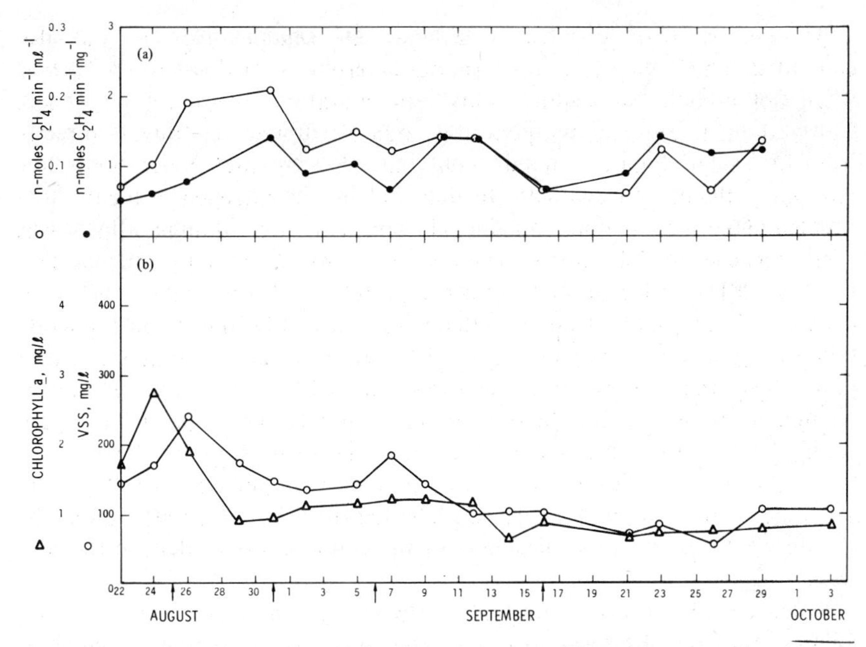

Fig. 3. *Anabaena* and *Nostoc* mixed culture grown on effluents from second-stage batch ponds. Arrows indicate the following changes in pond operation. August 25, semicontinuous operation begun with θ = 6 days and depth = 15 cm; August 31, θ = 7 days and depth = 18 cm; September 6, θ = 8 days and depth = 20 cm; September 16, 17, 18, pond not diluted, θ = 8 days thereafter. (a) Nitrogen-fixing capacity; (b) Pond density.

Green algae and an *Oscillatoria* sp. were always present to a small extent. During the beginning of September, the proportion of *Anabaena* began dropping. It fell below 5% by the middle of that month. The *Nostoc* fell below 50% of the total biomass shortly thereafter with the remainder made up by diatoms, *Oscillatoria*, and green algae, respectively. The pond was invaded by rotifers and crustaceans in early September. This experiment was typical in that the decline in density was slow. Diatoms were often found to invade these ponds initially attaching to the edges of clumps. Presumably, a source of fixed nitrogen was made available by the blue-green algae through excretion and/or cell lysis. Free NH_4^+-N was less than 75 ppb in these cultures, and 1–2 mg/liter of total PO_4-P were present in microstrained effluents from these ponds.

The slow disappearance of the blue-green algae was not indicative of photo-oxidative death, to which blue-greens are notably susceptible. A bioassay was performed on effluents from the second-stage growth ponds. Green algae did not proliferate, and no nutritional deficiency was discovered for growth of blue-greens. These blue-green algal ponds were run at ambient temperatures (11–18°C at 9 AM; 15–25°C at 4 PM), which was probably suboptimal.

Another strain of nitrogen-fixing algae, an *Anabaenopsis* sp., was also cultivated. This coiled alga has terminal heterocysts. If floats to the surface when not mixed, but requires only mild agitation to remain suspended. Cultivation, at ambient temperatures, was attempted in effluents directly from first-stage, high-rate ponds (containing 3–5 mg/liter NH_4^+-N); in low nitrogen effluents; in synthetic media; and in low nitrogen effluents fully supplemented with synthetic media. The ammonium-containing effluent was tried because in laboratory chemostats, a low-ammonium media (1.5 mg/liter NH_4^+-N) had little effect on growth of this alga. The *Anabaenopsis* in the outdoor pond diluted with first-pond effluents died quickly, followed by a bloom of green algae. The culture bleached completely, at a time when other *Anabaenopsis* cultures retained their pigments. However, pigment content was low before this in all ponds. The typical blue-green color of the cultures had faded to a brownish-green. A month later, all of the *Anabaenopsis* cultures died when a period of cloudiness was followed by a couple of days of sunny weather. The densities had dropped during the cloudy weather, and total bleaching of the cultures was evident in the sunshine.

Because of the slow growth of all of the blue-green algae and the low pigmentation of the *Anabaenopsis*, we tested the effect of increased temperature. Goldman [16] has indicated temperature as a determining factor in algal culture. Synthetic media was used in these experiments.

CULTIVATION OF *Anabaenopsis* ON SYNTHETIC MEDIA

Two 0.55-m^2 ponds were operated on one-quarter strength Allen and Arnon's [15] media supplemented with 0.2 g/liter $NaHCO_3$. The ponds were sparged with air/carbon dioxide equal to 99.7%/0.3% (v/v) to provide

TABLE VII

Summary of Data from Cultivation of an *Anabaenopsis* sp.[a]

DATE	Pond VSS mgℓ^{-1}	% Chlor. a	Harvest % day^{-1}	$\theta \propto$ 1/(avg. harv) days	Productivity, Total VSS, gm m^{-2} day^{-1}	Langleys day^{-1}	Photo. Eff. % Total	Temp. °C a.m./p.m.	N_2 Fixation cap. nm C_2H_4 min^{-1}mg^{-1}	Particulate Nitrogen, % of VSS
a.										
11/6-12/17/77	145	1.6	11.1	10.0	3.0	220	0.8	22/23	4.0	8.1
12/18-2/18/78	127	1.5	4.6	24.1	1.1	175	0.3	22/23	1.5	9.5
2/19-4/1	190	1.5	13.5	8.3	4.7	330	0.8	23/24	3.5	10.2
4/2-4/29	225	1.3	12.8	8.8	5.3	470	0.6	24/25	3.0	—
4/30-6/5	285	1.3	14.0	8.0	7.3	—	—	25/26	—	—
b.										
11/20-12/17	57	1.0	13.1	8.5	1.4	190	0.4	11/14	4.5	9.6
12/18-2/18	30	1.0	2.1	53.0	0.1	175	<0.1	11/13	1.5	10.0
3/5-4/1	180	1.3	10.2	11.0	3.3	370	0.5	18/21	2.8	—
4/2-4/29	150	1.2	21.8	5.1	6.0	470	0.7	24/26	3.6	—

[a] Grown on synthetic media in 0.55-m^2 ponds. pH was 8.5–10.0, depth was 20 cm, air/CO_2 was 99.7%/0.3%, sparging rate was 7 liter/min (= 100 g $CO_2/m^2/day^2$). (a) with heating, (b) without heating from 11/1977 to 2/1978 and with heating from 3/1978 on.

carbon, pH control, and mixing. Table VII shows results of six months of operation of these ponds. One was heated to the indicated temperatures while the other was run at ambient temperature for the first four months.

The *Anabaenopsis* grew continuously in the heated pond except during a period of particularly low insolution (100 langleys/day average) and high precipitation in January. The algae grew poorly in the unheated pond in November and not at all in December, January, and February. Its color changed from blue-green to brown within the first two weeks in the unheated pond. Chlorophyll *a*, allophycocyanin, and phycocyanin content were 1.0, 1.2, and 2.7% of the dry weight in this pond compared to 1.6, 3.9, and 7.3% in the heated pond. Despite the lower phycobiliprotein levels, particulate nitrogen (as a percentage of pond VSS) was slightly greater in the unheated pond. The slower growing algae were greater than 10% N, while the heated, faster growing algae were about 9% N. When nitrogen is not limiting to growth, blue-green algae are known to store it intracellularly under conditions adverse to growth.

Nitrogen-fixation capacities were measured using the acetylene reduction technique in the laboratory at 27°C under saturating light intensity. The lower temperatures did not reduce this capacity when both cultures were still growing. On a specific mass basis, in the unheated pond, the algae often exhibited a higher capacity. However, the results of the assays were found to be very dependent on the light intensities during the hours preceding the assay. Light would be more limiting in the heated pond due to the higher density, and, thus, internal levels of ATP and reductant would be lower.

Productivity was a function of temperature. A pond heated to 23–25°C produced one-third more biomass than a pond heated to 18–20°C. The unheated pond produced from 0–40% of the heated ponds. In November, the heated blue-green algal pond produced 3 $g/m^2 \cdot day$ at $\theta = 10$ days. From 19 February to 1 April, at $\theta = 8.4$ days, 4.7 $g/m^2 \cdot day$ of blue-green algae were produced. Production of green algae at this time was very variable—between 3 and 6 $g/m^2/day$ at similar dilution rates. Thus, productivities were comparable at long detention time and low insolation. The decrease in productivity due to the energy costs of fixing atmospheric nitrogen are difficult to determine, but some small reduction is likely.

To determine the extent to which productivity can be increased (at about 25°C), one pond was operated at a shorter detention time ($\theta = 5$ days, see Table VIIb). Productivity increased only about 10–20%. This increase is much less than would be expected with green algae. Green algal ponds operated at $\theta = 5$ days produced about 10 $g/m^2/day$ (when the eight-day, blue-green pond produced 7 $gm/m^2/day$), while those diluted with $\theta = 4$ days produced about 15 $g/m^2/day$.

Blue-green algae become photoinhibited at lower light intensities than green algae. Photosynthesis is also saturated at lower intensities. It may be that, in general, the light-saturated rate of photosynthesis is lower for blue-green algae. This would limit the maximal photosynthetic efficiency of blue-green algae relative to green algae at the high light intensities of outdoor cultivation. The differences would be most noticeable at low culture densities. On the other hand, lower maintenance requirements of blue-green algae minimize to some extent the decrease of net efficiency at higher densities. The temperature for optimum growth of the *Anabaenopsis* sp. has yet to be determined, but this genus is normally tropical. Thus, 25°C is most likely suboptimal for growth and productivity. Pigmentation is evidently very temperature dependent.

The data shown in Table VII were obtained in ponds that had 30 cm of freeboard. This shaded the ponds as much as 70% in afternoons and mornings and 20% at noon. Thus, the photosynthetic efficiencies and areal productivities listed are considerably lower than what would have been calculated taking this shading into account. However, the shading may have had the beneficial effect of lowering the average light intensity in the cultures as well as the time any particular cell spent in high light. This could have lowered photoinhibition and photo-oxidation particularly in the ponds operated at lower densities. Indeed, when the culture, described in Table VIIb was transferred to a pond with only 10 cm freeboard (data not shown), the culture density began to fall after a few days. It bleached quickly thereafter, indicating that at a density of 75–125 mg/liter, shading was necessary to prevent photo-oxidation. Thus, although this nitrogen-fixing alga (and many free-floating, nonclumping blue-greens) must be cultured at higher densities if grown under full sunlight, productivity of blue-greens at higher densities may be significantly greater than that of green algae. Longer detention time, nitrogen-fixing ponds may turn out to be substantially more productive when the shading is reduced. Experiments along these lines are under way.

SUMMARY AND CONCLUSIONS

The data presented represent an initial, limited attempt on a small scale to cultivate nitrogen-fixing blue-green algae on both chemically defined media and low nitrogen sewage pond effluents. The rates of blue-green algal biomass production were low compared to those of green algae. Nevertheless, it appears that cultivation of nitrogen-fixing blue-green algae is possible on sewage effluents where these algae could be used as a method of tertiary treatment.

These preliminary experiments have brought out some of the problems involved in outdoor cultivation of nitrogen-fixing blue-green algae. Unless cold-adapted strains can be isolated, use of these algae may be limited to regions with warmer climates. Growth on low nitrogen sewage effluents might require some nutrient supplementation, a factor which can be investigated in the laboratory. The endurance of most of the outdoor cultures allows optimism that these effluents are not inhibitory to growth and that supplementation will be minor.

The sensitivity to high light intensities is a factor that limits the maximal productivity of these organisms by requiring that they be cultivated at high density. At 25°C, we have attained 6–8 $g/m^2/day$ with an eight-day detention time (250 mg/liter) in the spring. This rate can undoubtedly be increased to about 12 $g/m^2/day$ in the summer and possibly twice that at optimal temperatures. At 10% N, 12 $g/m^2/day$ corresponds to 1.2 $g/m^2/day$ of nitrogen, which, if it can be sustained year round, extrapolates to approximately 5 metric tons of nitrogen per hectare per year. This rate would allow such systems to be considered in agricultural fertilizer production.

Production of sewage pond effluents suitable for the growth of nitrogen-fixing algae is feasible in one- or two-stage systems fed with sewage influents. In the two-stage system, levels of fixed nitrogen are reduced in a continuously diluted high-rate pond with the growth of green algae. These algae, and the remaining nitrogen, are removed in second-stage batch ponds where a growth or isolation process results in efficient algal settling. The year-round reliability of these systems must yet be established, but it appears that batch growth ponds are effective in spring and summer whereas isolation works during the remaining months. The relationship among mixing, nutrient limitation, flocculation, and settling in batch ponds is as yet unclear. It is also not known whether the relationships are the same in growth and isolation ponds. It may be possible that complete NH_4^+-N removal and production of rapidly settleable flocs can be achieved in a one-stage system by adjusting the sewage strength with effluent recycling in fast-mixed ponds.

We would like to thank Professor William J. Oswald for encouragement and support. This work was supported by the U.S. Department of Energy, Contract No. EY-76-S-03-0034-279.

References

[1] E. F. Gloyna, J. F. Malina, and E. M. Davis (Eds.), *Ponds as a Waste Treatment Alternative*, Water Resources Symposium No. 9, Univ. of Texas, Austin, 1976.

[2] B. L. Koopman, J. R. Benemann, and W. J. Oswald (unpublished).

[3] J. R. Benemann, J. C. Weissman, B. L. Koopman, and W. J. Oswald, *Nature*, *268*, 19–23 (1977).

[4] A. Watanbe, *J. Gen. Appl. Microbiol.*, *6*, 85 (1959).

[5] G. S. Venkateraman, *The Cultivation of Algae* (Indian Council on Agricultural Research, New Delhi, 1972).

[6] S. H. Pantastico and J. L. Gonzales, *Kalikasan Phollipp. J. Biol.*, *5*, 221 (1976).

[7] C. G. Golueke, W. J. Oswald, and H. B. Gotaas, *Appl. Microbiol.*, *547*, 551 (1957).

[8] W. J. Oswald and J. R. Benemann, "Fertilizer from algal biomass," RANN II Symposium, Washington, D.C., 1976.

[9] W. J. Oswald, *Developments in Ind. Microbiol.*, *4*, 112 (1963).

[10] G. Mackinney, *J. Biol. Chem.*, *140*, 315 (1941).

[11] *Standard Methods for the Examination of Water and Wastewater*, 13th ed. (Amer. Pub. Hlth. Assn., Washington, D.C., 1976).

[12] J. C. Weissman and J. R. Benemann, *Appl. Environ. Microbiol.*, *33*, 123 (1977).
[13] A. Bennett and L. Bogarad, *J. Cell Biol.*, *58*, 419 (1973).
[14] J. C. Weissman, D. Eisenberg, J. R. Benemann, and W. J. Oswald (unpublished).
[15] M. B. Allen and D. I. Arnon, *Plant Physiol.*, *30*, 366 (1955).
[16] J. C. Goldman, *Limnol. Oceanogr.*, *22*, 932 (1977).

Energy Recovery from Landfilled Solid Waste

F. B. DeWALLE
Department of Environmental Health, University of Washington, Seattle, Washington 98195

E. S. K. CHIAN
Department of Civil Engineering, Georgia Institute of Technology, Atlanta, Georgia 30332

INTRODUCTION

Most energy sources considered to be important in the current energy shortage constitute nonrenewable resources. While oil and coal are indirectly derived from solar energy through plankton and plant formation over long conversion periods, the current rate of depletion would classify these fuels as nonrenewable. Recently formed biomass and direct solar energy, however, are renewable resources. Only 1% of the 1.7×10^{11} tons of the biomass produced annually is currently being used for fuel and human usage [1]. It would seem possible to increase in this percentage by recycling biomass after its initial human usage. For example, biological matter present in municipal solid waste (newspaper, construction debris, vegetable scraps, etc.) could provide valuable additional energy resource.

Energy can be generated from municipal solid waste by direct combustion, pyrolysis, or by the formation of hydrogen and methane by microbial processes and the subsequent combustion of the gas. The latter alternative produces 630 kcal/mol carbohydrate degraded. The lignins, comprising 30–50% of woody plant material, are easily combustible but are not very amenable to microbial digestion. Thus, carbohydrates and plant waxes and fats are the major renewable substrates available for microbial conversion to methane.

The actual amount of energy recovery from the solid waste will depend on the efficiency of the conversion process used. The net energy yield, expressed as a percentage of the theoretical refuse energy content, ranges from approximately 10% for methane recovery to between 30 and 40% for pyrolysis. Heat recovery by incineration can yield as much as 60% of the theoretical energy [2].

While pyrolysis and incineration require elaborate processes, the production of methane from solid waste occurs in sanitary landfills without any sophisticated controls. The generated gas can be recovered from the fill through simple extraction wells maintained under a slight vacuum after which it is processed in a gas cleaning unit. Construction and operating

Biotechnology and Bioengineering Symp. No. 8, 317–328 (1978)

0572-6565/78/0008-0317$01.00

costs for such a landfill gas recovery system are considerably less than those for the pyrolysis and incineration processes.

It is therefore not surprising that gas recovery wells have been installed at several landfills. Schuyler [3] reported that pumping rates ranging from 12,234 m^3/day (300 ft^3/min) to 22,468 m^3/day (550 ft^3/min) were obtained from a 33.5-m-deep (110-ft) extraction well at the Palos Verdes landfill in Los Angeles County, California (a landfill receiving relatively wet solid waste). Based on a measured radius of influence of 76 m (250 ft) and an estimated dry density of 714 kg/m^3 (1200 lb/yd^3), the pumping rate of 13,050 m^3/day (320 ft^3/min) corresponds to a gas production rate of 30 ml/kg solid waste/day, with methane representing approximately 50% of the gas phase. The maximum flow rate observed was equivalent to 56 ml/kg/day.

In the Mission Canyon landfill (Los Angeles County, California), which receives relatively dry solid waste, 15 gas recovery wells, each approximately 30 m (100 ft) deep, recover 3260 m^3/day (80 ft^3/min), with methane and carbon dioxide representing approximately 70% of the gas phase.

In a test well at the Sheldon Arleta landfill, the Los Angeles Bureau of Sanitation [3] measured a pumping rate of 5909 m^3/day (145 ft^3/min), with 62% of the gas phase consisting of CO_2 and CH_4. At a radius of influence of 76 m (250 ft), a depth of 38 m (125 ft), and an estimated solid waste density of 595 kg/m^3 (1200 lb/yd^3), this pumping rate corresponds to a gas production rate of 22 ml/kg/day.

A substantial extraction rate, 2039 m^3/day (50 ft^3/min), with 78% of the gas phase consisting of CO_2 and CH_4, was reported for the relatively shallow Moutain View, California, landfill, which is only 12 m (40 ft) deep. With a radius of influence of 40 m (130 ft) and an estimated solid waste density of 595 kg/m^3 (1000 lb/yd^3), the extraction rate corresponds to a gas production rate of 45 ml/kg/day [4].

The above results indicate that it is technically feasible to recover methane from anaerobically decomposing solid waste in landfills. The rate and amount of gas produced, however, are uncertain, as is the influence of environmental factors such as moisture content, density, temperature, waste particle size, and exposure to air. Because systematic testing is best accomplished on a laboratory scale, the bench-scale study described here was initiated to evaluate the effects of these factors. Solid waste was placed in sealed containers, and the production and composition of the generated gases were measured.

MATERIALS AND METHODS

Eighteen 208-liter (55-gal) steel containers of the type shown in Figure 1 were used in the laboratory simulation. Each drum was lined with 10 polyethylene bags of 0.15-mm thickness. A cushion of construction sand was

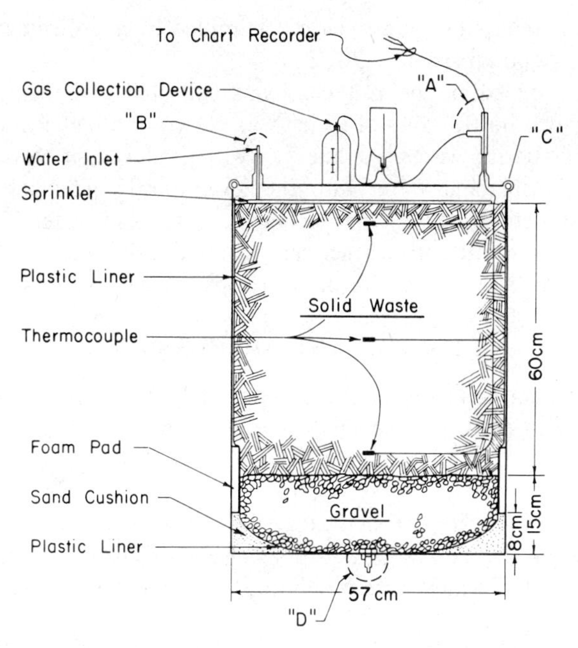

Fig. 1. Cross section of solid waste container: D = drum cross section, not to scale. All dimensions typical.

placed between the drum floor and the plastic liner, with a gradual slope to a height of about 8 cm at the periphery of the drum to funnel leachate toward a central drain. A 15-cm layer of class A gravel was placed inside the linear to provide for collection of the leachate and screening of the drain. Two fittings were installed on the top of each drum. One was used to apply water to the solid waste, and the second allowed for the collection of gases and provided access for the thermocouple wires. The drain installed at the bottom of the container was used to collect the leachate.

The containers was filled with from 55.0 to 80.5 kg (dry weight) of solid waste. They were then sealed and maintained under different environmental conditions. Seventeen of the containers were located in an insulated and air-conditioned room maintained at an average temperature of 17°C (62°F). One was located in an adjacent room maintained at an average temperature of 26°C (79°F).

The solid waste was obtained from the City Solid Waste Reduction Plant in Madison, Wisconsin. It had been collected by municipal employees in wards 6, 7, 13, and 15 on Thursday, January 15, 1976. In ward 6, located in downtown Madison, the values of the homes were generally less than \$10,000. Ward 7, also located in downtown Madison, is composed largely of old homes in the \$20,000 range, many of which are occupied by students. Wards 13 and 19 are both in suburban areas with homes in the \$50,000 and \$30,000 ranges, respectively. The solid waste is collected once a week and is brought to the reduction plant where it is milled by a Tollemache mill and shredded to a nominal size of 0.7–2.5 cm.

The gas produced in the test cells was collected with Mariotte flasks consisting of two bottles, one of which was placed at a higher level than the other and filled with water. As the gas was produced, it flowed into the upper bottle, displacing water into the lower bottle. The volume of gas produced was determined by measuring the change in water levels. Gases were collected intermittently rather than continuously, and the drums were kept completely sealed except for brief periods during gas measurement.

RESULTS AND DISCUSSION

The initial data collected in the present study have been presented previously by Chian et al. [5] and De Walle et al. [6] All containers produced gas, the amount varying between 0.09 and 6.0 liter/kg dry weight, as shown in Figure 2.

The effect of the different environmental variables can be deduced from the data in Figure 2. The largest amount of gas was produced in containers 6, 13, and 14, all of which were maintained at the highest moisture content.

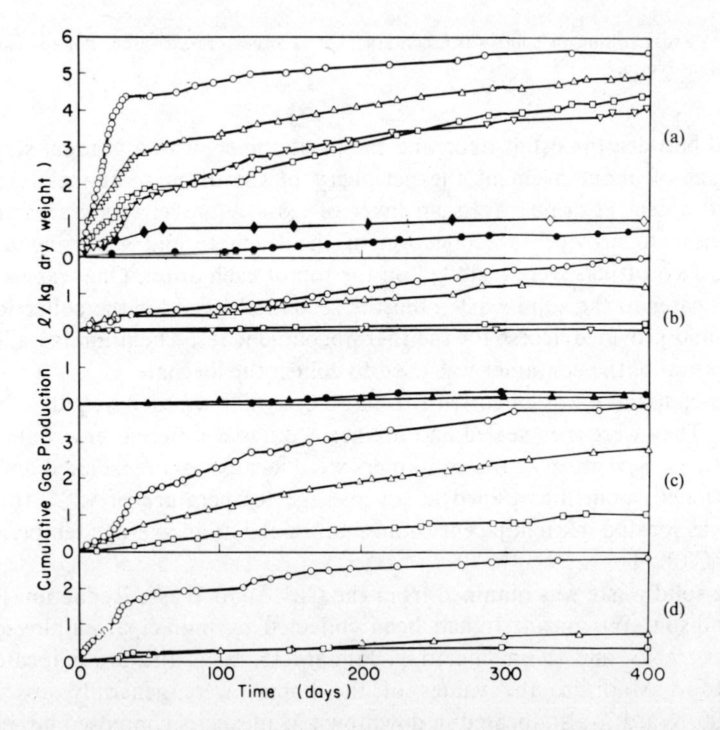

Fig. 2. Cumulative gas production of the enclosed solid waste as affected by different variables during steady-state testing. (a) (▽) cell 15; (○) cell 14; (△) cell 13; (◆) cell 10; (□) cell 6; (●) cell 7; (b) (●) cell 11; (□) cell 9; (△) cell 2; (▽) cell 1; (▲) cells 1, 7, 9; (c) (△) cell 18; (○) cell 15; (□) cell 4; (d) (○) cell 16, (□) cell 12; (△) cells 1, 2, 4, 5, 7.

Substantially lower quantities were produced at 60% and 36% moisture. The size of the solid waste is probably the second most important factor, as shown by the gas production from cells 4, 15, and 18. The moisture content of these three cells was maintained at 78%, but the cells contained waste of different sizes. Cell 18 contained 12.5-cm solid waste, and cell 4 contained 25-cm waste, whereas all other cells (including cell 15) contained 2.5-cm waste. Increasing the density of the solid waste tends to decrease the gas production, possibly because of a decrease in the effective surface area exposed to enzymatic hydrolysis. Temperature is shown to be the environmental variable having the least effect on gas production.

While the gas production data for the containers clearly showed the effects of the environmental factors, significant variability was found among individual containers maintained under identical conditions. The highest gas-production rate at the 36% moisture content, for example, was 265% of the average value, while the lowest rate was 14% of the average. At higher moisture contents the variability was less; the highest rate at 99% moisture was 113% of the average, and the lowest rate was 88% of the average. The small variation in the rates at 99% moisture is probably a result of the homogeneous distribution of moisture within the drums.

The results in Figure 2 further indicate that in cells having high gas-production rates, the gas is produced first at a high rate and then at a reduced rate which is a factor of 20–40 less than the initial rate. The high initial rates, which were sustained for up to 40 days, indicate that a major amount of the gas-producing solid waste fraction is readily and rapidly degradable, probably at zero-order rates. The shape of the other cumulative gas production curves generally do not show this stepwise pattern. Instead, they resemble more closely a parabolic function, possibly reflecting first-order reaction rates.

Gas analysis showed that the N_2 initially present at 79% was rapidly replaced by CO_2. The most rapid N_2 displacement was observed in the containers with the highest moisture contents, reflecting the CO_2 "bloom" often observed in landfills [7]. After approximately 50 days, hydrogen was found in the gas phase from 8 of the 15 sealed containers. The amount of hydrogen peaked after approximately 100 days and gradually decreased thereafter. Only one of the containers filled with large waste particles (cell 4) has, thus far, produced methane. The methane, which constituted 50% of the gas volume, appeared immediately in the container rather than following the typical sequence in which methane generation follows the CO_2 bloom [7]. The gas composition data, therefore, indicate that at the time of the most recent measurement most of the containers had just completed the acid hydrolysis phase characterized by CO_2 and H_2 generation. Methane fermentation apparently did not start in most containers within the initial 1-year period.

As the moisture content was found to be the most important variable during the intial 400-day monitoring period, this parameter was studied in greater detail during the following 300-day period. Six of the 15 containers received simulated rainfall at a rate of 25–50 cm/year to simulate transient-state conditions, as opposed to the steady-state conditions maintained during the first period. The resulting data are plotted in Figure 3, with day 0 representing the end of the first 400-day period and the beginning of the transient-state testing.

The results presented in Figure 3 show that the greatest gas production occurred in the containers that were brought to a 99% moisture content at the beginning of the first period (cells 6, 13, and 14). Addition of 25 cm/year of simulated rainfall to cells 13 and 14 immediately produced leachate, as the 99% moisture content was approaching field capacity. Adding a buffering substance, 25.5 g/liter of $NaHCO_3$, to cell 13 approximately doubled the rate of gas production, increasing it to 7 ml/kg/day as compared with the value of 3.5 ml/kg/day observed before adding the simulated rainfall and buffer. Cell 14 received the same amount of rainfall as cell 13 but, in addition, it received recirculated leachate to simulate conditions in a solid-waste landfill which uses leachate recirculation. The leachate collected at the bottom of cell 14 was sprayed on top of a simulated landfill cover to allow a portion of the leachate to evaporate. The leachate that drained through the simulated cover was then mixed with the simulated rainfall and added to the top of cell 14. The addition of leachate and rainwater to the solid waste did not result in a noticeable increase in gas production; that is, it remained at approximately 1 ml/kg/day. At 200 days after the start of the transient period, the gas production was gradually

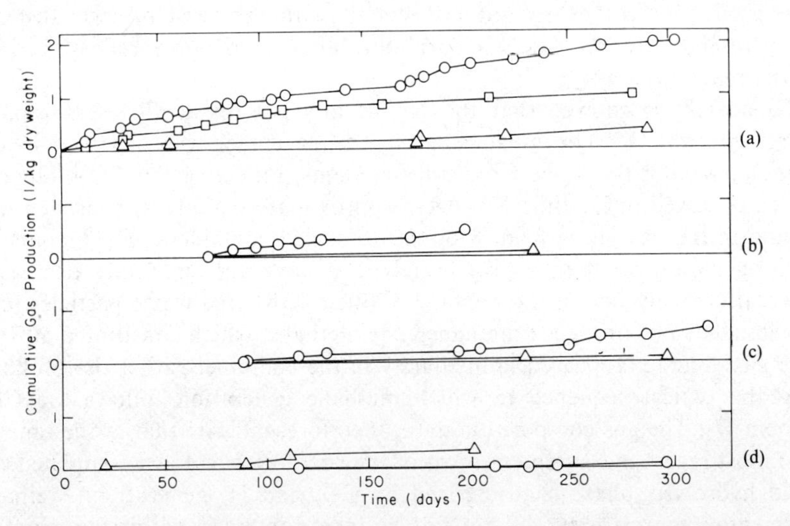

Fig. 3. Cumulative gas production of the enclosed solid waste during subsequent transient-state testing. (a) Initial moisture content 99%: (Δ) cell 14, (O) cell 13, (□) cell 6; (b) initial moisture content 36%: (O) cell 12, (Δ) cell 7; (c) initial moisture content 36%: (O) cell 9; (Δ) cell 1; (d) initial moisture content 36%; (Δ) cell 5; (O) cell 2.

increasing. Control cell 6, which was always maintained at a 99% moisture content, continued to produce gas at a gradually decreasing rate.

The gas produced to date by all three of these high-moisture containers is primarily carbon dioxide. However, cell 13 (the container receiving the buffer) is currently producing traces of methane in addition to carbon dioxide. The fact that the pH of the leachate collected from cell 13 was 5.7 as compared to 5.5 for container 14 indicates that a sufficiently high pH value is required to initiate methane fermentation.

Lesser quantities of gas were produced in the four cells in which the solid waste was initially maintained at 36% moisture. Addition of simulated rainfall at a rate of 25 cm/year was observed to generate leachate at the bottom of cell 12, containing high-density solid waste (485 kg/m^3 dry density), after a 145-day period, of cell 1 after 200 days, and of cell 2 after 247 days. Cell 9, which received rainfall at 50 cm/year, produced leachate after a 120-day period. The gas production in these cells generally started to increase before the solid waste had reached field capacity and before the leachate started to break through, indicating that a moisture content less than the field capacity is sufficient to produce increased gas production.

Before rainfall was added to cell 12, the container had ceased to produce any gas. Adding simulated rainfall caused gas production to resume after a 70-day "acclimation" period, indicating that the addition of water is capable of initiating gas generation. A similar observation can be made about the gas production in cells 1, 2, and 9. It is interesting to note that cell 9, which received the greastest amount of rainfall, did not produce twice as much gas as the containers receiving 25 cm/year, possibly indicating that the initial moisture content has a greater effect than the amount of moisture added at a later date. Additional research will be necessary to determine the relative importance of the initial moisture adjustment as compared to a gradual moisture increase.

The effects of environmental variables such as moisture content, temperature, size, density, and exposure to air, as evaluated in the present study, were compared with similar variables tested in other studies. The importance of moisture content in gas production, for example, is confirmed by the results of other studies [7–9]. As shown in Figure 4, those results may indicate that a logarithmically increasing gas production can be realized by a linear increase in moisture content. The plotted rates of the present study represent those volumes produced during the initial 20- to 50-day monitoring period when the highest rates were observed. The values given here, therefore, tend to be higher than those from other studies.

In the present study, methane was detected only in the container with the large solid waste (cell 4) and in the container receiving buffer (cell 13). Using shredded solid waste fractions, Merz [9] observed a maximum CH_4 content of only 0.9% despite the fact that a wide range of moisture contents were tested. Even in the large solid waste container, Rovers and Farquhar [7] noted a methane content of only 2.8%. Ramaswamy [10], however, often

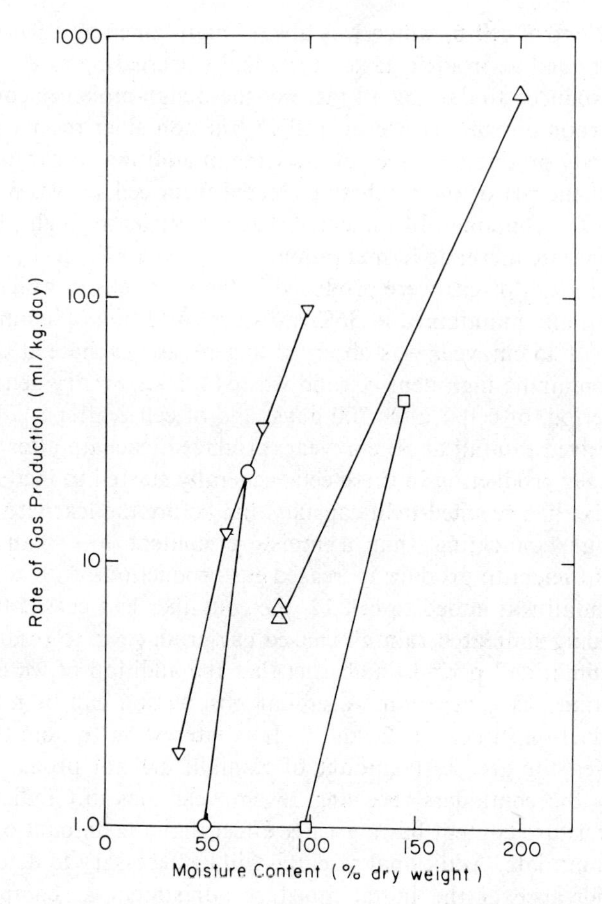

Fig. 4. Effect of moisture content on the rate of solid waste gas production as noted in different studies. (▽) This study; (○) Merz and Stone [8]; (△) Merz [9]; (□) Rovers and Farquhar [7].

noted stable methane fermentation as soon as 40 days after initiation of the tests. Careful examination of his data shows that methane fermentation does not start until the pH of the solid waste moisture increases beyond 5.0. The fact that methane fermentation was found only in that study may well be a result of the waste's high food content, which would have resulted in the release of sufficient amounts of ammonia during the degradation of the amino acids to counteract the pH decrease resulting from the release of free volatile fatty acids generated during the acid hydrolysis. The generation of methane in the container with the large solid waste may be a result of the relatively slow release of the volatile fatty acids, preventing the inhibition of the methane-generating bacteria. Thus pH, the concentration of free fatty acids, and buffer conditions seem to play a key role in the initiation of methane production.

The results in Figure 5 clearly indicate that an increasing moisture content results in a larger percentage of methane in the gas phase and a higher rate of methane generation. However, in several studies discussed earlier, no methane was observed at the highest moisture content. Thus, the way in which the moisture interacts with the solid waste may determine whether methane generation will or will not commence. It would therefore seem beneficial to add buffering substances such as dewatered anaerobic digester sludge, industrial alkaline sludge, septic tank pumpings, or lime sludges to initiate methane generation.

Using the temperature data obtained in the present study and those from Ramaswamy [10], it is possible to calculate the activation energy of the reaction, defined as:

$$\ln k_2/k_1 = (E/R)(1/T_1 - 1/T_2)$$

where k_1, k_2 are the reaction constants at temperature T_1 and T_2, E is the activation energy, R is the gas constant. The magnitude of the results shown in Figure 6 indicate that the rate of gas production, which parallels the hydrolysis of cellulose, is chemically controlled and not diffusion limited above a moisture content of about 75%. Substantially increased rates of gas production are also at moisture contents greater than 75%. Moisture

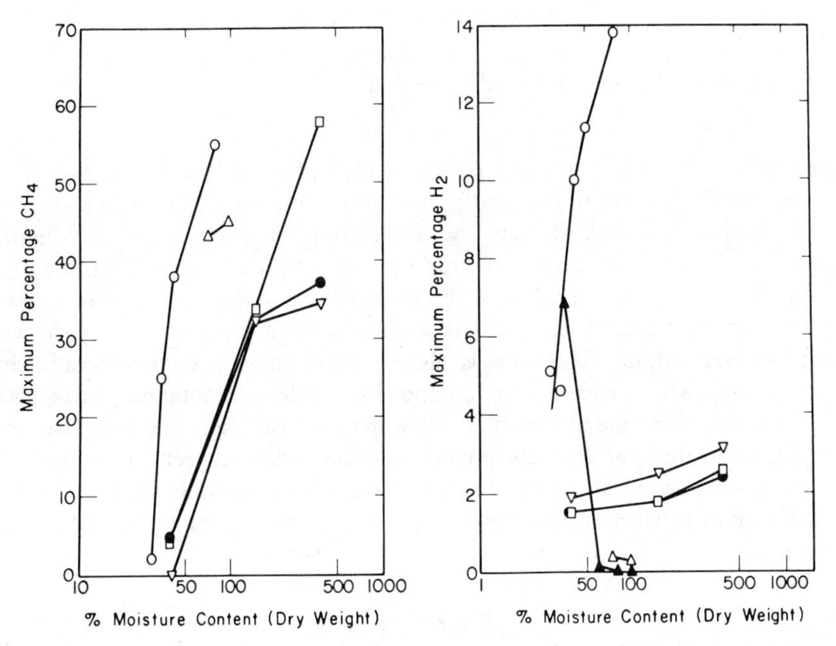

Fig. 5. Effect of moisture content on percentage of methane and hydrogen in the gas produced by enclosed solid waste. (□) Ramaswamy [10] at 35°C; (●) Ramaswamy [10] at 25°C; (▽) Ramaswamy [10] at 55°C; (○) Merz and Stone [8]; (△) Merz and Stone [8]; (▲) this study.

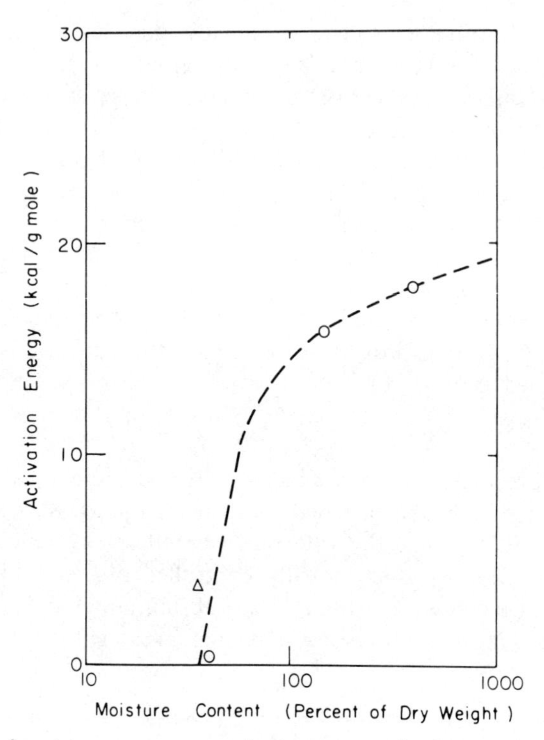

Fig. 6. Effect of moisture content on activation energy of solid waste placed at different temperatures. (Δ) This study; (O) Ramaswamy [10].

contents above 100% are not recommended because leachate will start to appear at the bottom of the landfill and may contaminate groundwater.

It should be realized that the generation of gas occurs only under anaerobic conditions; aerobic degradation of solid waste does not produce a large amount of gas because of the synthesis of biomass. It is also known that methane bacteria are obligate anaerobic, and exposure to oxygen will inhibit their action. This fact is clearly illustrated by data generated by Stone [11], who measured gas composition in leaky containers filled with solid waste and sludge mixtures. The data in Figure 7 indicate that the methane content of the gas phase decreases with respect to the carbon dioxide content when the percentage of nitrogen in the gas phase increases to 80% that is, when it becomes equal to the atmospheric content of this element.

CONCLUSIONS

The present study measured the rate and composition of gases released during anaerobic degradation of solid waste. The major gases observed were CO_2, H_2, and CH_4. Methane was produced in the container with the large

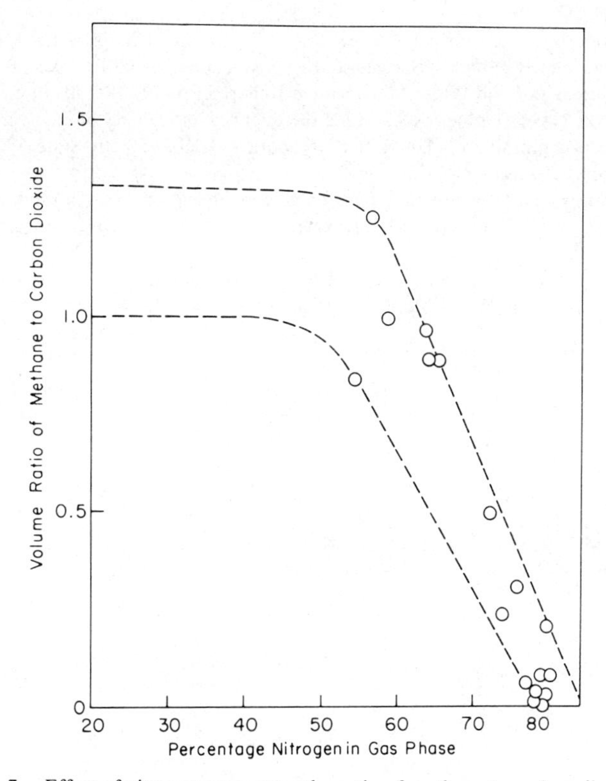

Fig. 7. Effect of nitrogen content on the ratio of methane to carbon dioxide.

waste particles and in the cell receiving buffering substances. The moisture content was the most important variable influencing the rate of gas production. It is recommended that landfills to be used for gas production be maintained at a moisture content above 75% but below 100%, or field capacity, to maximize gas production while holding the generation of leachate to an acceptable minimum, thereby preventing groundwater contamination.

References

[1] H. G. Schegel and J. Barnea, *Microbial Energy Conversion* (Pergamon, Oxford, 1977).
[2] W. C. Boyle, in *Microbial Energy Conversion*, H. G. Schegel and J. Barnea, Eds. (Pergamon, Oxford, 1977).
[3] R. E. Schuler, "Energy recovery at the landfill," paper presented at the 11th Annual Seminar on Governmental Refuse Collection and Disposal Association, Santa Cruz, Calif., 1973.
[4] R. A. Colona, *Solid Waste Manage.*, *19*, 90 (1976).
[5] E. S. K. Chian et al. in *Management of Gas and Leachate in Landfills*, S. K. Banerji, Ed. (U.S. Environmental Protection Agency, Research and Development Report EPA 600/9-77-026, Cincinnati, Ohio, 1977).
[6] F. B. DeWalle, et al. *J. Environ. Eng. Div., ASCE*, *104*, EE3, 415 (1978).

[7] F. A. Rovers, and G. J. Farquhar, *J. Environ. Eng. Div., ASCE*, *99*, 671 (1973).
[8] R. C. Merz and R. Stone, "Special studies of a sanitary landfill," U.S. Public Health Service, Bureau of Solid Waste Management Report EPA-SW 8R6-70, 1968.
[9] R. C. Merz, "Investigation to determine the quantity and quality of gases produced during refuse decomposition," University of Southern California to State Water Quality Control Board, Sacramento, Calif., 1964.
[10] J. N. Ramaswamy, "Nutritional effects on acid and gas production in sanitary landfills," Ph.D. thesis, Department of Civil Engineering, West Virginia University, Morgantown, W. Va., 1970.
[11] R. Stone, "Disposal of sewage sludge into a sanitary landfill," U.S. Environmental Protection Agency, Solid Waste Management Series SW-71D, 1974.

Pilot Plant Demonstration of an Anaerobic Fixed-Film Bioreactor for Wastewater Treatment*

R. K. GENUNG, D. L. MILLION, C. W. HANCHER, and W. W. PITT, JR.

Chemical Technology Division, Oak Ridge National Laboratory, Oak Ridge, Tennessee 37830

INTRODUCTION

Biological processes incorporated into conventional wastewater treatment schemes can generally be described either as stirred reactor systems, as in activated sludge or aerated lagoon processes, or as fixed-film systems, as in rotary disk contactors or trickling filter processes. Many of the treatment schemes developed during the part two decades combined forms of the stirred reactor and fixed-film systems. Most of these schemes concentrated on energy-intensive aerobic technology despite the potential energy-saving opportunities possible with anaerobic schemes through either reduced operating costs or the production of valuable by-products such as fuel gas. The limited research and development of anaerobic processes by the waste and wastewater treatment industries resulted from a lack of economic incentive to overcome such difficulties as long residence times and poor settling characteristics of anaerobic-produced solids. In addition, the industry has historically held the view that anaerobic fermentation processes are efficient only for wastewaters with high carbon concentrations (greater than 1000 ppm of total organic carbon) and only at temperatures above 25°C. Therefore, anaerobic processes are seldom considered for the treatment of low-strength, low-temperature wastewaters.

The increasing national concern of the 1970s for energy conservation and environmental protection provided new incentive for the development of innovative energy-saving technologies. The impact of this concern was accelerated in the area of pollution control by the passage in 1972 of Public Law 92-500, which required secondary sewage treatment for communities by 1977, certain farms and feedlots during 1978, and industries in general by 1983. As a result, it was projected that the increased treatment of municipal and industrial wastes would require twice as much energy in 1980 as was consumed in 1975 [1]. The need for new technologies which could

* Research sponsored by the Division of Building and Community Systems, U.S. Department of Energy, under contract No. W-7405-eng-26 with the Union Carbide Corporation.

Biotechnology and Bioengineering Symp. No. 8, 329–344 (1978)
Published by John Wiley & Sons, Inc. Not subject to U.S. copyright.

reduce the increases in costs and energy consumption in the near term was widely recognized.

In response to the need for major near-term decreases in the energy-intensiveness of contemporary technologies, Oak Ridge National Laboratory (ORNL) has been engaged in the development and demonstration of a pilot plant wastewater treatment facility based on an anaerobic, fixed-film bioreactor. The bioreactor employs a process which consists of attaching microorganisms to stationary packing material and passing liquid wastes upward through the unit for continuous treatment by biophysical filtration and anaerobic fermentation. The process has been demonstrated using municipal sewage with a bioreactor designed to process 500 gpd. The economic advantages of the process depend on the elimination of operating energy requirements associated with the aeration of aerobic-based processes and with the significant decrease of sludge-handling costs required with conventional activated sludge treatment systems. Methane can also be produced by the fermentation. With a view toward enhancement of technology transfer, the unit was designed during the summer of 1976 as a joint venture between ORNL and the Norton Company (Akron, Ohio). It was installed with the cooperation of the Norton Company and the City of Oak Ridge in the late fall of 1976. Since that time, operation has proceeded on a continuous basis. Downtime has been minimal, and the performance has been satisfactory except for minor mechanical problems.

BACKGROUND

Early investigation of upflow contact processes coupled with anaerobic treatment of wastewaters demonstrated that such approaches could be used to provide some gasification of the waste and some solids removal [2–4]. These processes depended primarily on the use of filtering devices but did provide a basis for pursuing the development of anaerobic bioreactors.

Selected recent references which are relevant to the development of anaerobic bioreactors for waste treatment are summarized in Table I. These begin with the pioneering work of Young and McCarty [5], who performed extensive laboratory studies of the anaerobic contact process. During these studies, they demonstrated the potential of upflow, fixed-film, anaerobic bioreactors for the treatment of wastewaters and the production of off-gases with methane contents as high as 75%. Their work involved high concentrations of synthetic feeds with low concentrations of suspended solids. Reactors were controlled at 25°C and typically operated with residence times of 36 hr in reactors of 1-ft^3 volumes. They measured chemical oxygen demand (COD) removals of greater than 90%. The effluents produced also had low contents of suspended solids; it was noted that the amount of solids that accumulated in the column would depend on the feed characteristics. These researchers commented that the performance of the bioreactor might improve if high-void-fraction packing was used. They also realized the

TABLE I

Selected References for Development of Anaerobic Bioreactors for Sewage Treatment

References	Reactor size (ft^3 packed volume)	Feed	Temp. (°C)	Type of packing	Feed strength (ppm COD)	Residence time (hr)	Loading (lb COD/day/1000 ft^3 of reactor)
"Anaerobic Filter for Waste Treatment," Young and McCarty Purdue Univ., Engr. Bull., Ext. Serv. 1967, No. 129 (Pt. 2), 559-574.	1	Synthetic feeds with low-conc. suspended solids	25	1- to 1-1/2-in.-diam quartzite stone $\varepsilon = 0.4$	1500-6000	4.5-72	27-212
"Anaerobic Digestion of Raw Sewage," W. A. Pretorius, Water Research, Pergamon Press, 1971, Vol. 5, pp. 681-687.	70	Raw sewage diluted with tap water	20	Layered stone and sand	500	24-45	15-30
"Bench-Scale Anaerobic Fixed-Film System Tests on a Low-Temperature, Low-Strength Synthetic Wastewater," Griffith and Compere, ORNL/TM-5267, Chem. Tech. Div., ORNL, in press.	2	Synthetic feed	20	1-in. stoneware Raschig rings $\varepsilon = 0.7$	650-1300	8	100-200
This paper.	200	Raw sewage	10-25	1-in. ceramic Raschig rings $\varepsilon = 0.7$	30-110	6-35	1.5-12

potential economic advantages of a treatment plant based on this bioreactor.

Pretorius [6] demonstrated the anaerobic digestion of raw sewage using a two-stage process, an agitated anaerobic digester for pretreatment followed by a multimedia biophysical filter. Using medium-strength sewage, Pretorius obtained a 90% reduction of COD in 24 hr at a temperature of 20°C; 60–65% of incoming solids accumulated in the digester and had to be removed periodically. The major part of the gasification was found to occur in the biophysical filter. More recent work by Pretorius [7] also demonstrated that since the anaerobic nature of his first stage, as opposed to aerobic processes, produced electron donors, an efficient denitrifying stage could be added to extend the treatment possible with the overall process.

Griffith and Compere [8] extended the work of Young and McCarty with detailed laboratory studies using columns with volumes of 2 ft^3, high-void-volume packing, and synthetic feeds. These studies were conducted with low-strength wastes at 20°C. Their work was given impetus by early nitrogen starvation tests conducted on columns; these tests showed that the decay coefficients of the columns were lower than would be expected from calculations and suggested the presence of a very stable biomass in the columns. These researchers reported reductions in biological oxygen demand (BOD) of greater than 80% for residence times of approximately 8 hr. Griffith and Compere also performed limited studies of these columns in the field using dilute raw sewage (30–100 ppm TOC) in the temperature range of 15–25°C. Results of these studies showed that the columns were capable of adapting to the low-temperature, low-strength feeds while retaining their ability to significantly reduce the pollutant concentration and to generate off-gas with a methane content approaching 70%. They also showed that the columns were capable of reducing the total suspended solids (TSS) content by more than 70%. The implied advantages of this flow process over conventional anaerobic digestion processes, which could require 10–30 days of holdup for equivalent treatment, were very significant.

The research projects described above demonstrated that upflow anaerobic, fixed-film bioreactors could clearly provide efficient treatment of high- or medium-strength wastewaters with off-gases having high methane contents. The residence times required were significantly lower than those required for conventional technology, and operating energy requirements were also anticipated to be greatly lowered [1]. Sludge removal requirements would be related to the rate of solids accumulation in the bioreactor; these rates were undefined but predictably low. Data indicated that such bioreactors would also efficiently treat low-strength, low-temperature wastewaters. The potential for developing a needed, near-term technology for wastewater treatment based on an anaerobic, fixed-film bioreactor led

ORNL to a pilot plant development and demonstration project for realistic evaluation of a total process.

DESCRIPTION OF PILOT PLANT

Reactor Description and Process Flows

The demonstration pilot plant is based on an anaerobic upflow (ANFLOW) bioreactor as shown in the flowsheet in Figure 1. The ANFLOW bioreactor, a cylindrical tank constructed of fiberglass, is 5 ft in diameter and 18.3 ft high; it contains 10 ft of packing (200 ft^3), which consists of ceramic Raschig rings. The bottom of the column is a 45° cone with a flanged outlet, and the top is flat with a view port. Nozzles for feed inlet and gas outlet extend through the tank wall, and two auxiliary nozzles are located near the top and bottom of the column. A 4-in. gate valve is installed on the cone flange. The column is surrounded by 4 in. of insulation; all external piping is insulated with electrical traces. There are thermocouple taps near the top and bottom of the packed section, a U-tube manometer tap at the top, and sampling taps at 1.0-ft intervals vertically along the packed section. An overflow weir and a collection trough in the top of the column are designed to remove effluent from the center of the tank.

The raw sewage stream is sampled immediately downstream of the comminutor used in the headworks of the Oak Ridge Waste Treatment Plant. The feed rate is controlled, with the bypass being returned to the main sewage flume. The heat exchanger is not routinely used but was installed for

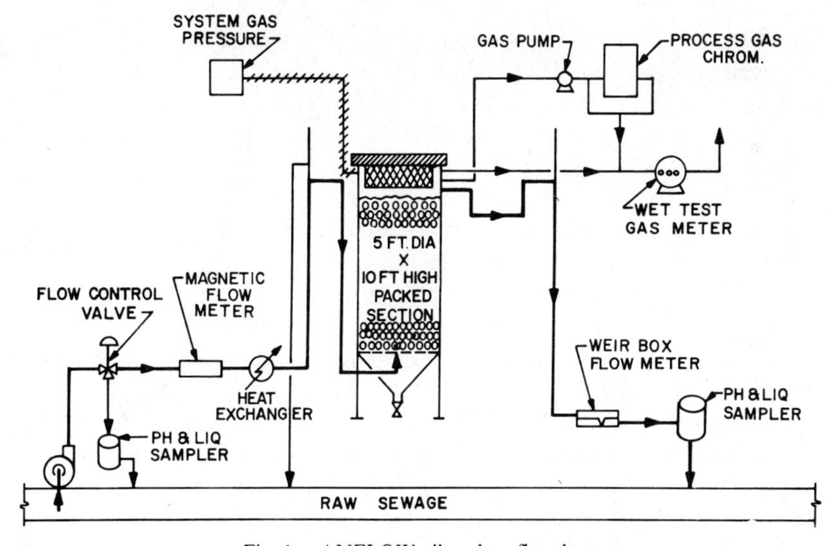

Fig. 1. ANFLOW pilot plant flowsheet.

contingency purposes, particularly since start-up was scheduled for November. Feed is pumped to a height above the column outlet and allowed to pass through the column by gravity flow; a hydraulic head of 2–4 in. is adequate to produce this flow. A liquid seal in the outlet line prevents gas from entering the column through the effluent-return vent line. A final weir box is monitored continuously before the effluent is returned to the main sewage flume. The off-gas from the column is sampled and analyzed by a process gas chromatograph; the total off-gas volume is measured by a wet-test meter.

OPERATION AND PERFORMANCE SUMMARY

The operating conditions experienced and performance parameters obtained during the pilot plant operation are summarized with monthly averages as histograms in Figures 2–7. Temperature variations of the raw sewage were seasonal, ranging from 10 to 25°C. Generally, column effluent temperatures closely followed the temperatures of the raw sewage except for the period January through April 1977, during which the preheater was used to prevent feed from freezing in the external pipelines. Neither inlet nor outlet pH levels have differed significantly from the value of 7 during the operating period. However, there have been pH spikes in the feed for as

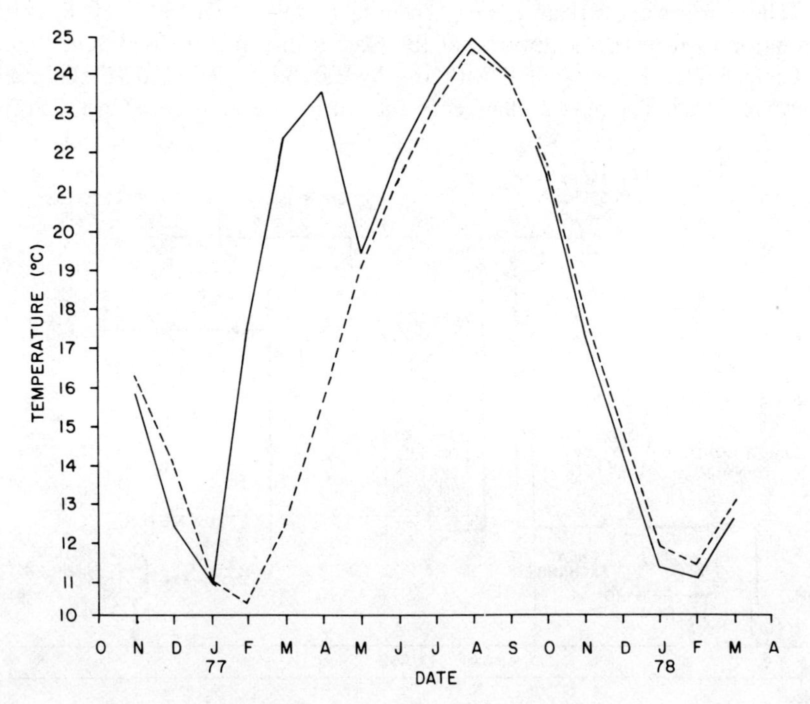

Fig. 2. History of operating temperatures at ANFLOW pilot plant: (——) outlet; (– – –) inlet.

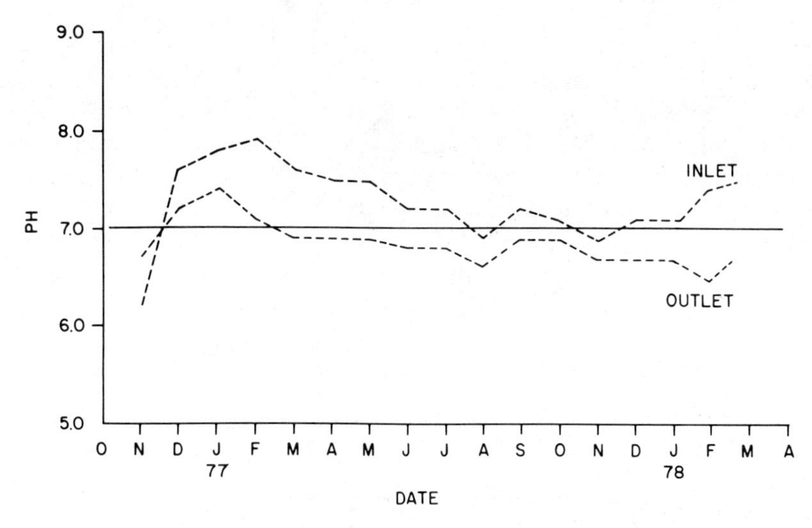

Fig. 3. Feed pH history for ANFLOW pilot plant.

long as 8 hr during which pH values reached lows of 3 and highs of 10; neither acidic nor basic spikes affected the effluent pH. TSS levels in the inlet have varied significantly with a maximum concentration of 210 ppm, but TSS levels in the outlet have remained at approximately 30 ppm. The percentage removal of TSS has consistently been greater than 70%.

BOD levels have varied from 60 to 180 ppm in the feed, with removals of approximately 50% (maximum of 62%) by the ANFLOW unit. TOC levels

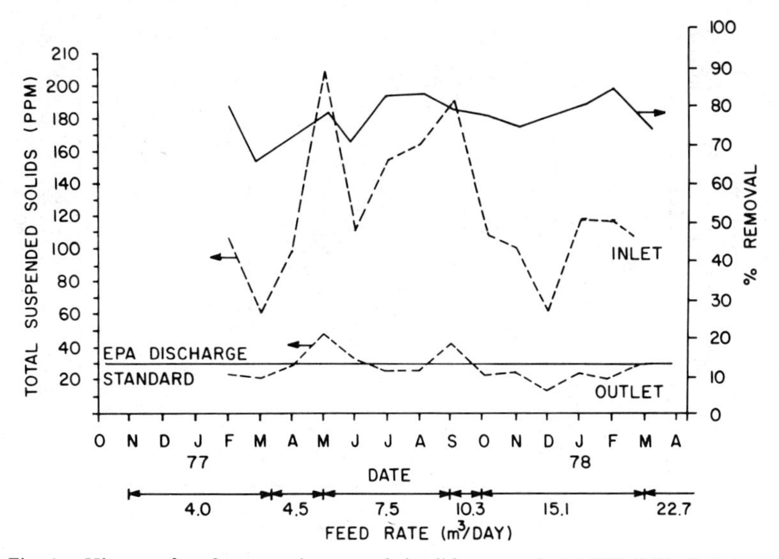

Fig. 4. History of performance in suspended solids removal at ANFLOW pilot plant.

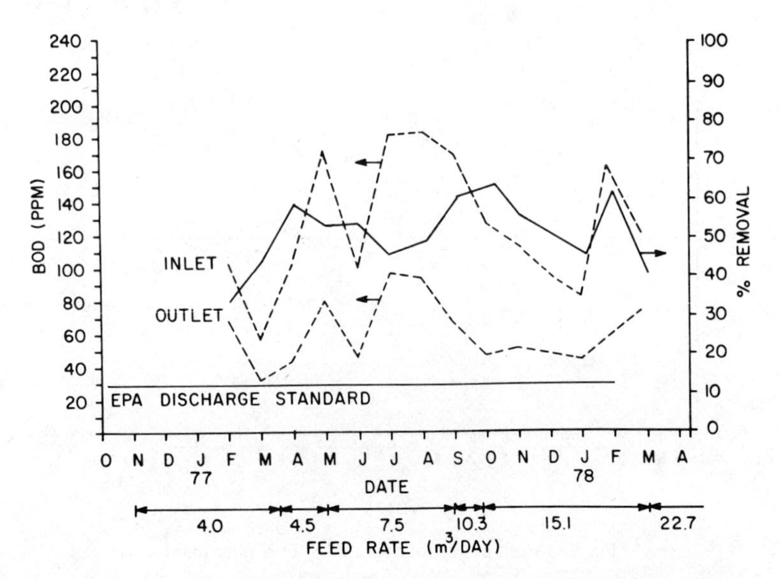

Fig. 5. History of performance in BOD reduction at ANFLOW pilot plant.

in the feed have varied from 30 to 120 ppm, with removals of approximately 60% (maximum of 80%) by the ANFLOW unit.

Gas production has ranged from 10 liter/day during the winter start-up months to 110 liter/day during October 1977. Methane content has reached a maximum of 72% in the off-gas. These rates have been defined for bench-scale units operated continuously over a 7-month period using the same

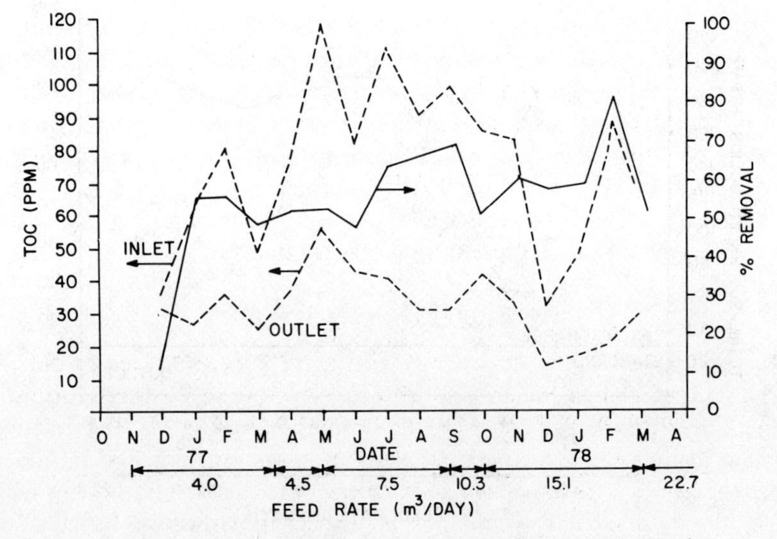

Fig. 6. History of performance in TOC reduction at ANFLOW pilot plant.

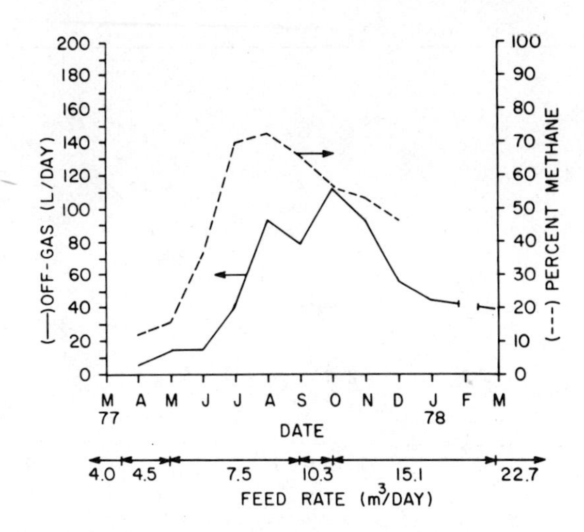

Fig. 7. History of gas production at ANFLOW pilot plant.

feed, packing, and cultures as used in the pilot plant bioreactor. The bench-scale units produced 2.6 liter of gas per day per ft^3 of reactor with a methane content of 70%; on this basis, rates of 500 liter of gas per day could conceivably be achievable in the pilot plant unit.

In order to establish baseline data, the process has been operated at constant feed rates ranging from 1000 to 6000 gpd. Routine stresses such as diurnal variations will be investigated before the project is concluded.

Removal Rates

Removal rates of BOD by the ANFLOW bioreactor are presented in Figure 8, which shows that removal rates can be readily predicted from the conventional expressions for reactor loading. Included with the results in Figure 8 are the dates for which the monthly averages were calculated. Midrange loading data have been obtained during various periods of pilot plant operation. The slope of the line defines a relatively constant 55% removal rate of BOD, indicating both the apparent limit of the current process design and the stability of that process thus far.

Effect of Temperature

The midrange loading data shown in Figure 8 were utilized to obtain a correlation of the averaged removal rates with the averaged operating temperatures during the corresponding months (see Fig. 9). For relatively constant loading rates, temperature changes in the range of 10–25°C do not affect the removal rates. Since all of these data fall within the mesophilic range, dramatic effects would not be expected. However, for a system designed to operate with minimal energy input at ambient temperatures,

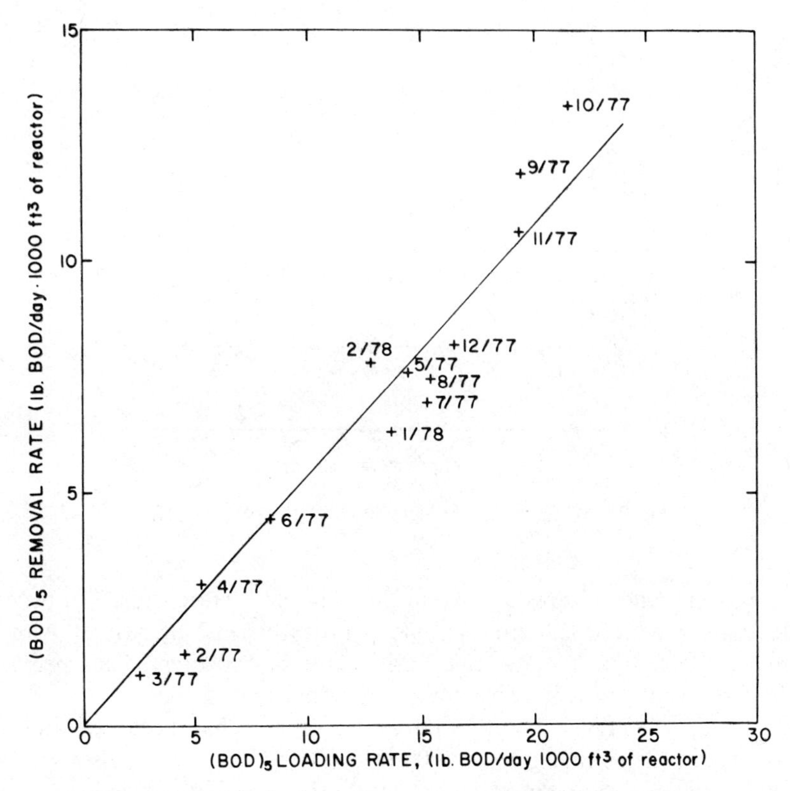

Fig. 8. BOD_5 removal rates at the ANFLOW pilot plant. Dates of monthly averages are shown with the data.

they do show an encouraging resistance to temperature effects in anticipated ranges of operation. The data may also indicate that the bioenergetics of the system are not highly temperature sensitive and, possibly, that the system has low energy requirements for survival. This would contribute to explanations for both biomass stability and the relatively low BOD conversion rate. It might also predict that biomass growth alone would be slow to cause column plugging.

Effluent Quality

Effluent from the ANFLOW pilot plant has consistently met EPA discharge standards with regard to pH level and suspended solids concentration; discharge standards for the BOD levels have also been met, but not regularly. Therefore, possible polishing steps for the effluent are being considered. Settled-effluent tests have shown very favorable results with BOD levels ranging from 3 to 22 ppm (the EPA discharge standard is 30 ppm). The results indicate that the ANFLOW process has been very successful in converting dissolved carbon to physically removable, suspended

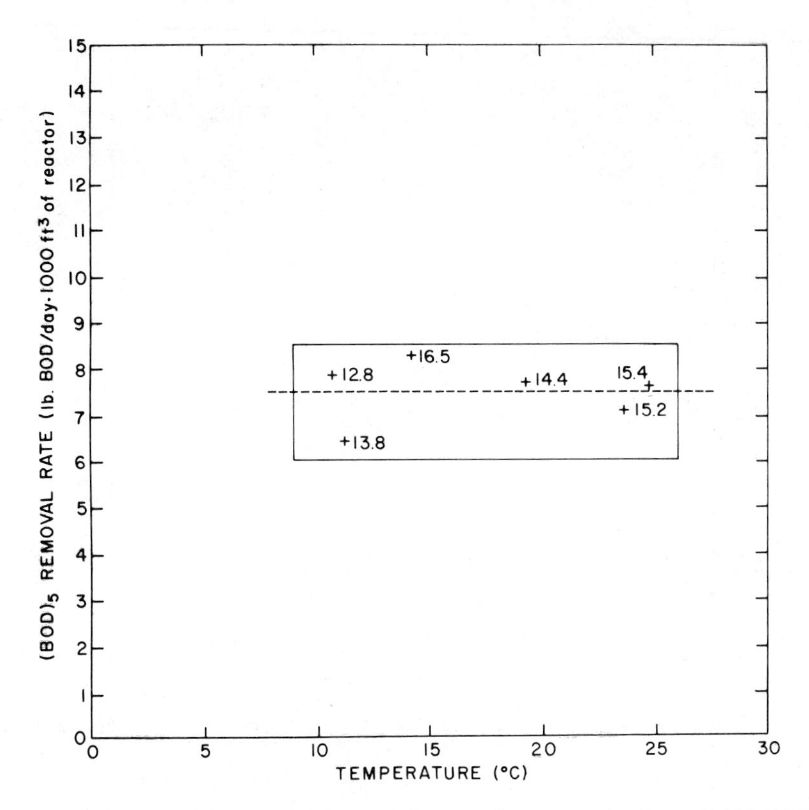

Fig. 9. Effect of temperature on BOD removal rates under constant loading at the ANFLOW pilot plant. Loading rates shown with the data are in units of lb BOD/day/1000 ft^3 of reactor.

carbon. To test this conclusion, effluents were filtered through 0.45-μm Millipore filters and analyzed. BODs of 7–9 ppm were obtained, showing very low concentrations of dissolved substances. Sand filtration tests have also been conducted with similar results. Therefore, it is reasonable to expect that the ANFLOW bioreactor followed by a sand filter will produce an EPA acceptable effluent.

Flow Distribution

The column residence time for wastewater being treated by the ANFLOW system was investigated by residence time distribution (RTD) tests performed at different flow rates. For each test a dye pulse (19 liter of 10 ppm fluorescein) was introduced into the column through the feed-line standpipe. The column effluent was sampled continuously and combined with a reagent development stream (0.5*M* NaOH, 0.005*M* EDTA) before being monitored by a fluorometer. The reagent development stream produced a basic pH that enhanced the fluorescence of the dye; the EDTA complexed dissolved metal ions to prevent precipitation in the flow cell at

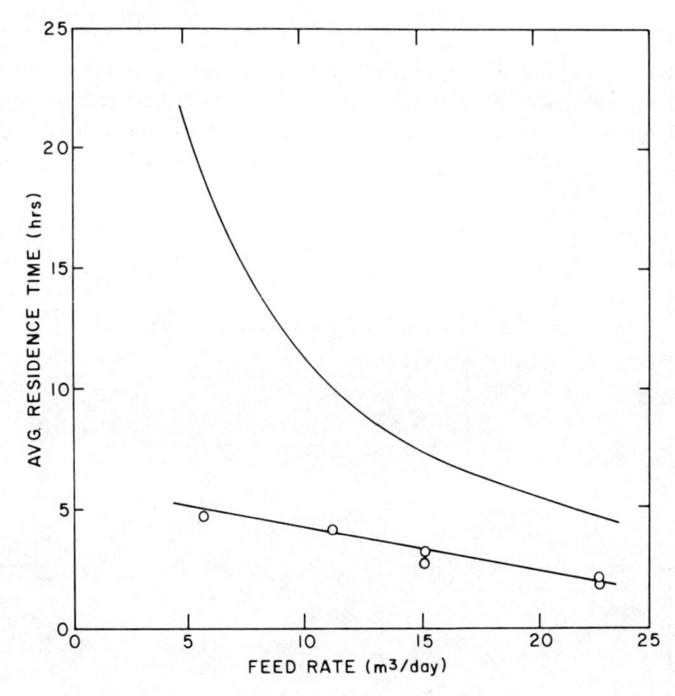

Fig. 10. Residence times of bioreactor at ANFLOW pilot plant. (——) theoretical (plug flow); (O–O) experimental.

the basic pH. Results of the RTDs, compared with plug-flow residence times, are shown in Figure 10. These data indicate that flow channeling is occurring in the column, causing particularly adverse effects at low flow rates. These effects could be the result of solids accumulation in the column or of nonoptimal design of the mechanical flow distribution devices in the column. A more complete answer will be possible when the column is dismantled at the conclusion of pilot plant operation (in late 1978).

TABLE II

Estimated Carbon Accumulation in the ANFLOW Unit

Operating period	Total carbon in feed[a] (kg)	Total carbon in effluent[a,b] (kg)		Accumulated carbon (kg)	
		Assumption 1[c]	Assumption 2[d]	Assumption 1[c]	Assumption 2[d]
Jan. 77 - Apr. 78	321.67	148.27	230.57	173.4	91.1

[a] Including both dissolved and suspended forms.

[b] Including both gas and liquid effluents.

[c] That no CH_4 dissolved in liquid.

[d] That liquid effluent saturates with dissolved CH_4 20 g CH_4/m^3 liquid.

Solids Accumulation

Data regarding solids accumulation in the column are obtained routinely and have been evaluated by general material balances. Accuracy is presently limited by measurements of gas production in this relatively large-scale system. However, Table II provides an estimate of the carbon accumulation to date.

Assuming that 1 g of carbon eventually represents 10 g of biomass and that a specific gravity of 1.1 can be used for this biomass, each 110 g of accumulated carbon will occupy 1 liter of column volume. For a void fraction of 0.7, the initial column void volume can be estimated as 3975 liter; therefore, it would require approximately 437 kg of accumulated carbon to fill the initial void volume. The present effect of solids accumulation can be

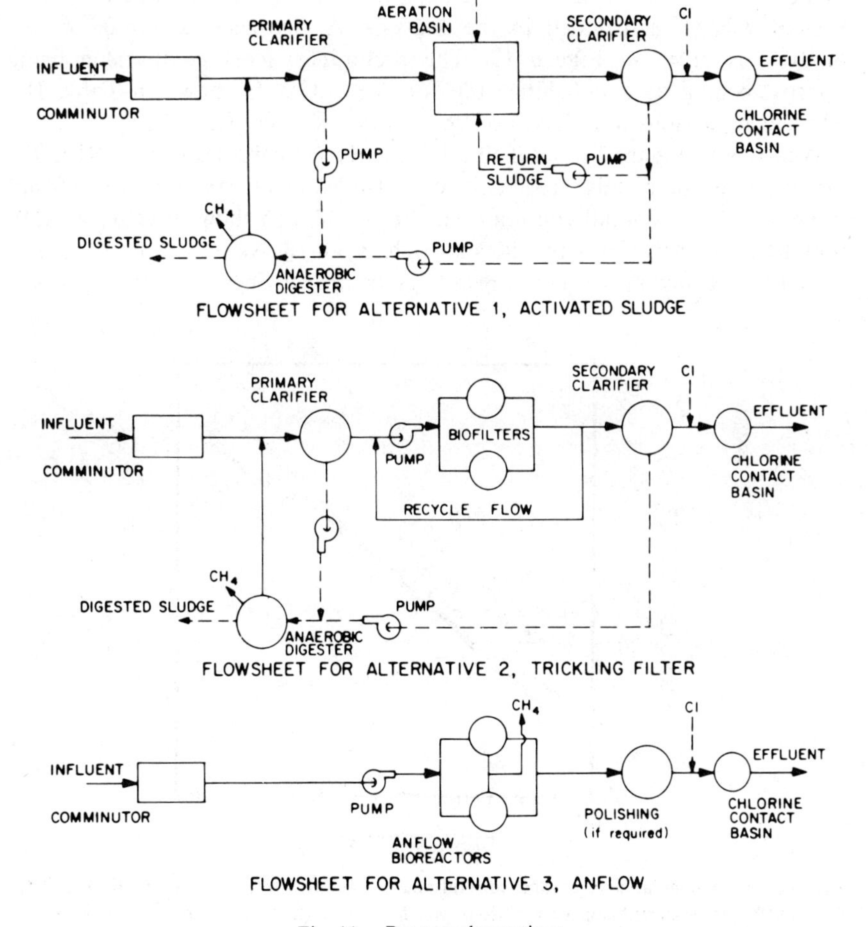

Fig. 11. Process alternatives.

m Table II. If the effluent stream contains no dissolved
roximately 40% of the initial void volume has been filled; if
ream is saturated with methane, approximately 21% of the
ume has been filled. Notably, these estimates occur after 16
pilot plant operation.

Short-term material balances are presently being performed on the unit for a more accurate appraisal. The actual extent of solids accumulation will be determined after the column has been dismantled.

ECONOMIC EVALUATION

The final economic evaluation will be performed after completion of the project. However, a preliminary comparison of the costs for ANFLOW, an activated sludge process, and a trickling filter process has been made on the basis of a plant capacity of 1 mgd. The flow schemes used for this development are shown in Figure 11. The capital cost, labor cost, and energy cost breakdowns were included in the analysis. A composite summary of the analysis is given in Figure 12. The preliminary cost comparison using microscreening as a polishing step for ANFLOW is shown in Table III. These comparisons were based on contemporary references [9–25].

Analysis of Figure 12 and Table III clearly indicates that the ANFLOW process is economically attractive for a 1-mgd plant capacity, a level that might serve many small communities. Figure 12 also shows that ANFLOW may be competitive for capacities as high as 5 mgd. As seen from Table III, ANFLOW's advantages come primarily from a savings in operating costs;

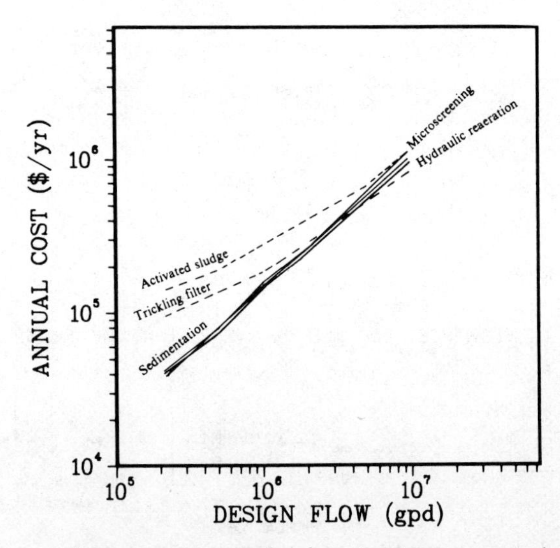

Fig. 12. Cost comparisons among conventional wastewater treatment alternatives and the ANFLOW process combined with various polishing steps (microscreening, hydraulic reaeration, and sedimentation): ANFLOW-media $10/ft^3.

TABLE III

Preliminary Cost Comparison*

Operation	Activated sludge		Trickling filter		ANFLOW	
	Cost	kWhr/day	Cost	kWhr/day	Cost	kWhr/day
Control house and site preparation	$150,000	60	$150,000	60	$100,000	60
Preliminary treatment	60,000	17	60,000	17	60,000	17
Primary sedimentation	400,000	41	450,000	41	-	-
Aerator, pumps, and blowers	300,000	580	-	-	-	-
Sludge handling	600,000	230	600,000	230	-	-
Biofilter	-	-	120,000[a]	240	900,000[b]	-
Final clarifier	400,000	31	400,000	31	-	-
Disinfection	40,000	4	40,000	4	40,000	4
Microscreen	-	-	-	-	40,000[c]	20
Total	$1,950,000	963	$1,820,000	623	$1,140,000	105

[a] One Norton trickling-filter, 38 ft diam × 22 ft height packed bed. Packing cost, $2.50/ft^3.

[b] Two ANFLOW units 50 ft diam × 10 ft height packed beds. Packing cost, $10/ft^3.

[c] If required.

* Plant capacity is 1 mgd.

ANFLOW consumes energy at approximately one-sixth the rate of the trickling filter process and at approximately one-ninth the rate of the activated sludge process.

CONCLUSIONS

Based on progress to date, it is obvious that significant treatment of dilute wastewaters at low temperatures can be obtained with an upflow, anaerobic, fixed-film bioreactor as demonstrated by the ORNL pilot plant. The bioreactor performance is stable after 16 months of operation and has been shown to be relatively insensitive to ambient temperature changes. Sludge removal requirements are still under evaluation, but preliminary indications suggest that the rate of biomass accumulation in the column is slow.

The quality of the effluent produced by the pilot plant at the present stage of development indicates the need for a polishing step; a physical method such as sand filtration will probably be adequate. The overall performance of the bioreator can probably be improved with better designs for mechanical flow distributors.

Gas production rates have not been defined as yet due to mechanical difficulties in sealing the large-scale unit, but rates in excess of 100 liter/day have been observed with methane contents as high as 70%. These rates are still under investigation.

A preliminary economic evaluation shows that the ORNL ANFLOW process is competitive at plant capacities of 1 mgd or less. The savings are primarily related to operating costs but occur in capital costs as well.

References

[1] J. R. McMillan, "Potential energy conservation utilizing bioconversion of municipal and indistrial wastes," study under contract with Economic Development Administration, Technical Assistance Center, The University of Tennessee, Knoxville, Tenn.

[2] J. B. Coulter, S. Soneda, and M. B. Ettinger, *Wastes*, *29*(4), 468 (1957).

[3] G. J. Schroepfer and N. R. Ziemke, *Sewage Ind. Wastes*, *32*(2), 164 (1959).

[4] G. J. Schroepfer and N. R. Ziemke, *Sewage Ind. Wastes*, *32*(6), 697 (1959).

[5] J. C. Young and P. L. McCarty, *Eng. Bull. Purdue Univ. Eng. Ext. Ser.*, *129*(2), 559–574 (1967).

[6] W. A. Pretorius, in *Water Research* (Pergamon, Oxford, 1971), Vol. 5, pp. 681–687.

[7] W. A. Pretorius, "Complete treatment of raw sewage with special emphasis on nitrogen removal," in *Advanced Water Pollution Research Proceedings of the 6th International Conference* (Pergamon, Oxford, 1973), pp. 685–693.

[8] W. L. Griffith and A. L. Compere, *Bench-Scale Anaerobic Fixed-Film System Tests on a Low-Temperature, Low-Strength Synthetic Wastewater* (ORNL/TM-5267, Oak Ridge National Laboratory, Oak Ridge, Tenn., in press).

[9] Battelle—Pacific Northwest Laboratories, *Evaluation of Municipal Sewage Treatment Alternatives* (Council on Environmental Quality, Washington, D.C., 1974).

[10] W. L. Berk, *Am. City*, *81*(9), 120 (1966).

[11] Black and Veatch Consulting Engineers, *Estimating Costs and Manpower Requirements for Conventional Wastewater Treatment Facilities* (U.S. Environmental Protection Agency, Washington, D.C., 1971).

[12] Brown and Caldwell Consulting Engineers, *Cost Curves for Basin Plans, Final Report of BCAC Subcommittee III-4*, 1974.

[13] J. J. Convery, "Treatment techniques for removing phosphorus from municipal wastewaters," presented at New York Water Pollution Control Association Meeting, New York, N.Y., Jan. 29, 1970.

[14] G. L. Culp, G. M. Wesner, and R. L. Culp, *Lake Tahoe Advanced Wastewater Treatment Seminar Manual*, 1975.

[15]Corr-Oliver, Inc., *Cost of Wastewater Treatment Processes* (Department of the Interior, Washington, D.C., 1968).

[16] Environmental Quality Systems, *Technical and Economic Review of Advanced Waste Treatment Processes* (Army Corps of Engineers, Washington, D.C., 1973).

[17] Hazen and Sawyer, Engineers, *Process Design Manual for Suspended Solids Removal* (U.S. Environmental Protection Agency, Office of Technology Transfer, Washington, D.C., 1975).

[18] J. A. Logan, W. D. Hatfield, G. S. Russel, and W. R. Lynn, *J. Water Pollut. Control Fed.*, *34*, 860 (1962).

[19] R. McNabey and J. Wynne, *Water Wastes Eng.*, *8*(8), 46 (1971).

[20] Metcalf and Eddy, Inc., *Report to National Commission on Water Quality on Assessment of Technologies and Costs for Publicly Owned Treatment Works* (Metcalf and Eddy, Inc., Boston, 1975).

[21] R. Smith, *J. Water Pollut. Control Fed.*, *40*(9), 1546 (1968).

[22] R. Smith, *Deisgn and Simulation of Equalization Basins.* (U.S. Environmental Protection Agency NERC-AWTRL report, Washington, D.C., 1973).

[23] R. Smith, *Cost Effectiveness Analysis for Water Pollution Control* (U.S. Environmental Protection Agency report, Washington, D.C., 1974).

[24] R. Smith and R. G. Eilers, *Cost to the Consumer for Collection and Treatment of Wastewater* (U.S. Environmental Protection Agency NERC-AWTRL report, Washington, D.C., 1970).

[25] R. Smith and W. F. McMichael, *Cost and Performance Estimates for Tertiary Wastewater Treating Processes* (U.S. Department of the Interior FWPCA-AWTRL report, Cincinnati, Ohio, 1969).

Use of Powdered Activated Carbon to Enhance Methane Production in Sludge Digestion

TIMOTHY McCONVILLE and WALTER J. MAIER
Environmental Engineering, Civil and Mineral Engineering Department, University of Minnesota, Minneapolis, Minnesota 55455

INTRODUCTION

Anaerobic digestion of sewage sludges is widely used for stabilizing organic materials to facilitate their ultimate disposal. It requires no extraneous energy input and actually produces small quantities of methane as a by-product; however, the major economic justifications for anaerobic digestion are associated with reductions of the total quantity of organic matter and improved dewatering characteristics. Sludge digestion is also beneficial for reducing potential health hazards and odor problems associated with land spreading (farmland, golf courses, parks). Alternative treatment and disposal processes are generally energy intensive because they involve use of chemicals for conditioning, vacuum filtration for dewatering, and incineration. As a result, alternative treatment costs have become increasingly more costly and have led to a renewed interest in the application of anaerobic digestion even in large-scale facilities. There is also a growing interest in maximizing extraction of methane for energy recovery and on making the sludges accepatble for land spreading in the vicinity of large urban centers.

Unfortunately, operating experiences with anaerobic digesters and their cost effectiveness has not been consistently good. Digesters are susceptible to malfunctioning due to shock loading and a variety of toxic substances. Malfunctioning manifests itself in terms of reduced gas production, reduced degradation of organic materials, and a simultaneous increase in acidity. Researchers have shown the need for pH control and the advantage of mixing. However, further process improvements are needed, particularly to reduce costs. Current design practice calls for exceedingly long detention times (upward of 60 days) and hence large capital expenditures, which lead to high operating costs. It follows that there is considerable incentive for finding procedures to increase rates of digestion.

This paper describes the results of a study aimed at improving both the performance stability and the rate of decomposition of anaerobic digesters by adding high surface area activated carbon. The objective was to evaluate the effects of adding high surface area activated carbon. The objective was to evaluate the effects of adding small doses of activated carbon on the rates

Biotechnology and Bioengineering Symp. No. 8, 345–359 (1978)

0572-6565/78/0008-0345$01.00

of methane generation, on the extent of decomposition of organic solids, and on the stability of the process in terms of pH control at different detention times, and to obtain a measure of the dewaterability of the residual sludge solids.

PREVIOUS EXPERIENCE WITH USE OF POWDERED ACTIVATED CARBON (PAC)

Activated carbon has long been recognized for its ability to absorb odoriferous compounds. A number of papers [1–7] have reported on the reduction of odors after using PAC. Odor is a difficult parameter to quantify, but reducing odors may result in savings on deodorant chemicals and fewer complaints. One plant using PAC reported savings of $5000 per year on chemicals for odor control. Research has shown that the addition of PAC stabilizes process performance. Digesters with PAC addition were able to absorb shock loads and could be operated at lower detention times without serious reduction in volatile solids conversion [8, 9]. PAC addition also enhanced gas production, and increased methane content of the gas has been reported. Early work with PAC addition showed that the dewaterability of sludge was improved [1, 2, 4, 7, 10]. Savings in the use of ferric chloride and polymer were reported and in a few cases the quantity of sludge to be dewatered was decreased after the use of PAC was initiated [6, 8, 9]. However, the results of previous studies are not conclusive because they did not quantify the effects of different PAC dosages.

APPARATUS AND PROCEDURES

The effects of carbon addition were tested in bench-scale digesters consisting of modified 4-liter bottles as shown in Figure 1. The liquid phase was maintained at 3.5 liter and was submerged in a 37°C water bath. The digesters were continuously mixed with a flat paddle attached to a sealed shaft rotating at approximately 60–100 rpm. Gas was collected and measured by displacing a weak sulfuric acid solution from precalibrated collection bottles. The gas collection system maintained a small positive pressure on the reactor, and changes in the rate of gas collection were monitored as indications of leaks. All reactors were fed once daily using a gravity feed system. Prior to feeding, sludge was withdrawn from the bottom of the reactor, using a vacuum line. Activated carbon was added to the feed in the form of a water slurry using a syringe. A high surface area, petroleum-base carbon (Amoco, PX-21*) was added in dosages ranging from 150 to 3000 mg/liter of sludge. A high surface area carbon was used because it was theorized that beneficial effects would be greater due to greater adsorption of organic substances and cells. Petroleum-base PAC was used because it was reported to have a low heavy metals content. The sludge was obtained

* Obtained from American Oil Co., Chicago, Illinois.

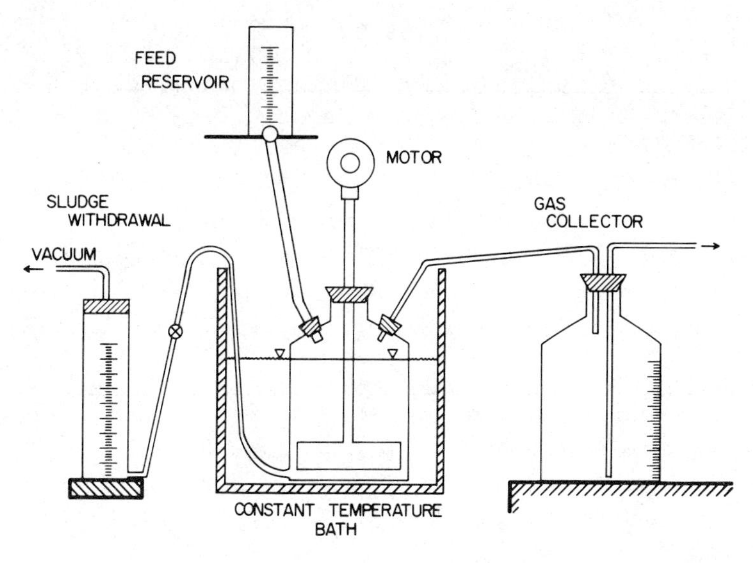

Fig. 1. Anaerobic digester apparatus.

from Anoka Wastewater Treatment Plant and stored in the refrigerator for periods of 1–2 weeks. Anoka has primary and secondary treatment that handles primarily domestic wastewaters. Volatile suspended solids (VSS) were measured using the procedures described in *Standard Methods* [11]. Methane and carbon dioxide concentrations in the gas phase were measured by gas chromatography [12]. Filterability of the sludges was measured using a capillary suction timer (CST). This instrument consists of an open-ended cylinder 1.8 cm in diameter and 2.5 cm high resting on chromatography paper. When the cylinder is filled with sludge, water sorbs into the paper under the influence of gravity and capillary action. The rate of wetting is determined using a digital timer to measure the time needed to increase the wetted diameter from 3.2 to 4.5 cm. This measurement has been found to correlate well with the filterability of sludges measured by vacuum filtration [13].

RESULTS

The effect of PAC addition was tested using five reactors operating at different dosages and at four hydraulic retention times. Operating data for each of the reactors at retention times of 15.2, 7.6, and 4.0 days are summarized in Table I. All the reactors dosed with PAC showed an increase in methane gas production. At a 15.2-day detention time, which is representative of a well-operating, unstressed digester, increases of 12 and 10% were measured in the digesters dosed at 150 and 3000 mg/liter. Lower gas production was observed at intermediate dose rates. The 150 mg/liter dosage gave the highest gas production at all detention times. The inter-

TABLE I

Effect of Powdered Activated Carbon[a]

	Carbon Dosage (mg/l)	0	150	750	1500	3000
a)	% VSS Reduction	50	52	50	51	52
	Avg Gas Prod. (l/day)	2.73	3.07	2.95	2.81	3.0
	PH	7.4	7.3	7.3	7.3	7.3
	CST (sec)	418	358	391	275	290
	% Methane	68	68	68	68	68
	% Increase in Gas Prod.	--	12	8	3	10
b)	% VSS Reduction	53	54	52	52	54
	Avg Gas Prod. (l/day)	4.63	5.22	4.71	4.08	5.03
	PH	7.2	7.2	7.2	7.2	7.2
	CST (sec)	699	522	607	586	507
	% Methane	70	70	70	70	70
	% Increase in Gas Prod.	--	13	2	-12	9
c)	% VSS Reduction	40	41	37	39	41
	Avg Gas Prod. (l/day)	1.7	6.0	2.3	3.9	5.7
	PH	6.0	7.0	6.5	6.5	6.9
	CST (sec)	994	597	707	695	494
	% Increase in Gas Prod.	--	253	35	130	235

[a] (a) 15.2-day retention time, 2.96% solids feed; (b) 7.6-day retention time, 3.2% solids feed; (c) 4.0-day retention time, 2.5% solids feed.

mediate dose rates showed a slight improvement over the control to which no PAC was added. At 3000 mg/liter the increase in gas production was approximately the same as with 150 mg/liter. At first this trend was thought to be a peculiarity of the individual reactors. However, by altering the dosage to different reactors, it was shown that the effect was indeed the result of PAC dosage and not the reactors. A similar effect was noted by Flower [1] who reported an optimum dosage at 200 mg/liter but did not determine the effects of larger doses of PAC.

Comparisons of the effects of PAC addition on gas production at different hydraulic retention times are summarized in Figures 2 and 3. At 15.2-day hydraulic retention time (HRT), cumulative gas production was only slightly larger than with the control whereas a 4-day HRT (Fig. 3) the effect was more pronounced. However, it should be noted that the total quantity of gas produced in the control at 4-day HRT was relatively low. Low gas production was probably due to incipient washout conditions at the short hydraulic residence time. Thus PAC addition appears to stabilize operations

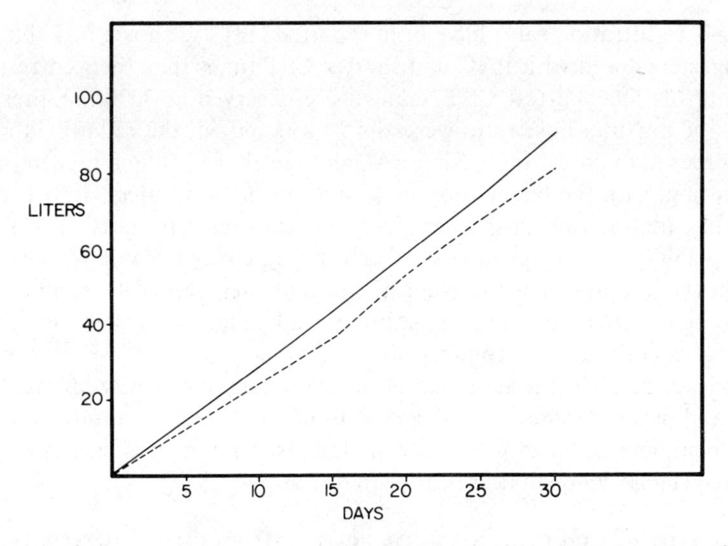

Fig. 2. Cumulative gas production, 15.2-day HRT (——) 150 mg/liter: (---) no carbon.

at short detention times. The VSS were measured in the feed sludge and in the digested sludges from each reactor. The effect of PAC on VSS reduction follows the same trend as gas production. The highest VSS reduction was observed in the reactors dosed at 150 and 3000 mg/liter PAC.

The filterability of the sludge solids was measured using a CST which measures the relative resistance of a sludge to dewatering. Sludges that have high filtration rates generally have low CST values, and correlations relat-

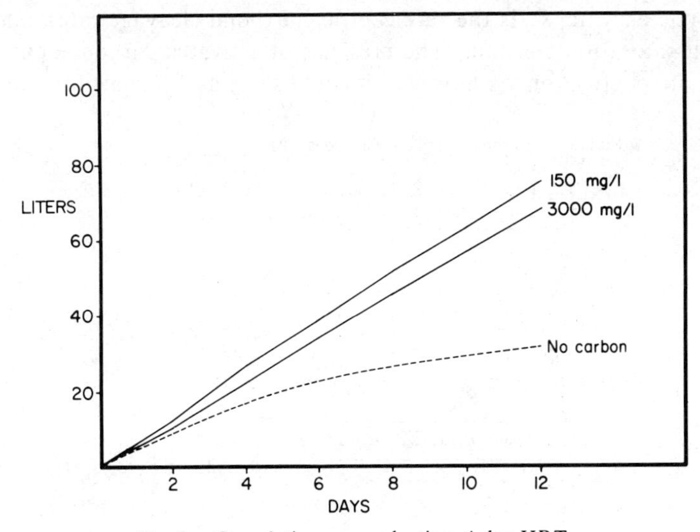

Fig. 3. Cumulative gas production, 4-day HRT:

ing CST to filtration rates have been reported [13]. As shown in Table I, all the digesters dosed with PAC had shorter CST times than their corresponding control. The shortest CST values were observed at 3000 mg/liter dose rates; 150 mg/liter dose rates were almost as good but the 750 mg/liter dose was somewhat poorer. Increasing PAC dosage above 750 mg/liter improved dewatering, with the largest dosage having the greatest effect. Improved filterability (dewatering characteristics) has two major benefits. The first is the possibility of savings in cost of dewatering chemicals due to the reduction in the amount of ferric chloride and polymer needed to condition the sludge before filtration. The second major advantage is realized in systems using a nonmixed or secondary digester where improved settling of the solids leads to a clearer supernatant and preferential retention of viable cell mass and organic solids. This increase in the solids retention time allows for more complete stabilization of the sludge resulting in smaller quantities of digested sludge solids that are easier to dispose of.

MODEL EVALUATION OF THE DIGESTION PROCESS

In order to gain insight on the mechanisms by which activated carbon affects digestion, the data were analyzed in terms of a simple mathematical model. The model considers three sequential coupled processes (Fig. 4), enzyme hydrolysis of polymeric materials into soluble (transportable) substrates, metabolism of soluble substrates into low-molecular-weight fatty acids by acid-forming bacteria, and conversion of fatty acids to methane and carbon dioxide by methanogenic bacteria. The VSS are assumed to consist of hydrolyzable polymeric constituents represented by P. Hydrolysis of polymers to form soluble, transportable substrate is described in terms of Michaelis-Menton enzyme kinetics where E is the concentration of hydrolytic enzyme, k_p is the rate coefficient (bond cleavage rate), and K_p is the half-saturation constant. The presence of activated carbon is unlikely to have a direct effect on k_p; however, it could serve as a surface for concentra-

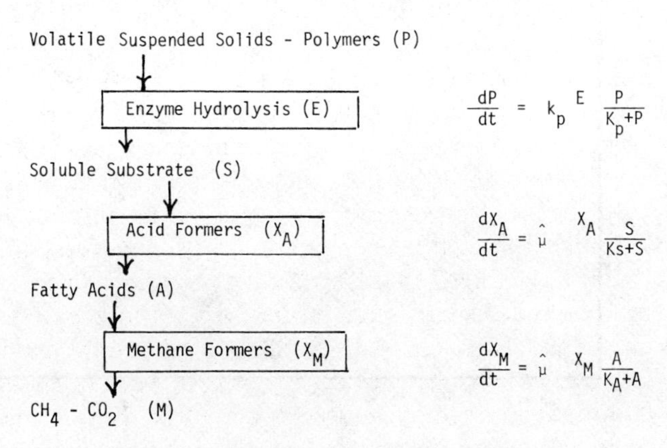

Fig. 4. Anaerobic decomposition sequence.

tion polymer and enzymes which would manifest itself as an appa[illegible] decrease in K_p.

Metabolism of soluble substrate (S) by acid forming bacteria (X_A) to produce fatty acids (A) is described in terms of Monod kinetics; $\hat{\mu}$ represents the maximum growth rate coefficient, and K_s is the half-saturation constant. Addition of activated carbon to this reaction system could serve to enhance rates of reaction by the same mechanism described above, namely, that substrate and microbial cells concentrate on the carbon surface thereby reducing the effective value of K_s. It is unlikely that $\hat{\mu}$ would be affected directly because internal cell metabolism is not influenced except insofar as the addition of PAC changes the chemical environment of the medium. However, if control of pH at more nearly optimum conditions or the adsorption and removal from solution of toxic substances changes the chemical environment of the medium, then it is possible that PAC addition would lead to more rapid growth and an apparent increase in $\hat{\mu}$.

Growth of methanogenic bacteria (X_M) is described in terms of Monod kinetics with $\hat{\mu}$ as the maximum growth rate coefficient and K_A as the half-saturation constant. The effect of adding PAC would be expected to have the same effects as described for the acid formers, namely, a reduction in the apparent value of K_A and the possible enhancement of growth rates due to pH control and removal of toxic substances from solution.

The coupled rate equations and material balances for a well-mixed digester are shown in Figure 5. The time rates of change of feed materials, products, and enzyme are described in terms of a mass balance around the digester of volume V and hydraulic flow rate Q. The first term in all but Eq. (7) represents inflow, the second term describes outflow, and subsequent terms describe the rates of formation, growth, or utilization. The rate of production of hydrolytic enzyme (E) is assumed to be proportional to the concentration of acid-forming bacteria present; α_1 is a proportionality coefficient. This model for characterizing enzyme production has not been evaluated for anaerobic systems but was found to be useful and gave reasonable results in correlating data for the aerobic decomposition of carbohydrate polymers [14]. Conversions of substrate into cell mass are defined in terms of yield coefficients; Y_A represents grams of acid formers produced per gram of soluble substrate consumed, and Y_M represents grams of methane formers produced per gram of acid consumed. Gas production is related to the utilization of fatty acids, and α_2 is the proportionality coefficient. It is recognized that sewage sludges consist of mixtures of polymers and suspended solids that are composed of different types of organic constituents, and it is likely that there exist correspondingly different hydrolytic enzymes and bacteria that mediate anaerobic digestion of different feed materials. However, the complexity of precise modeling is probably not warranted in the context of the present work because there is not enough information on either the feed materials or the rate coefficients of different microorganisms. The model calculations were therefore based

Polymer

$$V \frac{dP}{dt} = QP_o - QP - V\,k_p\,E\,\frac{P}{K_p+P}$$

Enzyme

$$2.\quad V \frac{dE}{dt} = QE_o - QE + V\alpha_1\,X_A$$

Soluble Substate

$$3.\quad V \frac{dS}{dt} = QS_o - QS - V\left(\frac{\hat{\mu}}{Y}\right)_A X_A \frac{S}{K_s+S} + V\,k_p\,\frac{E\,P}{K_p+P}$$

Acid Formers

$$4.\quad V \frac{dX_A}{dt} = QX_{Ao} - QX_A + V\,\hat{\mu}_A\,X_A \frac{S}{K_S+S}$$

Fatty Acids

$$5.\quad V \frac{dA}{dt} = QA_o - QA + V\left(\frac{\hat{\mu}}{Y}\right)_A X_A \frac{S}{K_S+S} - V\left(\frac{\hat{\mu}}{Y}\right)_M X_M \frac{A}{K_A+A}$$

Methane Formers

$$6.\quad V \frac{dX_M}{dt} = QX_{Mo} - QX_M + V\,\hat{\mu}_M\,X_M \frac{A}{K_A+A} - V\,X_M\,kd_M$$

Gas Production

$$7.\quad V \frac{dM}{dt} = -QM + V\,\alpha_2\left(\frac{\hat{\mu}}{Y}\right)_M X_M \frac{A}{K_A+A}$$

Fig. 5. Mass balances for well-mixed anaerobic digester.

on a single substrate–enzyme–bacterium association with the expectation of using model results to explain process variable effects but not to predict absolute values. The numerical values of the coefficients used for model calculations are shown in Table II. Most of the coefficients were obtained from review of the cited literature sources but are not intended to be all inclusive. The concentration of hydrolyzable polymer *P* was defined in terms of the VSS concentration and the degradable fraction coefficient α_3. The coefficient α_3 was estimated from digestion experiments on each sludge.

A series of model calculation were made to simulate steady state conditions holding inflow, outflow, and reactor volume constant; at steady state the time rate of change of all feed and product concentrations are equal to zero. The steady state solutions have been compared to the experimental results obtained using a daily feed-drawoff schedule as illustrated in Figures 6 and 7; model calculations are shown as solid lines. The results show that gas production increases with dilution rate (inverse of detention time) up to dilution rates of 0.2. In the absence of activated carbon, gas production drops off markedly at dilution rates above 0.2 whereas with 150 mg/liter PAC addition, gas production is consistently higher at all dilution rates and

TABLE II
Model Coefficients

			Source of Data
Enzyme Hydrolysis Rate Coefficient	k_p	10-500 $\frac{\text{mg VSS}}{\text{day}}$	(*)
Half Saturation Constant	K_p	100-1000 mg/l	(*)
Growth Rate Coefficient			
Acid Formers	μ_{max}	0.7 day^{-1}	(15)
Methane Formers	μ_{max}	0.4 day^{-1}	(15)
Monod Coefficients			
Acid Formers	K_A	0-5 mg/l	(15)
Methane Formers	K_M	0-500 mg/l	(15)
Death Rate Coefficient	kd_m	0.01 day^{-1}	(15)
Yield Coefficients			
Acid Formers**	Y_A	0.2 g/g	(15)
Methane Formers	Y_M	0.05 g/g	(15)
Enzyme Production Factor	α_1	0.25 g/g	(14)
Gas Yield/Unit of Acid	α_2	10^{-6} l/g	(*)

* Estimated by curve fitting of data at different residence times.
** Revised model includes fatty acid yield coefficient α_4.

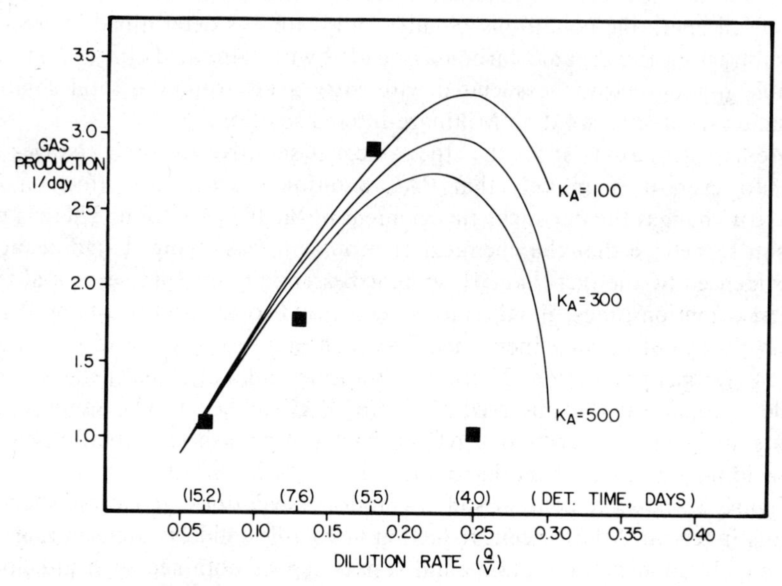

Fig. 6. Daily feed-drawoff schedule, no carbon. $\alpha_3 = 0.52$; $k_p = 2.2 \times 10^5$; $K_p = 750$.

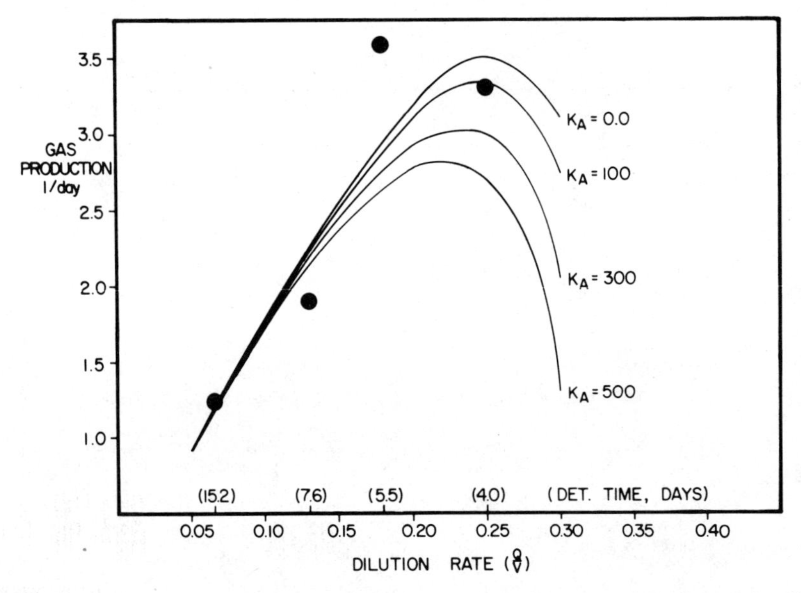

Fig. 7. Daily feed-drawoff schedule, 150 mg of carbon per liter: α_3 = 0.53; k_p = 2.2 × 10^5; K_p = 750.

remains high at dilution rates of 0.25. Model calculations for successively smaller values of K_A show a similar trend, for example, gas production levels are higher, and the drop-off occurs at higher dilution rates. K_s was not considered to be an important factor in accounting for the effects of PAC addition because soluble substrate concentrations were found to be low at all operating conditions. Soluble substrate was determined indirectly by subtracting the organic carbon associated with fatty acids from the total soluble organic carbon associated with fatty acids from the total soluble organic carbon measured on Millipore-filtered solutions.

Model calculations show that the system is sensitive to small changes in k_p. However, it is unlikely that PAC addition has a direct effect on k_p unless it changes the chemical environment of the bulk solution. There is no reason to believe that the chemical environment was changed significantly as evidenced by the fact that pH remained essentially the same except at the lowest detention times. Furthermore, removal of toxic constituents by PAC can be ruled out because increasing PAC concentrations above 150 mg/liter gave lower gas production. Therefore, it appears unlikely that changes in k_p could account for the observed effects of PAC addition. The same arguments apply to the effects of PAC on bacterial growth rate coefficients of the acid formers and the methane formers.

Figure 8 shows a comparison of model calculations and experimental results in terms of the percent reduction in volatile solids vs dilution rate. A reasonably good fit of experimental results can be obtained by a judicious choice of the magnitude of the coefficients. The effect of PAC addition on

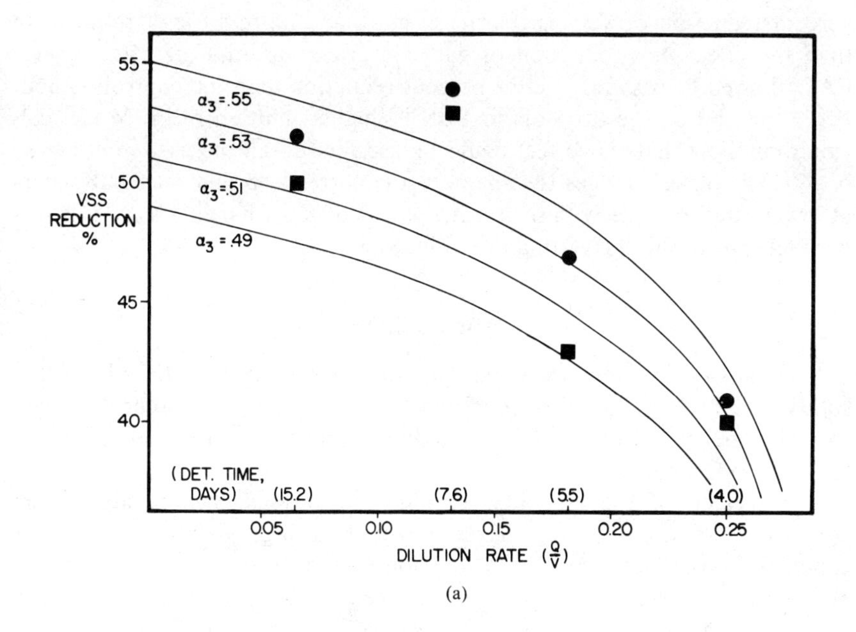

(a)

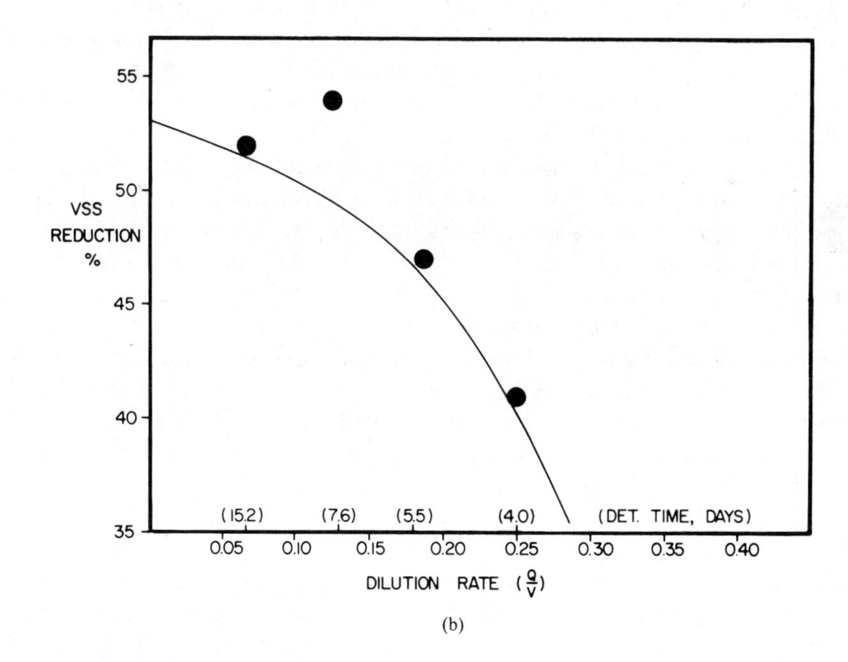

(b)

Fig. 8. Comparison of model calculations and experimental results in terms of the percent reduction in volatile solids vs dilution rate. (a) $K_p = 750$; $k_p = 2.2 \times 10^5$; (●) 150 mg carbon/liter; (■) no carbon; (b) 150 mg carbon/liter; $\alpha_3 = 53\%$; $k_p = 2.2 \times 10^5$; $K_p = 750$.

percent reduction of VSS is relatively small and more difficult to measure than the effect on production of gas. Neverless, addition of 150 mg/liter PAC showed consistently better percent reduction than the control without PAC and shifted the drop-off in VSS to higher dilution rates. Model calculations show that this effect could be ascribed to an increase in either α_3 or K_p. The latter changes the shape of the correlation line; washout occurs at lower dilution rates with increasing values of K_p. Changing α_3 results in a parallel shift of the correlating line (Fig. 8).

DISCUSSION

Results showing increased gas production and conversion of volatile solids with PAC addition are consistent with the findings of other investigators. The beneficial effects of PAC addition have been ascribed to several different mechanisms.

The surface of the activated carbon provides adsorption sites where substrate can accumulate, thereby providing high localized substrate concentrations. These areas of adsorption provide a more favorable growth environment for bacteria-substrate systems with large half-saturation constants. This environment would manifest itself as a lower half-saturation constant when small concentrations of PAC are added, as illustrated in Figures 6 and 7. Addition of larger concentrations of PAC ultimately provides too much surface, therefore diluting out the effect. Although this dilution would help to explain the poorer performance observed at intermediate PAC dosages, it does not provide a rational explanation of the beneficial effects of high PAC dosages (3000 mg/liter).

It has been reported that PAC addition acts like a buffer and prevents fluctuations in pH, thereby providing optimum conditions for growth. This does not appear to be a significant factor in this work because pH remained essentially constant in tests with and without PAC addition except at the lowest retention time.

PAC could sequester toxic substances. Removal of these substances from the bulk solution would be expected to increase gas production and conversion of volatile solids. However, the benefits would be expected to increase with PAC dosage, and a reversal at intermediate PAC dosage levels would be unlikely. Therefore, it seems improbable that sequestering of toxic substances is a significant factor in these experiments.

Preferential retention of active biomass and sorbed volatile solids on activated carbon particles could increase the solids residence time and lead to greater conversion without changing hydraulic residence time. Preferential retention of particles is likely to occur in large-scale digesters that are not completely mixed and where partially clarified supernatant is drawn off near the top of the tank. However, this was not the case in the laboratory digesters because the contents were well mixed and draw-off was from the bottom of the tank. Therefore, it must be concluded that preferential solids retention was not a factor in the laboratory digesters.

It has been suggested that PAC addition enhances the dispersion of hydrophobic substances into the slurry. Increasing PAC dosages would therefore be expected to increase the conversion of volatile solids and gas production. For example, fats or surfactants that otherwise float on the surface or accumulate in surface foams are more likely to be sorbed on the carbon surface and dispersed in the slurry. It was noted in the course of this work that there was directionally less foam accumulation in tests using high PAC dosage. However, the foam was not measured, nor was it analyzed; thus, it is not known whether the reduced foaming was a significant factor in increasing conversion of volatile solids and gas production.

Models of the anaerobic digestion process generally assume that the digester liquor is a homogeneous mixture. This assumption is probably valid in the case of soluble substrates that can be directly transported across cell membranes but may not be valid for the case of polymeric substrates. In such a system it is likely that localized aggregations of polymers, enzymes, substrates, and cells act as centers of biochemical activity to produce favorable microenvironments which differ from the bulk solution phase. Digestion of polymer requires the presence of extracellular hydrolytic enzymes to convert them into soluble-transportable substrates followed by the sequential conversion to acids and gas by two or more different species of bacteria. The coupling and attendant mass transfer processes of these three steps are not fully understood but it seems likely that micro-scale concentration gradients are important. The growth kinetics of aerobic myxobacteria-metabolizing cellulosic substrate are a case in point. It has been shown that cell-mass doubling time is inversely related to bacterial concentration when small quantities of inoculum are used [16] but no effect is found at higher bacterial concentration. It is theorized that the presence of aggregations of bacteria and attendant extracellular enzymes create a more favorable microenvironment for cell growth. By analogy, the higher conversion of volatile suspended solids obtained with 150 mg/liter PAC addition may be due to the formation of centers of biochemical activity with more favorable microenvironments than the bulk solution. This could be characterized in terms of a decrease in the effective value of K_s or K_p as discussed above. At higher PAC dosages the benefits would be smaller due to the dilution effect of additional sites. This is consistent with the observed fall-off in conversion at intermediate dosage rates. To account for the improved results at the highest PAC dosage rates, attention is drawn to the previously described effect of PAC addition on the dispersion of hydrophobic constituents. Increased dispersion of hydrophobic substrates contributes to the formation of favorable microenvironments in localized centers of biochemical activity. This is equivalent to increasing α_3, the fraction of degradable polymer. However, it appears likely that this effect would be most pronounced at higher PAC dosages where the effects of reduced foaming were most noticeable. Thus, the combined effects of enhanced dispersion of hydrophobic substances and optimization of environmental factors at localized sites of biochemical activity can be used to explain the observed effects of PAC addition.

If the proposed explanations are correct, it suggests that the benefits of PAC addition at large dosages will be strongly dependent on the chemical composition of the sludge whereas the improvement at lower dosages will be nearly the same for all sludges. This hypothesis should be checked experimentally. The question of using different types of activated carbon has not been explored. A high surface area material was used in this study because it was orginally assumed that this would enhance surface effects. However, the benefits for using high surface area PAC have not been established nor have the effects of using different particle size activated carbon been evaluated.

SUMMARY

Addition of small concentrations of powdered activated carbon to sewage sludges has been shown to have beneficial effects in anaerobic digestion. Gas production (65–60% methane) is increased by 10–15% at high detention times; larger increases were observed at low detention times. Increased gas production is mirrored by increased conversion of volatile suspended solids. PAC addition also improves the dewatering characteristics of the digested sludges and reduces potential odor problems. The latter benefits are particularly important in terms of facilitating the ultimate disposal of sludge solids. Dewatering characteristics improve with increasing PAC dosages whereas gas production is optimized at 150 mg/liter dosage rates.

The benefits for PAC addition are magnified at lower detention times. Stable operations with 40% reduction in volatile suspended solids were obtained with 150 mg/liter PAC dosage at a 4-day detention time. The control digester without PAC addition was on the verge of washout as evidenced by lower gas production, lower pH, and poor dewaterability. Thus PAC addition could be used to increase the capacity of existing digesters or save capital on new designs.

Analysis of the test data in terms of mathematical models indicates that the benefits associated with PAC addition may be the result of localized changes in bacterial-substrate-enzyme concentrations. It is postulated that PAC addition leads to the formation of biochemically active sites that enhance the coupling of the sequential conversion of polymeric substances into soluble substrates, fatty acids, and finally into gas. The surface properties of the activated carbon should therefore be tailored to providing optimal conditions for the formation of biochemically active sites.

We wish to thank Mr. C. Yee for his help in setting up the analytical procedures and Mrs. M. Henderson for her help in the preparation of the manuscript. This work was supported from funds provided by the Metropolitan Waste Control Commission and by the Minnesota Mininig and Manufacturing Company.

References

[1] G. E. Flower *et al.*, *Sewage Works J. 10*, 441 (1938).
[2] R. W. Haywood, *Sewage Works J. 9*, 785 (1937).
[3] M. Hunsicker, *Water Sewage Works 123*, 62 (1976).
[4] R. W. Kehr (Ed.), *Sewage Works J. 7*, 950 (1935).
[5] A. H. Rodgers, *Sewage Works J. 7*, 691 (1935).
[6] W. Rudolfs, *Sewage Works J. 9*, 207 (1937).
[7] C. L. Walker, *Sewage Works J. 9*, 207 (1937).
[8] A. D. Adams, *Water Sewage Works 122*, 46 (1975).
[9] T. Ventetuolo and A. D. Adams, *Deeds Data*, July 1976.
[10] G. J. Wiest, *Water Works Sewerage 88*, 235, (1941).
[11] *Standard Methods for the Examination of Water and Wastewater*, 13th ed. (APHA, New York, 1971).
[12] C. K. Yee, Anaerobic Digestion of Crop Residues for Methane Production, M.S. Thesis, University of Minnesota, 1977 (unpublished).
[13] R. C. Baskerville and R. S. Gale, *J. Institute Water Pollut. Con. 2* (1968).
[14] J. V. Maxham, Bacterial Growth on Polymeric Substrates in the Activated Sludge Process, Ph.D. Thesis, University of Minnesota, 1976 (unpublished).
[15] Metcalf and Eddy, Inc., *Wastewater Engineering*, text, 1972.
[16] M. Dworkin (private communication).
[17] C. F. Keefer and H. Kratz, *Water Sewage Works, 123*, 62 (1976).

Operation of a Fluidized-Bed Bioreactor for Denitrification*

C. W. HANCHER, P. A. TAYLOR,† and J. M. NAPIER†

Oak Ridge National Laboratory, Oak Ridge, Tennessee 37830

INTRODUCTION

Many commercial processes such as fertilizer production, paper manufacturing, and metal finishing, as well as several steps in the nuclear fuel cycle, yield nitrate-containing wastewaters, which are currently discharged because traditional recovery or disposal methods are economically unacceptable. Nitrate-containing wastewater causes stream eutrophication and can be a health hazard. In addition, the anticipated discharge limit (i.e., 10–20 g/m^3 NO_3^-) being considered by many states will not allow the continued release of these wastewaters. This program is primarily concerned with nitrate wastewater streams resulting from the nuclear fuel cycle but is also applicable to nitrate-containing wastewaters arising from other sources.

At the Oak Ridge National Laboratory, we have developed a fluidized-bed bioreactor process for denitrification of nitrate wastewaters containing up to 10,000 g/m^3 NO_3^-; higher levels can be successfully treated after appropriate dilution. The basic concept of the process is that denitrification bacteria are allowed to grow and attach themselves to 30/60 mesh coal particles, thus forming a stable-bacterial-population bed through which nitrate-containing wastewater can be pumped. The bacteria also need a carbon source, such as ethanol, and micronutrients for successful growth.

We will report a scale-up comparison between 5- and 10-cm-i.d. by 5-m-tall bioreactor systems. Also, operating data will be presented for a 10-cm-i.d. bioreactor functioning as a single unit and the 10-cm-i.d. bioreactor units connected in series. The results from these tests form the basis of the design of a pilot plant which consists of dual 20-cm-i.d. by 7-m-long bioreactors.

MATERIALS AND METHODS

Nitrate Waste Sources

It has been estimated that up to 2.5 million tons of dissolved nitrogen-bearing substances reach the surface waters of the United States annually

* Research sponsored by the Divison of Waste Management, U.S. Department of Energy under contract W-7405-eng-26 with the Union Carbide Corporation Nuclear Division.

† Y-12 Development Division, Y-12 Plant, P. O. Box Y, Oak Ridge, Tennessee 37830.

Biotechnology and Bioengineering Symp. No. 8, 361–378 (1978)
Published by John Wiley & Sons, Inc.

[1, 2]. The nitrogen waste discharged directly from industrial installations is estimated to be about 20% of the total, or 500,000 tons/year. Much of this nitrogen pollution is present as dissolved nitrates at high concentrations. The majority of industrial nitrate pollution comes from industrial sources such as metal finishing, fertilizer production, and paper manufacturing. Liquid effluents from the nuclear fuel cycle (Fig.1) also contribute significantly (~5000 tons/year) to the total nitrate pollution problem [3].

Many aqueous nitrate waste streams are generated by the nuclear fuel cycle (see Fig. 2). Disposition of the nitrate (recovery, conversion, or discharge) will be governed by the economics of the process technologies which may be applied. In situations where nitrate recovery is not feasible, the reduction of nitrous oxides to nitrogen gas (chemically or biologically) appears to be the only acceptable long-range solution.

We recognize that attempts have been made to evade the liquid waste nitrate problem by volatilization into a gaseous effluent. This practice, although expedient in the short term, will probably not meet EPA restrictions in the future. Thus it appears that, in cases where nitrate cannot be economically recovered, biological denitrification is the preferred choice of processes because the end-product nitrogen gas, carbon dioxide, and biomass are ecologically acceptable; in addition, it is more economical than chemical reduction.

Biological Denitrification

Biological denitrification, as referred to in this paper, is the biological reduction of nitrate or nitrite and carbon to gaseous molecular nitrogen and carbon dioxide [4]. It commonly takes place in soil under anaerobic conditions by the various strains of facultative anaerobic bacteria, such as *Pseudomonas denitrificans* or *Pseudomonas stutzeri*, that are responsible for recycling nitrogen compounds to the atmospheric molecular nitrogen pool. The reaction requires a source of carbon, which has usually been provided in the form of various alcohols and acetates. The rate of denitrification is dependent on nitrate concentrations and the type of carbon substrate supplied as well as on other operating parameters such as the pH and temperature of the system. With ethanol as the carbon source, the reaction may be written in unbalanced form as:

$$NO_3^- + C_2H_5OH \rightarrow CO_2 + N_2 + H_2O + OH^- + X\,C_5H_7O_2N$$

It has been observed that the molar ratio of carbon consumed to nitrogen (as nitrate) reacted is about 1.3 to 1.5. The composition of the biomass may be approximated as $C_5H_7O_2N$. The biomass yield is roughly 0.1 g/g nitrate consumed.

The denitrification bacteria have been successfully allowed to grow and attach themselves to small solid supports, thereby forming a stable bacteria-population bed through which the nitrate wastewater can be processed [5].

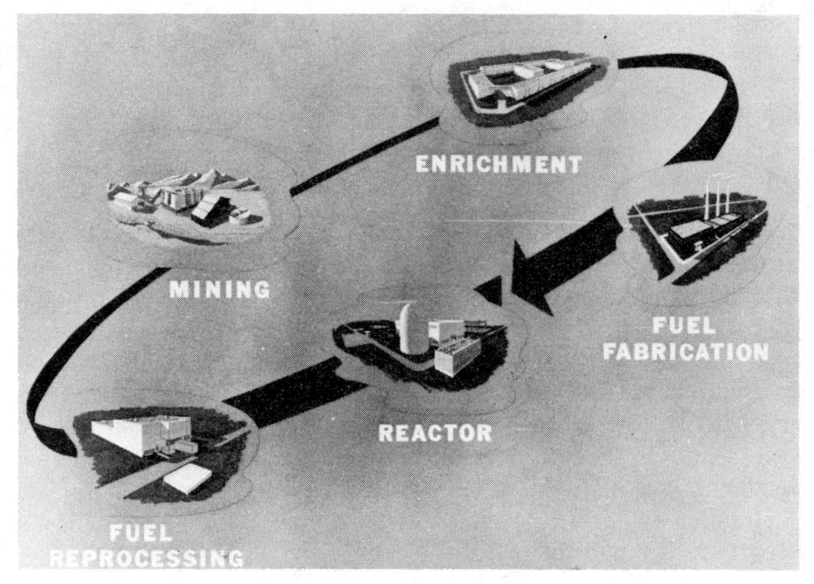

Fig. 1. Nuclear fuel recycle.

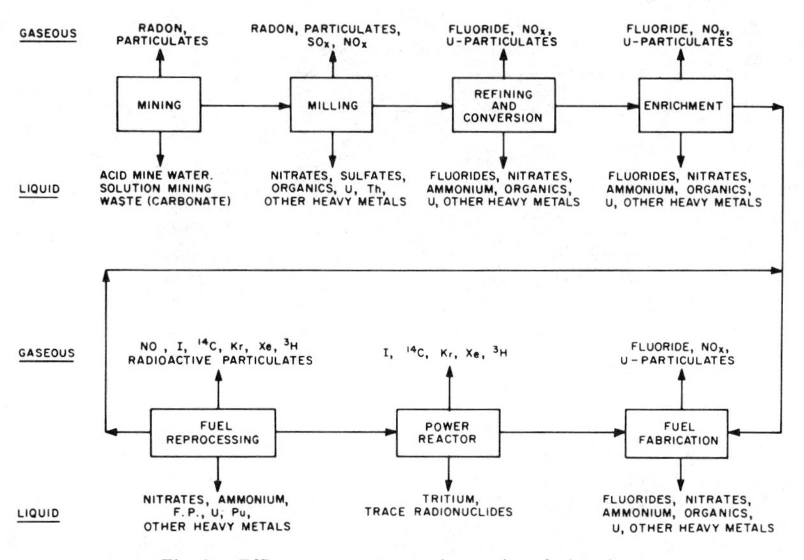

Fig. 2. Effluent waste streams in uranium fuel cycle.

Denitrification bacteria have been successfully attached to 30/60 mesh coal (density = 1.5 g/cm³) and sand (density = 2.2 g/cm_3) particles. Coal particles, which have a lower density than the sand granules, have been used for most of the tests since they will fluidize at a lower liquid flow rate and hence provide a longer contact time for a given bioreactor length.

Feed Solution

In the tests reported here, we selected a feed composition representing nitric acid wastewater that had been neutralized with ammonium hydroxide (Table I). Ethanol was added as a carbon source at a carbon-to-nitrogen (C/N) ratio of 1.5. Ethanol was used rather than the less costly methanol because it could be easily determined in the presence of other soluble carbon compounds by the ethanol Calbiochem kit [6]. Sodium phosphate was added directly to the feed as it entered the column to eliminate phosphate precipitation problems in the concentrated feed. The 20% nitrate feed concentration solution was diluted to the desired nitrate level with water before it was fed to the bioreactor.

Bacteria Inoculum

The culture used for seeding the columns was taken from a stock of frozen bacteria that had been prepared for starting the Y-12 Plant denitrification reactors. These bacteria were grown in 55-gal drums; however, since the drums were open to contamination, the bacteria are almost certainly a mixed population.

TABLE I

Concentrated Feed Solution

30% NITRATE	
HNO_3	300 g/liter
NH_4OH	120 g/liter
ETOH	200 g/liter
$MgSO_4$	1 g/liter
Dow-Antifoam "A"	1 g/liter
Trace metal solution	0.1 g/liter
TRACE METAL SOLUTION MIX (g/liter)	
H_3BO_3	1
$ZnSO_4 \cdot 7H_2O$	0.4
$NH_4MO_7O_{24} \cdot 4H_2O$	0.2
$MnSO_4 \cdot 7H_2O$	0.25
$CuSO_4 \cdot 5H_2O$	0.45
$FeSO_4 \cdot 7H_2O$	0.25
KI	10
Fe-chelate	200

Analytical Methods

The analyses for nitrate, nitrite, and ethanol are performed using a Gilford 2400 spectrophotometer. The nitrate concentration is determined by measuring the absorbance at 220 nm [7] and then subtracting the reading at 275 nm from the 220-nm result to correct for interference by dissolved organics. Nitrite samples are combined with sulfanilic acid, EDTA, and naphthylamine hydrochloride to produce a red azo dye. The intensity of this dye is measured by absorbance at 520 nm. Ethanol is measured by using a commercial blood alcohol analysis kit from Calbiochem [6]. In this method, ethanol is enzymatically oxidized to acetaldehyde while nicotinamide adenine dinucleotide (NAD) is reduced to NADH. The concentration of NADH is determined by measuring the absorbance at 340 nm. Three standards, whose concentrations span the range of the sample concentrations, are used for calibration in all of these methods. The concentration of total organic carbon (TOC) is measured with a Beckman 915-A carbon analyzer.

Biomass-on-Coal Determination

The biomass-covered particle is gently fluidized and washed with water to remove any nonattached biomass. The sample is air-dried for 24 hr at 105°C, cooled in air, and then weighed. The dried sample is subsequently soaked in 4*M* NaOH for 4 hr to remove the attached biomass. The cleaned particles are water-washed, dried at 105°C for 25 hr, cooled, and reweighed. The biomass is reported in terms of weight percent on a dry basis. The biomass is 58% carbon, based on the biomass formula $C_5H_7O_2N$.

Experimental Equipment

Bioreactors. The 5- and 10-cm-i.d. Pyrex glass bioreactors used in this series of experiments were similar in design. Each bioreactor consisted of a tapered bottom followed by five samplers and five cylinders alternately spaced, the top disengaging taper, and the liquid gas separating section (Fig. 3). The sections are described in detail, as follows:

(1) A tapered bottom, 0.5 cm i.d., enlarging to 5 or 10 cm i.d. over a 15-cm length for flow distribution.
(2) A 2-cm-thick Plexiglas sampler for liquid and solid samples.
(3) A 60-cm-long straight cylindrical section.
(4) A top solid disengaging section tapering from either 5 to 10 cm or 10 to 15 cm i.d. and 60 cm long.
(5) A top liquid-gas disengaging section 10 or 15 cm i.d. and 60 m long.

The liquid fluidizing rates were 800 to 1000 ml/min for the 5-cm-i.d. bioreactor and 4000 ml/min for the 10-cm-i.d. bioreactor.

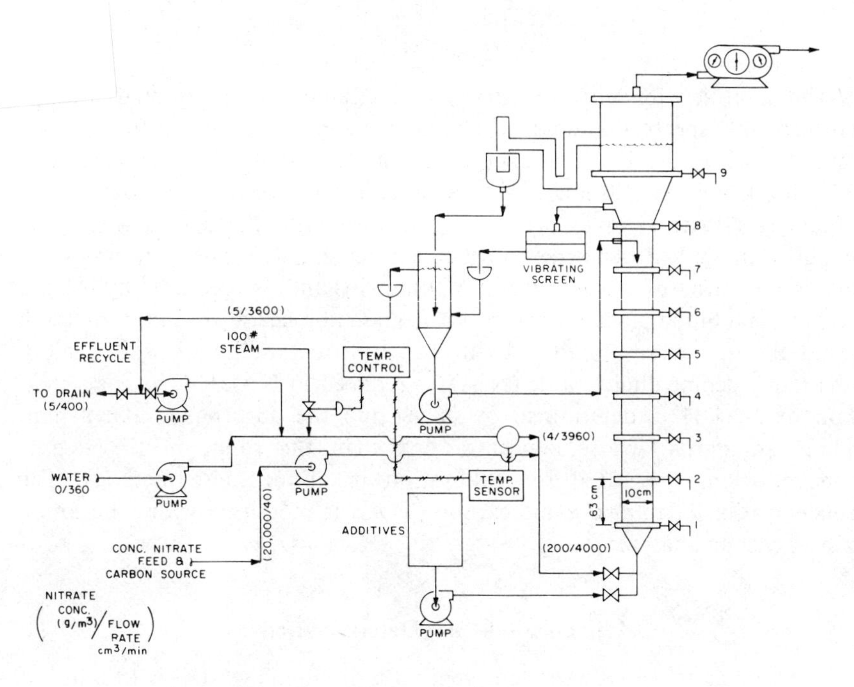

Fig. 3. Bioreactor system with temperature control and recycle capability.

Two identical 10-cm-i.d. bioreactors piped in series were used in the dual bioreactor tests (Fig. 4). The liquid feed rate was 4000 ml/min. The inter-reactor feed pump rate was controlled so as to maintain a constant liquid level in the surge tank between the two bioreactors.

Pumps. The pumps used in these experiments were Milton Roy (Milton Roy Co., Flow Control Div., Ivyland, PA 18974) positive-displacement piston pumps, which are fabricated of 316 stainless steel.

Biomass Control

The excess biomass generated in the denitrification reaction must be removed periodically for stable bioreactor operation. The amount of biomass to be removed is theoretically 0.1 g/g NO_3^- decomposed. Operating experience has indicated that a biomass loading of no more than 5 to 10% dry weight is satisfactory. Overgrowth of a biomass has the following detrimental effects on the reactor operation:

(1) The reactive surface area and, in turn, the denitrification rate are reduced.

(2) Nitrogen gas that is formed during the denitrification r
trapped, causing the particle to float and wash out of the biore
(3) Particles tend to become sticky,and fluidization is diffic

When the density of the overgrown biomass-covered parti
the particle will migrate to the top of the bioreactor. The less-dense overgrown particle can be removed from the top of the bioreactor, mechanically cleaned, and returned to the top of the bioreactor. The clean particle will sink, and biomass growth will continue.

Our biomass control system consists of a Sweco vibrating screen equipped with a 30-mesh (0.50-mm diam) screen. Liquid and solid are pumped from the top of the bioreactor, vibrated through the 30-mesh screen to remove most of the attached bacteria, and then returned to the bioreactor

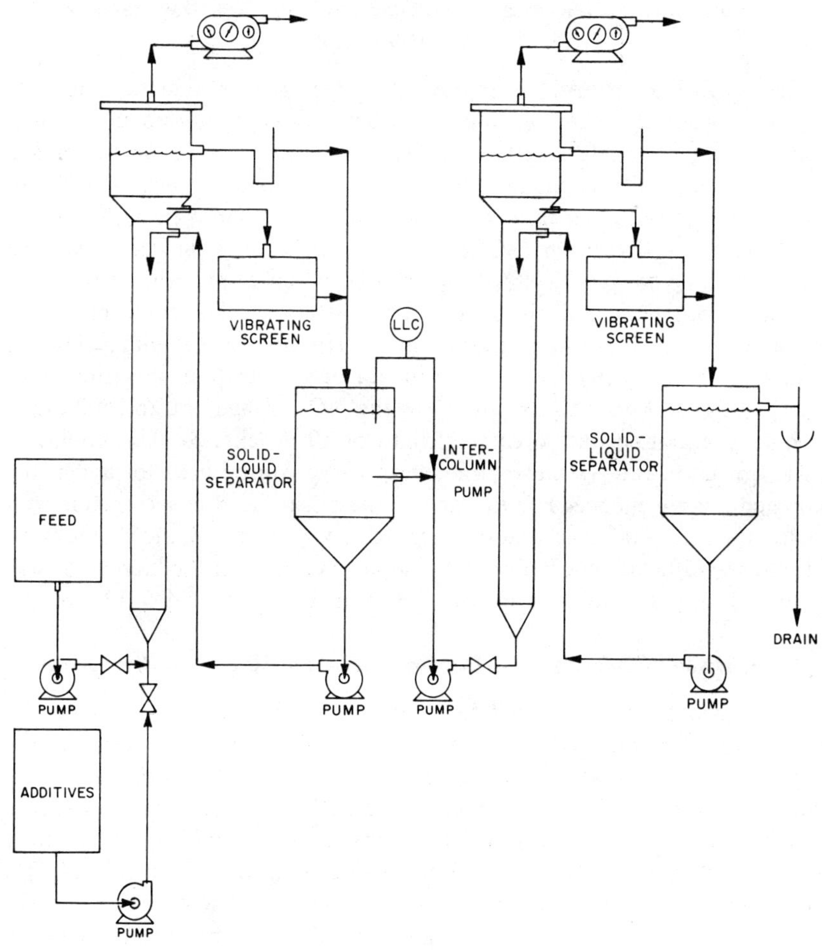

Fig. 4. Dual fluidized-bed bioreactor system.

about 10 to 20 cm below the removal point (see Fig. 3). The unattached biomass is discharged with the bioreactor effluent for later disposal by one of a number of conventional processes.

RESULTS

Many of the nitrate-containing industrial and DOE nuclear fuel cycle waste streams are very concentrated, some with nitrate contents as high as 20 or 30% (300,000 g/m^3 NO_3^-), as compared with the maximum operating range of 10,000 g/m^3 NO_3^- for biological denitrification. This makes it necessary to dilute the wastewater in some manner before it is fed to the bioreactor.

Effect of Nitrate Concentration on Denitrification Reaction Rates in a 5-cm-i.d. Bioreactor

The operating rationale is to introduce feed with the highest possible nitrate concentration to the bioreactor but have a near-zero nitrate discharge in a reasonable length of bioreactor. The 5-cm-i.d. bioreactor has been operated with various inlet NO_3^- concentrations ranging from 200 to 7500 g/m^3 with all other process variables held constant as possible [Standard Operations Condition (SOC)]. This series of tests indicated that the denitrification rate [kg N(NO_3^-)/day-m^3, based on empty reactor volume], increased with increased nitrate feed concentrations. The curve of nitrate concentration vs bioreactor length was constructed by superimposing the results of these denitrification tests (nitrate concentration at various bioreactor heights) "head to toe" to cover the NO_3^- range of 0 to 7500 g/m$_3$, producing a composite bioreactor length of 19 m (Fig. 5). This empirical nitrate reaction rate vs bioreactor length curve can be used to predict the bioreactor performance. For example if the nitrate feed concentration to a bioreactor was 5000 g/m^3 and the bioreactor was operated under SOC conditions, the effluent concentration would be 2000 g/m^3 if the bioreactor was 5.2 m in length and liquid residence time in the reactor would be 13 min.

Comparison of Operating Performance of 5- and 10-cm-i.d. Bioreactors

The 5- and 10-cm-i.d. bioreactors were operated under similar conditions of feed concentrations, pH, temperature, unit cross section flow rate (m^3/m^2), and bacterial loading. Both units were operated with NO_3^- feed concentrations of 200, 500, 1000, 1500, and 2000 g/m^3. As shown in Table II, the denitrification rates for the two units show no significant differences. Figure 6 shows the data for a typical set of tests using a feed with a NO_3^- concentration of 200 g/m^3. The 10-cm-i.d. unit was somewhat easier to operate because its larger diameter allowed gas bubbles to rise through the fluidized solids with less interference, due to wall resistance.

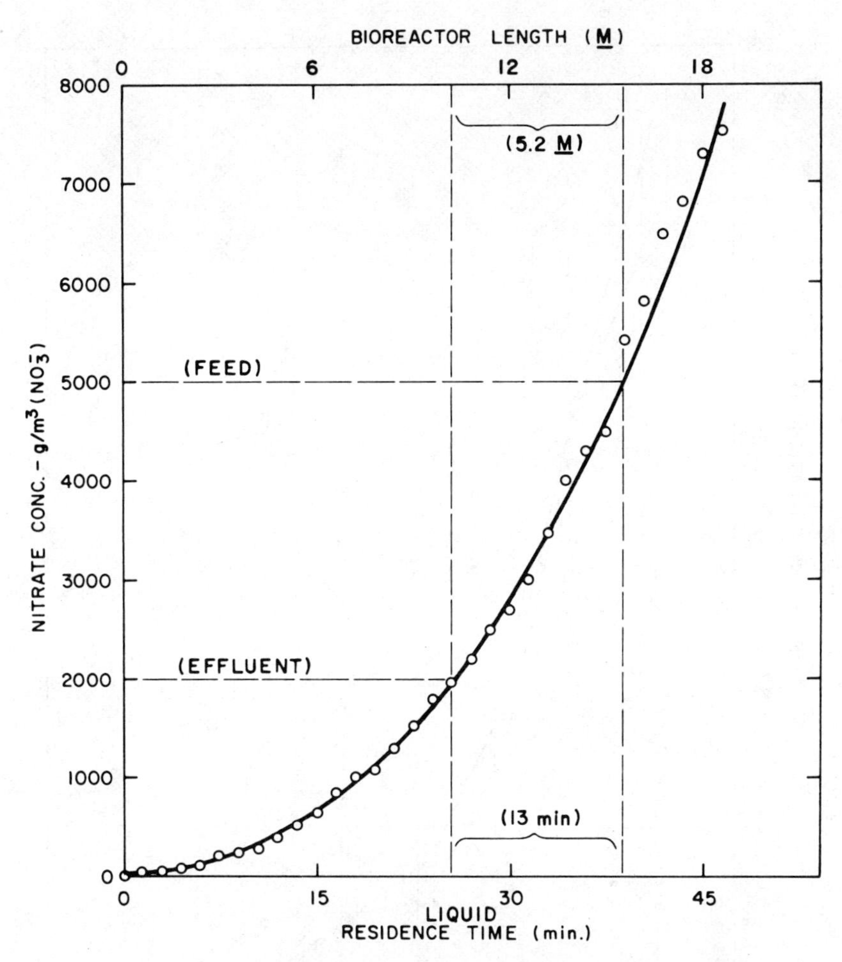

Fig. 5. Design curve for residence time requirements based on 5-cm-i.d. bioreactor.

TABLE II

Comparison of Denitrification Rate for 5- and 10-cm-i.d. Bioreactors

FEED/EFFLUENT CONCENTRATION (NO_3) g/m^3		DENITRIFICATION RATE [kg N(NO_3)/day-m^3]	
5 cm I.D.	10 cm I.D.	5-cm-I.D.	10-cm-I.D.
210/5	230/5	7.8	8.6
490/0	505/0	22.2	18.8
950/0	980/300	26.6	25.5
-----	1500/185	----	50.3
2300/500	------	49	----

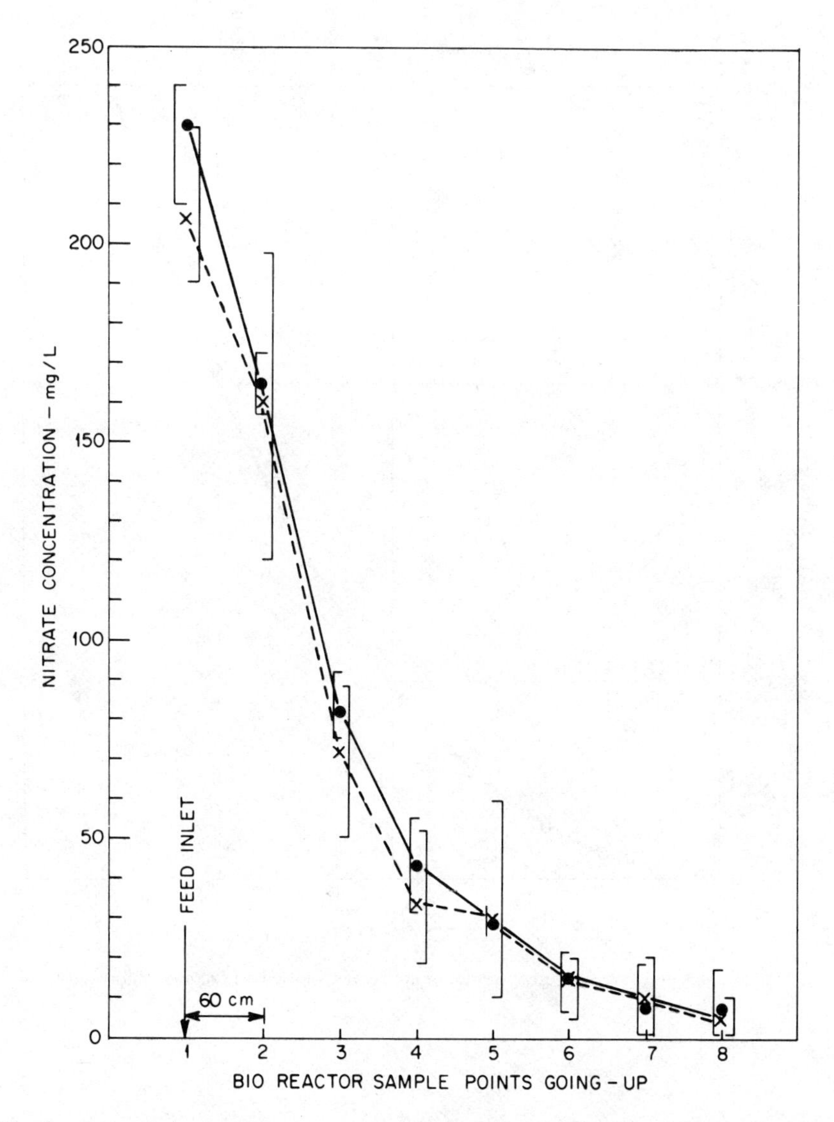

Fig. 6. Comparison of 5-cm-i.d. and 10-cm-i.d. bioreactors (nitrate feed concentration; π200 g/m³). (—●—) 10-cm-diam, 8.6 g N/D-2 (4 tests); (--×--) 5-cm-diam, 7.8 g N/D-L (3 tests).

10-cm-i.d. Dual Bioreactors in Series Test

The two 10-cm-i.d. by 6-m-long bioreactors were operated in series to test the denitrification rates and the mechanical features of staged bioreactors.

The feed for the test contained ~2000 g/m³ NO_3^- with a C/N ratio of 1.5 The average ethanol carbon usage (C/N) ratio was 1.2. When the nitrate concentration profile along the length of each bioreactor in series was plotted along with 5-cm-i.d. empirical nitrate reaction rate vs bioreactor length, good agreement of the data was obtained (see Fig. 7).

The fluidization pump for the second bioreactor operating with a liquid-level controller maintained a constant and correct flow of bioreactor 1 effluent for feed to bioreactor 2. The biomass control system, which used the Sweco filter to remove part of the attached biomass from the overloaded particle, controlled the biomass on the particles in the bioreactor to near the desired value of not more than 8–10 wt% (dry basis).

Recycle Mode of Operation

If the cost or quantity of dilution water is limiting, it can be minimized by using effluent with a near-zero nitrate concentration to dilute the

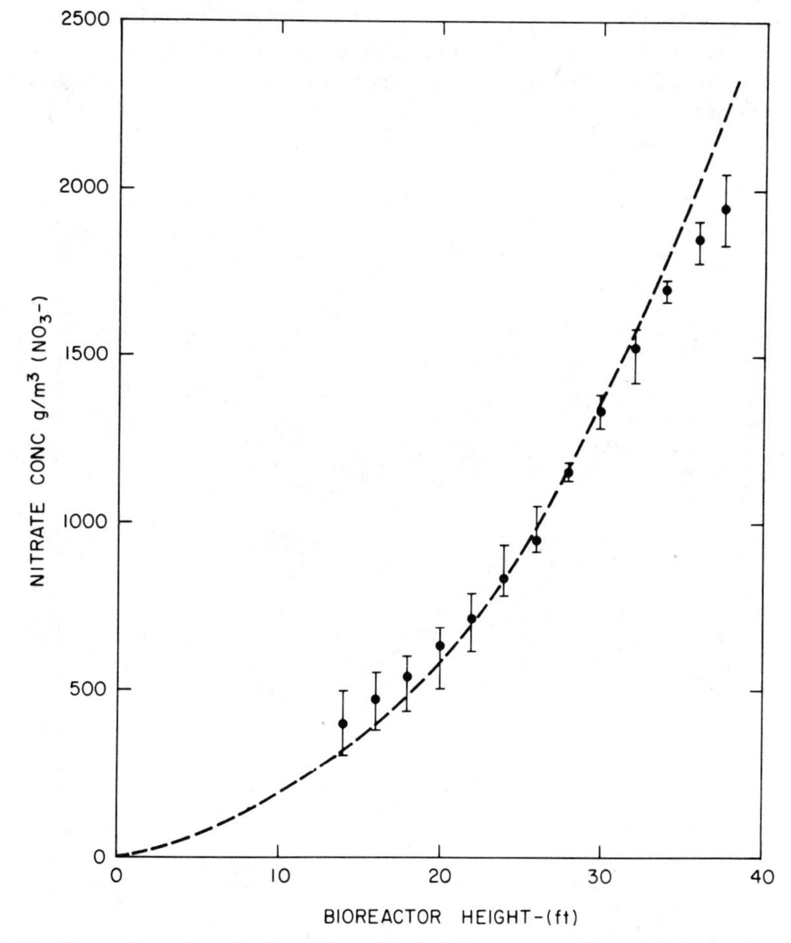

Fig. 7. Comparison of results from 10-cm-i.d. dual bioreactor and 5-cm-i.d. design curve. (I) Range and average values; (– – –) composite 5-cm i.d. bioreactor curve.

concentrated feed solution; this practice would also reduce the amount of treated effluent to be discharged. The 10-cm-i.d. bioreactor has been operated with a 9/1 recycle ratio and an inlet NO_3^- concentration of 200 g/m^3. Effluent solution was pumped to the bottom of the column at the rate of 3600 cm^3/min, where it was combined with fresh feed introduced at the rate of 400 cm^3/min before entering the column.

A typical nitrate concentration profile for the column is shown in Figure 8. The denitrification rate and the average nitrate concentration were calculated for each 63-cm section of the column. These results are presented in Figure 9. The average denitrification rate for the entire column was 6.1 kg $N(NO_3^-)$/day-m^3.

One disadvantage of this mode of operation is that any unreacted component such as the major cation (Na or NH_4) will be present at the same concentration as in the concentrated feed. The denitrifying bacteria will grow satisfactorily in the presence of Na and NH_3 up to 1*M* concentration; therefore, the nitrate of the concentrated feed cannot exceed about 70,000 g/m^3 NO_3 (7% NO_3). With other minor impurities, care must be taken to prevent buildup to a toxic level. The C/N ratio must also be moni-

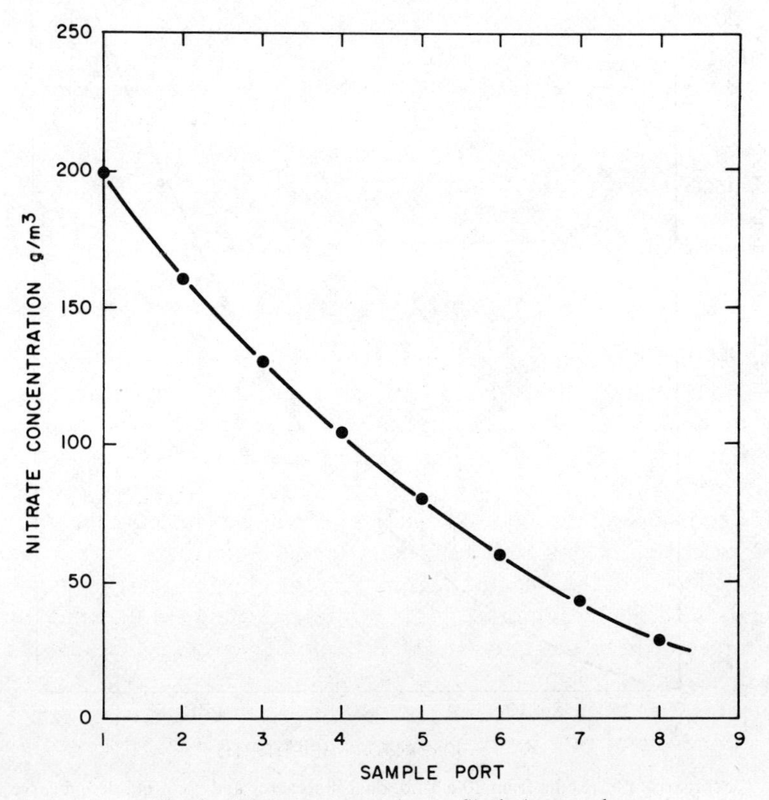

Fig. 8. Nitrate concentration profile during recycle.

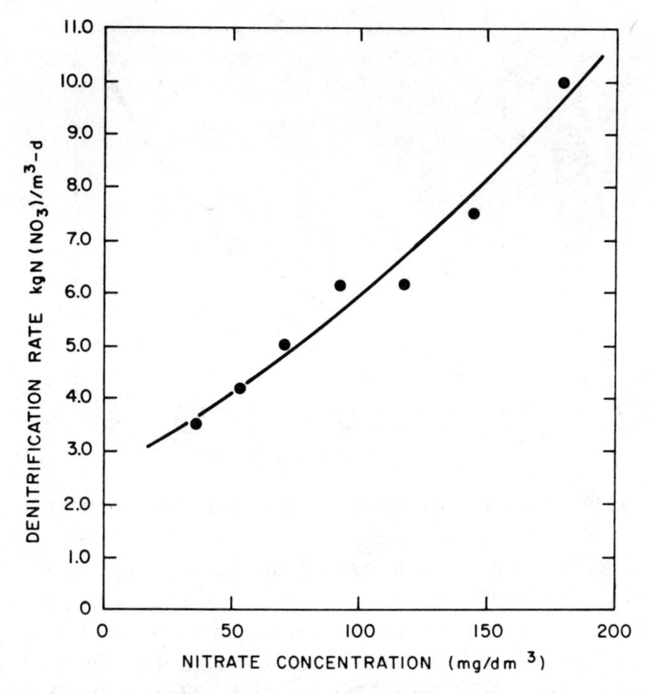

Fig. 9. Effect of feed nitrate concentration on denitrification rate in recycle.

tored closely since any unused organic carbon is returned to the bottom of the column, where it remains and concentrates. In many instances, however, the extra care required by this mode of operation will be more than compensated for by the reduction in dilution-water requirements.

Rate versus Temperature

The effect of temperature on denitrification rate has been measured in the 10-cm-i.d. column. In the first set of experiments, a heat exchanger was used to control the inlet temperature to the column. The maximum temperature achievable with this design was 22°C. The inlet NO_3^- concentration was kept at 1000 g/m³ for this set of experiments. At the lower inlet temperatures, the combination of heat generated by the bacteria and convective heat flow into the column caused an increase in temperature along the length of the column. To minimize the effect of these temperature changes, the denitrification rate in the first 62-cm section of the column was used for comparing the rates at the various temperatures. The results of these experiments, summarized in Figure 10, show that increasing the temperature has a dramatic effect on the denitrification rate.

A second set of rate vs. temperature experiments was performed three months later. The inlet NO_3^- concentration for these tests was 500 g/m³. A direct steam injection system was installed to control the inlet temperature.

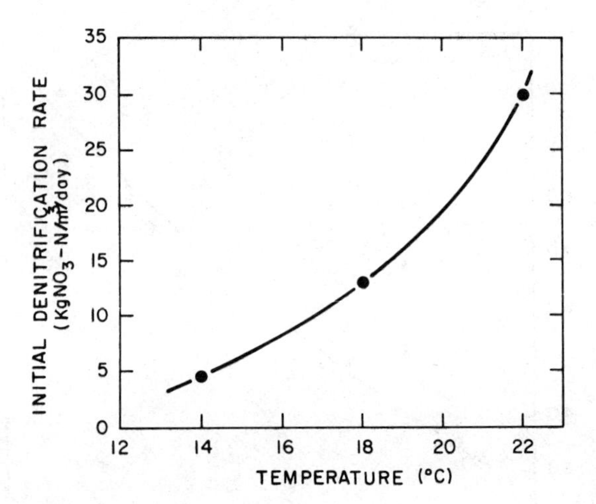

Fig. 10. Effect of operating temperature on initial denitrification rate.

The denitrification rate was measured at 22, 27, and 32°C. At these temperatures, the heat generated by the bacteria was approximately balanced by the convective heat loss from the column; thus the temperature remained reasonably constant. No increase in rate was observed when the temperature was increased from 27°C to 32°C. The results obtained from this set of experiments are shown in Figure 11, where the rates are plotted

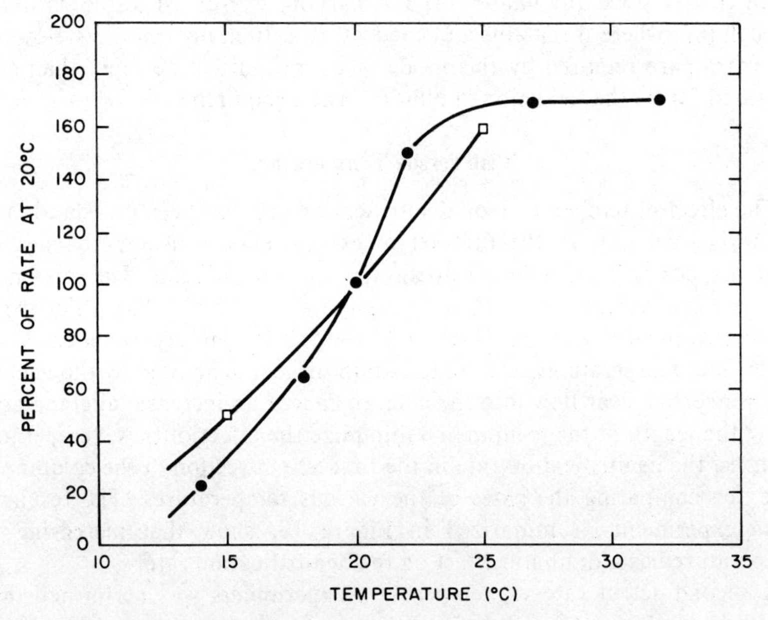

Fig. 11. Effect of operating temperature on denitrification rate. (—●—) Experimental; (—□—) published (Ref. 1).

as a percentage of the rate at 20°C in order to reduce the effects of the different nitrate concentrations in the two sets of experiments. A compilation of other published results is shown in Figure 11 for comparison [8]. These data suggest that supplemental heating above 22°C will probably be uneconomical.

Biomass Control and Distribution

The higher the nitrate concentration in the bioreactor, the higher the denitrification rate and the faster the biomass buildup on the fluidizing particles. Our present mode of operation is to remove and partially clean the dense, overgrown, biomass-covered particles which migrate to the top of the bioreactor, typically 5–10% (dry weight) reduced by one-half. The cleaned particles are returned to the bioreactor where their density causes them to fall through the fluidized bed to the lower zones of the bioreactor. The small part of the biomass growth that is not attached to particles is swept out of the bioreactor with the effluent. Soluble organic compounds, such as butyric acid, that are excreted by the biomass and dissolved carbon dioxide are discharged with the effluent along with excess ethanol used as carbon growth source. We attempted to control the carbon growth source utilization rate so that it matched the nitrate consumption; however, the varying biomass loading, which in turn changes the carbon usage rate, usually makes it necessary to provide a small excess.

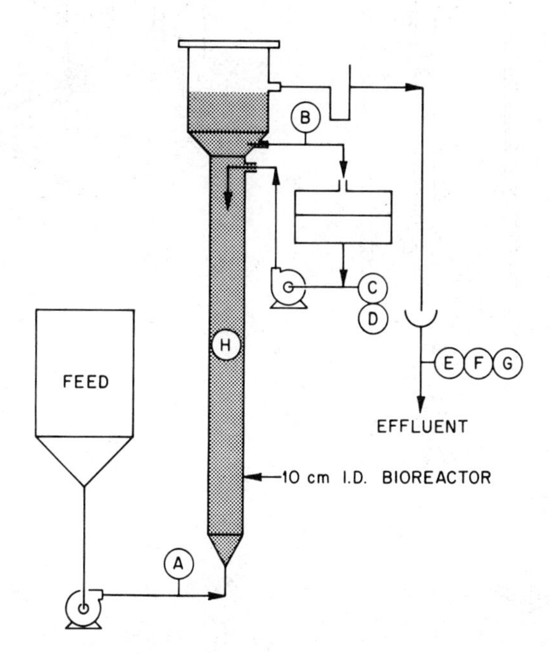

Fig. 12. Distribution of organic carbon.

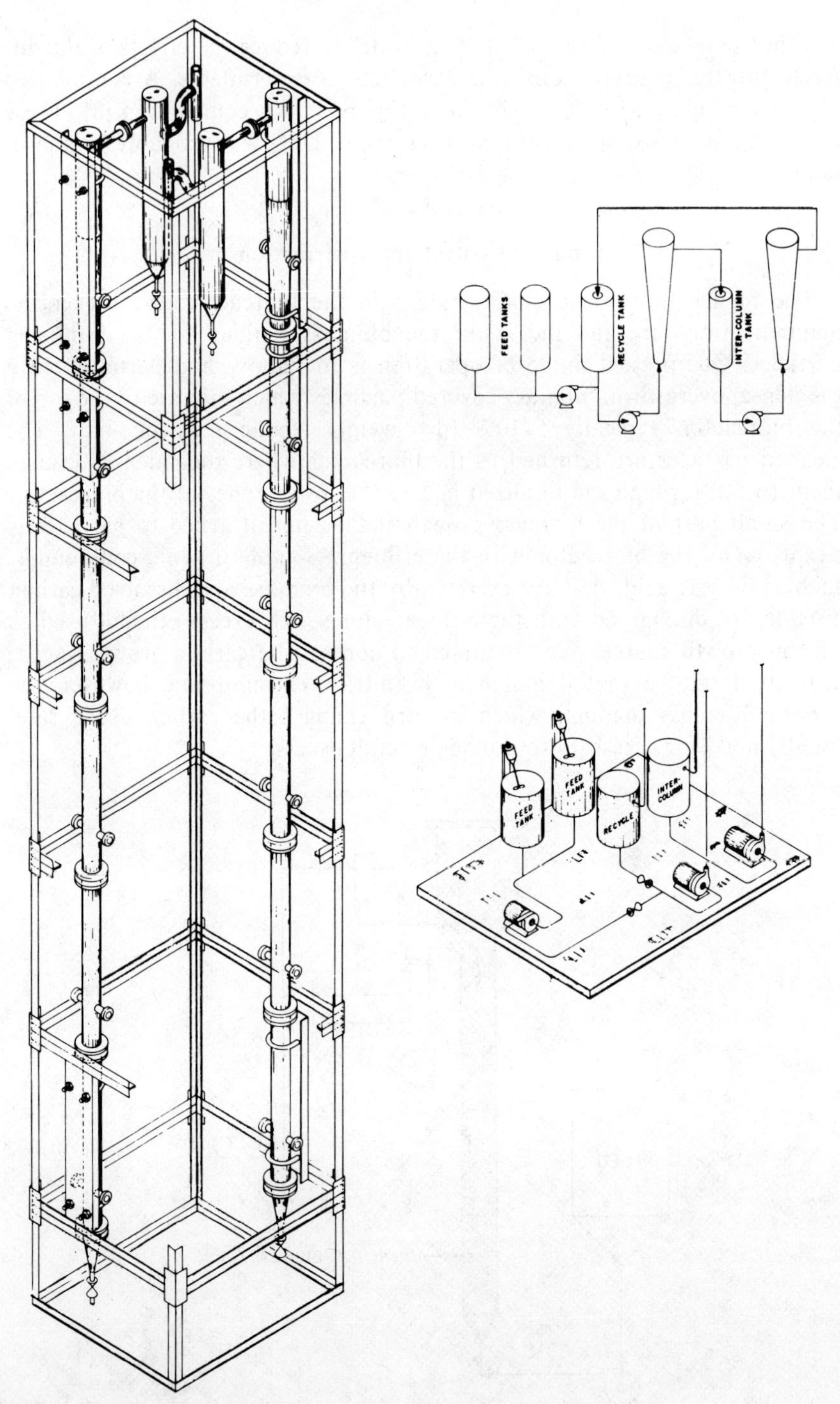

Fig. 13. Conceptual design of demonstration pilot plant for biological wastewater treatment.

TABLE III

Typical Organic Carbon Flowsheet Balance[a]

STREAM	FLOWSHEET SYMBOL (FIG.14)	CARBON TYPE	CARBON CONC (g/m^3)	CARBON QUANTITY (g/DAY)
FEED	A	ETOH	200	1150
EFFLUENT	D	SOLUBLE	25	160
	F	ETOH	40	240
	G	BIOMASS	20	130
		DIS. CO_2	100	660
			TOTAL	1190
BIOREACTOR	H	10 cm ID x 6 m BIOREACTOR CONTAINED 20 L OF SETTLED SOLIDS-AT 5% BIOMASS LOADING = 1600 gm OF CARBON		
TO SWENCO SCREEN	B	ATTACHED BIOMASS (~50% OF BIOREACTOR/DAY)		950
FROM SWENCO SCREEN	C	UNATTACHED BIOMASS		330

[a] 500 g/m^3 NO_3^- Feed concentration test.

The effluent will contain the following carbon sources: (1) biomass removed from particles; (2) biomass not attached to particles; (3) soluble organic biomass waste by-product; (4) carbon dioxide reaction gas; and (5) excess organic feed nutrient. Figure 12 and Table III show typical levels of organic carbon and biomass distribution in a 10-cm-i.d. bioreactor. The carbon concentration of the effluent can be reduced to discharge levels using standard, available sanitary sewage procedures. The removal of the filterable biomass is one possible solution, but since it is a minor component of the whole carbon discharge problem, we chose to treat the whole mixed effluent stream. We believe that if the bioreactor biomass was controlled at a 1 or 2% dry weight, the carbon discharge problem would be minimized.

CONCLUSIONS

The purpose for the preceding development program was to design a mobile denitrification pilot plant that could be used in a number of DOE and commercial nuclear fuel cycle operations to test the applicability of this biological denitrification process to their individual nitrate wastewaters. The results indicate that the scale-up of the bioreactor from 5 cm i.d. to 10 cm i.d. is directly proportional. The tests with the single and dual 10-cm-i.d. by 6-m-tall bioreactors corresponded satisfactorily with the 5-cm-i.d. bioreactor empirical curve of nitrate concentration vs. bioreactor length. Therefore, we have a basis for predicting the performance of our mobile pilot plant.

The mobile pilot plant will contain two 20-cm-i.d. by 7-m-long bioreactors operated in series (see Fig. 13 and the appendix). For near-zero NO_3^- effluent (5 to 10 g/m_3), the feed will have a maximum concentration of 4000 g/m^3 and will be introduced at a flow rate of 16 liter/min. The temperature will be regulated between 22 and 30°C.

APPENDIX

The specifications of the mobile denitrification pilot plant are as follows:
Bioreactors: two operated in series (20 cm i.d. by 7.3 m long).
Operating conditions:

Nitrate concentration	feed-4000 g/m^3 max effluent-5 liter g/m^3
flow rate	16 liter/min (0.84 cm/sec)
temperature	22–30°C
pH	feed-6.5–6.8, effluent, 8.1–8.5

The authors wish to thank C. D. Scott and W. W. Pitt for the program technical direction and G. B. Dinsmore and J. E. Miner, our laboratory technicians, for their dedication to details that lead to good results.

References

[1] L. Landner, *Proceedings of Conference on Nitrogen as a Water Pollutant* (LAWPR Specialized Conference, Copenhagen, Denmark, 1975), Vol. 1.
[2] S. E. Shumate II, C. W. Hancher, and G. W. Strandberg, "Biological processes for environmental control of effluent streams in the nuclear fuel cycle," *Proceedings of Waste Management and Fuel Cycle '78 Symposium* (University of Arizona Press, in press).
[3] *Evnironmental Survey of the Nuclear Fuel Cycle*, prepared by the U.S. Atomic Energy Commission (AEC), Fuels and Materials Staff of the Directorate of Licensing (dated November 1972).
[4] C. D. Scott and C. W. Hancher, *Biotechnol. Bioeng.*, *18*, 1393 (1976).
[5] C. D. Scott, C. W. Hancher, and S. E. Shumate II, "A tapered fluidized bed ad a bioreactor," *Proceedings of the 3rd International Conference on Enzyme Engineering*, 1975.
[6] U.S. patent 3,926,736.
[7] *Standard Methods for the Examination of Water and Wastewater*, 13th ed. (American Public Health Association, Washington, D.C., 1971).
[8] *EPA Process Design for Nitrogen Control Manual*, October 1975.

Biological Treatment of Aqueous Wastes from Coal Conversion Processes*

J. A. KLEIN and D. D. LEE
Oak Ridge National Laboratory, Oak Ridge, Tennessee 37830

INTRODUCTION AND BACKGROUND

The abundance of this country's coal resources, relative to those for crude oil and natural gas, has prompted both the Federal Government and industry to undertake an intensive effort to develop processes for converting this solid form of energy into substitute gaseous and liquid fuels. Environmental and economic reasons will compel developers of coal conversion processes to maximize the reuse of water while reducing pollutant levels in aqueous effluents. Data on the amenability of aqueous coal conversion wastes to treatment need to be obtained in order to ensure that water supply and quality are not adversely affected by the developing coal conversion industry.

Some of the most difficult wastewater problems encountered today are inherent in the coal conversion industry. These include high concentrations of phenolics, ammonia, and sulfur compounds, as well as trace elements (heavy metals) and polynuclear aromatic hydrocarbons (PAHs).

Although wide variations exist in the composition of coal conversion wastewaters, depending on the type of process, operating conditions, and nature of the coal used, generalizations can be formulated. The major constituents of a representative wastewater, and their concentrations, are listed in Table I [1]. The high phenol and ammonia concentrations are typical of coal conversion process wastewaters, regardless of whether they are condensation liquors or scrubber waters. Also listed are the concentration ranges expected to be specified in future Federal regulations for a variety of components. Standards of related industries such as coking and petroleum refining were used to develop these values [1–3] since there are as yet no regulations for the coal conversion industry. As shown, the levels of all the components listed will exceed the limits of the anticipated regulations. A survey of the literature reveals a number of presently available and proposed technologies for the cleanup of wastewaters with compositions

* Research sponsored by the Division of Environmental Control Technology, U.S. Department of Energy, under contract W-7405-eng-26 with the Union Carbide Corporation.

Biotechnology and Bioengineering Symp. No. 8, 379–390 (1978)

TABLE I
Composition of a Typical Coal Conversion Wastewater

Component	Concentration in wastewater ($\mu g/cm^3$)	Limits of anticipated regulations ($\mu g/cm^3$)
Phenol	6,000	0.03-0.3
NH_3	10,000	0.8 -5.0
H_2S	1,000	0.02-0.2
CN^-	100	0.02-0.1
SCN^-	500	---
PAH	10	---
TOC	20,000	BOD 4-30 COD 20-350

similar to those indicated in Table I [1]. As expected, they represent a wide range of removal efficiencies, final residual concentration levels, and costs.

Efforts to obtain relative economic costs of various environmental control technologies for the treatment of aqueous wastes from coal conversion processes have been completed [1]. Six processes have been identified as potentially promising in this area: ozonation, adsorption, biological degradation, solvent extraction, membrane processes, and coagulation-flocculation.

Ozonation, a fairly recent entry into the American wastewater treatment field, offers certain advantages as a method of coal conversion wastewater cleanup. For example, recent work has shown that reasonable reductions in phenolic and PAH contents can be achieved without the production of questionable toxic or carcinogenic compounds [4, 5]. However, phenols are oxidized in discrete steps via simple oxidation products to CO_2 and H_2O, with each step requiring larger amounts of ozone. Present indications are that the complete oxidation of coal conversion wastewaters cannot be achieved economically with ozone [4, 6]. If a small amount of organic needs to be treated, as in a final polishing step, or only a small amount of degradation is required, as would be the case when only the phenolic structure is to be cleaved and not completely oxidized to carbon dioxide and water, then ozonation may be preferred.

Adsorption on activated carbon or other sorbents (coal conversion char or high-surface-area coals) appears to have two possible uses: (1) treatment and recovery of concentrated wastes and (2) "polishing" of more dilute wastes for reuse or discharge. The first application appears to suffer from problems associated with moving and regenerating large quantities of activated carbon, whereas the second application will have much smaller demands. In addition, coal conversion char shows a potential for sorbing wastes, and the spent char could be burned or gasified, thereby eliminating the regeneration process and recovering the fuel value of the dissolved wastes [7].

Biological oxidation shows a very high potential for application to hydrocarbonization wastewater. Numerous experimenters have demonstrated that the concentration of phenol, the principal contaminant in aqueous coal conversion wastes, can be reduced by up to 99.9% to achieve levels as low as 20 ng/cm^3 [8, 9]. PAH levels appear to be relatively unaffected by current biological methods and, in fact, may be somewhat increased for those biological units that require long residence times.

Bioreactors of widely different configurations are available for biological degradation. Representative examples are: fluidized beds, trickling beds, activated sludge, packed beds, biological lagoons, oxidation ditches, rotating biological filter, overland purification, and soil purification. Such systems are susceptible, to varying degrees, to environmental shocks. The long residence time requirement is the major disadvantage for activated sludge processes and biological lagoons. Advantages which are common to all the bioreactor types mentioned above include relatively high availability factors and low maintenance and operating costs, typically $5 to $10 per thousand gallons of wastewater [1].

Solvent extraction has attracted the attention of investigators for many years. Various solvents have been investigated for use in phenol removal in both typical solvent extraction systems and dual solvent processes [10, 11]. Although high distribution coefficients have been reported for many of these systems, Lurgi's Penolsolvan Process appears to be the only one that has been commercialized. Even though the solvent for the contactor in this case is chosen for its high distribution coefficient for phenol, other types of organics will also be stripped out of the aqueous phase. The Phenolsolvan Process should remove 99.5% of the monohydric phenols but only 15–60% of other organics and phenols [12]. Solvent extraction appears to offer an advantage only when a moderate amount of reduction in phenol content is needed in order to allow an additional waste treatment system to operate or when a salable crude phenol can be produced.

Another area of interest includes membrane processes such as reverse osmosis, ultrafiltration, and electrodialysis. In reverse osmosis and ultrafiltration, separation is achieved by passing the water through a membrane, leaving the concentrated solute behind; conversely, in electrodialysis, the solute passes through the membrane, leaving the purified solvent behind. These membrane processes, although considered expensive, may be useful in upgrading wastewater for inplant reuse. Although ultrafiltration and reverse osmosis have shown phenol reductions of 50 to 95% [13, 14], the resulting levels are still in excess of anticipated regulations. In addition, a concentrated organic stream will be produced and require a suitable disposal method.

Some investigators have shown that as much as 80% of the chemical oxygen demand of some wastewater streams may be contained in particles colloidal or larger in size, while up to 50% of the total organic carbon in the

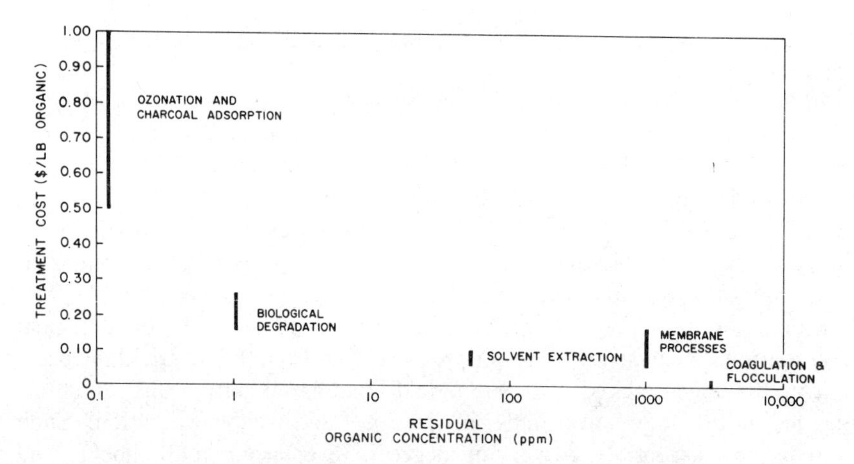

Fig. 1. Economic costs of environmental control technologies for aqueous coal conversion wastes.

effluent of biological treatment plants may be colloidal [15]. The coagulation and flocculation of these particles could lower dramatically the demand on downstream treatment steps.

The general consensus seems to be that biological degradation will give the lowest cost per thousand gallons of wastewater or per pound of material degraded at a reasonable removal efficiency. The advantageous position of biological degradation [1] is indicated in Figure 1, where treatment cost is plotted versus residual organic levels.

PROPOSED INTEGRATED FLOWSHEET

Figure 2 shows a proposed flowsheet that takes advantage of several of the above removal techniques for the complete treatment of coal conversion wastewaters. The processing scheme includes physical, chemical, and biological treatment of the waste stream. Wastewater from the coal conversion process enters a surge tank where the pH is adjusted for H_2S removal in a stripping tower. The H_2S off-gas is subsequently sent to a sulfur recovery facility. The pH of the water stream is then adjusted and the ammonia stripped by air or steam to a concentration of less than 100 ppm. After readjustment of the pH, the waste stream is sent first to a solvent extraction step for phenol recovery and finally to the bioreactor surge tank as feed for the bioreactor.

The bioreactor oxidizes phenolic compounds, thiocyanates, and other organics, with additional air or oxygen and a nutrient supplement being required. Excess biomass and solid support that are carried overhead are sent to a solids treatment facility for recycle. Residual ammonia oxidation and denitrification are completed in two additional biological columns; the sludge is routed to the solids treatment facility and the water to a carbon adsorption tower. The activated carbon tower removes the remaining biologically refractory phenolics and, possibly, some PAHs. Fresh or regenerated carbon is periodically added to the carbon adsorber, and the spent material is sent to the bioreactor for use as a bacteria support medium. The wastewater than flows successively to a settling basin for removal of suspended solids and to an ozonation reactor where the remaining organics, including any PAHs, are oxidized to CO_2. The final, purified water is pH adjusted, if necessary, and discharged. The solid sludges produced at various locations are treated to regenerate the activated carbon by a wet oxidation process or some similar method.

In an effort to maintain the low cost generally associated with biological degradation and yet reduce the long residence time requirements of activated sludge and biological lagoon systems normally used for industrial wastewaters, Oak Ridge National Laboratory's Chemical Technology Division has been involved in a program for the development of a tapered fluidized-bed bioreactor that can be adapted to the treatment of coal conversion wastewater [16].

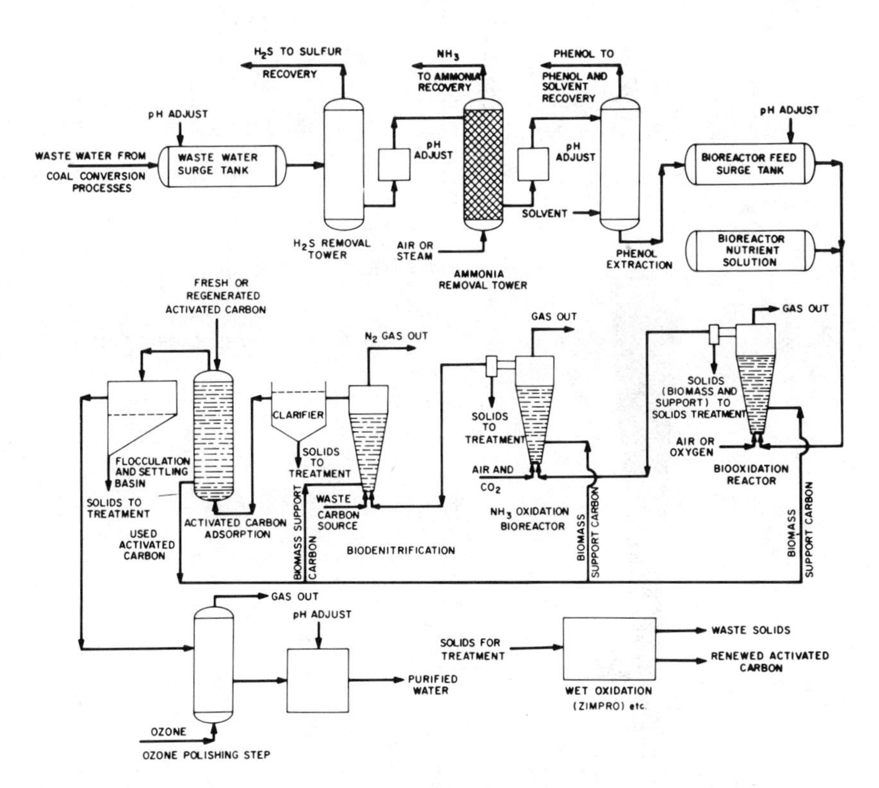

Fig. 2. Proposed integrated flowsheet for treatment of aqueous coal conversion wastes.

Description of Tapered Fluidized-Bed Bioreactor

This tapered fluidized-bed bioreactor system, shown in Figure 3, consists of a tapered (2.54–7.62 cm) section containing a solid support to which bacteria can become attached. The solid support, which is generally a substance such as coal or sand of about 30/60 mesh particle size, is fluidized by the flow of the waste stream to be treated. The tapered bed permits a wide range of fluidizing conditions and allows for expansion of the bed as biomass coats the particles. In addition, the tapered section produces few large eddies and tends to minimize backmixing, especially at the entrance point for two-phase systems.

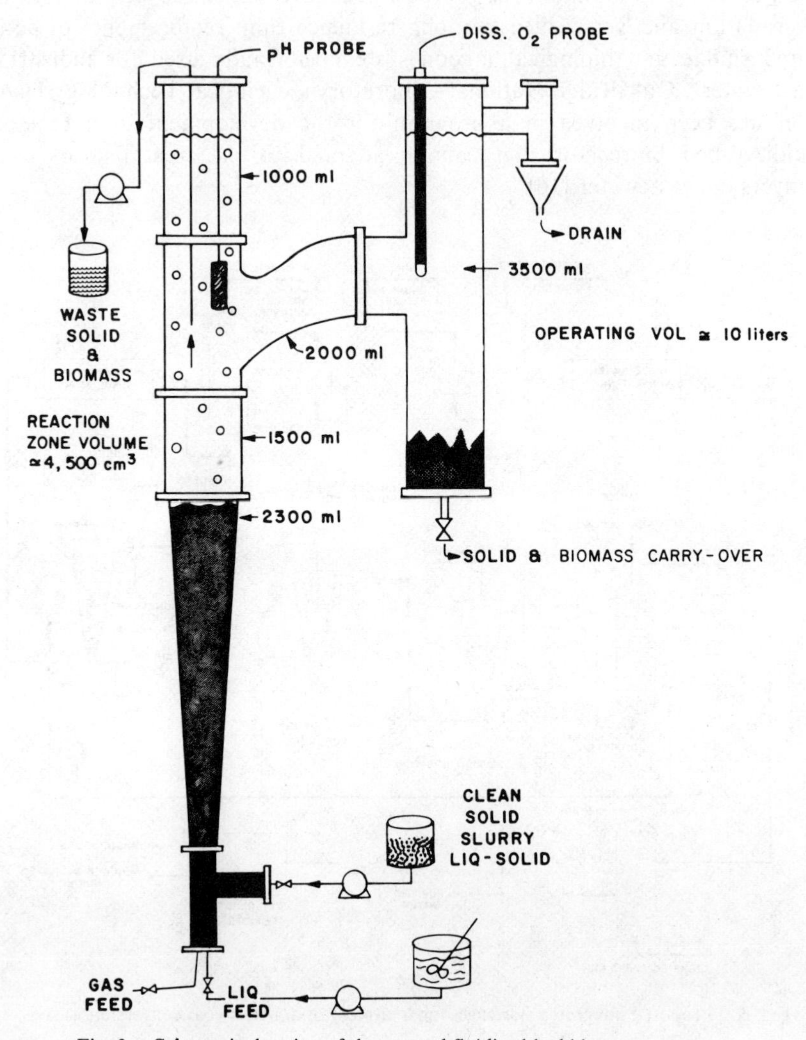

Fig. 3. Schematic drawing of the tapered fluidized-bed bioreactor system.

The chief advantage of the fluidized-bed system is the high bacteria loading per unit reactor volume [17], which is made possible by the large surface area available on the small particles. Also, the overloaded support particles can be easily removed and fresh particles added to the reactor while operation continues. The bench-scale operating systems are shown in Figures 4–6. Figure 4 is a photograph of a single 105-cm-long fluidized bed, with a 2.54 to 7.62-cm fluidized section, containing about 1000 cm^3 settled 30/60 mesh anthracite coal particles. Figure 5 depicts a two-stage system of two reactors connected in series, while Figure 6 illustrates a 10.2-cm-diam cylindrical reactor, 4.50 m high, with a 10.2- to 15.2-cm tapered section 0.61 m long on top. The latter reactor holds about 40 liters of fluidized medium (about 12–15 liter settled volume of coal). In each case, the reactor operates

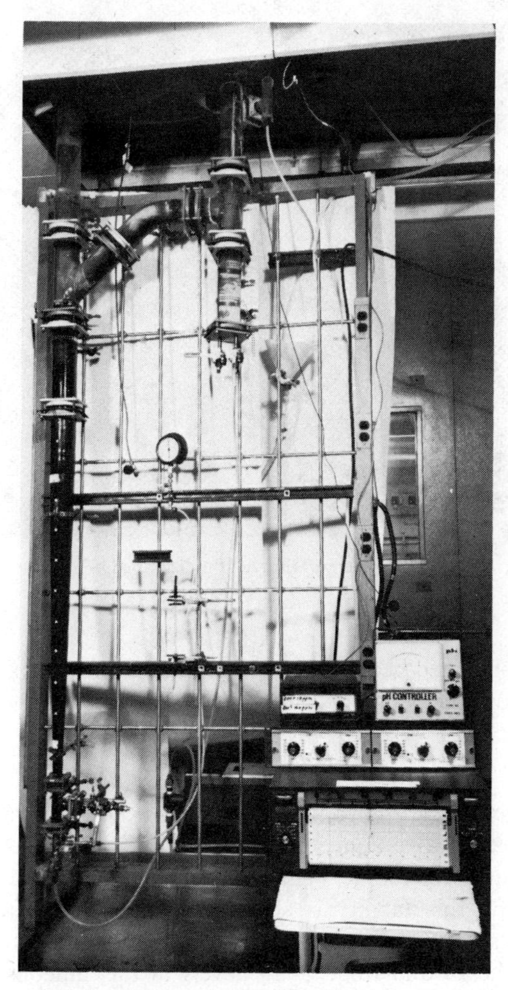

Fig. 4. Single tapered fluidized-bed bioreactor.

Fig. 5. Two-stage system of tapered fluidized-bed bioreactors.

as a three-phase fluidized bed and the oxygen needed for metabolism is supplied by sparged oxygen which is admitted to the bottom of the reactor.

Results of Operation

The bench-scale system shown in Figure 4 has been in operation for more than a year, treating the aqueous waste produced by the ORNL coal hydrocarbonization bench-scale facility. The composition of the scrubber water produced by this facility is shown in Table II. Some preliminary results indicate that the tapered fluidized bed in a single contacting stage can reduce the phenol and thiocyanate levels in the aqueous waste to ≤ 1 $\mu g/cm^3$ and 1–5 $\mu g/cm^3$, respectively, at rates up to 5–10 kg phenol/m^3

reactor volume per day, and 0.5–1 kg thiocyanate/m^3 reactor volume per day, at a flow rate of 500 cm^3/min.

It is expected that, in general, a pretreatment of the scrubber water is required. This consists of ammonia stripping and suspended solids precipitation. The pH is adjusted to that required for the bacterial action, 6.0–7.5, phosphate ion at 10–20 $\mu g/cm^3$ is added, and the material is fed to the reactor either in a prediluted form (in which tap water is used to dilute the scrubber water) or by using waste recycle to dilute the feed to the maximum phenol concentration, 200 $\mu g/cm^3$, that the bacteria (Phenobac, Polybac Corp.) can tolerate and achieve maximum rates. If phenol recovery is practiced (via liquid-liquid extraction), dilution may not be required.

Fig. 6. Single 10.2-cm-diam tapered fluidized-bed bioreactors.

TABLE II

Analysis of Scrubber Water Generated by the ORNL Coal Hydrocarbonization Facility

Component	Concentration ($\mu g/cm^3$)		
	High	Low	Average
Phenols	10,000	3000	6900
Thiocyanate	600	200	380
Ammonia	7,000	1000	3600
Phosphate	50	3	20
Total carbon	20,000	3800	9300
Carbon dioxide	2,000	1000	1250
Organic carbon	18,000	2800	8150
pH	10.1	8.5	9.5

The larger reactor (Fig. 6) has been operated with a synthetic feed mixture containing phenol at concentrations up to 200 $\mu g/cm^3$, thiocyanate at up to 100 $\mu g/cm^3$, and nutrients (<1 $\mu g/cm^3$ trace metals), plus phosphate and nitrogen at 10 and 50 $\mu g/cm^3$, respectively. Feed rates ranged as high as 20 kg phenol/m^3 reactor volume/day, and flow rates up to 4000 cm^3/min were used. Both 30/60 mesh coal and 60 mesh alumina were used in the reactor in separate runs. The alumina, which is much denser than the coal, gave better performance during operation of the three-phase bed since it did not get carried (overhead) out of the reactor as rapidly as coal when loaded with bacteria. Synthetic feed with a phenol concentration of over 100 $\mu g/cm^3$ was fed at rates up to 4000 cm^3/min, and the waste phenol level was less than 10 $\mu g/cm^3$. Sampling points in the lower sections of the reactor showed phenol removal rates as high as 30 kg/m^3 day. Thiocyanate was degraded at rates up to 1 kg/m^3 day.

In the fluidized-bed bioreactors for phenol degradation, the total soluble organic carbon content is reduced an average of 95%, while phenol is reduced an average of $>99.5\%$, using the trace metal–phosphate-supplemented hydrocarbonization scrubbing water as a reactor feed. The concentrations of other organic components (PAHs, xylenols, etc.) are also decreased to some extent, depending on the chemical species involved and the residence time available in the reactor.

Ongoing Work

The objective of our present work is to characterize the effectiveness of various biological treatment systems, including the tapered fluidized-bed bioreactor, for the cleanup of aqueous coal conversion wastes. Since the tapered fluidized-bed bioreactor concept is still in the developmental stage, an integrated procedure is being formulated to test the efficiency of this treatment technique. This plan includes the Biology, Environmental Sciences, Analytical Chemistry, and Chemical Technology Divisions of ORNL and is aimed at developing procedures for characterizing the efficiency of environmental control methods for any coal conversion aqueous effluent. Untreated, biotreated, and final polishing treatment material from ORNL's hydrocarbon experiment will be analyzed and tested for both acute and mutagenic toxicity.

Future work will include utilizing "low cost per unit material removed" biological methods as well as a higher-cost, but more-efficient, alternative cleanup method to remove the remainder of the refractory materials as described in our proposed flowsheet.

CONCLUSIONS

A tapered fluidized-bed bioreactor has been adapted to degrade actual hydrocarbonization wastewater. Relatively low residual organic levels at high removal rates per unit volume have been achieved. The use of this type of bioreactor in conjunction with more efficient alternative control technologies may offer real economic advantages and result in extremely low levels of refractory organics.

References

[1] J. A. Klein and R. E. Barker, *Assessment of Environmental Control Technology for Coal Conversion Aqueous Wastes* ORNL/TM-6263 (July, 1978).

[2] EPA, *Effluent Guidelines and Standards for Iron and Steel Manufacturing Point Source Category*, 39FR24114 (June 28, 1974).

[3] EPA, *Effluent Guidelines and Standards for Petroleum Refining*, 39FR16560 (May 9, 1974); 39FR32614 (September 10, 1974); 40FR21939 (May 20, 1975).

[4] H. R. Eisenhauer, *Water Res.*, *5*, 467 (1971).

[5] G. S. Sforzolini, A. Savito, S. Monarca, and M. N. Lollini, *Decontamination of Water Contaminated with Polycyclic Aromatic Hydrocarbons (PAH) I. Action of Chlorine and Ozone on PAH Dissolved in Doubly Distilled and De-ionized Water, Ig. Mod.*, *66*(3),309 (1974).

[6] R. E. Rosfjord, R. B. Trattner, and P. N. Cheremisinoff, *Water Sewage Works*, *123*, 96 (1976).

[7] A. J. Forney, G. E. Johnson, A. DeGalbo, W. P. Haynes, and D. A. Green, *Internal Quarterly Technical Progress Report*, *April–June 1975*, PERC/QTR-75/2; *July–Sept. 1975*, PERC/QTR-75/3; *Oct.–Dec. 1975*, PERC/QTR-75/4; *Jan.–March 1976*, PERC/QTR-76/1; *April–June 1976*, PERC/QTR-76/2; and *July–Sept. 1976*, PERC/QTR-76/3.

[8] J. Barker and R. Thompson, *Biological Removal of Carbon and Nitrogen Compounds from Coke Plant Wastes*, EPA-R2-73-167, Washington, D.C., 1973.

[9] G. J. Capestrany, J. McDaniels, and J. L. Opgrande, *J. Water Pollut. Control Fed.*, *49*(2), 257 (1977).

[10] J. P. Earhart, K. W. Won, H. Y. Wong, J. M. Prausnitz, and C. J. King, *Chem. Eng. Prog.*, *73*(5), 67 (1977).

[11] L. I. Volkova, A. M. Kunin, N. N. Amagaeva, G. V. Nenast'eva, A. Ya. Tkachenko, and L. N. Lapina, *Tr. Kalinin. Politekh. Inst.*, *13*, 20 (1972).
[12] M. R. Beychok, "Coal gasification and the phenolsolvan process," presented at the American Chemical Society, Division of Fuel Chemistry, 1974 Annual Meeting; Vol. 19, No. 5, p. 85.
[13] W. A. Duvel and T. Helfgott, *J. Water Pollut. Control Fed.*, *47*, 57 (1975).
[14] S. L. Klemetson, "Treatment of phenolic wastes," presented at the Conference on Environmental Aspects of Fuel Conversion Technology III, Hollywood, Fla., Sept. 13–16, 1977.
[15] D. F. Bishop, L. S. Marshall, T. P. O'Farrell, R. B. Dean, B. O'Connor, R. A. Dobbs, S. H. Griggs, and R. V. Villiers, *J. Water Pollut. Control Fed.*, *39*, 188 (1967).
[16] D. D. Lee and C. D. Scott, "A tapered fluidized-bed bioreactor for treatment of aqueous effluents from coal conversion processes," presented at the 70th National AIChE Meeting, New York, N.Y., Nov. 13–17, 1977.
[17] J. S. Jeris, R. W. Owens, R. Hickey, and F. Flood, *J. Water Pollut. Control Fed.*, *49*, 816 (1977).

Acclimation and Degradation of Petrochemical Wastewater Components by Methane Fermentation

W. LIN CHOU and R. E. SPEECE
Drexel University, Philadelphia, Pennsylvania 19104

RASHID H. SIDDIQI
College of Petroleum and Minerals, Dharan, Saudi Arabia

INTRODUCTION

Methane fermentation has two unique advantages in wastewater treatment. The organic pollutant is almost quantitatively converted to CH_4, a high quality energy resource, and for most soluble industrial pollutants there is only negligible synthesis of excess biomass [1]. These are exceptionally desirable characteristics of a wastewater treatment process. Approximately 11×10^6 BTU of CH_4 is produced per ton of chemical oxygen demand (COD) destroyed. Synthesis of excess biomass is normally 20 to 100 lb of dry mass per ton of COD destroyed. By contrast, aerobic biological treatment processes consume approximately 8×10^6 BTU (1000 kWhr) per ton of COD removed and synthesize approximately 1000 lb of excess biomass per ton of COD removed. Disposal of the biomass synthesized during aerobic treatment is costly and energy intensive.

Methane fermentation has proven to be reliable as well as cost and energy effective for the treatment of domestic sludges and certain industrial wastewaters such as from packing plants [2–3]. Yet methane fermentation has traditionally been restricted to treatment of municipal wastewater sludges and is often not considered to be applicable to industrial wastewater treatment. The area of industrial wastewater treatment may, however, offer the greatest potential for exploitation of the significant advantages inherent in methane fermentation. Petrochemical production in the United States is approximately 1×10^{11} lb per year [4] and approximately 0.5 to 2.0% of the product is lost to the wastewaters during production [5]. This demonstrates the magnitude of the petrochemical wastewater problem—5×10^8 to 2×10^9 lb per year of pollutants. Degradation of petrochemical wastewaters by methane fermentation has been noted by various workers [6–9]. Specifically this paper is intended to demonstrate the wide variety of petrochemicals which are amenable to methane fermentation treatment.

Biotechnology and Bioengineering Symp. No. 8, 391–414 (1978)

0572-6565/78/0008-0391$01.00

Toxicity and Acclimation

A petrochemical may initially exhibit toxicity to an unacclimated population of methane-fermenting bacteria, but with acclimation the toxicity may be greatly reduced or disappear. In addition the microorganisms may develop the capacity to actually degrade compounds which showed initial toxicity. Inducible enzymes may be synthesized and/or a shift in the predominant population may result allowing metabolism to occur.

The purpose of cross acclimation is to shorten the length of time required to start a new digester for different industrial wastes. The possible reasons cross acclimation might work are (a) the enzyme system for utilizing a new organic compound is relatively not substrate-specific as long as the new compound has a similar structure to the substrate to which the bacteria are acclimated and (b) the enzyme systems required to metabolize a new organic compound may already exist within the bacteria, that is, the enzymes may be constitutive or they may have been induced for some reason. Weng and Jeris [10] reported that glutamic acid–acclimated cultures could utilize acetic, propionic, measaconic, α-ketoglutaric and pyruvic acids and convert them to methane and carbon dioxide. Healy and Young [11] studied the cross acclimation of potential aromatic products from the heat treatment of lignin. They indicated that the relative length of time required for acclimation depended upon the substrate structure and the source of the inoculum. They suggested that a microbial population acclimated to a particular aromatic substrate can be simultaneously acclimated to other aromatic substrates which have similar arrangements of side group on the aromatic ring. Ludzack [12] reviewed the cross acclimation of aerobic cultures. He summarized from available data that organisms acclimated to alcohols appear to be adequate for assimilation of corresponding aldehydes and acids with the possible exception of formaldehyde and isobutyraldehyde. Phenol acclimated organisms appear suitable for a wide variety of materials such as cresols. Catechol and hydroquinone were the major exceptions.

Objectives

A number of aspects are important in the design of methane fermentation processes for the treatment of petrochemicals.

(a) Toxicity of chemical structural characteristics to unacclimated methane fermentation systems.

(b) The degree to which acclimation reduces toxicity to methane fermentation systems.

(c) Long-term acclimation and cross acclimation to identify those petrochemicals which are degraded by methane fermentation.

(d) Kinetics associated with the methane fermentation of various petrochemicals.

The first aspect is covered in another paper [13]. Only the third aspect is covered in this present paper. With acetic acid being the major precursor of methane, the major objective of this study was to determine whether bacteria developed on acetate substrate could metabolize other organic compounds. Therefore a series of experiments was carried out to assay the degradability of the compounds in Table I. Cultures developed on benzoic acid was also used to study the possible acclimation to other aromatic compounds such as phthalic acid, aniline, and phenol.

EXPERIMENTAL PROCEDURES

Warburg Respirometer Study Procedure

Warburg flasks of 150-ml volume containing the reaction mixture, before adding the seed, were attached to manometers. A train was set up to purge the flasks of oxygen. The end of the train was connected to an anaerobic seed collection bottle and a dispensing pipet. The flasks were flushed for 1½ hr with a mixture of CO_2 and N_2 at flow rates of 0.5 and 1.5 scfh respectively. Fifty milliliter of inoculum from a stock digester was transferred anaerobically to the prepared Warburg flask, and 10 mg of acetate and 25

TABLE I

List of Petrochemicals Assayed

1.	Acetic Acid	$CH_3.COOH$	29.	Maleic Acid	$HOOC.CH:CH.COOH$ (cis)
2.	Propanal	$CH_3.CH_2.CHO$	30.	Sorbic Acid	$CH_3CH:CH.CH:CH.COOH$
3.	Propanol	$CH_3.CH_2.CH_2OH$	31.	Hydroquinone	OH-⌬-OH
4.	2-Propanol	$CH_3.CHOH.CH_3$	32.	Crotonaldehyde	$CH_3.CH:CH.CHO$
5.	Propylene Glycol	$CH_3.CHOH.CH_2OH$	33.	Crotonic Acid	$CH_3CH:CH.COOH$
6.	Glycerol	$CH_2OH.CHOH.CH_2OH$	34.	Acrylonitrile	$CH_2:CH.C{\vdots}N$
7.	Propionic Acid	$CH_3.CH_2.COOH$	35.	Ethyl Acrylate	$CH_2:CH.COOCH_2.CH_3$
8.	Butyric Acid	$CH_3.(CH_2)_2.COOH$	36.	Phenol	⌬-OH
9.	Isobutyric Acid	$(CH_3)_2.CH.COOH$	37.	Succinic Acid	$HOOC.CH_2CH_2COOH$
10.	Acrylic Acid	$CH_2:CH.COOH$	38.	Fumaric Acid	$HOOC.CH:CH.COOH$ (trans)
11.	Methyl Acetate	$CH_3.COO.CH_3$	39.	Catechol	⌬(-OH)(-OH)
12.	Ethyl Acetate	$CH_3.COOC_2H_5$	40.	Nitrobenzene	⌬-NO_2
13.	Acetaldehyde	$CH_3.CHO$	41.	Resorcinol	⌬(-OH)(-OH)
14.	Acrolein	$CH_2:CH.CHO$	42.	1-Butanol	$CH_3.(CH_2)_2CH_2OH$
15.	Acetone	$CH_3.CO.CH_3$	43.	Sec-Butyl Alcohol	$CH_3.CH_2.CHOH.CH_3$
16.	Methyl Ethyl Ketone	$CH_3.CO.CH_2CH_3$	44.	Tert-Butyl Alcohol	$(CH_3)_3COH$
17.	3-Chloro-1,2-Propandiol	$CH_2OH.CHOH.CH_2CL$	45.	Butanal	$CH_3(CH_2)_2.CHO$
18.	1-Amino,2-Propanol	$CH_3.CHOH.CH_2NH_2$	46.	Valeric Acid	$CH_3.(CH_2)_3.COOH$
19.	1-Chloropropane	$CH_3.CH_2.CH_2CL$	47.	Vinyl Acetate	$CH_3COOCH:CH_2$
20.	1-Chloropropene	$CH_2:CH.CH_2CL$	48.	4-Aminobutyric Acid	$NH_2.CH_2.(CH_2)_2$-$COOH$
21.	Adipic Acid	$COOH.(CH_2)_4.COOH$	49.	Sec-Butylamine	$CH_3.CHNH_2.CH_2.CH_3$
22.	Hexanoic Acid	$CH_3.(CH_2)_4.COOH$	50.	Lauric Acid	$CH_3.(CH_2)_{12}.COOH$
23.	Benzoic Acid	⌬-COOH	51.	Formic Acid	$H.COOH$
24.	Phthalic Acid	HOOC-⌬-COOH	52.	Methanol	CH_3OH
25.	Ethyl Benzene	⌬-$CH_2.CH_3$	53.	Formaldehyde	$H.CHO$
26.	Aniline	⌬-NH_2			
27.	2-Chloropropionic Acid	$CH_3.CHCL.COOH$			
28.	Glutaric Acid	$HOOC.(CH_2)_3.COOH$			

mg of the compound to be assayed for cross acclimation were added to the flask through the side arm using a hypodermic syringe at the beginning of the test. Response of the organisms over periods of up to 30 days was observed.

Hungate Serum Bottles Technique

The Hungate serum bottle technique [14] was substituted for the Warburg Respirometer technique later in this study due to its simplicity and reliability. The serum bottles were filled with water and the water was displaced with an inert gas mixture of CO_2 and CH_4. A 50-ml inoculum was injected into the serum bottle along with 100 mg of acetate and 25 mg of test compound. Gas production was monitored and subsequently injections of the pure test compound or pure acetic acid were made with a microliter syringe as needed.

Petrochemicals listed in Table I were all injected into serum bottles to yield initial concentrations of 500 mg/liter for the first six injections. If metabolism was indicated, the injections were increased to doses of 1000 mg/liter in the seventh injection and thereafter. When metabolism was not indicated by stoichiometric gas production, acetate was subsequently injected to assay the viability of the culture.

Long-Term Feeding Acclimation Studies

In contrast to the Warburg and serum bottle studies where no biomass was wasted from the system and only minute amounts of pure compounds were added occasionally, a series of long-term feeding acclimation studies were conducted. These studies were conducted in mixed reactors operating on a 20-day hydraulic retention time (HRT) schedule and in submerged anaerobic upflow filters of 2- to 10-day HRT. Both of these systems were inoculated with the acetate enriched culture. Daily addition of an inorganic salt solution shown in Table II, plus acetate and the test petrochemical compound were made to these systems. Initially the test petrochemical concentrations were low, approximately 500 mg/liter in the feed, but these concentrations were increased as stated later, as metabolism was observed. The control parameter was COD or TOC reduction.

Petrochemicals 1–26 as shown in Table I were acclimated on the 20-day HRT completely mixed systems while compounds 27–49 were acclimated in the anaerobic filters.

Inoculum Source

A separate system was continually maintained to provide an acetate enriched culture to serve as an inoculum for all of these studies. This inoculum source consisted of a 400-liter vessel which was intermittently stirred each hour and continuously fed a solution of the inorganic salts listed in

TABLE II

Concentration of Inorganic Salts in Feed Solution

Constituent	Concentration in Feed mg/l
NH_4 Cl	400
KCl	400
$MgSO_4.6H_2O$	400
$FeCl_2.6H_2O$	40
$CoCl_3$	4
$(NH_4)_2HPO_4$	80
Cysteine	10
KI	10
Na hexameta phosphate	10
$MnCl_2$	0.5
$NH_4V_2O_3$	0.5
$ZnCl_2$	0.5
$Na_2MoO_4.2H_2O$	0.5
H_3BO_3	0.5
$NaHCO_3$	As needed to maintain alkalinity about 3000 mg/l as $CaCO_3$

Table II. The hydraulic retention time was 60 days. It was initially started with a small seed of well-digested domestic sludge. Subsequently, it has been fed only acetate for a period of several years. The suspended solids concentration was approximately 1000 mg/liter and essentially all of biological origin due to the prolonged operation on acetate substrate. This inoculum source provided somewhat of a uniform, enriched acetate culture for use over the prolonged period of these studies, even though it is realized that microbial population dynamics still occur under such "steady state" conditions.

RESULTS AND DISCUSSION

Cross Acclimation and Long-Term Acclimation

As stated earlier, acetate is the major precursor of methane in an anaerobic digester. Microbial cultures were developed on acetate substrate for these studies. Due to the population dynamics of these enriched cultures, the length of time required for a petrochemical to be acclimated varied; however, the results of replicates were the same. Therefore, the results obtained from these studies should be considered somewhat qualitatively.

TABLE III

Petrochemicals Metabolized by Enriched Methane Cultures

Acetaldehyde	Ethyl Acrylate	Phenol
Acetone	Formic Acid	Phthalic Acid
Adipic Acid	Fumaric Acid	Propanal
1-Amino-2-Propanol	Glutaric Acid	Propanol
4-Amino-Butyric Acid	Glycerol	2-Propanol
Benzoic Acid	Hexanoic Acid	Propionic Acid
Butanol	Hydroquinone	Propylene Glycol
Butyraldehyde	Isobutyric Acid	Resorcinol
Butyric Acid	Maleic Acid	Sec-Butanol
Catechol	Methanol	Sec-Butylamine
Crotonaldehyde	Methyl Acetate	Sorbic Acid
Crotonic Acid	Methyl Ethyl Ketone	Succinic Acid
Ethyl Acetate	Nitrobenzene	Tert-Butanol
		Valeric Acid
		Vinyl Acetate

Table III is a list of all of the petrochemicals which were metabolized by the enriched acetate cultures after acclimation.

The responses of the acetate cultures cross acclimated with various petrochemicals were observed in terms of gas production in serum bottles. Figures 1, 2, and 3 represent typical results found in cross acclimations and degradations of assayed compounds. The figures show the cumulative gas production vs time. The arrows in these figures indicate the time at which the petrochemicals were introduced. The first injections and the specific amount of the compound injected appear on the top of arrows in these figures. Subsequent injections were of the same mass unless otherwise indicated.

Table IV shows the petrochemicals which were cross acclimated to acetate culture in serum bottles and the length of time required for the acetate cultures to start metabolizing these cross-fed compounds. The percent of substrate of these compounds removed and their average removal rates at the end of the acclimation period are also included in the table. The total gas production of the culture divided by the theoretical gas production for the respective compound gives the percent of utilization. Theoretical gas yields were obtained according to Buswell's equation:

$$C_nH_aO_b + \left(n - \frac{a}{4} - \frac{b}{2}\right) \rightarrow \left(\frac{n}{2} - \frac{a}{8} + \frac{4}{b}\right) CO_2 + \left(\frac{n}{2} + \frac{a}{8} - \frac{b}{4}\right) CH_4$$

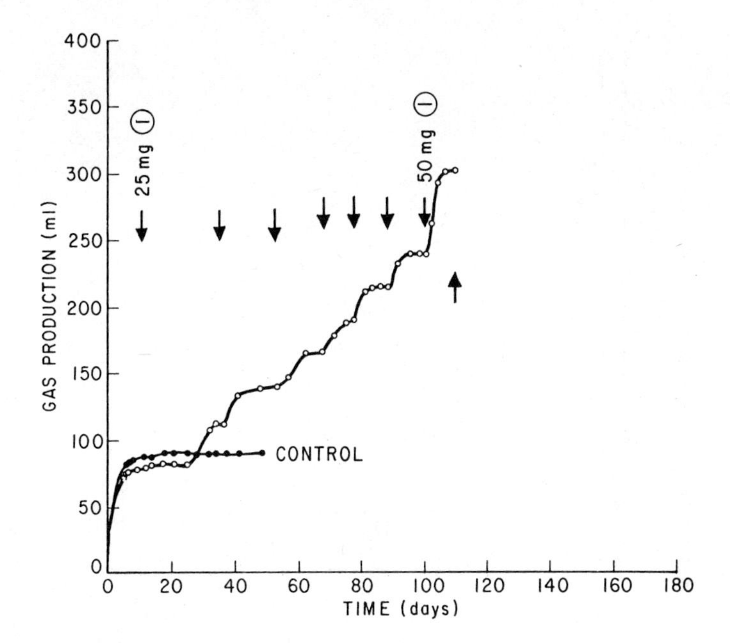

Fig. 1. Cross acclimation of acetate culture to acetone.

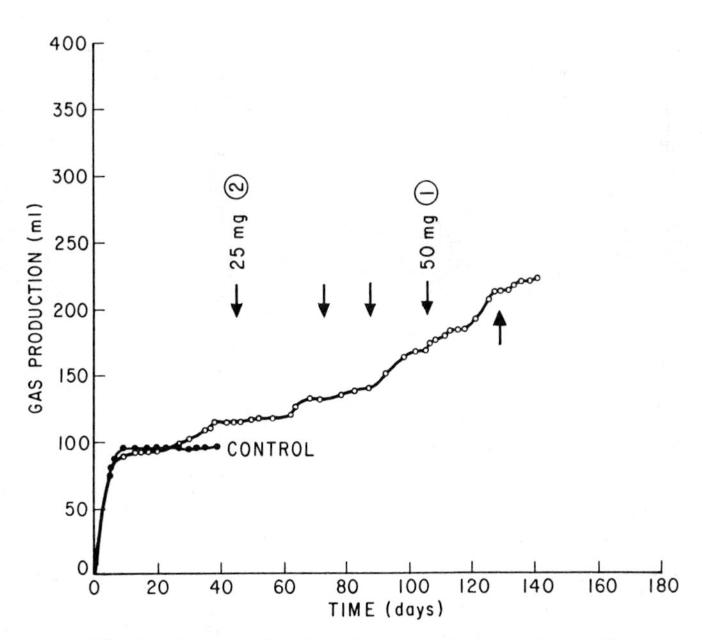

Fig. 2. Cross acclimation of acetate culture to adipic acid.

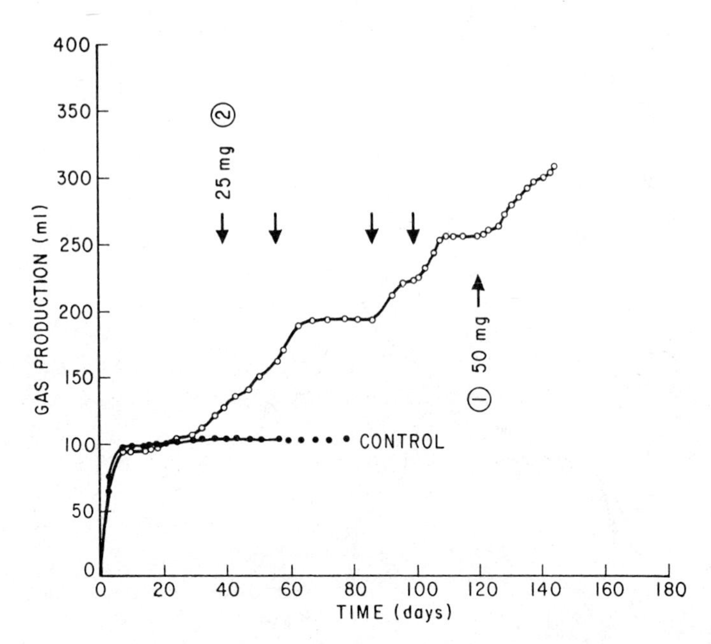

Fig. 3. Cross acclimation of acetate culture to phenol.

TABLE IV

Petrochemical Removal Rates Observed in Serum Bottles

Compounds	Lag Time (Days)	Percent Substrate Removed (%)	Removal Rate (mg/l/Day)
1-Amino, 2-Propanol	9	65	22
Sec-Butyl Alcohol	14	100	42
Phenol	7	98	42
Phthalic Acid	14	90	50
Adipic Acid	10	82	50
Valeric Acid	3	100	72
Methyl Ethyl Ketone	8	100	72
Maleic Acid	10	95	74
Glutaric Acid	10	100	90
Propionic Acid	2	100	90
Butanol	4	100	100
Propanol	4	100	110
Succinic Acid	10	93	110
Hexanoic Acid	6	100	112
Propylene Glycol	4	100	125
Acetone	5	100	125
Fumaric Acid	13	86	125
Vinyl Acetate	3	100	125
Sorbic Acid	10	76	142
Butyraldehyde	7	99	142
Methanol	6	66	182
2-Propanol	4	100	200
Glycerol	8	90	200
Benzoic Acid	11	82	200
Crotonic Acid	2	100	200
Isobutyric	3	100	250
Methyl Acetate	3	100	250
Ethyl Acetate	3	100	250
Butyric Acid	3	100	284
Formic	4	89	286

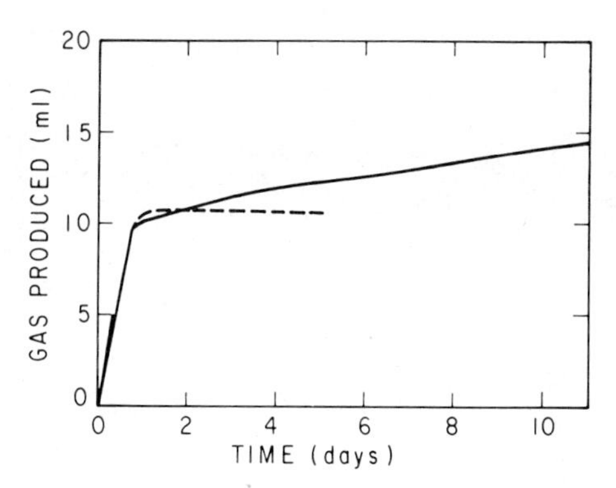

Fig. 4. Acclimation of benzoate culture to aniline: (---) control; (——) aniline.

The benzoate cultures decomposed phthalic acid and aniline without a noticeable lag period and both had relatively the same gas production rates (Figs. 4, 5). Phthalic acid required 14 days for acclimation by acetate cultures, and aniline was not acclimated during the assay period by acetate cultures. It appeared that the structural characteristics play an important role in determining the length of lag period required for acclimation.

Contrary to the previous findings, the benzoate culture required 23 days to acclimate to phenol although the chemical structure of phenol is similar to benzoic acid. It took 30 days lag period for the acetate culture to start using phenol (Figs. 3 and 6).

Compounds which were not acclimated to acetate cultures during the period of the experiment can be grouped into two categories: (1) Compounds which were toxic to unacclimated acetate cultures at 500 mg/liter; 3-chloro-1,2-propandiol, 2-chloro-propionic acid, 1-chloro-propane, 1-chloro-propene, acrolein, crotonaldehyde, acrylic acid, ethyl acrylate, nitrobenzene, formaldehyde, lauric acid, and acrylonitrile. (2) Compounds which were neither toxic to unacclimated acetate cultures nor available to the cultures are ethyl benzene and aniline.

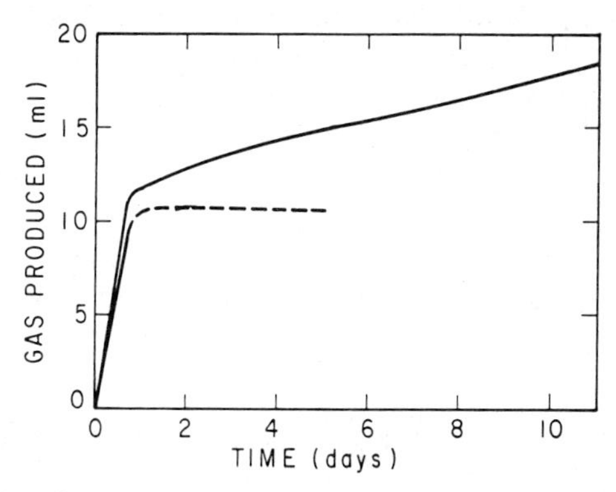

Fig. 5. Acclimation of benzoate culture to phthalic acid: (---) control; (——) phthalic acid.

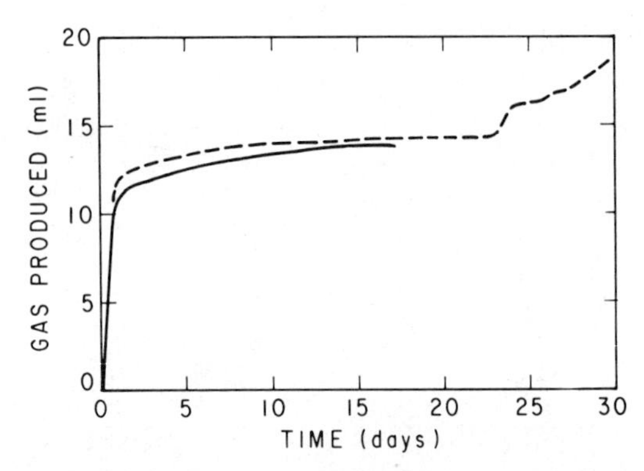

Fig. 6. Acclimation of benzoate culture to phenol. (——) Control; (---) phenol.

The toxicity of a substrate which was cross-fed to an acetate culture seemed to affect the length of a lag period for the culture to start metabolizing this substrate significantly. Results showed that the toxic compounds 2-chloropropionic acid, 3-chloro-1,2,-propandiol, formaldehyde, and acrylic acid were not utilized by acetate cultures. But non-toxic compounds with similar structures such as propionic acid, propylene glycol, butyraldehyde, and crotonic acid were all metabolized within four days.

The structural characteristics of compounds also appeared to influence the length of lag period for cross acclimation. Data suggested that

(1) Non-toxic aliphatic compounds containing carboxyl, ester, or hydroxyl groups were all acclimated readily by acetate cultures. Butyric acid, valeric acid, propanol, propylene glycol, ethyl acetate, and vinyl acetate were all degraded within 4 days.

(2) Toxic compounds with carbonyl functional groups or double bonds all acclimated within a short period of time, although these compounds were usually toxic to unacclimated acetate cultures. For example, butyraldehyde and sorbic acid required 7 and 10 days for acclimation respectively.

(3) Compounds with amino functional groups showed difficulty in acclimation. Aniline, 4-amino-butyric acid, and sec-butylamine were not acclimated in this study. 1-amino-2 propanol was the only compound acclimated, but it degraded at an extremely slow rate.

(4) Compounds with dicarboxylic groups required a longer length of time to be acclimated as compared to compounds with one carboxylic group and the same carbon chain length, for example, succinic acid, adipic acid, and phthalic acid all had longer lag periods as compared to butyric acid, hexanoic acid, and benzoic acid respectively.

(5) The position of functional group also affected the length of lag period for acclimation. In the case of the hydroxyl group, primary butanol required 4 days for acclimation, and secondary butyl alcohol required 14 days, while tertiary butyl alcohol was not acclimated in the serum bottles.

It has been noted that acetate cultures had to exhaust the acetic acid as a carbon and energy source before they would start to utilize the cross-fed compound. Studies showed that 30 days were required for an acetate culture to start metabolizing butyric acid if acetic acid was present during the period of acclimation. It only took 3 days to acclimate if butyric acid was added after the acetic acid was exhausted. This phenomenon was also observed in acclimation studies of propanol, 2-propanol, propylene glycol, methyl acetate, and methyl ethyl ketone.

Healy and Young [11] studied the cross acclimation of aromatic compounds with similar structures and suggested that a microbial population acclimated to a particular substrate is often simultaneously acclimated to other substrates with a very similar structure. The results of our study are in general agreement. Microbial cultures acclimated to a particular compound were simultaneously acclimated to other compounds which had the same functional group. With the hydroxyl group, cultures originally acclimated to 2-propanol metabolized propylene glycol and propanol immediately. Propylene glycol cultures also metabolized propanol, 2-propanol, and glycerol without delay. Glutarate-acclimated cultures metabolized adipic acid and fumaric acid with no lag. For the carbonyl group, cultures acclimated to methyl ethyl ketone also could metabolize acetone with no lag. In addition to the length of lag period required for acetate cultures to start metabolizing test compounds, the acclimation characteristics in terms of utilization efficiency are shown in the acclimation studies conducted in anaerobic filters and completely mixed reactors and discussed later.

Compounds listed in Table I except formic acid, lauric acid, formaldehyde, and methanol were also acclimated and metabolized either in completely mixed reactors or anaerobic filters. Reactors which were fed 1-Cl-propane, 1-Cl-propene, ethyl benzene, and aniline had to be discontinued since these compounds were extremely insoluble and collected on the liquid surface in the reactor, making it impossible to estimate their rates of degradation under the experimental conditions. Table V gives the utilization efficiency of substrates which acclimated in completely mixed reactors after 90 days of acclimation. The utilization efficiency is obtained from a mass balance of each respective substrate.

At the end of 90 days acclimation, acetate, glycerol, butyric, and isobutyric acids, methyl acetate, ethyl acetate, acetaldehyde, methyl ethyl ketone, and adipic acid acclimated very well with high utilization efficiency (above 66%). Propanol, 2-propanol, propylene glycol, propionate, acrolein, acetone, hexanoic, and benzoic acids had degradation ranging from 33% to 66%. Propanal, acrylic acid, 3-chloro-1,2-propandiol, and 1-amino-2-propanol acclimated poorly (below 33% of utilization).

The anaerobic filters proved to be greatly superior to the completely mixed reactors as acclimation systems. It was found that some petrochemicals could not be acclimated and metabolized in the completely mixed reactors because the 20-day Solid Retention Time (SRT) tended to allow washout of the bio-

TABLE V

Utilization Efficiency of Petrochemicals 1–24 During Acclimation Period in Completely Mixed Reactors (Day 70 to 90)**

	Petrochemical*	%Utilization**
1.	Acetic Acid	87
2.	Propanal	26
3.	Propanol	41
4.	2-Propanol	56
5.	Propylene Glycol	53
6.	Glycerol	66
7.	Propionic Acid	47
8.	Butyric Acid	97
9	Isobutyric Acid	99
10.	Acrylic Acid	21
11.	Methyl Acetate	96
12.	Ethyl Acetate	96
13.	Acetaldehyde	97
14	Acrolein	42
15	Acetone	55
16.	Methyl Ethyl Ketone	77
17.	3-Chloro-1,2-Propandiol	26
18.	1-Amino,2-Propanol	26
21.	Adipic Acid	67
22.	Hexanoic Acid	55
23.	Benzoic Acid	40
24.	Phthalic Acid	0

* The serial numbers are the same as those assigned to these compounds in Table I.

** Digesters were acclimated to 10,000 mg/liter substrate concentration in the feed—20-day HRT completely mixed reactors with no solids recycle.

mass. However, the exceptionally long SRT inherent to the anaerobic filters allowed retention of the biomass over the prolonged acclimation periods so that very high utilization efficiencies were eventually possible. Table VI shows the utilization efficiencies of petrochemicals 27–41 after 110 days of acclimation. The 2-Cl-propionic acid and acrylonitrile filters acclimated to 2000 mg/liter substrate with only 14% and 17% utilization, respectively, at the end of 110 days. The remaining filters all acclimated well to their respective substrates with at least 80% utilization, as shown in Table VI. Petrochemicals 42 to 49 were also acclimated in anaerobic filters, but for only a 52-day period. Table VII shows the utilization efficiency for these compounds. At the end of 52 days of acclimation, butanol acclimated to 95% utilization efficiency, and tert-butanol and sec-butylamine also acclimated with 73% and 66% utilization. Sec-butanol, valeric acid, vinyl acetate, and 4-amino-butyrate had good acclimation with utilization efficiencies of at least 87% at the end of 52 days.

It should be pointed out that many compounds which were not acclimated in serum bottles were metabolized in anaerobic filters or completely mixed reactors. Therefore a compound not being acclimated in the present study does not necessarily mean that it cannot be acclimated; rather it is a matter of time required and techniques of acclimation. For instance, propanal which

TABLE VI

Utilization Efficiency of Petrochemicals 27–41 in Anaerobic Filters (110 Days)

	Chemicals	Eff. Conc. mg/l	% Util.
27	2-Cl-Propionate	1560	14
28	Glutaric Acid	70	96
29	Maleic Acid	80	96
30	Sorbic Acid	110	94
31	Hydroquinone	280	85
32	Crotonaldehyde	100	95
33	Crotonic Acid	70	96
34	Acrylonitrile	1500	17
35	Ethyl Acrylate	100	95
36	Phenol	90	95
37	Succinic Acid	90	95
38	Fumaric Acid	100	95
39	Catechol	120	93
40	Nitrobenzene	350	81
41	Resorcinol	90	95

* The serial numbers are the same as those assigned to these compounds in Table I.

was not acclimated by acetate cultures in serum bottles was acclimated in a Warburg flask after 10 days of lag period. It was also acclimated in a completely mixed reactor and had 54% removal during the steady state period when the reactor was operated on 20 days hydraulic detention time. Acrylic acid was reported to be metabolized by an independent investigator.

Acetaldehyde was a toxic compound but it had a high utilization efficiency from the start (85%). It did not suffer adverse effects as the substrate concen-

TABLE VII

Utilization Efficiency of Petrochemicals 42–49 in Anaerobic Filters (52 Days)

	Petrochemicals	Eff. Conc. mg/l	% Util.
42	Butanol	80	95
43	Sec-Butanol	110	93
44	Tert-Butanol	400	73
45	Butyraldehyde	270	82
46	Valeric Acid	110	93
47	Vinyl Acetate	140	91
48	4-NH_2-Butyrate	200	87
49	Sec-Butylamine	510	66

* The serial numbers are the same as those assigned to these compounds in Table 1.

tration increased sharply in the second 30 day period, and had 97% utilization efficiency at the end of 90 days acclimation. Acrolein as stated earlier was found to be an extremely toxic compound to unacclimated acetate cultures, but with acclimation showed 42% utilization and there were indications that over 90% utilization is possible with proper acclimation. Propanal, 1-amino-2-propanol, and 3-Cl-1,2,propandiol acclimated poorly by comparison with acrolein and acetaldehyde.

Table VI shows that the mildly toxic compounds such as glutaric, maleic, and fumaric acids, hydroquinone, catechol, resorcinol, ethyl acrylate, and phenol all acclimated very well in the anaerobic filters. The utilization efficiency also improved with the increasing length of acclimation time. Filters receiving crotonaldehyde and nitrobenzene, considered to be extremely toxic compounds with unacclimated acetate cultures, both acclimated very well with 95% and 81% utilization respectively. The results in Table VII also show that the toxic compound vinyl acetate had good acclimation with utilization efficiency of 82% at the end of 52 days acclimation. Chou et al. [13] reported that the remaining activities of unacclimated acetate culture which were treated with 500 mg/liter of crotonaldehyde and nitrobenzene were 6% and 0% respectively. The unacclimated acetate culture treated with 1000 mg/liter of vinyl acetate had a remaining activity of 31%.

Structural Effect on Acclimation Characteristics and Degradation Rates

1) *Type of functional group.* During the acclimation, the results show that compounds with ester, hydroxyl, and carboxyl groups acclimated better than compounds having a carbonyl, amino, or chloro group. For example, 4-carbon compounds butanol, butyric acid, and ethyl acetate had about the same degree of degradation (about 95%) while 4-amino-butyric acid had 87% utilization and methyl ethyl ketone 77% (Table IV). The structural characteristics of compounds also appeared to affect the degradation rates of these compounds. Results of the study conducted in serum bottles showed butanol, butyric acid, and ethyl acetate had degradation rates of 100 mg/liter/day, 284 mg/liter/day, and 250 mg/liter/day respectively; this contrasted with 4-amino-butyric acid which was not degraded at the end of 120 days and with methyl ethyl ketone which had a 72 mg/liter/day degradation rate. Of the aromatic compounds, phenol, nitrobenzene, and benzoic acid had utilization of 95%, 81%, and 40%, respectively, in anaerobic filters. In the serum bottles, phenol had the highest utilization (98%) and lowest degradation rate (42 mg/liter/day) among the aromatic compounds studied. Benzoic acid had 82% utilization and 50 mg/liter/day degradation rate, while nitrobenzene was not acclimated by acetate cultures in serum bottles.

2) *Chloro group.* Two chloro-substituted compounds were assayed for acclimation characteristics, that is, 3-Cl-1,2-propandiol and 2-Cl-propionate. Neither compound acclimated well in either system. The utilization efficiency of 3-Cl-1,2-propandiol was consistently lower than propylene glycol

throughout the acclimation period. 2-Cl-propionate acclimated slightly in the anaerobic filter, but had a lower utilization efficiency than propionate which acclimated in a stirred reactor. These results are in agreement with earlier findings that chloro-substituted compounds were not metabolized by acetate cultures in serum bottles. However, the toxicity was reduced with acclimation.

3) *Amino group:* Three amino-containing compounds were studied and all showed adverse effects on acclimation: 1-amino-2-propanol had consistently lower utilization efficiency than 2-propanol; 4-amino-butyrate had lower utilization than butyric acid, 87%, vs 97% respectively; Sec-butylamine also had lower utilization (66%). In addition, the amino group reduced the degradation rate and utilization efficiency as shown in Table IV. 1-amino-2-propanol had 65% utilization and 22 mg/liter/day degradation rate as compared to 2-propanol which was 100% utilized and had a 200 mg/liter/day degradation rate.

4) *Position of the Functional Group.* 2-propanol had a higher utilization efficiency (56%) than propanol (41%) in the stirred reactors. Sec-butanol had almost the same utilization as butanol (93%, 95%, respectively) at the end of 52 days acclimation. Results obtained from studies conducted in serum bottles also showed that 2-propanol had a higher degradation rate (200 mg/liter/day) than propanol (110 mg/liter/day) with the same utilization efficiencies (100%) at the end of 150 days acclimation. However, sec-butanol had a lower degradation rate (42 mg/liter/day) than *n*-butanol (100 mg/liter/day). Although hydroquinone was the least toxic compound among the tree isomers, hydroquinone acclimated with somewhat lower utilization (85%) at the end of the acclimation period as compared to its isomers, for example, catechol and resorcinol which had utilizations of 93% and 95% respectively. It appears that structural characteristics play a more important role in acclimation than the toxicity characteristics. Results found in the studies of serum bottles also support this suggestion. In the case of ester compounds, the position of the double bond in vinyl acetate and ethyl acrylate is different but their relative toxicities are essentially the same [13]. However, vinyl acetate was acclimated by acetate cultures in three days and had a degradation rate of 125 mg/liter/day with 100% utilization as opposed to ethyl acrylate which was not acclimated by the end of the study (130 days).

5) *Odd and even number carbon compounds.* Within the same system, the odd- and even-numbered carbon compounds did not affect acclimation in the case of esters. The results showed that methyl acetate and ethyl acetate had similar utilization efficiency throughout the whole acclimation period and the same utilization (100%) at the end. This also held true for degradation rate and utilization efficiency (250 mg/liter/day and 100% respectively) obtained from the results of studies conducted in serum bottles. In the case of carbonyl containing compounds, propanal acclimated in a stirred reactor with 20% utilization while acetaldehyde had 97% in the same system and butanal acclimated in a filter with 82% utilization. However, the even-numbered carbon

compound methyl ethyl ketone had a lower degradation rate than the odd-numbered carbon compound acetone (72 mg/liter/day and 125 mg/liter/day respectively) with the studies conducted in serum bottles. As for the carboxyl-containing compounds, propionic acid had a lower utilization of 47% as compared with even-numbered carbon compounds, for example, acetic and butyric acids which had 87% and 97%, respectively. However, odd carbon length valeric acid had a higher utilization of 93% than the even carbon hexanoic acid at 67%. This result may be attributed to the fact that valeric acid was acclimated in the anaerobic filter while hexanoic acid was acclimated in a stirred reactor. The surfactant nature of hexanoic acid may also have affected its utilization rate. In the serum bottles, the degradation rates of odd-numbered carbon compounds propionic acid and valeric acid (90 mg/liter/day and 72 mg/liter/day respectively) were found lower than even-numbered carbon compounds, butyric acid and hexanoic acid (284 mg/liter/day and 112 mg/liter/day respectively), although all of them had 100% utilization at the end of study.

6) *Increasing the multiplicity of a given functional group*. From the results of anaerobic filters and mixed reactors, there is no clear relationship between the acclimation characteristics and the increase of a given functional group during the period of acclimation. However, the increase of a given functional group appears to decrease the rate of degradation. In the case of the carboxyl group, compounds containing dicarboxyl groups had lower degradation rates than compounds having only one carboxyl group. Succinic acid, adipic acid, and phthalic acid had degradation rates of 110, 50, and 50 mg/liter/day, respectively, as opposed to butyric acid, hexanoic acid, and benzoic acid which had 284, 112, and 200 mg/liter/day, respectively. However, the 5-carbon compound valeric acid had a degradation rate of 72 mg/liter/day while that for glutaric acid was 90 mg/liter/day. The three compounds propanol, propylene glycol, and glycerol showed that degradation rate increases with the number of hydroxyls, for example, 110, 125, and 200 mg/liter/day, respectively.

7) *Configuration*. One pair of compounds with a different configuration was compared here, for example, fumaric acid and maleic acid. The utilization efficiency of fumaric acid was lower than maleic acid (125 mg/liter/day vs. 74 mg/liter/day) in the serum bottles at the end of 150 days acclimation.

Since all petrochemicals were not acclimated in one type of system, the acclimation period and the concentration level to which the microorganisms were acclimated were also different. Therefore, the acclimation characteristics of these petrochemicals in terms of utilization efficiency during the period of acclimation should not be directly compared on a quantitative basis.

Toxin Acclimation and Acetate Utilization

The presence of some petrochemicals prevented the enriched acetate cultures from metabolizing acetate initially. Then after an acclimation period, the cultures regained their ability to metabolize acetate even though the toxin was not metabolized. This is shown for ethyl acrylate, 1-chloro-propane, and nitrobenzene in Figures 7–9 at 500 mg/liter concentrations of the toxins.

When acetate and the above toxic compounds were added to an enriched acetate culture initially, the only gas production was from CO_2 production from the pure acetic acid addition. Then after 30 to 60 days, the full stoichiometric quantity of CH_4 was produced, but with no metabolism of the toxic compound. Subsequent addition of acetic acid showed metabolism without delay, indicating that the acetate cultures were no longer inhibited by the toxins.

Phenol Column

An 8-liter liquid volume-packed anaerobic filter was fed continuously with phenol as the sole carbon source along with the inorganic nutrients listed in Table II for four months. The concentration of phenol and feeding volume were increased gradually in order for the enriched acetate culture to have ample time to acclimate. The filter was fed 4 liter per day with a phenol concentration of 4100 mg/liter as COD for a month and a half. The average effluent concentration and gas production daily were 1400 mg/liter as COD and 5.5 liters of gas per day. The methane content of gases produced from the filter was 70%. The mass balance on a COD basis showed 66% reduction.

Effect of SRT on Acclimation and Resistance to Toxic Slugs

The anaerobic filter was proven to be an inherently stable system capable of tolerating and eventually acclimating to substances exhibiting initial

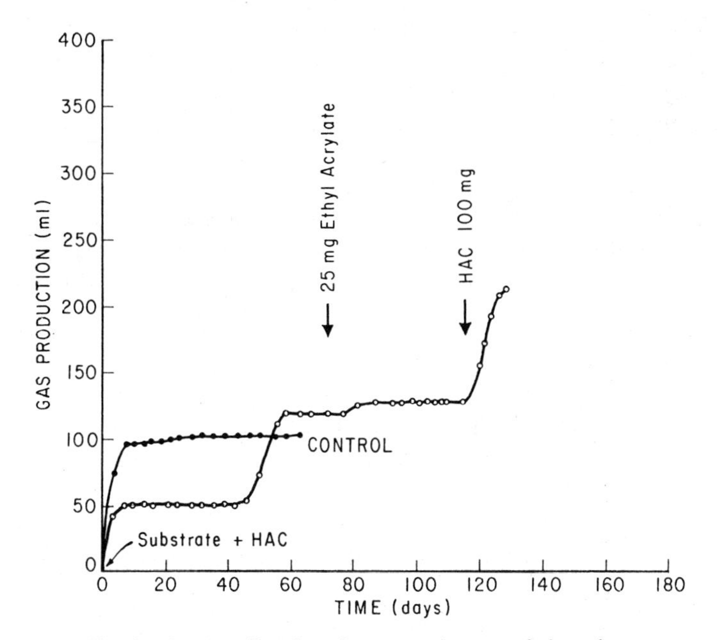

Fig. 7. Cross acclimation of acetate culture to ethyl acrylate.

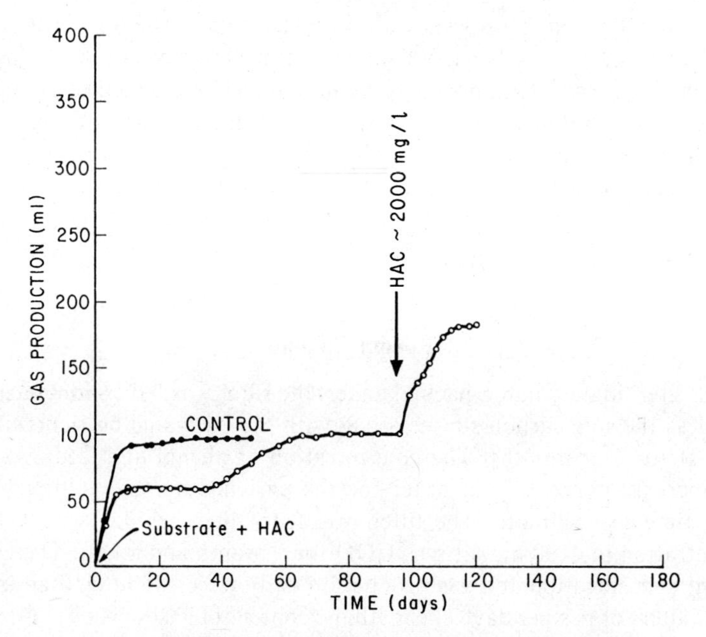

Fig. 8. Cross acclimation of acetate culture to 1-chloro-propane.

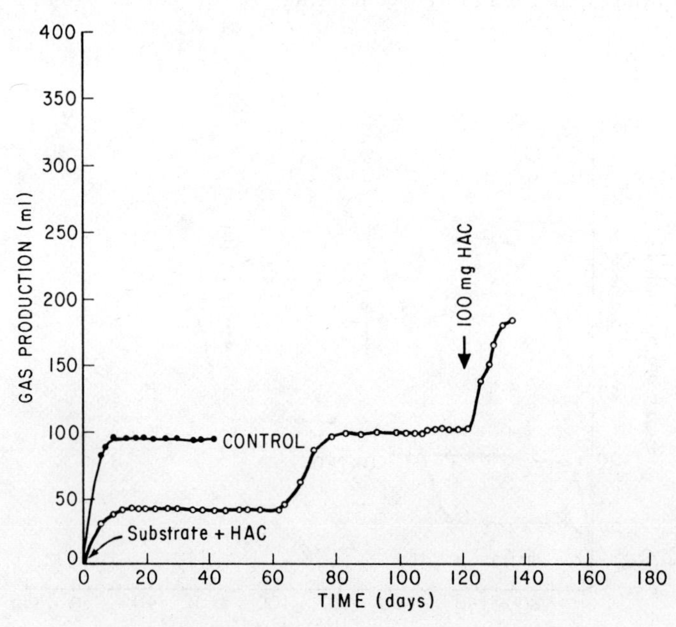

Fig. 9. Cross acclimation of acetate culture to nitrobenzene.

toxicity. At 35°C, the methane bacteria fermenting acetate to methane have a generation time of three to five days under favorable environmental conditions. Under unfavorable environmental conditions, this generation time would be extended. Conceivably, cell division may cease for a period of time in the presence of a toxin. This points up the fundamental importance of maintaining extended SRT to ensure process stability in the presence of adverse environmental conditions. For instance a completely mixed reactor with no recycle having a 20-day SRT could have digestion failure if a toxin was introduced even though it has more than ample SRT under favorable conditions. To illustrate this point 100 mg/liter of formaldehyde (after addition to the reactor) was added to completely mixed digesters having 10-day and "infinite" SRT. The "infinite" SRT digester was fed pure glacial acetic acid with a microliter syringe and had no wasting of contents. The 10-day SRT digester ceased gas protection after four days and had not recovered after 30 days. The "infinite" SRT digester had a temporary reduction in gas production to 50% of the control but gas production returned to 100% of the control within six days as shown in Figure 10.

Anaerobic filters were fed 3200 mg/liter of acetate on a one-day hydraulic detention time and were purposely subjected to toxic slugs of petrochemicals, disinfectants, sulfides, and heavy metals. Their response was observed on the basis of gas production and effluent COD. They proved to be very "forgiving" and stable when subjected to unusually high concentrations of these toxins.

Formaldehyde concentrations of 75 mg/liter produced 50% inhibition to unacclimated, enriched acetate cultures, but Figure 11 shows that the filter could be subjected to 500 mg/liter slugs for up to four days with little effect.

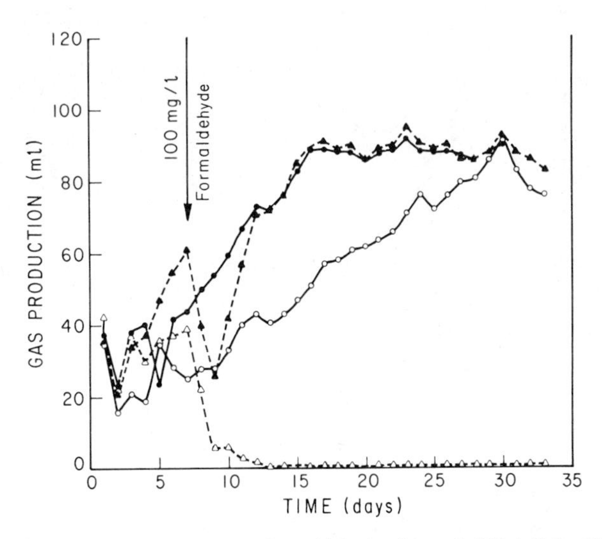

Fig. 10. Effect of SRT on response to formaldehyde. Control: (●) infinite DT; (○) 10-day DT. 100 mg/liter formaldehyde added, (▲) infinite DT, (△) 10-day DT.

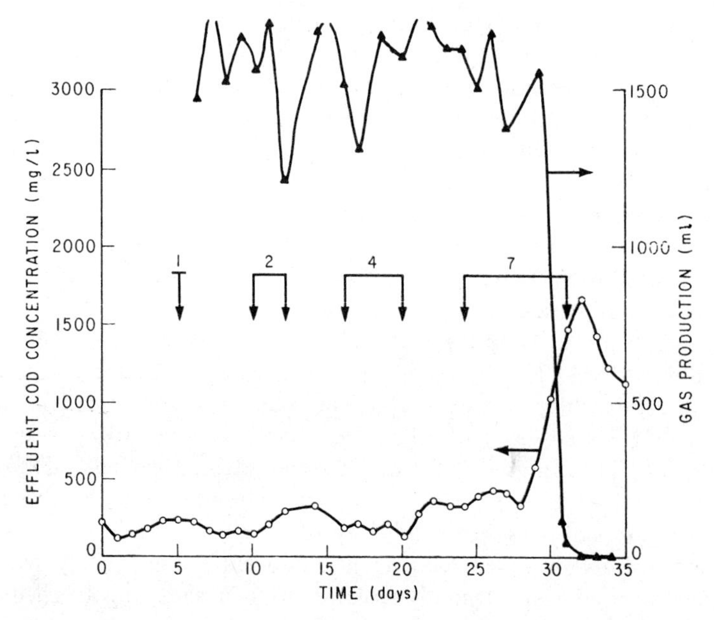

Fig. 11. Response of anaerobic filter to toxicant: 500 mg/liter formaldehyde.

Seven days continuous feed at this dose caused gas production to cease. More gradual acclimation may have permitted continuous exposure with eventual metabolism.

Dinitrophenol is an "uncoupling" agent of oxidative phosphorylation in aerobic systems. It took 500 mg/liter of 2,4-dinitrophenol to substantially retard methane production as shown in Figure 12. Acrolein slugs were

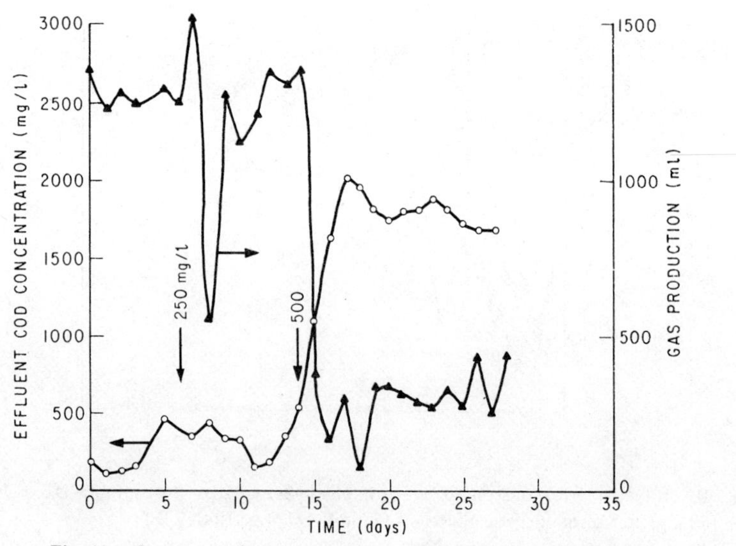

Fig. 12. Response of anaerobic filter to toxicant: 2-4-dinitrophenol.

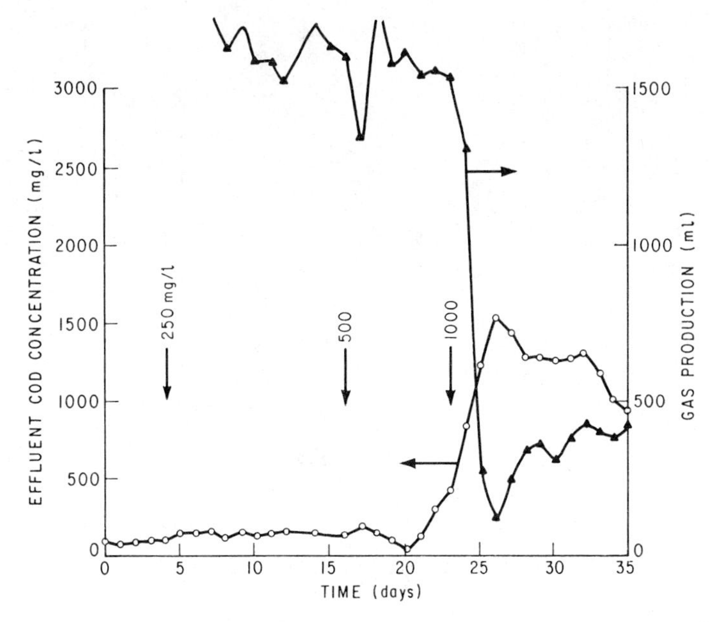

Fig. 13. Response of anaerobic filter to toxicant: acrolein.

tolerated up to 500 mg/liter (Fig. 13). Quarternary ammonium compounds are bacteriocidal agents. Figure 14 shows that 100 mg/liter had little effect on the methane production from a filter.

In petrochemical plants, spills or deliberate discharge to the plant wastewaters of inorganic materials such as sulfides or heavy metal catalysts occur. Figures 15–17 show the effect of very high concentrations of sulfides,

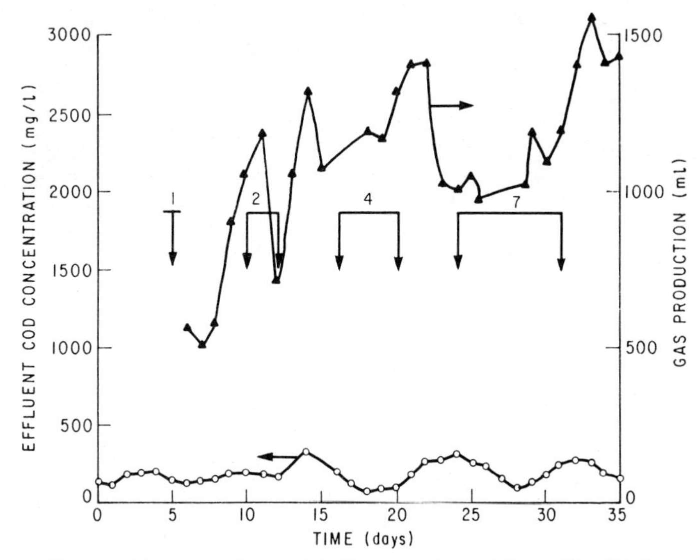

Fig. 14. Response of anaerobic filter to toxicant: 100 mg/liter QAC.

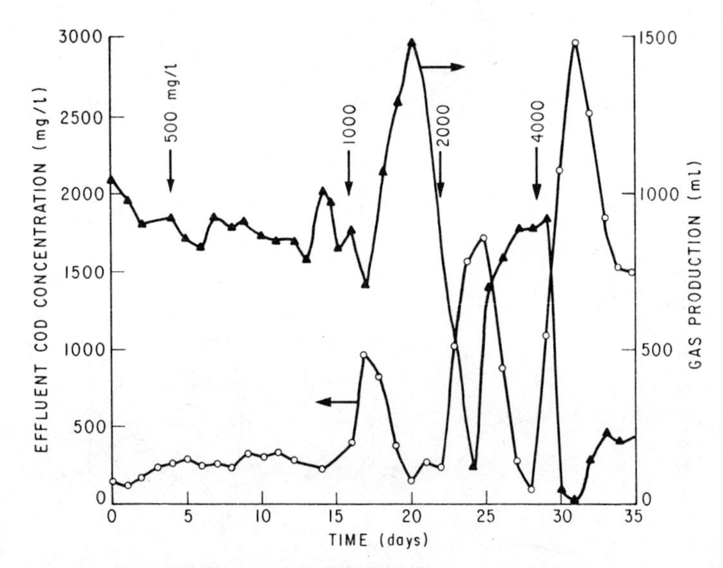

Fig. 15. Response of anaerobic filter to toxicant: S^{2-}.

copper, and nickel. The filters were able to rapidly recover from a slug of 2000 mg/liter of sulfide, but 4000 mg/liter of sulfide retarded gas production for a prolonged period. Copper at a concentration of 50 mg/liter had little effect on gas production, as shown in Figure 15. In a related experiment 250 mg/liter of copper was added for 14 consecutive days to a filter with no effect on gas production. Sulfate reduction across the filter would have been able to

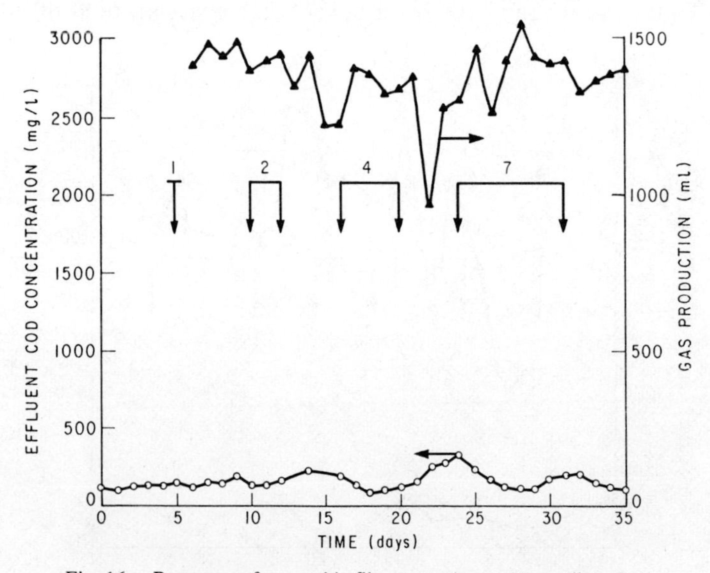

Fig. 16. Response of anaerobic filter to toxicant: 50 mg/liter Cu^{2+}.

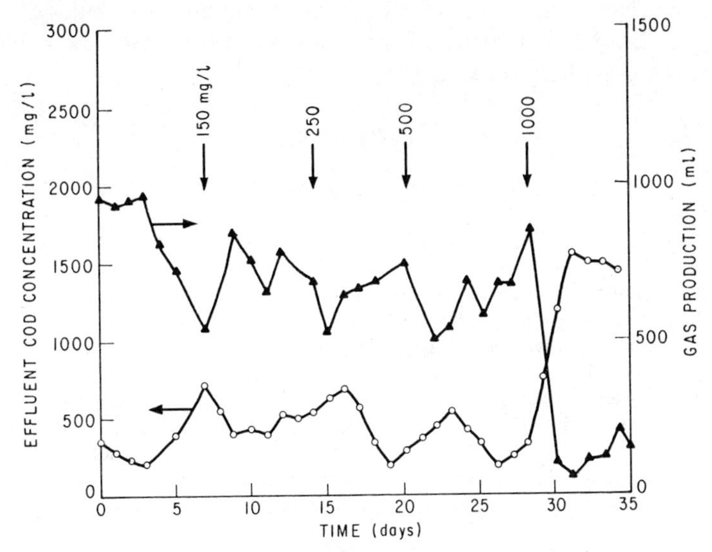

Fig. 17. Response of anaerobic filter to toxicant: Ni^{2+}.

precipitate only 40 mg/liter of copper. Slugs of $NiCl_2$ at concentration up to 500 mg/liter as Ni^{2+} had only a slight momentary effect on gas production. However, 1000 mg/liter severely retarded gas production. These concentrations were so high that the nickel was not completely dissolved.

CONCLUSIONS

All the results stated above regarding the toxic and structural effects on acclimation characteristics and degradation rates indicate that:

(1) The toxicity of a compound affects the acclimation and degradation of the compound.

(2) The longer the acclimation time, the better the degradation.

(3) Compounds with chloro, amino and carbonyl groups have an adverse effect on acclimation and degradation.

(4) The position of the functional group has a significant effect on the acclimation with respect to the length of lag time, utilization rate, and degradation rate of a compound.

(5) Odd- and even-numbered carbon compounds had no effect on the length of lag period required for acclimation. However, they had an effect on the utilization efficiency and degradation rate of a compound.

(6) Dicarboxyl functional groups increased the length of time required for acclimation and decreased the degradation rate of a compound as compared to the same chain length compounds with a single carboxyl group.

(7) Anaerobic filters proved to be superior systems to completely mixed reactors for the purpose of acclimation and stability, although the microorganisms in the completely mixed reactors were not subjected to as

high a concentration of toxicity (due to the immediate dilution) as the organisms in the anaerobic filters, because there were lower biomass concentrations in the stirred reactors.

References

[1] R. E. Speece and P. L. McCarty, *Proceedings of First International Conference on Water Pollution Research*, London (1962).

[2] J. D. Swanwick, D.C. Shurben, and S. Jackson, *J. Inst. Water Pollut. Control*, *6*, 2–24 (1969).

[3] Cecil Lue-Hing, (personal communication, 1974).

[4] G. T. Austin, *Chem. Eng.*, *96* (August, 1974).

[5] N. L. Nemerow, *Industrial Water Pollution—Origins, Characteristics and Treatment* (Addison—Wesley, Reading, Mass. 1978).

[6] J. M. Gossett, J. B. Healy, W. Owen, D. C. Stuckey, L. Y. Young, and P. L. McCarty, *Heat Treatment of Refuse for Increasing Anaerobic Biodegradability*. NSF/RANN/SEE/AER-74-17940-A01/PR/7612, Technical Report No. 212, Civil Engineering Department, Stanford University Stanford, California (November 1976).

[7] J. C. Hovious, R. A. Conway, and C. W. Ganze, *J. Water Pollut. Control Fed.*, *45*, 71 (1973).

[8] J. S. Jeris, Y. Chen, T. Chi, Y. Su, C. Weng, and R. Schneeman, *The Biochemistry of Anaerobic Digestion*, Federal Water Pollution Control Administration, Research Grant 17070 DFK, Final Report (1970).

[9] J. Chmielowski, A. Grossman, and I. Wegrzynowska, *Z. Nauk. Politech. Slask., Inz. Sinitaria*, *6*, 97 (1964).

[10] C. N. Weng and J. S. Jeris, *Water Res.*, *10* (1976).

[11] J. B. Healy and L. Y., Young, in *Heat Treatment of Refuse for Increasing Anaerobic Biodegradability*, Gossett, *et. al.* (Eds.) NSF/RANN/SE/AER-74-17946-A01/PR/76/2, Technical Report No. 212., Civil Engineering Department, Stanford University, Stanford, California (1976).

[12] F. J. Ludzack and M. B. Ettinger, *J. Water Pollut. Control Fed.*, *32*, 1173, (1960).

[13] W. S. Chou, R. E. Speece, R. H. Siddiqi, and K. McKeon, *Ninth International Conference on Water Pollution Research*, Stockholm (1978).

[14] T. C. Miller and M. J. Wolin, *Appl. Microbiol.*, *27*, 985 (1974).

Biodegradation of High-Level Oil-in-Water Emulsion*

R. M. DAVIS†
Union Carbide Corporation, Nuclear Division, Oak Ridge Y-12 Plant, Oak Ridge, Tennessee 37830

INTRODUCTION

Metalworking industries, particularly those engaged in machining of metals, use large quantities of special coolants and lubricants for cutting, grinding, and drilling metal surfaces. These metal machining operations require constant application of fluids which cool friction-heated surfaces and provide sufficient lubrication to prevent oxidation. Where cooling is of paramount importance, this is commonly accomplished through the use of petroleum oils emulsified in water. Stable oil-in-water emulsions are formulated by using emulsifiers such as alkyl sulfonates, naphthenic and fatty acid soaps, and other surface-active agents.

During machining operations, cooling emulsions are subjected to contamination by waste oils used to lubricate machine surfaces, metallic particulates, and microflora. Bacteria will attack the emulsified oils and emulsifying agents. This results in partial deemulsification with odors emanated typical of anaerobic decomposition.

Partially deemulsified and contaminated coolants lose their effectiveness in machining operations and must be discarded. The discarded coolant can result in a significant waste disposal problem.

At the Oak Ridge Y-12 Plant, we have used a stirred-bed, aerobic bioreactor process for the biological degradation of waste machining coolants. This process is an extended activated sludge process. Since the cooling fluids are mainly aliphatic organics and water, mineral salts such as nitrates and phosphates must be added to complete the nutrient requirements for the bacteria.

MATERIALS AND METHODS

Biological Degradation

Biological degradation, as referred to in this paper, is the decomposition of organic matter through the aerobic biological cycle to carbon dioxide

* Research sponsored by the Department of Energy under contract with the Union Carbide Corporation, Nuclear Division.

† Y-12 Development Division, Oak Ridge Y-12 Plant, P. O. Box Y, Oak Ridge, Tennessee 37830.

Biotechnology and Bioengineering Symp. No. 8, 415–422 (1978)

0572-6565/78/0008-0415$01.00

gas, water, plus biomass. In this process, bacteria utilize organic compounds as a source of energy and carbon for the synthesis of new cellular material. Inorganic elements, such as nitrogen, phosphorus, and other trace elements, are vital to cell synthesis. For organisms that utilize energy fermentation for cell synthesis from an organic oxidation-reduction reaction, oxygen must be supplied. In the work presently discussed, the organic carbon source for the organic oxidation reaction is the petroleum oil and emulsifiers present in the machining coolant emulsions.

Many investigators have shown that bacterial utilization of hydrocarbons in most types of oils can be induced [1–6]. More than one hundred species of yeasts, fungi, and bacteria have been found to oxidize one or more hydrocarbons. These microorganisms are widely distributed in oil field soils, sump waters, and coastal areas [1, 2]. Other investigators [3–5] have found that petroleum-cutting oil emulsions were good breeding solutions for microbial populations, with one predominant species in oil emulsions being pseudomonads [4]. Laboratory experiments have demonstrated effective microbial oxidation and removal of oil from water [6] and suggest the use of microbial action for disposing of oil wastes specifically [7]. Other investigations [8, 9] have successfully degraded lubricating and vacuum pump oils along with oil-in-water emulsions in field plots in which natural soil bacteria were utilized as inoculum.

Waste Coolant Sources

Waste machine coolants used in this study were from several different manufacturers. Coolants were collected from individual machine sumps and transported to storage tanks for pumping into the bioreactor. Although the specific coolant formulations are proprietary, a qualitativc laboratory analysis of some of the constituents in the wastes are included in the Appendix.

A test of biodegradation could be done on individual machining coolants. From a practical view, however, it would be impossible, logistically and economically, to maintain each waste as a separate stream.

Feed Solutions

Machining coolants when mixed for use contain from 1 to 5 vol % oil. After use in machining operations, these fluids are contaminated by oils used to lubricate machine surfaces and by fine metal particulates that fall from the metal workpieces. These discarded fluids are mixed with different coolants from other machines for disposal.

Table I lists chemical analyses of a typical coolant mix ready for disposal. As shown in the table, organic concentrations are measured by both chemical oxygen demand (COD) and total organic carbon (TOC). In later reported operational data, only COD figures are provided. This is due to faster analysis times and repeatabilities experienced in the COD analyses vs TOC analyses.

TABLE I
Analysis of Waste Coolant[a]

SPECTROGRAPHIC ANALYSIS
(Semiquantitative analysis (reported as %)

Ag <0.001	Al 0.001	Au <0.001	B 0.0001	Ba 0.001
Be <0.0001	Bi <0.0001	Ca 0.04	Cd <0.0001	Co 0.0001
Cr 0.0002	Cs <0.001	Cu 0.001	Fe 0.002	Hf <0.001
K 0.002	Li 0.0002	Mg 0.0006	Mn 0.0001	Mo 0.0001
Na 0.02	Nb <0.001	Ni 0.0001	P 0.005	Pb 0.0009
Rb <0.0003	Sb <0.0001	Si 0.002	Sn 0.0001	Ta <0.0003
Th <0.0001	Ti 0.0001	U <0.006	W 0.0002	Zn 0.0004
Zr 0.0001				

[a] COD = 50000 mg/liter (6000–10000 range); TOC = 30000 mg/liter; pH = 10.5.

In initial laboratory experiments on biodegrading machining coolants, additional nutrient sources such as calcium nitrate and potassium phosphates were added to the coolant feeds. Operational difficulties were experienced using this method in that the added salts have a tendency to partially deemulsify the emulsions. Since best biodegradation can be expected to proceed with a larger water-to-oil interface, later experiments were conducted using a separate nutrient addition. Sufficient amounts of nutrients are added so as not to be deficient or growth limiting.

Bacteria Inoculum

The biological reactor was originally seeded from an operating, municipal-activated sludge plant. Identification of the bacteria now present in the reactor has not been performed.

Analytical Methods

The measurements of soluble COD volatile suspended solids, nitrate, and phosphate concentration were performed according to *Standard Methods for the Examination of Water and Wastewater* [10]. Other performance measures were pH and temperature measurements, settleable solids concentrations, and dissolved oxygen concentrations. Hydrogen ion concentrations were measured using a Photovolt portable pH meter. Dissolved oxygen concentrations were measured using a YSL Model 54A dissolved oxygen meter.

Experimental Equipment: Bioreactor

A concrete vessel with a capacity of 157 m^3, a length of 11.7 m, a width of 9.75 m, and an operational depth of 1.38 m was used (Fig. 1). The reac-

Fig. 1. Bioreactor (concrete vessel).

tor was equipped with a 28-m^3 feed tank with a timer-controlled centrifugal pump which was used to feed coolant solutions on a periodic basis into the bioreactor. The reactor was also equipped with a separate biological settling chamber with a timer-controlled sump pump. Aeration and mixing was accomplished by using four 5-HP floating aerators. A flow sheet of the process is shown in Figure 2.

RESULTS

Operational Rational

The objective of the machining coolant degradation facility is to reduce as efficiently as possible organics into carbon dioxide gas before release. An effective biological removal process demands careful monitoring and control of a number of environmental conditions, including pH, mixed liquor temperature, dissolved oxygen concentration, and nutrient concentrations.

pH Control

Influent machining coolants (Table I) have relatively high pH, reaching levels of 10–10.5. Experience has shown that, when the bioreactor is functioning properly, pH levels of the mixed liquor in the reactor will range from 7.2 to 8.5. This is within the limits normally accepted for heterotrophic microorganisms.

Temperature Control

During most of the year, the mixed liquor temperature in the bioreactor will be allowed to follow closely the ambient temperature. During freezing weather, steam must be added to the reactor to prevent freezing of the reactor solution.

Dissolved Oxygen Control

Dissolved oxygen concentrations are monitored regularly and are perhaps the best indicators of overloading of the reactor. Best conditions exist at dissolved oxygen concentrations of 2 μg/g or greater.

Nutrient Concentrations

Since nitrogen and phosphate concentrations are low in the waste stream, these nutrients are added batchwise to the reactor. Mineral salt additions of calcium nitrate and potassium phosphate are controlled by laboratory analysis and are added so as to be in excess and not rate limiting to biological growth. Lower limits for nitrate and phosphate in the reactor are 60 and 10 mg/liter, respectively.

Biomass control. Biological growth in the growth in the bioreactor is largely controlled by available organic substrate. Biological growth and substrate utilization is also strongly dependent on available dissolved oxygen in the mixed liquor.

When an excess amount of organic carbon is added to the bioreactor or when a low dissolved oxygen concentration occurs, a condition will exist in which the bacterial mass or floc will rise to the surface. This condition is referred to as "bulking" of the reactor. The "bulking" condition can also exist when a lack of nutrients such as nitrate or phosphate occurs.

During normal operations, the bacterial floc will settle when not agitated. Mixed liquor from the bioreactor exits into a settling chamber as shown in Figure 1. Periodically, the settled biomass from the chamber is pumped back into the bioreactor. Some wasted cells from the settler flow out with the effluent.

Bacterial concentrations in the reactor-mixed liquor are routinely analyzed for mixed liquor volatile suspended solids. Highest recorded levels

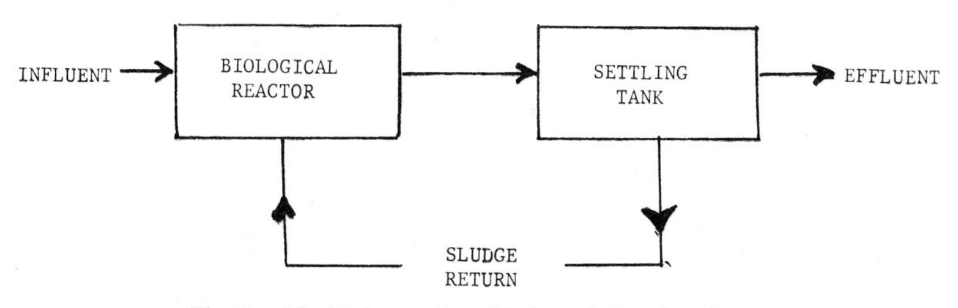

Fig. 2. Machining coolant biodegradation flowsheet.

TABLE II

Reactor Performance 1977–1978

Month	COD Added (kg)	COD in Effluent (kg)	Average Daily Feed (COD in g liter^{-1} day^{-1})	Reactor Efficiency (%)
JULY	325.5	7.09	0.11	97.9
AUGUST	687.9	11.1	0.20	98.4
SEPTEMBER	1705.0	30.2	0.50	98.8
OCTOBER	3156.9	83.7	0.89	97.4
NOVEMBER	2015.0	86.0	0.80	95.6
DECEMBER	462.1	36.9	0.40	92.0
JANUARY	1786.6	105.3	0.74	94.1
FEBRUARY	3100.7	69.2	1.44	97.8
MARCH	2738.5	61.7	1.20	97.8

in the reactor have been 14,000 mg/liter. Average concentrations in the reactor during normal operation are at levels of 8000–9000 mg/liter.

Bioreactor efficiency. Efficiency in the process is measured by monitoring reduction of chemical oxygen demand. Table II lists monthly average amounts of feed and reactor efficiencies. As can be seen in the table, feed rates have a low of 0.11 to a high of 1.44 g of COD per liter of reactor solution per day. Efficiencies in COD removal have remained above 92% for nine months of continuous operation. During the months of November, December, and January, the lowest efficiencies were noted and can be related to cold temperatures experienced during that quarter. Although

TABLE III

Spectrographic Analysis of Effluents

Semiquantitative analysis (reported as %)				
Ag <0.0001	Al 0.0003	Au <0.0001	B 0.0003	Ba 0.00001
Be <0.0001	Bi <0.0001	Ca 0.002	Cd <0.0001	Co <0.0001
Cr <0.0001	Cs 0.0006	Cu <0.0001	Fe 0.0003	Hf <0.0001
K 0.003	Li 0.0001	Mg 0.0006	Mn <0.0001	Mo <0.0001
Na 0.01	Nb <0.0001	Ni 0.003	P <0.001	Pb <0.0002
Pd <0.0001	Rb <0.0002	Sb <0.0001	Si 0.002	Sn <0.0001
Ta <0.0002	Th <0.0003	Ti <0.0001	U <0.0001	V <0.0001
W 0.0002	Zn 0.0001	Zr <0.0001		

TABLE IV
Spectrographic Analysis of Reactor Sludge

Semiquantitative analysis (reported as %)				
Ag 0.006	Al 0.2	Au <0.002	B 0.005	Ba 0.008
Be <0.0001	Bi <0.001	Ca 2	Cd <0.004	Co <0.0005
Cr 0.002	Cs <0.06	Cu 0.004	Fe 0.3	Hf <0.002
K 0.1	Li 0.02	Mg 0.1	Mn 0.004	Mo <0.0002
Na 0.01	Nb <0.002	Ni 0.02	P 0.5	Pb <0.001
Pd <0.0002	Rb <0.02	Sb <0.004	Si 1.5	Sn 0.001
Ta <0.02	Th <0.004	Ti 0.03	U <0.03	V 0.0005
W <0.008	Zn <0.008	Zr 0.001		

steam was added during this period, there were periods when mixed liquor temperatures dropped to 10°C or less.

Effluent quality. A number of factors can influence effluent quality and include (1) excess organic feed, (2) excess mineral nutrients, (3) soluble organic waste by-product, and (4) biomass washed from the reactor.

While average removal efficiencies in the bioreactor have been quite high, the average soluble COD in the effluent during the nine months of operation has been approximately 900 mg/liter. Analyses of the effluent have shown that the organic components in the effluent were, in large part, a mixture of organic acids and sulfonated vegetable oils. Both are present in the coolant samples; the organic acids can also be biologically produced as a cellular by-product.

While the soluble chemical oxygen demand is efficiently lowered in the process, it is known that a significant and variable amount of oil may be occluded with the biological solids. This, along with the fact that large amounts of biomass are produced, make it highly desirable to have the effluent receive post-treatment. Using standard sanitary wastewater treatment procedures, it is possible to lower effluent values to point-source discharge limits. Alternately, it may be possible to include also an anaerobic treatment for the effluent, since there would be an adequate supply of nitrate and phosphate ions, plus remaining organics which should be amenable to this decomposition process.

As shown in Table I, there are metals which might build up and pose a problem. Analyses of effluents and reactor sludge (Tables III and IV, respectively) suggest that metals tend to be retained by the biological floc, and at the operational pH ranges, this is not at all surprising.

CONCLUSIONS

It has been demonstrated that oil-in-water emulsions used for machining coolant can be effectively biodegraded. Future studies should include operation at several levels of steady-state operation to determine kinetics of biological species involved and, thus, determine optimum operating conditions.

It is evident in the operation of the reactor that post-treatment of effluents will be required to meet possible point-source discharge limits as imposed by NPDES.

APPENDIX I

Chemical constituents of machining coolants are

triethanolamine	tall oil fatty acids
diethanolamine	alkali soaps
sodium nitrate	alkyl phenol polyglycol ether
petroleum sulfonates	polyalkaline glycol
boric acid	polyethylene glycol
mineral oil	

References

[1] C. E. Zobell, *Bacteriol. Rev.*, *10*, 1–59 (1946).
[2] C. E. Zobell, *Adv. Enzymol.*, *10*, 443–486 (1950).
[3] M. Lee and A. C. Chandler, *J. Bacteriol.*, *41*, 373–386 (1941).
[4] H. Pivnick and F. W. Fabian, *J. Bacteriol.*, *70*, 1–6 (1955).
[5] N. D. Duffett, S. H. Gold, and C. L. Weirich, *J. Bacteriol.*, *45*, 37–38 (1943).
[6] J. B. Davis, *Petroleum Microbiology* (American Elsevier, New York, 1967), pp. 350–375.
[7] J. B. Davis and R. L. Raymond, U.S. patent 3,152,983 (1964).
[8] C. B. Kincamon, *Demonstration of Oily Waste Disposal by Soil Cultivation Process*, Prepublication EPA R-2-72-110 (Environmental Technology Series, Washington, D.C., 1972).
[9] F. E. Clark et al., *Disposal of Oil Wastes by Microbiol Assimilation*, Y-1934, May 16, 1974 (NTIS, Union Carbide Corporation, Nuclear Division, Oak Ridge Y-12 Plant, Oak Ridge, TN 37830).
[10] *Standard Methods for the Examination of Water and Wastewater*, 14th ed. (American Public Health Ass., Washington, D.C., 1975).

Biomimetic Approaches to Artificial Photosynthesis

JOSEPH J. KATZ and MICHAEL R. WASIELEWSKI
Chemistry Division, Argonne National Laboratory, Argonne, Illinois 60439

INTRODUCTION

There is very general agreement about the desirability, indeed the urgency, of developing methods for the utilization of solar energy. As fossil fuels become depleted, there is increasingly strong pressure to turn to solar energy as our only truly renewable source of energy. Even though there appears no reasonable possibility of foregoing the use of essentially limitless nuclear fission and fusion energy in the foreseeable future, prudence dictates a pluralistic approach to our energy problems, and this means effective use of solar energy to the maximum extent dictated by social, economic, and technological considerations. Social factors, in particular, provide a very strong impetus to exploit solar energy in any and all ways available to us.

At this writing, solar heating and the direct conversion of sunlight to electricity are the most advanced technologies for solar energy conversion. It is a reasonable assumption that both of these techniques for solar energy conversion will achieve large scale use in the not too distant future. However, there are other less familiar and less widely discussed but in the long run perhaps more broadly useful approaches to solar energy conversion. One such approach, which is the subject of this paper, is the utilization of solar energy for chemical purposes rather than as a heat source or in competition with various nuclear technologies for the production of electricity. Green plants and certain bacteria, in fact, use solar energy in precisely this way. These organisms use the energy of sunlight to effect the synthesis of the great array of organic compounds (e.g., carbohydrates, proteins, lipids) characteristically produced by green plants. The process by which green plants use solar energy for chemical purposes is called photosynthesis, and it is one of the objects of this paper to survey some of the salient features of artificial photosynthesis as a general approach to solar energy conversion.

Biomimetic Technologies

We refer to processes that mimic the essential features of similar processes in living organisms as *biomimetic* (from the Greek *bios*, life, and *mimesis*, imitation). Biomimetic technologies are abstractions from a living

Biotechnology and Bioengineering Symp. No. 8, 423–452 (1978)

0572-6565/78/0008-0423$01.00

model. As our understanding of the intimate details of the ways in which living organisms, for example, fix nitrogen, or carry out chemical reactions by enzymatic catalysis, or manufacture polymers and fibers, the prospects for technological advances of industrial significance based on the methods used by living organisms for similar ends become ever more realistic goals. All the important biological processes that may ultimately serve as models for biomimetic technologies are solar energy powered. Solar energy conversion by a biomimetic approach may well be among the earliest of biomimetic technologies that may attain practical significance. Features of biomimetic technologies in general and solar in particular have been discussed previously by Katz et al. [1] and by Katz [2].

Solar energy conversion by a biomimetic technology has a distinct appeal. The feasibility of solar energy conversion for chemical purposes has been demonstrated on the largest scale. For several billion years photosynthetic bacteria, blue-green algae, and green plants have obtained the energy necessary for survival by photosynthesis. Indeed, all life as we know it depends on plant photosynthesis, for only photosynthetic organisms can exist exclusively on inorganic precursors. All forms of life other than plants require preformed organic compounds containing carbon–hydrogen bonds for continued existence, and these are produced only by photosynthetic organisms using the energy of sunlight as the necessary energy source. In plant photosynthesis we have a highly successful solar energy conversion process that has been evolved and refined over a period of billions of years. Plant photosynthesis because of its demonstrated success should certainly qualify as an attractive model for solar energy conversion.

Efficiency of Photosynthesis

The objection that plant photosynthesis is an inherently inefficient process is sometimes raised. Thus, Bolton [3] estimates the maximum solar energy storage efficiency of photosynthesis is $9.5 \pm 0.8\%$, and Boardman [4] concludes that maximum short-term growth rates of high-yielding crops only represent solar energy conversion efficiencies of 2.7–4.6%. Boardman's estimate is based on the ratio of biomass produced to incident solar energy. It should be recognized, however, that biomass/solar energy ratios are not necessarily a proper measure of solar energy efficiency. Plants are highly organized entities, the elaboration of which must require a large amount of energy. The amount of energy required to produce an intricate piece of wooden furniture or a wood carving cannot be deduced from its heat or combustion, which would hardly be different from an equal weight of wood. Growing plants transpire huge amounts of water, and energy for this must be included in the energy budget for photosynthesis. Plants must also expend energy in one way or another for survival. They operate what must be highly energy-intensive reproductive systems, and they must pay a price not only to survive long periods of dormancy, but also the (circadian) disappearance of the solar energy source during the night. All of these activities must clearly be at the expense of overall photosynthetic efficiency as gauged only by biomass production.

Judged by the efficiency with which a photon is converted to an electron and a positive hole (reducing and oxidizing capacity respectively), the light-energy conversion step itself in plant photosynthesis is highly efficient, even though the efficiency of photosynthesis as judged by biomass production is low. The best available experimental evidence indicates that conversion of one photon produces one electron in the photoreaction centers of both green plants [5] and photosynthetic bacteria [6]. The quantum yield of 1 for the light-conversion event argues for essentially 100% efficiency for this step. Although green plants and blue-green algae use light quanta over the entire spectral range from blue through the red, only that fraction of the energy of a photon corresponding to a 700 nm photon in green plants and blue-green algae, and an 865 nm photon in photosynthetic bacteria is actually used in electron production. Thus, in a sense, photosynthetic organisms degrade any photon they absorb to that of a 700 nm photon (green plants, blue-green algae) or 865 nm (purple photosynthetic bacteria). We should not be too hasty in assuming, however, that the conversion of more energetic photons to the energy of 700 nm or 865 nm photons is necessarily a sign of inefficiency. We do not know to what ends the "wasted" portion of the more energetic photons may be put to by the photosynthetic organism.

What may be optimum for the plant overall may not coincide with the maximum production of biomass, primarily because of the energy requirements of the plant for survival. Consequently, replication of the highly efficient light-conversion step *in vitro* may increase the net efficiency of solar energy utilization to a substantial degree. Not only can the energy requirements for growth, reproduction, and dormancy be evaded, but light-energy conversion itself can probably be made more effective. For example, a synthetic photoreaction center might be designed to use blue light more efficiently. Contemporary photosynthetic organisms are the beneficiaries of billions of years of evolution, but they also pay a certain price in the constraints imposed by their origins and in the evolutionary paths accessible to them. The photosynthetic pigments, the chlorophylls in particular, seem not to have undergone significant evolution; evidence from fossil remains seems to suggest that chlorophyll *a* was involved in photosynthesis from the earliest times. That the chlorophylls did not evolve into a large family of compounds (there are no more than five chlorophylls in the vast number of green plants and blue-green algal species now in existence) with spectral characteristics tailored to the enormous range of ecological niches occupied by photosynthetic organisms may not indicate that the earliest chlorophylls were incapable of improvement. Rather, the failure of the chlorophylls to experience structural divergence may be a consequence of the biosynthetic pathways available to the organisms. As pointed out by Broda [7], the properties of the photosynthetic machinery cannot "be as good as if it had been designed rationally without any constraint due to evolutionary history." Implicit in any biomimetic technology is improvement over nature. Because a biomimetic technology is sharply focused on a single limited objective, in contradistinction to a living organism for which survival and reproduction are paramount goals, any single objective of a biomimetic technology should ultimately be capable of achieving a higher intrinsic efficiency than its prototype *in vivo*.

Energy Storage in Photosynthesis

For an intermittent energy source such as solar energy, it has become evident that efficient and economical energy storage is probably indispensible to a practical solar energy system. Without energy storage, stand-by electrical power will be required to mitigate the adverse effects of extended periods of cloudy weather characteristic of much of the temperate zone as well as the periods of darkness every night. Plant photosynthesis copes with the intermittent nature of solar energy admirably. Solar energy used for photosynthesis is stored as energy-rich chemical compounds, e.g., carbohydrates, which then are available to the plant as an energy source as needed. Energy conversion and energy storage are automatically coupled in photosynthesis by the very nature of the *in vivo* solar energy conversion process. Energy storage must be regarded as a positive feature of any solar-energy conversion scheme in which light energy is used to synthesize chemical compounds.

Biomimetic vs Biomass Technologies

Biomimetic approaches to solar-energy conversion are directed to the development of completely synthetic (inanimate, abiotic systems for light energy conversion that mimic the light energy conversion process *in vivo*). They thus differ in a basic way from biomass production by plants, or (photovoltaic) devices (even if they use chlorophyll as the working element) for direct conversion of light to electricity or for photolysis of water. Schemes for solar energy conversion using immobilized algae or chloroplast fragments, valuable though they may turn out to be, are likewise not biomimetic in the sense we use the term here.

The extent to which we can successfully mimic a complicated *in vivo* process *in vitro* clearly must depend on the extent of our understanding of the process in nature. Only for those situations where the fundamental strategy of the living organisms has been grasped are there reasonable prospects for success. Although many aspects of plant and bacterial photosynthesis, among these being some of the most basic, are still obscure, nevertheless enough has been learned about chlorophyll function in photosynthesis to make it possible to mimic the essential features of the light-energy conversion step in the laboratory. We can now consider the recent advances in the understanding of chlorophyll function in light-energy conversion in plant and bacterial photosynthesis that are the point of departure for the development of biomimetic technologies for solar-energy conversion.

PHOTOSYNTHETIC UNIT (PSU) IN PHOTOSYNTHESIS

Photosynthesis is commonly defined as the reduction of carbon dioxide (CO_2) by electrons abstracted from water. The addition of electrons to CO_2 (reduction) forms organic compounds containing carbon–hydrogen bonds, and the abstraction of electrons from water (oxidation) forms molecular oxygen. The conversion of CO_2 to organic compounds and the simultaneous oxidation of water to oxygen are reactions that do not proceed spontaneously. An input of energy is required, and the energy needed to drive the reactions is provided by light (symbol $h\nu$). Thus, photosynthesis ("synthesis aided by light") converts purely inorganic substances (CO_2, H_2O, Mg^{2+}, NH_3, $SO_4{}^{2-}$, etc.) to organic matter, using the energy of light to effect the conversion. The light energy is trapped so to speak in the synthesized organic matter (or hydrogen) and can be recovered at will.

It is convenient to divide considerations of photosynthesis into (a) light energy conversion in which photons are captured and chemical oxidizing and reducing capacity are generated; (b) the complex of chemical reactions associated with photosynthesis that are involved in the synthesis of the multitude of organic compounds routinely produced by plants, and (c) the chemical reactions involved in oxygen production in green plant photosynthesis. The primary light-conversion event is the chief topic of concern in this paper. The reducing power generated in the primary event is used to form the reducing agent NADPH (reduced nicotinamideadenine dinucleotide phosphate). NADPH is then used for reductive biosynthesis, the so-called dark reactions of photosynthesis. These will not be further considered here because they follow long after the initial photochemistry is completed. As for the oxygen side of photosynthesis, the present lack of information makes discussion of it in a biomimetic context impossible.

Chlorophyll has long been recognized to be the primary photoacceptor in photosynthesis and to be the principal agent in the light conversion step itself. The other photosynthetic pigments, i.e., the auxilliary chlorophylls, phycobilins, carotenoids, and xanthrophylls, have important functions in plant and bacterial photosynthesis. Chlorophyll (Chl) *a* in green plants and bacteriochlorophyll (Bchl) *a* in most purple photosynthetic bacteria are particularly important chlorophylls. All organisms that carry out photosynthesis with the evolution of molecular oxygen contain without exception chlorophyll *a*. Our discussion of the primary photochemical events in photosynthesis, therefore, will be in terms of these two chlorophylls, keeping in mind at the same time the important part played by the auxilliary chlorophylls and the other photosynthetic pigments in light harvesting and energy transfer in photosynthesis.

Chlorophyll function in light energy conversion is a cooperative phenomenon. Almost 50 years ago, Emerson and Arnold [8, 9] recognized that a large number of chlorophyll molecules act in concert in the conversion of a single photon to an electron and a positive hole. The concept of the photosynthetic unit (PSU) has evolved from this original suggestion. In an early form, the PSU was conceived by E. Katz [10] as a solid-state device, and the chlorophyll molecules were considered to be highly ordered in a two-dimensional crystalline array. From solid-state considerations, the chlorophyll crystal would then be expected to have an electronic conduction band. Electrons and positive holes generated by light energy conversion would be mobile in a "crystal" of chlorophyll and would migrate in such a way as to effect charge separation. The terms in which the light conversion event was analyzed were essentially the solid-state description developed for the transistor [10]. Rabinowitch [11] has summarized the arguments against such a picture of the photosynthetic unit. Although the experimental evidence for cooperative behavior of the chlorophyll molecules in the PSU is good, there is no evidence to support the existence of electronic conduction bands and hole migration in the PSU. In its early forms, all of the chlorophyll molecules in the unit were considered to be identical in structure and function, and the properties of the PSU were then those resulting from the order in the array. However, it has been known that chlorophyll occurs in the photosynthetic apparatus in a variety of forms characterized by different spectral properties [12]. The contemporary versions of the photosynthetic unit still postulate cooperative behavior by the chlorophyll, but rely on well-differentiated species of chlorophyll to account for charge separation in light conversion.

The chlorophyll molecules in the PSU have at least two well-defined functions. The large majority of the chlorophyll is used for collecting light energy, and this kind of chlorophyll is referred to as bulk, light-harvesting, or antenna chlorophyll. Antenna chlorophyll is organized into a structure for absorbing electromagnetic radiation and then transferring the excitation energy to other specialized chlorophylls in a photoreaction center, where the light-energy conversion event actually occurs. A chlorophyll molecule that captures a photon is excited to a higher-energy excited state. The excitation energy then migrates through the chlorophyll antenna until it is trapped in the photoreaction center. The details of the energy-transfer process are under very active discussion at this writing. Energy transfer in photosynthesis has been considered either in terms of the migration of excitons in a crystal lattice [13], or, in an alternative formation, proceeding by resonance transfer by the mechanism described by Förster [14]. The antenna chlorophyll molecules have electronic transition spectra that are red-shifted relative to those of monomer chlorophyll, i.e., the energy level of the first excited state of a chlorophyll molecule in the antenna is at lower energy than in monomer chlorophyll. Whereas monomer chlorophyll *a* absorbs maximally at 660–670 nm, depending on the solvent system, antenna chlorophyll in green plants has its absorption maximum in the red near 678 nm. Antenna chlorophyll is often referred to as P680, the P standing for pigment. Whether the chlorophyll in the antenna has one or several molecular structures is still unresolved, nor is the reason for the red shift clear. Majority opinion at this time holds that antenna chlorophyll is a chlorophyll–protein complex, and that the red shift is a consequence of chlorophyll–protein interactions [15]. Another view of antenna structure assigns the red-shift to chlorophyll–chlorophyll aggregates [16], largely on the grounds that *in vitro* chlorophyll films [17] and chlorophyll polymers [18] have red-shifts very comparable to those of antenna chlorophyll. The suggestions that chlorophyll–protein or chlorophyll–chlorophyll interactions can provide the observed red shift are not mutually exclusive and both have merit. We suspect that the model of antenna structure that ultimately emerges will combine valid features of both views. As information on the structure and function of antenna chlorophyll *in situ* is sparse, antenna structure in a biomimetic context will not be further pursued.

The situation with respect to photoreaction-center chlorophyll is considerably more satisfactory. What is known about antenna structure and function is strictly by inference, as there are as yet no experimental techniques that allow energy transfer to be observed in normal photosynthesis. However, photoreaction-center function *in situ* in photosynthetic organisms in the course of normal photosynthesis can be monitored both by visible absorption spectroscopy (optical transients) and by electron paramagnetic resonance (EPR) spectroscopy. Further, highly successful methods have been developed for the isolation of bacterial photoreaction centers that have all of the optical, redox, and EPR properties characteristic of the photoreaction center in the intact organism. This of course makes it possible to study the composition of the photoreaction centers and to observe the light-energy coversion event in a greatly simplified system. Although photoreaction centers of similar utility have not as yet been isolated from green plants, the considerable similarities between bacterial and green plant photoreaction-center function *in vivo* make the findings with bacterial photoreaction-center preparations highly relevant to green plant photosynthesis.

The *in vivo* photoreaction-center chlorophylls in green plants are red-shifted relative to the antennas. The first excited states of the photoreaction-center chlorophylls in green plants are at lower energy than those of the antenna, thus enabling the excitation energy captured in the antenna to be trapped by the reaction centers. The photoreaction-center chlorophylls in green plants and in photosynthetic bacteria are designed P700 and P865, respectively. Chlorophyll absorbing light at these wavelengths is identified as photoreaction-center chlorophyll because reversible changes in light absorption occur at these wavelengths under conditions of active photosynthesis. Under strong illumination, photo-bleaching (decrease in absorption) is observed at 700 nm in green plants [19] and near 865 nm in bacteria [20]. The photo-bleaching is fully reversible, light absorption being restored in weak light. In the intact organisms, photo-bleaching occurs to only a very small extent (e.g., about 1% of the chlorophyll is involved); this is the basis for the conclusion that the great majority of the chlorophyll, perhaps 99%, is engaged in light harvesting, and that only a small number of specialized in chlorophyll molecules are involved in the light conversion step itself. In isolated bacterial reaction center preparations, however, complete and reversible bleaching occurs at 865 nm. The observations of optical transients thus permit direct observation of photoreaction-center activity.

Concomittant with the photo-bleaching of P700 and P865 is the appearance of an EPR signal characteristic of an unpaired electron. Most chemical compounds contain an even number of electrons, which are paired in groups of two. Removal of one electron (or the addition of an electron) results in formation of a compound with an odd number of electrons and in which one electron is unpaired. Compounds containing an unpaired electron (spin) are called free radicals. Electrons are paramagnetic, and therefore a compound containing an unpaired electron is paramagnetic. EPR is an extremely sensitive tool for the detection of free radicals, and the unpaired electron can readily be detected by an EPR experiment. Commoner and his colleagues [21, 22] were the first to detect the formation of chlorophyll free radicals in photosynthesis. The optical transients and the formation of free radicals have been interpreted in terms of photooxidation of the light-excited photoreaction-center chlorophylls:

$$\mathrm{P700} + h\nu \rightarrow \mathrm{P700}^{\dagger} + \epsilon$$
$$\mathrm{P865} + h\nu \rightarrow \mathrm{P865}^{\dagger} + \epsilon$$

Support for this interpretation comes from very careful studies that have shown that for each photon that reaches the trap, one P700 (P865) is bleached, and one free radical is produced [5, 6]. The quantum yield of 1 for conversion of a photon to an electron is the basis for the conclusion stated earlier in this paper that light energy conversion in photosynthesis is a highly efficient process. The ejection of the electron (the electron is of course not free as written in the equation) from the photoreaction-center leaves an entity containing a single unpaired electron. The kinetics of photo-bleaching and the formation of $\mathrm{P700}^{\dagger}$ are identical within experimental error. The photooxidized reaction center is written as a cation primarily because of important *in vitro* experiments on oxidized porphyrins [23], but in the intact photosynthetic organisms there is no evidence that bears on whether the photooxidized center carries a positive charge; for example, interaction of $\mathrm{P700}^{\dagger}$ with a hydroxyl group, OH^-, would leave a neutral entity but one that is still a free radical.

Figure 1 shows a highly schematic representation of a PSU and the course of events when a light photon is absorbed in the PSU. The structure of the antenna is not specified. The antenna is excited by the capture of a photon, and the excitation energy (in the form of electronic excitation to higher energy states) is transferred to the photoreaction-center. One of the

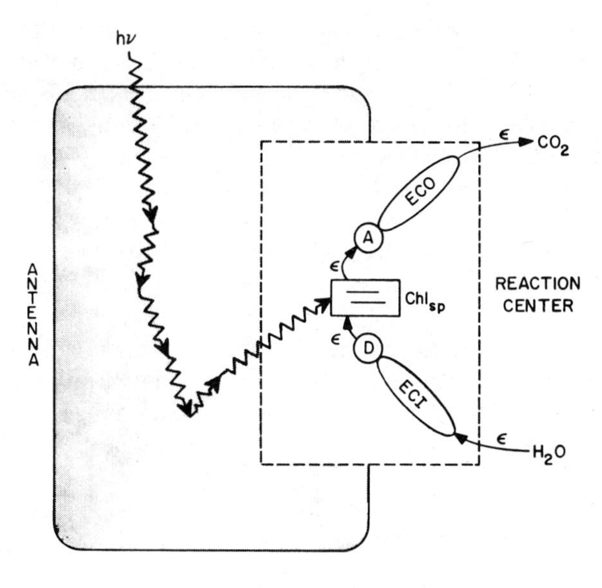

Fig. 1. Highly schematic photosynthetic unit. ε, electron; ECI, electron conduit in; ECO, electron conduit out; Chl_{sp}, the photoreaction center; A, the primary acceptor; D, the primary donor to Chl_{sp}^{+}. Antenna is excited by the capture of a photon by the antenna. Excitation energy is transferred to the Chl_{sp} where charge separation occurs. Note that only one photosystem reaction center is shown in this highly simplified representation.

important functions of the antenna is to degrade the energy of whatever photon is absorbed (the photon may range in color from blue to red) to that of a photon with the energy of a 700 nm photon. Only energy quanta of this size can be trapped in the reaction center. Once the photoreaction center is excited, photooxidation occurs by ejection of an electron. This is putting it loosely, for there is no reason to suppose that free electrons (in the sense of electrons in a metallic conductor) are present in the photosynthetic apparatus. Ejection of the electron proceeds by transfer to an electron acceptor. Subsequent electron transfer is by a number of transfer agents arranged in a bucket brigade that passes the electron along until the ultimate reductant, NADPH, is formed. The dark reactions of photosynthesis, powered by NADPH, then form carbohydrate, protein, lipid, and so forth. The dark reactions presumably proceed in the aqueous regions of the chloroplast well removed from the photosynthetic unit. The primary acceptor and the subsequent electron-transport chain comprise the electron conduit out. Another electron-transport chain functions as a conduit to bring in an electron abstracted from water (green plants) or some organic compound (purple photosynthetic bacteria) that restores the oxidized photoreaction center to its resting condition, ready for another light-conversion event. The proximate donor to the oxidized photoreaction center is generally considered to be a cytochrome. The electron transfer conduits in and out of the photoreaction center are of the utmost importance in PSU function. It is very important that a high degree of directionality be assured in electron transport. If an electron ejected from the photoreaction center returns to its origin, a photon is wasted. Reversal of electron transfer anywhere in the electron-transfer chains likewise means a loss in efficiency. There is good evidence to support the view that electron transfer is highly efficient under normal forward photosynthesis and that the graduation in redox properties and the geometrical arrangement of the electron transfer agents relative to each other is precisely specified to assure efficient electron transfer.

The PSU, then, can be considered to be an electron pump powered by light. Low-energy electrons are abstracted from water and raised by light to an energy level at which they can be used to reduce CO_2. The electrons are pumped uphill, and the energy required for this purpose is supplied by the energy of sunlight. The chemical reactions associated with photosynthesis, the oxidation of water and the reductive biosyntheses of CO_2, take place elsewhere, well removed from the apparatus that generates the necessary oxidizing and reducing capacity for powering the chemical reactions of photosynthesis. In green plants, the presumption is that the PSU is situated in the bilayer lipid membrane of the thylakoids, the small units of the chloroplast that house the chlorophyll. The thylakoid membrane provides an hydrophobic environment for the photosynthetic light conversion apparatus. In the view of Anderson [24], both the antenna and photoreaction center chlorophyll are embedded in the intrinsic membrane proteins, whereas Katz et al. [25] suggest that the photoreaction center chlorophyll may be closely associated with protein but that the antenna chlorophyll may exist for the most part in the hydrocarbon region of the lipid bilayer. The weight of the available evidence at this time supports an intimate association of the chlorophyll with the membrane intrinsic proteins. Photosynthetic bacteria do not have organized photosynthetic organelles such as the chloroplast, and the PSU of these organisms is presumably sited in a photosynthetic membrane. The exact spatial relation between chlorophyll, protein, and lipid in the photosynthetic membrane remains one of the more pressing problems in photosynthesis research.

SOME CHLOROPHYLL PROPERTIES RELEVANT TO ITS FUNCTION

For the purposes of a biomimetic technology, the nature of the PSU must ultimately be describable on the molecular level. A necessary preliminary to such an effort is an understanding of the ways in which the structural formula of chlorophyll can be translated into function. Physical chemical studies on chlorophyll over the past 15 years have yielded important insights into the functional possibilities inherent in the anatomy of the chlorophyll molecule. These new perspectives on the relation between chlorophyll structure and function result mainly from the application of infrared [26] proton and ^{13}C magnetic resonance [27], electron paramagnetic resonance [28], and electronic transition [29] spectroscopy. Two aspects of these studies are particularly relevant in the present context. These are the coordination properties and coordination interactions of chlorophyll, which largely determine the molecular architecture of chlorophyll *in vivo*, and the properties of chlorophyll–water aggregates, which serve as the paradigm for the photoactivity of chlorophyll.

All of the chlorophylls in nature are chelate compounds of magnesium. They all contain a central magnesium (Mg) atom coordinated to the four pyrrole nitrogen atoms of the chlorophyll macrocycle (Fig. 2). In the structure shown in Figure 2, the Mg has a coordination number of 4, that is, the Mg participates in 4 chemical bonds (to the four nitrogen atoms). Spectroscopic investigations, however, clearly indicate that Mg in chlorophyll with coordination number 4 is coordinatively unsaturated. Coordinative unsaturation implies that the Mg is electron deficient; it effectively carries a positive charge. Because of the electron deficiency, there is a strong driving force for the Mg to interact with electron donor molecules (nucleophiles) that have electrons not otherwise involved in chemical bonding and thus available for donation. Compounds containing O, N, and S have lone pairs of electrons that can serve as effective electron donors to Mg. The coordination unsaturation of the central Mg atom and the driving force to insert nucleophiles into the axial position(s) of the Mg have interesting structural

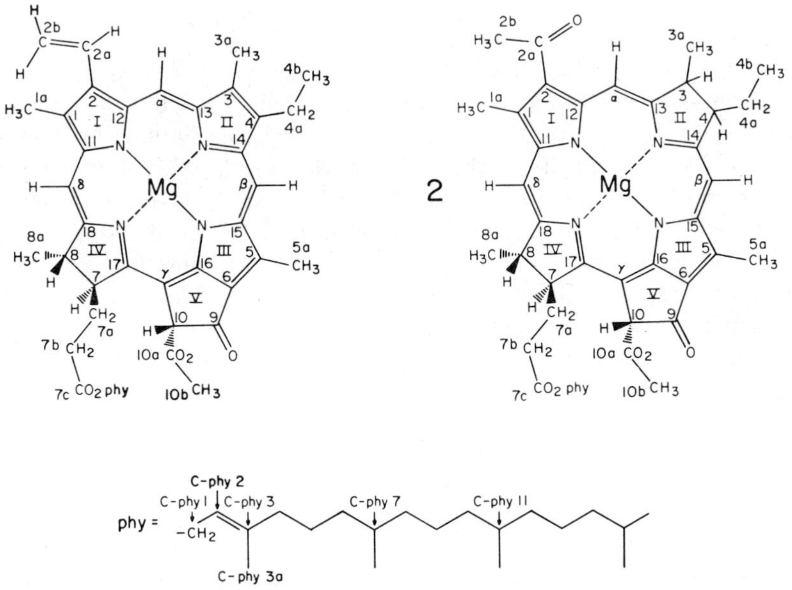

Fig. 2. Structure and numbering of green plant [chlorophyll *a*(**1**)] and purple photosynthetic bacteria [Bacteriochlorophyll *a*(**2**)] chlorophylls.

ramifications. Chlorophyll dissolved in a polar, nucleophilic solvent such as acetone, diethyl ether, pyridine, or other typical Lewis base solvents, will occur as monomeric chlorophyll, with one or two molecules of solvent in the axial Mg positions. Whether the Mg has the coordination number of five (one additional ligand) or six (both axial positions occupied by solvent ligands) depends on the basicity of the solvent. If, however, carefully purified chlorophyll is dissolved in a nonpolar, non-nucleophilic solvent, such as benzene or toluene from which adventitious nucleophiles such as water have been removed, then the chlorophyll is confronted with the problem of how to deal with the coordination unsaturation of its central Mg atom. Nature, however, has provided for such a contingency. The chlorophyll molecule itself has five oxygen atoms distributed between two ester functions and a keto-carbonyl function. The later, present in ring V (Fig. 2), is a completely adequate electron donor by virtue of the lone pair of electrons that it bears. Thus, in circumstances where extraneous nucleophiles are not available, one chlorophyll via its keto C═0 group can act as donor to the Mg of another. A keto C═0---Mg interaction between two chlorophyll molecules forms a dimer, and repetition of the interaction in an unfavorable solvent such as the aliphatic hydrocarbon *n*-octane can lead to the formation of chlorophyll polymers, $(Chl)_n$, in which 20 or more chlorophylls form an aggregate. Chlorophyll–chlorophyll aggregates formed by keto C═0---Mg coordination interactions have spectral properties similar to those of antenna chlorophyll in green plants, a resemblance first noted by Livingston [30] many years ago. It is the spectral properties of self-aggregated chlorophyll that are the basis for advancing chlorophyll oligomers and polymers as models for antenna chlorophyll.

Interaction of chlorophyll with monofunctional ligands such as acetone or pyridine leads inevitably to monomeric chlorophyll species. If the ligand is bifunctional, i.e., has two donor centers, then the coordination interaction can crosslink chlorophyll to other chlorophyll molecules[31]. For our purposes the most interesting bifunctional ligand is water. The oxygen atom in H_2O has a lone pair of electrons that can be used for coordination to Mg. The coordinated water molecule has two hydrogen atoms that can form hydrogen bonds with the oxygen atoms of another chlorophyll molecule, and thereby crosslinking them. Repetition of this interaction forms very large aggregates of macroscopic dimensions. Strouse and coworkers [32–34] have determined the crystal structure of ethyl chlorophyllide $a \cdot 2H_2O$, a compound which is a close derivative of chlorophyll; this derivative can be crystallized because it lacks the bulky phytyl chain present in chlorophyll itself. The structure of the ethyl chlorophyllide$\cdot 2H_2O$ crystal provides convincing confirmation of the coordination and hydrogen-bonding interactions postulated above for the chlorophyll–water aggregates.

Chlorophyll–water aggregates have some very interesting and unusual properties [35–38]. Conversion of a chlorophyll *a* oligomer to a chlorophyll–water aggregate [39] causes a very large red-shift from 680 nm (the oligomer) to 740 nm (the chlorophyll *a*–water adduct). The chlorophyll *a*–water aggregate is probably not an important physiological species, but bacteriochlorophyll *a*–water aggregates mimic with remarkable fidelity the visible absorption spectrum of intact *Rhodopseudomonas spheroides*, a purple photosynthetic bacterium popular in photosynthesis research [25].

The red shift observed in chlorophyll–water aggregates more or less automatically generates interest because chlorophyll *in vivo* is red-shifted. However, the red-shift is not the only, or even the most important, reason for an interest in these aggregates. Of all of the chlorophyll species prepared in the laboratory, only the chlorophyll–water adducts are photoactive by the EPR criterion [40]. Irradiation with red light of a chlorophyll *a*–water aggregate (P740) (suspended in *n*-octane or some other aliphatic hydrocarbon) results in the formation of an EPR signal characteristic of a free radical. This photosignal is Gaussian, has a *g*-value of 2.0026 (indicative of a "free" electron), and decays rapidly in the dark. The photooxidized P740 aggregates are likely to be π-cation free radicals ($P740^{\dot{+}}$) similar to those postulated for $P700^{\dot{+}}$ *in vivo*. It is a detailed consideration of the properties of the P740 free radical signal that has provided the basis for the interpretation of the EPR signal [41] and the structure [42] of photooxidized P700 *in vivo*.

CHLOROPHYLL SPECIAL PAIR

The EPR signal of monomeric Chl $a^{\dot{+}}$ obtained by chemical oxidation is about ~9 G wide. The EPR signal from $P700^{\dot{+}}$ is ~7 G wide. Despite this widely noted discrepancy, the *in vivo* EPR signal was long assigned to Chl $a^{\dot{+}}$. That the *in vivo* linewidth is about 40% narrower than that of *in vitro* monomeric Chl $a^{\dot{+}}$ was explained as a consequence of a "biological environment." The photo-EPR signal obtained from P740 is unusually narrow, about 1 G in width. This very narrow linewidth is surprising at first sight for an unpaired electron in such a large molecule as is chlorophyll, which contains many protons whose magnetic interactions with the unpaired electron are expected to broaden the EPR signal. However, the chlorophyll–water aggregate has many fully equivalent chlorophyll molecules, and an unpaired electron, given sufficiently rapid electron or spin migration, can occupy many equivalent sites. The process of spin migration (effectively electron delocalization) is known to cause a collapse to very narrow EPR signals when spin migration or delocalization occurs. The unpaired spin is delocalized, that is, occupies to varying extents all equivalent positions in all of the chlorophyll molecules comprising the aggregate. The linewidth in situations where spin delocalization occurs is in fact a function of the number of chlorophyll molecules that share the unpaired electron. A rigorous analysis indicates that the linewidth of monomer Chl $a^{\dot{+}}$ will be narrowed by the factor $N^{-1/2}$ when shared by N equivalent chlorophyll molecules. A chlorophyll–water aggregate containing a 100 or so chlorophyll molecules will have a cation free radical linewidth narrowed to about one-tenth that of the monomer, that is, 1 G. An extension of this analysis to the anomalous linewidth of P700† proceeds along the following lines. Assume the line narrowing of P700† relative to monomer Chl $a^{\dagger}$ is due to a process of spin-sharing or delocalization of the unpaired electron over more than one chlorophyll molecule. Over how many chlorophyll molecules would an unpaired electron have to be delocalized to account for a 40% narrowing? In the relation $N^{-1/2}$, which relates the linewidth of N-mer to that of the monomer [41], N equals 2 for a 40% narrowing. The linewidth *in vivo* in all photosynthetic organisms containing chlorophyll *a* or bacteriochlorophyll *a* can be accounted for with considerable precision by the assumption that the unpaired electron in P700† (or P865†) is shared by two chlorophyll molecules. In all green plants and in purple photosynthetic bacteria that contain bacteriochlorophyll *a* that have been examined the EPR linewidth can be rationalized by the assumption that the primary electron donor in photosynthesis is a pair of chlorophyll molecules. Two photosynthetic bacteria, *Thiocapsa pfenningii* and *Rhodopseudomonas viridis* have still a different chlorophyll, bacteriochlorophyll *b*. In these organisms, the oxidized reaction center has an EPR linewidth characteristic of a monomer rather than a special pair. A possible explanation for the anomaly is a highly asymmetric environment for the special pair, which results in spin localization on one of the bacteriochlorophyll *b* molecules in the special pair. The conclusions from EPR spectroscopy have been strengthened and refined by the application of the more refined techniques of endor spectroscopy [42] and by investigations on the triplet state of *in vivo* photoreaction centers [43]. Except for the particular chlorophyll from which it is constituted the photoreaction center of both green plants and purple photosynthetic bacteria appears to be constructed in essentially the same fashion. To differentiate the photoreactive pairs of chlorophylls from chlorophyll dimers formed in nonpolar solvents by keto C=O---Mg interactions, we think it preferable to refer to the donor chlorophylls as special pairs, Chl_{sp} or $Bchl_{sp}$, for green plant and purple photosynthetic bacteria respectively.

MODELS OF CHLOROPHYLL SPECIAL PAIRS

We can now consider how the Chl_{sp} can be converted into a biomimetic device. Since the first model of the Chl_{sp} was advanced [44, 45], a number of variants have been suggested [46–49]. In all of the models, two chlorophyll molecules are arranged in parallel fashion, but the geometry and symmetry vary. The original model was asymmetric, i.e., the two chlorophylls are not fully identical, one acting as donor, the other as acceptor in the coordination interactions that crosslink them. Other models have translational symmetry [47], and the models of Fong [49], Shipman et al. [46], and Boxer and Closs [50] have C_2 symmetry. A valid model for P700 or P865 must meet the following criteria: the model must account in a reasonable way for the red shift in the visible absorption spectrum, and it must account for a properly narrowed EPR signal when it undergoes one electron oxidation to a π-cation free radical. The model of Shipman et al. [46] and of Boxer and Closs [50] appears to meet these requirements, completely for P700 and at least in part for P865.

In the Chl_{sp} model of Shipman et al. [46] and Boxer and Closs [50] (Fig. 3), the two chlorophyll macrocycles are parallel, and their transition dipoles are also parallel. In the Shipman et al. [46] model, the two chlorophyll molecules are held together by two bifunctional molecules which may be but are not necessarily identical. The ligand should contain an O, N, or S atom with electron lone pairs that can be used as a donor for a coordination interaction to Mg. Attached to the donor atom must be an hydrogen atom that can form hydrogen bonds. Examples of typical nucleophiles that function in the desired way are water, HOH; ethanol, CH_3CH_2OH; *n*-butylamine, $C_4H_9NH_2$; or ethanethiol, CH_3CH_2SH. The electron lone pair on the O, N, or S atom of these ligands is coordinated to the Mg atom of one of the chlorophylls and a hydrogen bond is formed to the keto C═O function of the other chlorophyll molecule. Which of the oxygen functions is used for hydrogen bonding is of critical importance. The use of the keto C═O function as shown in Figure 3 brings the macrocycles to a π–π stack-

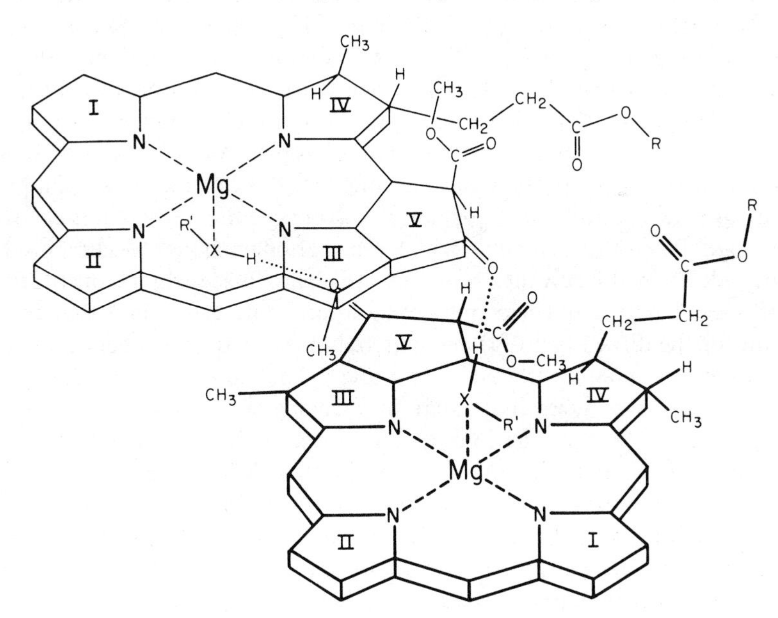

Fig. 3. Chl_{sp} model of Shipman et al. [46]. Nucleophiles of the general type R′XH, where X = O, NH, or S, and R′ = H, or alkyl can act as crosslinking agents. Water, HOH, or nucleophilic groups present in protein side-chains could act to form the Chl_{sp}. Crosslinking agent *in vivo* is unknown. Because the keto C═O functions are used for hydrogen bonding, the two macrocycles are at their van der Waals radii, 3.6 Å.

ing distance of 3.6 Å. The two π systems come just into contact at their van der Waals radii at this distance, and thus optimum π overlap is achieved. Further, in this structure, the Q_y electronic transition dipoles are parallel. The use of the carbomethoxy C═O group for this purpose, as proposed by Fong [49], positions the two macrocycles at a distance of ~6 Å, and the transition dipoles are at a 60° angle to each other. The distance between the macrocycles is far too large to permit the π–π overlap essential for spin sharing, and the 60° angle between transition dipoles does not permit an interaction between the transition dipoles sufficiently strong to provide the necessary red shift. Exciton calculations [51] indicate that the structure of Figure 3 is compatible with an optical shift to 700 nm [46].

A particularly interesting feature of the model proposed by Shipman et al. [46] is the variety of nucleophiles that can be used to form it. The geometrical arrangement in Figure 3 is sufficiently open to allow even bulky nucleophilic groups to function as crosslinking agents. Nucleophilic groups such as —OH, —NH_2, or —SH are present in protein side-chains, and from studies with models we conclude that the hydroxyl group of serine, threonine, or tyrosine, the amino group of lysine, the imino group of arginine, or the sulfhydryl group of cysteine can all be used to form the Shipman et al. [46] or Boxer and Closs [50] chlorophyll special pair. Possible protein participation in Chl_{sp} organization is of considerable interest in the light of current activity in the isolation of chlorophyll–protein complexes [52] and photoreaction centers. The latter contains several polypeptides as essential components. The possibility then exists that a bend in a peptide chain could position two nucleophilic groups at such a distance as to make formation of a chlorophyll special pair possible. As no covalent bond formation between the protein and the chlorophyll is involved in special pair formation, extraction with organic solvents will easily detach the chlorophyll from the protein. If a conformational change occurs in the polypeptide from the solvent treatment, it is easy to see why reconstitution of the special pair could become a very improbable event, and this could account for the difficulty so far encountered in reconstituting bacterial reaction centers or photosynthetic membranes. The possible existence of a family of Chl_{sp} structures that range in nucleophile composition from two water molecules to combinations of one water molecule and one protein side-chain nucleophile, or combinations of two different protein nucleophiles thus presents itself. All of these would have essentially the same optical and spin-sharing properties. However, their redox properties might be significantly different, a situation that conceivably might be pertinent to the concept of multiple photoreactions in photosynthesis.

Even though a number of different models for the chlorophyll special pair have been advanced, there is no convincing evidence from *in situ* experiments that allows at this time a definitive choice between them. However, the model of Shipman et al. [46] and of Boxer and Closs [50] has shown itself more readily translated into reality than the other variants. We now describe the laboratory synthesis of such a special pair.

SYNTHESIS OF BIOMIMETIC CHLOROPHYLL SPECIAL PAIRS

Chlorophyll *a* adducts with ethanol have been prepared that successfully mimic the optical and EPR properties of the green plant photoreaction center [53]. In a concentrated chlorophyll *a* solution in toluene containing ethanol in the molar ratio 1:1.5 the chlorophyll is in the form of monomer Chl *a*·ethanol, a short-wavelength chlorophyll species absorbing maximally at 668 nm. There is only a slight tendency for two monomer Chl *a*·ethanol units to assemble themselves into the Shipman et al. [46] structure, primarily because of entropy considerations. Ordering and immobilizing two separate chlorophyll molecules into a special pair involves a loss in entropy sufficient to prevent spontaneous ordering into special pairs. Ordering can be facilitated by strong cooling, which has the effect of making the solvent more hostile and so increasing the tendency to special pair formation. Cooling to 175°K forms a chlorophyll species absorbing maximally in the red at 702 nm. Infrared studies show that in the low-temperature species the free (uncoordinated) keto C=O absorption peak which absorbs at 1697 cm^{-1} is shifted to 1686 cm^{-1}. This peak is assigned to a keto C=O---Mg coordination interaction; the other carbonyl functions in the chlorophyll are not affected by the temperature change. The low temperature form has the red-shift anticipated from the Shipman et al. [46] model, and its keto C=O functions are known to be involved in the formation of the low temperature form. The photooxidized low temperature species has a narrowed EPR signal indistinguishable from that of $P700^{+}$ in the alga *Chlorella vulgaris*, and thus can spin share.

While this system can be considered a successful biomimetic replication of P700, and the system certainly has its uses, nevertheless it is not particularly convenient to work with. Cryogenic conditions are required, the assembled special pair must be handled in an organic glass or in solution at very low temperature, and, most importantly, other chlorophyll species are likely present. Judging from the visible spectra, special pairs must be the predominant species at 175°K, but there are reasons to suppose that short linear stacks (chlorophyll linked by ethanol into stacks resembling those of the Strouse [32] ethyl chlorophyllide *a*·$2H_2O$ crystal structure) may also be present in these systems. While this is not serious for some purposes, it could be a problem in other, as for example, in fluorescence studies on energy-transfer studies, where a species that may be present at very low concentrations may make interpretation of an experiment very difficult. Other approaches to a synthetic chlorophyll special pair that can be formed and manipulated at ambient temperatures thus are required.

The severe entropy requirements for special pair formation from two monomer chlorophyll molecules can be overcome by joining two chlorophyll macrocycles by a covalent link. The link must be chosen so that the macrocycles are sufficiently close to allow intramolecular folding but at the same time the link should not be so long that it encourages intermolecular interactions. The link must be flexible, and new nucleophilic groups should not be introduced into the linked dimer to avoid the possibilities of new coordination interactions competitive with those leading to folding. A link consisting of an ethylene glycol unit, —O—CH_2CH_2—O—, formed by esterification at the propionic acid side chain in position 7 of ring IV of chlorophyll turns out to have the necessary linkage properties. Boxer and Closs [50] have used an identical strategy to link two pyrochlorophyll *a* (chlorophyll *a* in which the carbomethoxy group at position 10 of ring V has been replaced by H) macrocycles. This synthesis is rather less demanding than the linkage of the intact chlorophyll *a* macrocycles that still retain their temperature-sensitive carbomethoxy groups, because the Mg lost from the macrocycles during the synthetic procedure can be reintroduced in the pyrochlorophyll macrocycle by a high-temperature procedure [54]. A high temperature reaction cannot be used with chlorophyll *a* because the carbomethoxy group is lost in the reinsertion of the Mg.

Cross-linking can be accomplished by selective hydrolysis of the ester functions to remove the phytyl residue at the propionic side-chain. Re-esterification with a bifunctional alcohol such as ethylene glycol then serves as a link to joint the two chlorophyll macrocycles. Both chlorophyll *a* [55] and bacteriochlorophyll *a* [56] have been converted to linked dimers in this way. Details of the synthetic procedures can be found in these reports. For the reinsertion of Mg, the procedure introduced by Eschenmoser et al. [57] has proved satisfactory for chlorophyll *a* but for bacteriochlorophyll *a*, Wasielewski [58] has devised an improved procedure for the reinsertion of Mg.

The synthetic procedures used for preparing the linked dimers were devised to assure retention of the carbomethoxy group at position 10. The presence of the carbomethoxy group at position 10 imposes stringent steric requirements in any special pair configuration where two chlorophyll molecules are brought into close proximity in a face-to-face orientation. Although the linked pyrochlorophyll *a* dimer has many useful features because of the increased stability that it acquires by the elimination of the carbomethoxy groups and the consequent loss of the reactive β-keto ester grouping in ring V, it cannot reflect the steric complications arising from the presence of the carbomethoxy group in ordinary chlorophyll *a*. The synthesis of linked dimers of both chlorophyll *a* and bacteriochlorophyll *a* thus permits the investigation of a category of problems that cannot be studied directly in the pyrochlorophyll *a* dimer, useful though it may be.

The linked chlorophyll *a* dimer in its open configuration is shown in Figure 4. A particularly useful reagent for the esterification of the pheophorbides (Mg-free derivatives of chlorophyll with a free propionic acid side-chain at position 7) with ethylene glycol is benzotriazole *N*-methane sulfonate plus a base in tetrahydrofuran solvent. The structure of the product was established by proton magnetic resonance spectroscopy and by mass spectroscopy of the Mg-free dimer. Present in a solution of the linked chlorophyll *a* dimer is an equilibrium amount (~15%) of an optical isomer designated as *a*′ [59]. The equilibrium forming the optical isomer is rapidly

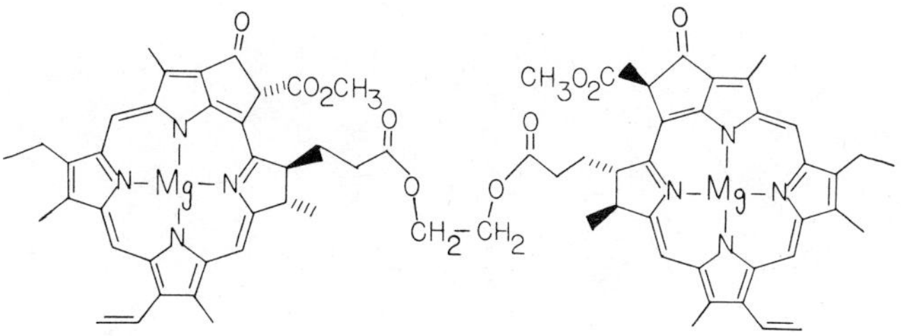

Fig. 4. The covalently linked chlorophyll *a* dimer (bis(chlorophyllide a) ethylene glycol diester) in its open configuration (in a polar medium).

established when any chlorophyll molecule with an intact β-keto ester system is dissolved in polar (basic) solvents. The stereoisomerization at position C-10 in the linked dimer results in a mixture consisting of 73% of the *a-a* dimer (the configuration at C-10 in chlorophyll in nature), 25% *a-a'* linked dimer, and about 2% of the linked *a'-a'* species. This is the origin of the line-broadening and the small satellite peaks that can be observed in the proton magnetic resonance spectrum of the open form of the chlorophyll *a* linked dimer.

The visible absorption spectrum of bis(chlorophyllide *a*) ethylene glycol diester (the systematic name for the covalently linked chlorophyll *a* dimer) in a polar solvent such as diethylether or pyridine is indistinguishable from that of monomer chlorophyll *a* itself under identical conditions [29]. In the open configuration, then, the two macrocycles in the linked dimer are independent and show no π-system interactions that would produce a shifted electronic transition spectrum.

Chlorophyll *a* dissolved in a nonpolar solvent undergoes self-aggregation to form dimers and higher oligomers via keto C=O---Mg interactions. These chlorophyll aggregates are moderately red-shifted. The linked chlorophyll *a* dimer undergoes a similar aggregation, probably involving an intramolecular keto C=O---Mg coordination interaction. This absorption spectrum of this species is red-shifted relative to its open configuration and absorbs maximally in the red at 677 nm; the regular dimer, (Chl $a)_2$, absorbs at 674 nm. In the open configuration, the linked dimer is self-aggregated by formation of an internal keto C=O---Mg bond.

Addition of water or ethanol to the internally self-aggregated, linked dimer in CCl_4 causes a marked spectral change. The band intensity at 677 nm is very much diminished, and a new strong absorption band at 697 nm appears. The 697 nm species is assigned to a folded linked dimer with the configuration shown in Figure 5. The residual absorption at 677 nm can be assigned to the *a-a'* dimer. The intensity of the 697 nm band relative to the 677 nm band is about 3:1. From studies with space-filling models it is easy to show that whereas both the *a-a* and *a'-a'* linked dimers can assume the structure of Figure 5, the *a-a'* dimer cannot fold because of the magnitude

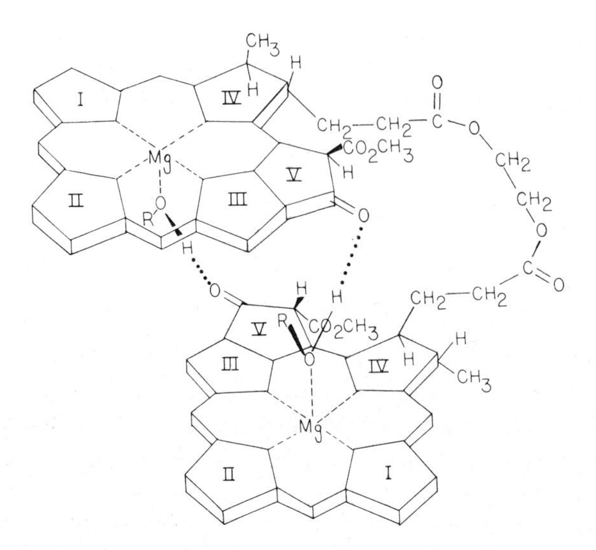

Fig. 5. Covalently linked chlorophyll dimer in its folded configuration. Chemical link is seen at the upper right. Except for the link, the structure is identical with that of Shipman et al. [46] shown in Figure 3.

of the steric compression of the carbomethoxy group demanded in the linked dimer in the folded configuration. Regardless of the direction in which folding occurs, a carbomethoxy is always positioned between the two macrocycles in the folded configuration. The relative intensities of the 677 and 697 nm bands are roughly consistent with the anticipated presence of nonfolded *a-a′* dimer in the system. The open *a-a′* dimer would of course be expected to exist as the internally coordinated dimer with an absorption maximum essentially the same as the corresponding *a-a* dimer. It is interesting to note that in the linked bis(pyrochlorophyllide *a*) ethylene glycol diester of Boxer and Closs [50] that complete conversion to a 697-nm-absorbing form occurs under similar solvent concentration, and nucleophile conditions. The structure shown in Figure 5 is essentially the same as the model suggested by Shipman et al. [46] shown in Figure 3 except for the link in the upper right quadrant of the figure. Experimental support for the assignment comes mainly from ^{1}HMR spectroscopy, which supports the view that the folded pyrocholorohyll *a*, chlorophyll *a*, and bacteriochlorophyll *a* linked dimers have the C_2 symmetry and the orientation of Figure 3. Both the linked pyrochlorophyll *a* and the chlorophyll *a* dimers, then, have the optical properties required for a biomimetic representation of P700.

The photochemical and spin-sharing properties of the linked chlorophyll *a* dimer in its folded configuration are also compatible with its proposed role as a biomimetic photoreaction center. In the presence of iodine and red light, complete bleaching of the 697 nm absorption occurs with 30 sec.

From these and other experiments by cyclic voltametry the redox potential of the folded linked chlorophyll *a* dimer is estimated to be close to 0.5 V. The oxidation potential of P700 from spinach chloroplasts is estimated by Kok [60] to be 0.43 V. The redox behavior of the folded chlorophyll dimer thus resembles that of *in vivo* P700 despite major differences in environment. Finally, irradiation of the folded chlorophyll *a* dimer in the presence of iodine yields an EPR signal with a linewidth of 7.54 G, which corresponds closely to the linewidth predicted for an unpaired spin delocalized over two chlorophyll molecules. The observed linewidth in this experiment is very similar to the EPR signal from $P700^{\dagger}$ in *Chlorella vulgaris*.

The properties of P700 and those of the folded chlorophyll *a* dimer are compared in Table I. In terms of optical, redox, and spin-sharing properties the synthetic biomimetic Chl_{sp} is a remarkably faithful imitation of its natural precursor.

The synthesis of a covalently linked bacteriochlorophyll *a* dimer (bis(bacteriochlorophyll *a*) ethylene glycol diester) has also been accomplished [56], and the identity of the product verified by magnetic resonance and mass spectroscopy. Because the equilibrium concentration of the unnatural diastereoisomer at position 10 is much lower in bacteriochlorophyll *a* than in chlorophyll *a*, interpretation of the proton magnetic resonance spectra is considerably facilitated. The ^{1}HMR spectra offer quite conclusive evidence that the dimer in CCl_4 or benzene solution in the presence of ethanol or water is folded into the configuration of Figure 3. The optical properties of the folded bacteriochlorophyll *a* dimer are, however, anomalous. The covalent dimer in dry benzene has its absorption maximum in the red at 780 nm, with a shoulder at 812 nm. This spectrum is very similar to that of self-aggregated $(Bchl\ a)_n$. When a suitable nucleophile is added to the benzene solution by saturating it with water, the 780 nm absorption maximum is replaced by a very broad band centered at 803 nm. This band has significant intensity almost to 900 nm, which allows for the possibility of the existence of several species. The shift in the visible absorption spectrum is accompanied by a sharpening of the ^{1}HMR lines, which can be interpreted as indicative of a single species in solution, or, alternatively, as the result of a mobile, rapid equilibrium among a number of species. The folded bacteriochlorophyll *a* dimer does not have the optical properties of P865, but the presence of an 865 nm absorbing species cannot be entirely excluded because of the unusually broad absorption peak.

A possible explanation for the failure of the folded dimer to absorb at 865 nm may be the absence of additional bacteriochlorophyll and bacteriopheophytin molecules in our laboratory system. It is known that bacterial photoreaction-center preparations contain at least four bacteriochlorophyll and two bacteriopheophytin molecules. It may be the interaction of the additional bacteriochlorophyll and bacteriopheophytin molecules with the $Bchl_{sp}$ that provides the red shift to 865 nm. From the ^{1}HMR evidence, neither of the acetyl functions in our folded dimer is used for coordination

TABLE I

Comparison of P700 and P865 with Synthetic Photoreaction Centers

Species	λ_{max} (nm)	ΔH_{pp} [a] (gauss)
In vivo P700	700	7.0[b]
Bis(chlorophyllide *a*)ethylene glycol diester		
open	662[c]	(9.3)[g]
folded	697[d]	7.5[h]
Bis(pyrochlorophyllide *a*)ethylene glycol diester[e]		
open	666	
folded	696	
Chlorophyll *a*·ethanol @ 175 K[f]	702	7.1[i]
In vivo P865	865	9.6
Bis(bacteriochlorophyllide *a*)ethylene glycol diester		
open	760[c]	(12.8)[g]
folded	803[d]	10.6[j]

[a] EPR peak-to-peak linewidth of the oxidized species.
[b] Oxidized with $FeCl_3$ or I_2 in methanol solution.
[c] In benzene solution containing $>0.1M$ ethanol.
[d] In benzene or CCl_4 containing water of $<0.1M$ ethanol.
[e] Data of Boxer and Closs [50].
[f] Formed in $0.1M$ Chl *a* in toluene containing $0.15M$ ethanol.
[g] Linewidth of monomer Chl $a^{\dagger}$ or Bchl $a^{\dagger}$.
[h] Photo-oxidized with I_2 in H_2O-saturated CCl_4.
[i] Photo-oxidized at 96°K with tetranitromethane (100mM) in a toluene solution containing 1mM Chl *a*, 1.5 mM methanol.
[j] Photo-oxidized with tetranitromethane in wet benzene.

interaction. The acetyl groups have good donor properties and could interact with additional bacteriochlorophyll molecules by either acetyl C=O---Mg interactions or by hydrogen bonding to form a species red-shifted to 865 nm. An unpaired spin is then delocalized over only two of the four bacteriochlorophyll molecules, but the red shift would then result from the interaction of all four.

In terms of photooxidation and spin sharing, however, the folded bacteriochlorophyll *a* dimer mimics P865 very well. Irradiation of a water-saturated benzene solution of bis(bacteriochlorophyllide *a*) ethylene glycol diester containing iodine with red light results in the complete bleaching of the 803 nm band and the appearance of a new absorption peak at 1150 nm; photobleached P865 has an absorption maximum at 1250 nm, [61] which is usually assigned to the π-cation radical $Bchl_{sp}^{+\cdot}$. Oxidation of the folded bacteriochlorophyll *a* linked dimer yields a free radical with an EPR linewidth of 10.6 ± 0.3 G as compared to the ~ 13 G linewidth observed for monomeric Bchl $a^{+\cdot}$. The folded dimer thus exhibits the property of spin-sharing and is able to delocalize an unpaired electron to the extent required by a valid model for P865. A comparison of the optical and EPR properties of the folded dimer with those of P865 is given in Table I.

In these covalently linked dimers we now have available synthetic systems that have the spin-sharing properties so characteristic of *in vivo* photoreaction-center chlorophyll, and in the chlorophyll *a*–linked dimer, a chlorophyll species that can in its folded configuration mimic the optical properties of green plant P700 as well.

BIOMIMETIC LIGHT-MEDIATED ELECTRON TRANSFER

The essential feature of light-energy conversion in photosynthetic organisms is the use of the energy of light quanta for electron transfer from the photoreaction center. The process of light-mediated electron transfer results in an electron that can be used as a reducing agent in chemical synthesis, and a positively charged entity (the photooxidized-reaction center) that constitutes oxidizing capacity that is likewise used for chemical reaction. Now that a variety of photo-active chlorophyll species have been prepared in the laboratory, we can proceed to an examination of the utility of these species in light-assisted electron transfer. Such an effort has been in progress for the past several years at Argonne National Laboratory under the title of the "Synthetic Leaf Project [1, 2]." We can now describe the extension of these studies to our biomimetic chlorophyll special pairs.

Our initial experiments on light-mediated electron transfer used the photoactive P740 chlorophyll–water adduct. Films of P740 were deposited on semipermeable or metallic supports and the electron transfer properties were then studied in a small two-compartment cell arranged in such a way that the film of the photoactive chlorophyll separated the two halves of the cell (Fig. 6). One compartment contained an aqueous solution of an electron donor, the other compartment an electron acceptor likewise in aqueous solution. Examples of suitable electron acceptors are dichlorophenylindophenol (DCPIP), potassium ferricyanide [$K_3Fe(CN)_6$], or anthraquinone sulfonic acid (AQS), and suitable electron donor compounds are ascorbic acid (Asc), cysteine (Cys), or tetramethylphenylenediamine (TMPD). Two platinum electrodes inserted into the donor and acceptor solution are used to monitor the electrical response of the cell.

In a typical experiment, a film of a photoactive chlorophyll species is deposited by evaporation on a thin platinum foil. The photoactive species may be the P740 chlorophyll–water adduct prepared by hydration of a

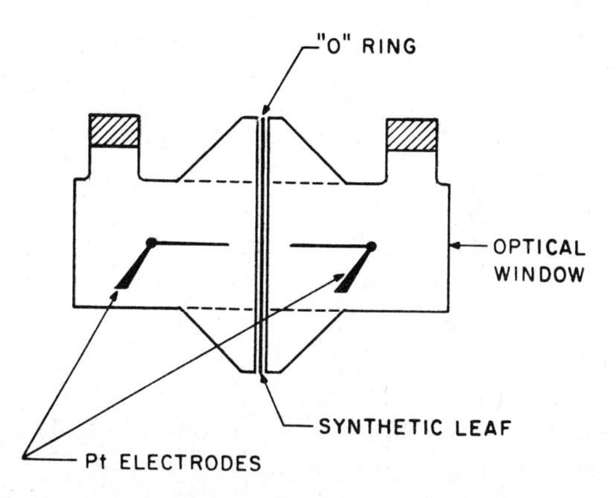

Fig. 6. Schematic synthetic leaf cell for studying light-mediated electron transfer. Two compartments are separated by a semipermeable membrane or a metal foil on which a film of photoactive chlorophyll is deposited. Two compartments contain buffered aqueous solutions of an electron donor and an electron acceptor. Window at the end of the cell is optically flat and is used for light irradiation. Two platinum electrodes are used for monitoring the electrical response of the cell.

chlorophyll oligomer in octane solution, or a solution of the folded linked dimer. The film of P740 or folded dimer is then clamped into the cell so that the photoactive chlorophyll is in direct contact with the electron donor solution. The cell is then inserted into a circuit containing an electrometer and x–y recorded to measure the electrical consequences of light irradiation by illumination through the optical flats at the ends of the cell (Fig. 7). A

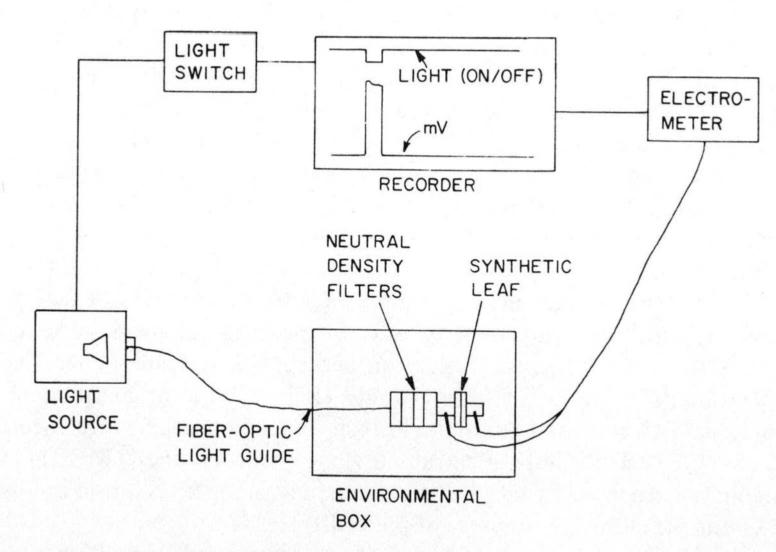

Fig. 7. Block diagram of circuit used for monitoring photoresponse of photoactive chlorophyll species.

typical photoresponse is shown in Figure 8. The presence of a photoactive chlorophyll species (i.e., photoactive by the EPR criterion) is essential. The dependence of the photosignal on the chlorophyll species present is shown in Figure 9, which shows the photoresponse of the P740 species (photoactive) as compared to that of P680 (not photoactive). It can be seen that in the absence of a photoactive chlorophyll species the response of the cell is minimal. The absence of either donor or acceptor, likewise greatly diminishes the photoresponse. The magnitude of the response is also a sensitive function of the metal substrate used (Table II). The largest photovoltages are observed when aluminum foils are used, but these have rather low reproducibility. There appears to be a correlation for noble metal substrates between the magnitude of the photoresponse and the work function of the metal. For a given pair of donor and acceptor, the photo-EMF produced is in the order Ag (3.67 eV) > Pt (4.09 eV) > Au(4.82 eV). The work functions in electron volts are shown in parentheses. The magnitude of the voltages developed in the cell is also strongly dependent on the donor and acceptor used, with the donor having by far the dominant influence on the size of the voltage produced by light irradiation.

The P740 system has only a limited lifetime of the order of several hours as a result of photodegradation of the chlorophyll. To probe some of the phenomenological questions raised by our synthetic leaf experiments and to

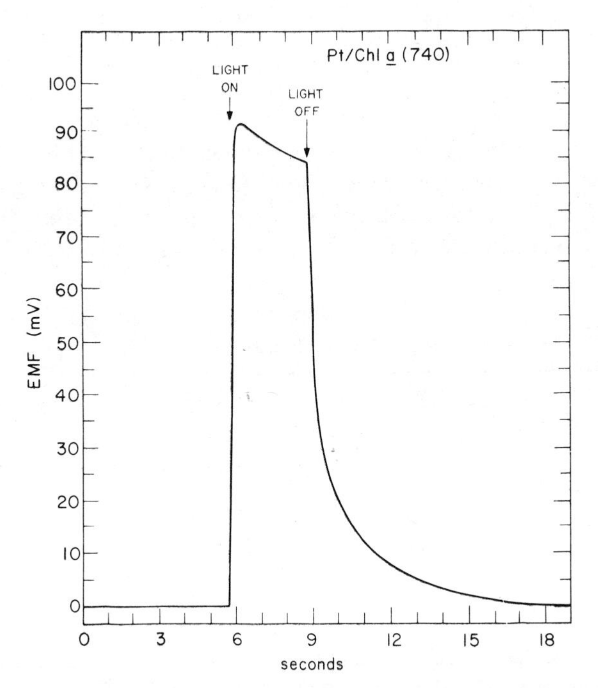

Fig. 8. Typical photoresponse elicited on irradiation of P740. The P740 film is supported on a platinum foil. Donor: sodium ascorbate; acceptor: anthraquinone-2-sulfonic acid, sodium salt.

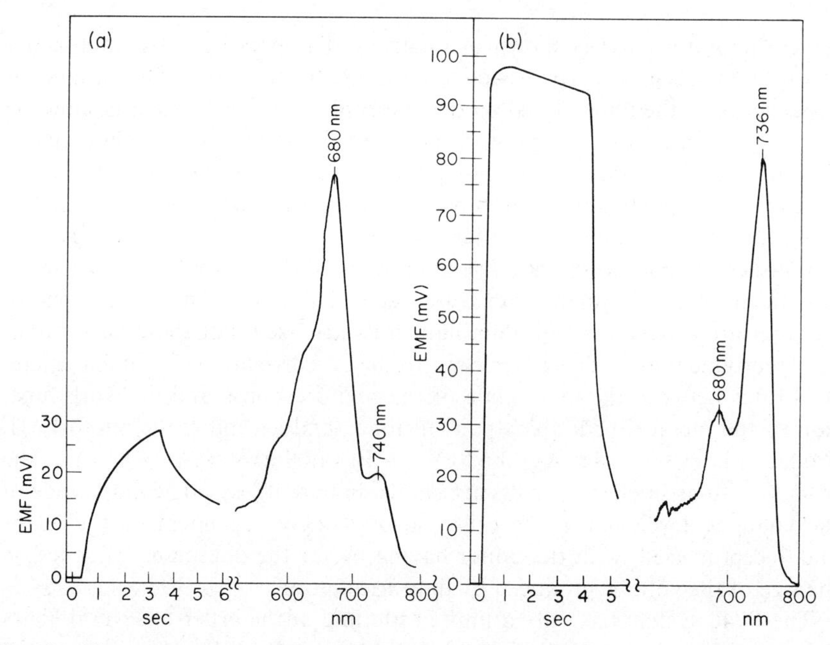

Fig. 9. Photoresponse of chlorophyll–chlorophyll (P680) and chlorophyll–water (P740) adduct. In each half of the figure is shown the photoresponse and the absorption spectrum of the chlorophyll species irradiated. (Chl *a*)$_n$ P680 species is not photoactive, whereas the P740 species shows a strong photoresponse. Significant photo-emf is observed only in the presence of a photoactive chlorophyll species, an electron donor, and an electron acceptor.

design a photogalvanic system with adequate stability over long periods of time, we have used the covalently linked dimers in their folded, photoactive configuration. An improved cell for these studies is shown in Figure 10. The linked dimer in its folded configuration is applied as a thin film on the platinum disc working electrode. This electrode is placed in a buffered solu-

TABLE II

Photoresponse of P740 as a Function of Metal Support

Metal	Donor[a]	Acceptor[b]	ΔEMF (mv)
Pt	Asc	$K_3Fe(CN)_6$	142
Pt	Asc+TMPD	$K_3Fe(CN)_6$	35
Al	Asc+TMPD	DCPIP	393
Au	Asc+TMPD	$K_3Fe(CN)_6$	0
Au	H_2O	H_2O	1

[a] Asc, ascorbate, TMPD, tetramethylphenylenediamine.
[b] DCPIP, 2,6-dichlorophenolindophenol.

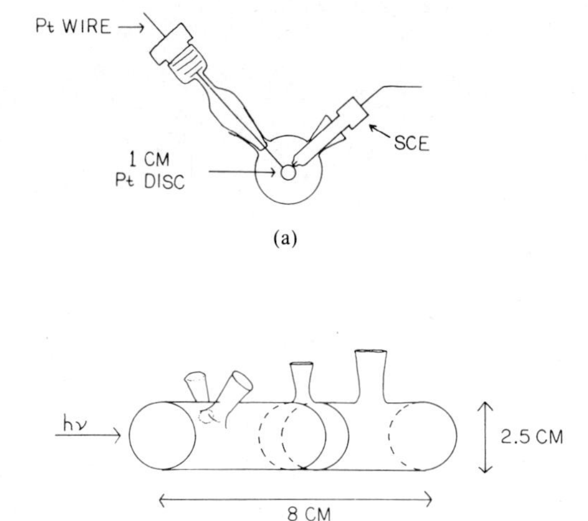

Fig. 10. Improved cell for light-mediated electron transport. Photoactive chlorophyll species is applied to the platinum disc, which is the working electrode. Cell compartments contain buffered aqueous media. Working electrode is potentiostated relative to a saturated calomel electrode. Third platinum wire serves as a counter electrode. (a) Endview; (b) side view.

tion, and its voltage controlled by a potentiostat at zero potential relative to a saturated calomel electrode. A third platinum wire serves as a counter electrode. The photocurrents produced on irradiation can be recorded both as a function of light intensity and of the potential applied to the cell. In this fashion the fundamental characteristics of current production can be measured.

The synthetic linked dimer is very much more stable to photodegradation than is P740. A properly prepared cell using the linked dimer as the photoactive species exhibits a substantial photogalvanic effect over a period of months. A typical photoresponse evoked from the folded dimer is shown in Figure 11. From this we can conclude that the ability of the biomimetic Chl_{sp} to act as an electron donor when excited by light is reasonably well established.

FUTURE DIRECTIONS

While we consider the successful development of the synthetic Chl_{sp} to be progress in the direction of a useful biomimetic technology for solar energy conversion, a great deal remains to be done before our goal can be reached. The chlorophyll special pair is only a portion of a naturally occurring photoreaction center. The electron transport conduits in and out of the Chl_{sp} make an indispensible contribution to the efficiency of light energy conversion, for it is the graduated redox properties and the precise relative orientations of the members of the transport chain that assure the very high degree

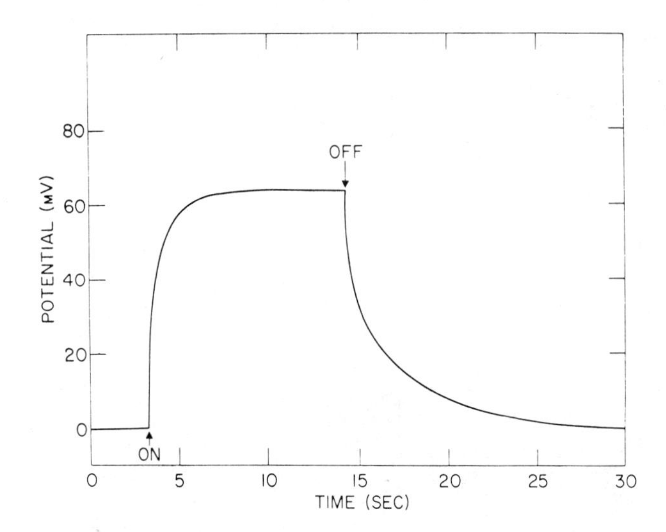

Fig. 11. Photo-emf produced by light irradiation of a synthetic Chl_{sp}. Asc, ascorbate, an electron donor; AQS, anthraquinone sulfonate, an electron acceptor. Lifetime of a cell with the artificial Chl_{sp} is greatly extended relative to P740. $h\nu \rightarrow$ ASC| Dimer.| AQS.

of directionality required for efficient operation. For a practical biomimetic solar energy conversion unit, it will be essential to provide for the highly directional electron flow that is needed. On the molecular level ordinary metallic electron conductors cannot be used, as metal of molecular dimensions is no longer a metallic conductor. An exercise in molecular electronics will therefore be necessary, and a biomimetic approach that seeks to simulate the electron transport chain in chloroplasts or mitochondria may well be indicated. Perhaps the most challenging of the problems to be solved is how to extract electrons (oxidize) from water. Although a fairly clear and detailed picture of the reducing side of photosynthesis has been developed, the oxidizing side of photosynthesis is still an enigma. Despite the voluminous literature, there is no concensus even about the general principles by which water is oxidized to oxygen, let alone agreement about details. Until the strategy used by green plants to oxidize water is understood and the photooxidation of water can be accomplished, no solar conversion process for chemical purposes is likely to be viable.

It is very likely that a biomimetic solar energy process in its early stages will be limited by our understanding of the prototype in nature. Consequently a fundamental prerequisite for success is a reasonably detailed understanding of plant and bacterial photosynthesis in all of its important aspects. It is probable that no radically new scientific principles will need to be elaborated to achieve the required level of understanding. What is needed is a comprehensive attack and a commitment that ensures continuity of effort. There is every reason to suppose that such a program of research and development will provide the basis necessary for a biomimetic solar energy conversion technology.

The experiments on light-mediated electron transfer with P740 were carried out by Dr. Thomas R. Janson. Work performed under the auspices of the Division of Basic Energy Sciences of the Department of Energy.

References

[1] J. J. Katz, T. R. Janson, and M. R. Wasielewski, *1977 Karcher Symposium on Energy and The Chemical Sciences*, S. D. Christian and J. J. Zuckerman (Eds.) (Pergamon, Oxford, England, 1978), pp. 31–57.

[2] J. J. Katz, in *Proceedings of the Fifth Energy Technology Conference*, R. F. Hill (Ed.) (Government Institutes, Inc., Washington, D.C., 1978), pp 499–510.

[3] J. R. Bolton, in *Proceedings of the Fourth International Congress on Photosynthesis* (Reading, England, 1977).

[4] N. K. Boardman, *Living Systems as Energy Converters* (North Holland, Amsterdam, 1977), pp. 307–318.

[5] J. T. Warden and J. R. Bolton, *J. Am. Chem. Soc. 95*, 6435 (1973).

[6] C. A. Wraight and R. K. Clayton, *Biochim. Biophys. Acta 333*, 246 (1973).

[7] E. Broda, *Int. J. Hydrogen Energy 3*, 119 (1978).

[8] R. Emerson and W. Arnold, *J. Gen. Physiol. 15*, 391 (1931–1932).

[9] R. Emerson and W. Arnold, *J. Gen. Physiol. 16*, 191 (1932–1933).

[10] E. Katz, in *Photosynthesis in Plants*, J. Franck and W. E. Loomis (Eds.) (The Iowa State College Press, Ames, Iowa, 1949).

[11] E. I. Rabinowitch, *Photosynthesis* (Interscience Publishers, New York, 1956), Vol. 2, Pt. 2.

[12] C. S. French, *Proc. Natl. Acad. Sci. USA 68*, 2893 (1971).

[13] D. L. Dexter and R. S. Knox, *Excitons* (Interscience, New York, 1965).

[14] Th. Förster, in *Modern Quantum Chemistry*, O. Sinanoglu (Ed.) (Academic Press, New York, 1965), p. 93.

[15] J. P. Thornber, *Ann. Rev. Plant. Physiol. 26*, 127 (1975).

[16] T. M. Cotton, A. D. Trifunac, K. Ballschmiter, and J. J. Katz, *Biochim. Biophys. Acta 368*, 181 (1974).

[17] A. A. Krasnovskii, in *Progress in Photosynthesis Research*, H. Metzner (Ed.), 1969, p. 709.

[18] K. Ballschmiter, K. Truesdell, and J. J. Katz, *Biochim. Biophys. Acta 184*, 604 (1969).

[19] B. Kok, *Nature 179*, 583 (1957).

[20] L. N. M. Duysens, W. J. Huiskamp, J. J. Vos, and J. M. van der Hart, *Biochim. Biophys. Acta 19*, 188 (1956).

[21] B. Commoner, J. J. Heise, and J. Townsend, *Proc. Natl. Acad. Sci. USA 42*, 710 (1956).

[22] B. Commoner, J. J. Heise, B. B. Lippincott, R. E. Norberg, J. V. Passoneau, and J. Townsend, *Science 126*, 57 (1957).

[23] D. C. Borg, J. Fajer, R. H. Felton, and D. Dolphin, *Proc. Natl. Acad. Sci. USA 67*, 813 (1970).

[24] J. M. Anderson, *Biochim. Biophys. Acta 416*, 191 (1975).

[25] J. J. Katz, W. Oettmeier, and J. R. Norris, *Phil. Trans. Royal Soc. London B 273*, 227 (1976).

[26] J. J. Katz, R. C. Dougherty, and L. J. Boucher, in *The Chlorophylls*, L. P. Vernon and G. R. Seely (Eds.) (Academic Press, New York, 1966), p. 185.

[27] H. Scheer and J. J. Katz, in *Porphyrins and Metalloporphyrins*, K. Smith (Ed.) (Elsevier, Amsterdam, 1975), p. 399.

[28] J. T. Warden and J. R. Bolton, *Acct. Chem. Res. 7*, 189 (1974).

[29] L. L. Shipman, T. M. Cotton, J. R. Norris, and J. J. Katz, *J. Am. Chem. Soc. 98*, 8222 (1976).

[30] R. Livingston, *Quart. Rev. (London) 14*, 174 (1960).

[31] J. J. Katz, in *Inorganic Biochemistry*, G. L. Eichhorn (Ed.) (Elsevier, Amsterdam, 1973), Vol. 2, p. 1022.

[32] C. E. Strouse, *Proc. Natl. Acad. Sci. USA 71*, 325 (1973).
[33] C. E. Strouse, *Proc. Inorg. Chem. 22*, 159 (1976).
[34] H.-C. Chow, R. Serlin, and C. E. Strouse, *J. Am. Chem. Soc. 97*, 7230 (1975).
[35] K. Ballschmiter and J. J. Katz, *Biochim. Biophys. Acta 256*, 307 (1972).
[36] K. Ballschmiter, T. M. Cotton, H. H. Strain, and J. J. Katz, *Biochim. Biophys. Acta 180*, 347 (1969).
[37] K. Ballschmiter and J. J. Katz, *Nature 220*, 1231 (1968).
[38] K. Ballschmiter and J. J. Katz, *J. Am. Chem. Soc. 91*, 2661 (1969).
[39] J. J. Katz and K. Ballschmiter, *Angew. Chem. 80*, 283 (1963); *Angew. Chem. Int. Ed. Engl. 7*, 286 (1968).
[40] J. J. Katz, K. Ballschmiter, M. Garcia-Morin, H. H. Strain, and R. A. Uphaus, *Proc. Natl. Acad. Sci. USA 60*, 100 (1968).
[41] J. R. Norris, R. A. Uphaus, H. L. Crespi, and J. J. Katz, *Proc. Natl. Acad. Sci. USA 68*, 625 (1971).
[42] J. R. Norris, H. Scheer, M. E. Druyan, and J. J. Katz, *Proc. Natl. Acad. Sci. USA 71*, 4897 (1974).
[43] H. Levanon and J. R. Norris, *Chem. Rev. 78*, 185 (1978).
[44] M. Garcia-Morin, R. A. Uphaus, J. R. Norris, and J. J. Katz, *J. Phys. Chem. 73*, 1066 (1969).
[45] J. J. Katz and J. R. Norris, *Current Topics Bioenergetics 5*, 41 (1973).
[46] L. L. Shipman, T. M. Cotton, J. R. Norris, and J. J. Katz, *Proc. Natl. Acad. Sci. USA 73*, 1791 (1976).
[47] J. J. Katz, J. R. Norris, L. L. Shipman, M. C. Thurnauer, and M. R. Wasielewski, *Ann. Rev. Biophys. Bioeng.* (to be published).
[48] J. J. Katz, L. L. Shipman, and J. R. Norris, *Ciba Foundation Symposium on Chlorophyll Organization and Energy Transfer in Photosynthesis* (North Holland Publishing Company, Amsterdam, in press).
[49] F. K. Fong, *Proc. Natl. Acad. Sci. USA 71*, 3692 (1974).
[50] S. G. Boxer and G. L. Closs, *J. Am. Chem. Soc. 98*, 5406 (1976).
[51] L. L. Shipman, J. R. Norris, and J. J. Katz, *J. Phys. Chem. 80*, 877 (1976).
[52] J. P. Thornber and R. S. Alberte, in *The Enzymes of Biological Membranes*, A. Martonosi (Ed.) (Plenum Press, New York, 1976), Vol. 3, p. 163.
[53] T. M. Cotton, P. A. Loach, J. J. Katz, and K. Ballschmiter, *Photochem. Photobiol.* (to be published).
[54] S. J. Baum, B. F. Burnham, and R. A. Plane, *Proc. Natl. Acad. Sci. USA 52*, 1439 (1964).
[55] M. R. Wasielewski, M. H. Studier, and J. J. Katz, *Proc. Natl. Acad. Sci. USA 73*, 4282 (1976).
[56] M. R. Wasielewski, U. H. Smith, B. T. Cope, and J. J. Katz, *J. Am. Chem. Soc. 99*, 4172 (1977).
[57] H.-P. Isenring, E. Zass, K. Smith, H. Falk, J.-L. Luisier, and H. Eschenmoser, *Helv. Chim. Acta 58*, 2357 (1975).
[58] M. R. Wasielewski, *Tetrahedron Lett. 16*, 1373 (1977).
[59] J. J. Katz, G. D. Norman, W. A. Svec, and H. H. Strain, *J. Am. Chem. Soc. 90*, 6841 (1968).
[60] B. Kok, *Biochim. Biophys. Acta 48*, 527 (1961).
[61] P. L. Dutton, K. J. Kaufmann, B. Chance, and P. M. Rentzepis, *FEBS Lett. 60*, 275 (1975).

Cell-Free Agriculture: The Concept and Its Initial Implementation*

CONSTANTIN A. REBEIZ and MAARIB B. BAZZAZ
Department of Horticulture, University of Illinois, Urbana-Champaign, Illinois 61801

I. INTRODUCTION

Present projections suggest that by the end of this century the world population would amount to 6 to 7 billion and by the year 2100 to 12 to 16 billion people [1]. Considering the rate of land degradation in the United States, the arable land per person available throughout the world, by the year 2100, will drop from the present 0.9 acre per capita to 0.15 to 0.2 acre per capita [1]. Since there are already insufficient land and energy resources to feed adequately the present world population of 4 billion, the world community is bound to experience a very dramatic food shortage in the foreseeable future. For example, a world population of 6 to 7 billion is predicted in less than 25 years; to feed this number, food production must be doubled, which would involve a three-fold increase in energy consumption by agriculture [1]. In view of the depletion of the world reserves of fossil fuels, such an additional expenditure of energy, even if feasible, is likely to drive the price of food beyond reach of a considerable fraction of the world population. More importantly it is not certain that it will be possible to double the world food production on the available arable land. The built-in limitations of whole plants, the limited availability of arable land, the environmental restraints placed on modern agricultural practices, and weather uncertainties seem to argue against success.

In the future the imminent food shortage may be accompanied by a "materials" shortage [1]. For example, in addition to chemicals of various kinds, almost all of the synthetic fibers now produced are derived from hydrocarbons [2]. Indeed the present nonenergy usage of fossilized photosynthetic products (i.e., petroleum, natural gas, and coal) amounts to 6–7% of the total fossil fuel usage and is bound to keep on rising [2]. Therefore, altogether and independently of an energy crisis, there appears to be an equally important food and materials (reduced carbon) crisis in the foreseeable future.

It is with such considerations in mind that in 1974 we proposed the con-

* This work was supported by funds from the Illinois Agricultural Experiment Station.

Biotechnology and Bioengineering Symp. No. 8, 453–471 (1978)

0572-6565/78/0008-0453$01.00

cept of cell-free agriculture as a means of meeting the future food and energy demands of the expanding world population [3].

II. CONCEPT OF CELL-FREE AGRICULTURE

Cell-free agriculture was visualized as "the massive production of food by man-made photosynthetic membranes in the absence of cellular entities and without nuclear cytoplasmic control and interference. These membranes which are the heart of the system are envisioned as being highly specialized structures tailored to meet three well-defined and limited goals: self-maintenance, self-perpetuation, and high photosynthetic rates. The highly specialized molecular architecture and limited tasks assigned to these membrane populations are expected to result in very high food-making capacities." The reduced carbon-making capacity of those membranes was visualized to reside in turn in the conversion of solar energy, carbon dioxide, and water into simple or complex carbohydrates via the process of photosynthesis [3]. The carbohydrate(s) could then be processed for direct consumption by humans, for use in feed concentrates, or as a renewable resource for petrochemicals as outlined by Calvin [2].

It now appears that the anticipated high "reduced carbon"–making capacity of such a system may accrue from several considerations. First, cell-free agriculture may be practiced on nonarable land such as the Sahara, South African, and Southwestern United States deserts. These regions are of high year-round solar intensity of about 250 W/m^2 and are ideally suited for the quantum conversion process [2]. At present only one part in 10,000 of the solar energy which strikes the earth is used in the process of photosynthesis, and agricultural photosynthesis (i.e., useful photosynthesis) is only a very small percentage (about 1 part in one thousand) of the average daily photosynthesis [4]. Second, because the energy output of cell-free agriculture would not be siphoned away into making leaves, stems, and roots, this process is expected to be more efficient than that of conventional plants [3]. For example, even in a highly efficient crop plant such as corn, the food energy in the harvested crop is generally less than half of the calories of the sunlight captured by the entire plant [5]. The rest of the captured calories are not readily utilizable and are returned to the soil for a slow and wasteful oxidation. Third, all plants operate much below the theoretical maximal energy conversion efficiency of the photosynthetic electron transport system (PETS) which is in the range of 10 to 12% of the total incident solar energy impinging on the photosynthetic apparatus [6]. Indeed, the average net photosynthetic productivity in conventional agriculture is at best in the range of 0.1 to 0.4% [6]. This is due to extrinsic factors (availability of H_2O, CO_2, and inorganic nutrients; ambient temperature; weather uncertainties, etc.) and intrinsic rate-limiting dark step(s) of the PETS [6]. Both the extrinsic and intrinsic factors could be more easily manipulated in a controlled cell-free system than in an intact plant. For example, it may be possible, in the distant future, to populate the reactor with (bio)engineered photosynthetic membranes having a smaller photosynthetic unit size (i.e., with more photoreactive centers per unit surface area) and an increased turnover rate of the PETS. Photosynthetic unit size and PETS turnover rates are presently considered to be the major reason for the low intrinsic efficiency of whole plants as compared to the maximum theoretical efficiency of the PETS [6].

In an effort to implement the concept of cell-free agriculture, an intensive research effort was undertaken. It consisted of (1) the assembly of a short-lived experimental photosynthetic reactor populated with "natural" chloroplasts, (2) the continued in-depth investigation of the chlorophyll biosynthetic pathway, and (3) the continued investigation of the assembly of photosynthetic membranes *in vitro*. The last two objectives were aimed at gaining enough understanding of the greening process to eventually replace the natural chloroplasts in the reactor, by man-made photosynthetic membranes. In this paper we wish to describe the assembly of the experimental photosynthetic reactor and some of the techniques that were developed for monitoring the degradation, maintenance, and repair potential of the cultured chloroplasts.

III. ASSEMBLY OF THE EXPERIMENTAL PHOTOSYNTHETIC REACTOR

The photosynthetic reactor was assembled from various modules marketed by Bio-Rad Laboratories, 220 Maple Avenue, Rockville Centre, New York, and was populated with mature *Cucumis sativus* chloroplasts. The physical layout of the various components is depicted in Figure 1. The biofiber module consisted either of a biofiber-50 minibeaker or a biofiber-50 beaker. Biofiber-50 are perforated hollow-bore cellulose fibers with a nominal molecular weight cutoff of about 5000 (Fig. 1(a)). The minibeaker and beaker had a fiber bundle volume (total volume of the bore of the fibers) of 0.4 and 4 ml and a jacket volume (space outside the fibers) of 8 and 100 ml, respectively. In the presence of the medium in which the chloroplasts were suspended (suspension medium) the biofibers swell and come into intimate contact with the chloroplast suspension. After introduction of the chloroplast suspension into the jacket volume, a carrier was pumped through the fibers with the help of the peristaltic pump module (Fig. 1(c)). During its passage through the fibers, solute exchange took place between the carrier and the chloroplast suspension through the huge surface area of the fibers. To enhance this exchange, the chloroplast suspension was stirred very gently (Fig. 1(b)). In all the preliminary work reported here, the carrier had the same composition as the suspension medium and consisted of 0.5 *M* sucrose, 0.2 *M* Tris-HCl, pH 7.7. After passing through the fibers, the carrier was collected at the outlet at atmospheric pressure or under a slight vacuum (Fig. 1(e)). The rate of exchange between the chloroplast suspension

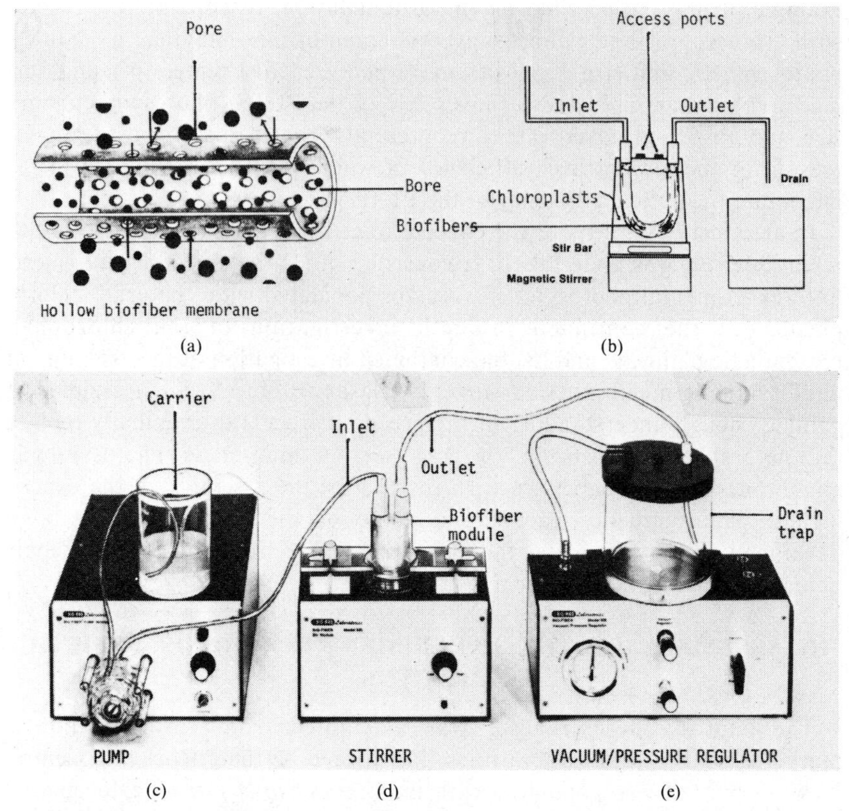

Fig. 1. First row: schematic view of a hollow biofiber (a); of a biofiber-50 beaker (b). Second row: physical layout of the assembled photosynthetic reactor, pump module (c); biofiber-50 beaker with stirrer (d); vacuum/pressure regulator (e).

and the carrier was controlled by their contact time, and the latter was controlled by the rate of pumping of the carrier. The rate of pumping was continuously adjustable from a few ml to several liter/hr. During operation of the reactor, it was possible to sample the chloroplasts via one of the access ports. In this respect the beaker set-up was easier to manipulate than the minibeaker and allowed the sampling of the chloroplasts without interruption of the pumping operation. As the solutes diffused through the fiber walls and were carried to the drain, the chloroplasts, which were much larger than the pores of the fibers, were excluded from the bore of the latter and from the carrier stream. However, after a few hours of operation during which the chloroplasts degraded, solubilized chlorophyll (Chl[s]) started to appear in the efflux stream.

The ultimate aim of assembling the photosynthetic reactor was to achieve a substantial collection of photosynthates for prolonged periods of time without disturbing the photosynthesizing chloroplasts. It is already well established that active chloroplasts which are incubated with ^{14}C–$NaHCO_3$

excrete mainly glycerate 3-P, and dihydroxy acetone-P as well as smaller amounts of fructose 1, 6-dip, sedoheptulose 1, 7-dip, and glycollate into the incubation medium [7]. The major problem encountered so far in the operation of the photosynthetic reactor resided in the instability of the isolated chloroplasts. The latter degraded very rapidly, and their photosynthetic efficiency decayed dramatically within an hour. Because the prolonged maintenance of isolated chloroplasts is a problem shared by both cell-free agriculture and the generation of H_2 gas from water (biophotolysis) by cell-free systems [6], a considerable effort was devoted to prolonging the life-span of the cultured chloroplasts. In turn, this process necessitated the description of analytical techniques for the early monitoring of the degradation of the cultured plastids and for the evaluation of their degree of maintenance. Since the prolonged maintenance of isolated chloroplasts is bound to be linked to repair phenomena taking place in the plastids, the repair potential of the cultured chloroplasts was also investigated. These techniques and phenomena will be described in the ensuing sections.

IV. BIOPHYSICAL AND BIOCHEMICAL DETECTION OF STRUCTURAL MAINTENANCE AND DEGRADATION IN MATURE UNFORTIFIED CHLOROPLASTS

In the past decade, several laboratories have studied the aging of chloroplasts *in vitro* [8–14], and more recently a great deal of interest was expressed in the survival of isolated chloroplasts *in vitro* [3, 15–18]. In all these studies the decay of various photosynthetic reactions and the disappearance of Chl were the main markers that were utilized in monitoring the maintenance or degradation of the isolated chloroplasts or of subplastidic particles.

Although the use of photosynthetic parameters to monitor the plastid maintenance and/or degradation conveys a good picture of the functional state of the isolated organelles, they do not constitute a direct probe of the structural alterations that do take place *in vitro* and which may precede the irreversible decay of photosynthetic efficiency. On the other hand, the disappearance of Chl, a spectrophotometric marker that has been widely used for the same diagnostic purposes as the photosynthetic markers, is thoroughly inadequate for the early detection of chlorophyll degradation. For example, although the use of absorption spectrophotometry to monitor the amount of chlorophyll *a* and *b*, which is extractable in organic solvents, is satisfactory when the ratio of Chl *a* to Chl *b* is less than 6.0 [19], this technique is not suitable for detecting and determining small amounts of Chl degradation products in the presence of large amounts of Chl. Indeed it now appears that the formation of small amounts of chlorophyllides (Chl that has lost its phytol) and pheophorbides (chlorophyllides that have lost their Mg) signal the beginning of chloroplast degradation before the disappearance of Chl becomes evident.

It thus appears that the successful retardation of chloroplast degradation would involve, among other things, the early detection and prevention of irreversible degradative processes. This in turn underlies the need for routine, sensitive, and specific techniques for the monitoring of these early degradative events. Some of those techniques currently in use in our laboratory are described below.

A. Biophysical Detection of Degradative Events in Isolated Chloroplasts

In chloroplasts the state of the Chl constitutes undoubtedly a marker of chloroplast disorganization even more sensitive than electron microscopic visualization. Chlorophyll is indeed a major structural and functional component of the chloroplast; it constitutes 5–8% of the dry weight of the chloroplasts [20] and possesses remarkable spectroscopic properties that are readily detectable by a variety of spectroscopic techniques [21]. Although the molecular organization of Chl *in situ* is not exactly known, an overwhelming body of evidence indicates that *in situ*, its arrangement is intimately interwoven with the molecular architecture of the other components of the photosynthetic membranes (i.e., thylakoids) [22]. Perhaps the best evidence of the organized state of chlorophyll in the thylakoids is expressed by the four to six Chl *a* forms and the two Chl *b* forms that are detectable in photosynthetic membranes [22–26], by their differential orientation and association in the thylakoids [27–37], and by their functional heterogeneity. The latter is best evidenced by the unequal distribution of the various chlorophyll forms among photosystem I (PSI) and photosystem II (PSII) of the thylakoids [23, 31]. Because the various Chl forms yield either Chl *a* or Chl *b* upon extraction in organic solvents, their multiple spectroscopic forms *in situ* are attributed either to different intermolecular associations of the Chl chromophores with one another or to different associations of the Chl with the lipoprotein components of the thylakoids [32]. This heterogeneous state of the Chl *in vivo* and the remarkable intrinsic spectroscopic properties of the Chl chromophore should therefore result in detectable changes in the *in situ* spectroscopic properties of the various Chl forms whenever changes in the molecular organization of the thylakoids do occur [33–36].

In conclusion, the state of Chl *in situ* should constitute a sensitive spectroscopic probe of chloroplast disorganization or maintenance. The former should be evidenced by changes in the *in situ* spectroscopic properties of the Chl while maintenance should be reflected by the preservation of those same properties.

1. Changes in the in vivo Absorption Spectra of Isolated Chloroplasts

The *in vivo* absorption spectra of higher plant chloroplasts recorded at room temperature are characterized by a broad red absorption band between 630 and 720 nm [23]. This red absorption band is presently interpreted in terms of six Chl *a* species (Chl *a* 662, 670, 677, 683, 691, and 704, where Chl *a* 662 refers to the Chl *a* species with an absorption maximum at 662 nm, etc.) and two Chl *b* forms (Chl *b* 640 and 650) [23]. In the past, several investigators have used *in vivo* absorption spectra of isolated chlorplasts to monitor the effect of various treatments on the state of the Chl *in situ* [33, 37, 38]. In our laboratory we have observed that readily detectable alterations of the red absorption band of the cultured chloroplasts constituted a rapid and sensitive marker of the onset of chloroplast degradation. These alterations of the red absorption band were very readily observed in the difference absorption spectrum (the absorption spectrum of aging chloroplasts from which the absorption spectrum of the freshly isolated chloroplasts was subtracted) of the isolated chloroplasts. The onset of chloroplast degradation was manifested by a preferential decrease in the absorption of the long-wavelength Chl species and by an enhancement in the absorption of the short-wavelength Chl species. For example, in the 1 hr minus 0 hr difference spectrum of the cultured chloroplasts (Fig. 2), the loss of long-wavelength Chl absorption after 1 h of incubation is manifested by the formation of a negative absorption band between 730 and 682 nm with negative absorption maxima at 724, 707, and 687 nm, and by a positive band between 682 and 650 nm with positive absorption maxima at 671 and 660 nm and a shoulder at 679 nm (Fig. 2). The molecular phenomena underlying these absorption changes are still ill understood and will have to await the thorough understanding of the molecular architecture of the thylakoid membranes.

2. *Changes in the Fluorescence Emission of Isolated Chloroplasts at 77 K*

At the temperature of liquid N_2, mature chloroplasts exhibit a three-banded fluorescence spectrum with emission maxima at 683–686 nm (i.e.,

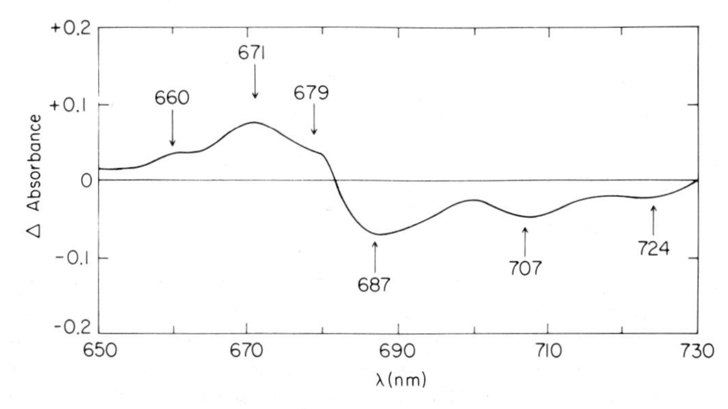

Fig. 2. Tracing of the difference absorption spectrum (1–0 hr) of unfortified chloroplasts incubated at 28°C in 0.5*M* sucrose 0.2*M* Tris-HCl, pH 7.7. The spectrum was recorded at room temperature with an Aminco dual beam spectrophotometer model DW-2, operated in the split beam mode. Arrows point to wavelengths of interest.

F686), 693–696 nm (F696), and at 735–740 nm (F740) (Fig. 3(a)) [25]. The emission at F686 nm is attributed to chlorophyll *a*, which absorbs at about 670 nm (Chl *a* 670) the emission at F696 nm to Chl *a* 677 and the emission at F740 nm to Chl *a* 704 [26]. Alternatively, it is suggested that the fluorescence emitted at F696 nm originates mainly from the photosystem II (PSII) antenna Chl *a*, that emitted at F740 nm originates primarily from the PSI antenna Chl *a*, and that emitted at F686 nm arises from the Chl *a* of the light-harvesting chlorophyll-protein complex [39].

Changes in the fluorescence emission ratio at 740 nm to that at 696 nm

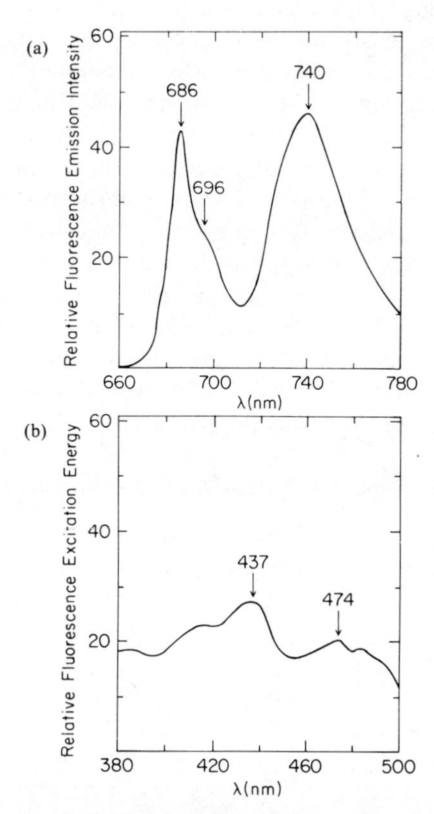

Fig. 3. Tracings of the fluorescence emission and excitation spectra of freshly prepared mature chloroplasts at 77 K. Unfortified chloroplasts were isolated in 0.5*M* sucrose, 0.2*M* Tris-HCl, pH 7.7. Unfortified chloroplasts are plastids to which none of the exogenous cofactors which were previously shown to be required for Chl biosynthesis *in vitro*, were added [61, 62, 64, 65]. Spectra were recorded on a Perkin–Elmer spectrofluorometer model MpF-3, equipped with a corrected spectra accessory [20, 72] in 67% glycerol [73]. Emission spectrum was recorded at an excitation slit width of 6 nm and an emission slit width of 3 nm. Excitation spectrum was recorded at an excitation slit width of 3 nm and an emission slit width of 6 nm. (a) Fluorescence emission spectrum elicited by a 440 nm excitation. (b) Soret excitation spectrum recorded with the emission monochromator positioned at 686 nm (i.e., at F686). Scale: 0.33 $\times$ ordinate. Arrows point to wavelengths of interest.

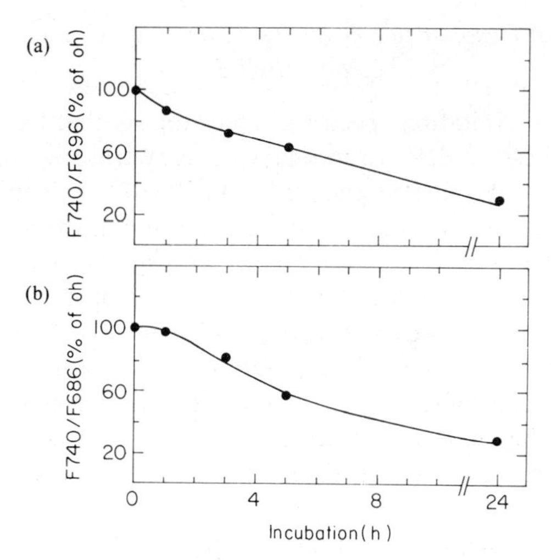

Fig. 4. Changes in the fluorescence emission ratios [F740/F696 (a) and F740/F686 (b)] at 77 K after various periods of incubation of the unfortified chloroplasts. Latter were incubated as described in Fig. 2. Fluorescence emission amplitudes at F740, 696, and 686 nm were obtained from the fluorescence emission spectra of the chloroplasts which were recorded as described in Fig. 3(a).

(i.e., F740/F696, where F740 represents the fluorescence emission amplitude at 740 nm, etc.), as well as changes in F740/F686, have been used by several researchers to probe a variety of chloroplastic events such as (a) changes in the distribution of excitation energy between PSI and PSII (40–42) and (b) preferential changes in the cross section of the absorbing pigments of the two photosystems[43–45].

We have observed that in the presence of Mg^{2+}, the onset of chloroplast degradation is readily manifested by a steady decrease in the F740/F696 and F740/F686 fluorescence emission ratios at 77 K (Fig. 4). Because the values reported in this figure represented the ratios of two values taken from the same spectrum, slight variations in the absolute emission magnitude from one preparation to another did not affect the reported ratios. The addition of Mg^{2+} to the isolated chloroplasts was found to be necessary to avoid the enhancement of fluorescence emission from PSI due to leakage of Mg^{2+} from the incubated chloroplasts. It is presently acknowledged that the loss of divalent cations from the chloroplast enhances the excitation energy distribution to PSI [40]. This in turn tends to obscure the detection of the early decrease in the F740/F696 and F740/F686 ratios. The drop in these fluorescence emission ratios in the presence of Mg^{2+} in degrading chloroplasts can be interpreted in terms of the decrease of the absorption cross section of PSI and the concomitant increase in that of PSII as was reported in Figure 2. This in turn results in a diminished energy transfer from PSII to PSI which manifests itself by a drop in the F740/F696 and F740/F686 fluorescence emission ratios.

3. Changes in Resonance Energy Transfer from Chlorophyll b to Chlorophyll a

The fluorescence excitation spectra of chloroplasts that have been cooled down to 77 K indicate a definite resonance energy transfer from Chl *b* and carotenoids to the Chl *a* of PSI and PSII [36, 46–49]. This in turn provides invaluable information about the spatial relationship of these components in the thylakoid membranes. Indeed resonance energy transfer can only occur among chromophores that exhibit the right absorption and emission properties and that are positioned within definite distances from one another. Energy transfer from Chl *b* to Chl *a* occurs with nearly 100% efficiency in mature chloroplasts [50] that have maintained their molecular organization. As the chloroplasts degrade and as their molecular architecture undergoes disorganization, it stands to reason to expect this resonance energy transfer to become less efficient.

The relative resonance energy transfer from Chl *b* to the Chl *a* of PSII was determined from corrected soret excitation spectra recorded at F686 nm, at 77 K (i.e., at the emission maximum of the Chl *a* of the light-harvesting Chl-protein complex). A typical, corrected soret excitation spectrum exhibiting energy transfer from Chl *b* to Chl *a* is depicted in Figure 3(b). The energy transfer from Chl *b* to Chl *a* was manifested by the obvious absorption of light by Chl *b* at 474 nm, which was detected when the emission monochromator was positioned at the emission maximum of Chl *a* at 686 nm (Fig. 3). Were it not for the transfer of the absorbed quanta of light from Chl *b* to Chl *a*, no light absorption at 474 nm would have been observed. To eliminate apparent changes in the absolute energy transfer amplitudes, caused by slight variations from one preparation to another, the energy transfer amplitudes from Chl *b* to Chl *a* at 474 nm (E474) were expressed as a ratio (E474/E437) of the energy transfer amplitude at 474 nm to the soret excitation of the Chl *a* at 437 nm of the same sample. In Figure 5, the values of the relative resonance energy transfer from Chl *b* to the Chl *a* of PSII at 77 K (i.e., E474/E437) are presented for various periods of incubation of the isolated chloroplasts. The relative energy transfer dropped from 100% before incubation to about 50% after 5 hr of incubation (Fig. 5). These results were corroborated by more involved determinations at room temperature of the relative quantum yield of fluorescence of Chl *a* at 685 nm (F685) sensitized by Chl *b* absorption. This relative quantum yield dropped from 100% before incubation to 95 and 72% after 2 and 5 hr of incubation respectively (M. B. Bazzaz and C. A. Rebeiz, unpublished).

The decrease in the efficiency of energy transfer from Chl *b* to Chl *a* as a function of incubation time was further confirmed by the gradual appearance of Chl *b* fluorescence at 665 nm. Indeed Chl *b* is nonfluorescent *in vivo* as it transfers its excitation energy with 100% efficiency to Chl *a* [50]. As a consequence of this efficient energy transfer, Chl *b* fluorescence is

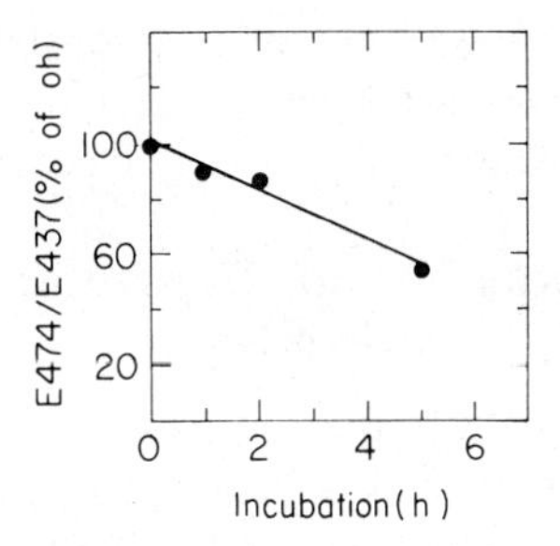

Fig. 5. Changes in the soret excitation ratio E474/E437 at 77 K after various periods of incubation of the unfortified chloroplasts. This ratio also expresses the relative resonance energy transfer from Chl *b* to Chl *a*. Chloroplasts were incubated as described in Fig. 2. Soret excitation amplitudes at E474 and E437 nm were obtained from the soret excitation spectra of the chloroplasts, which were recorded at F686 nm as described in Fig. 3(b).

not observed in freshly isolated chloroplasts (Fig. 6). It was observed, however, that as the incubation of the isolated chloroplasts proceeded *in vitro*, the onset of chloroplast degradation was accompanied by the appearance and gradual increase in the fluorescence emission of Chl *b* at 666 nm (Fig. 6). In other words during *in vitro* incubations a fraction of the

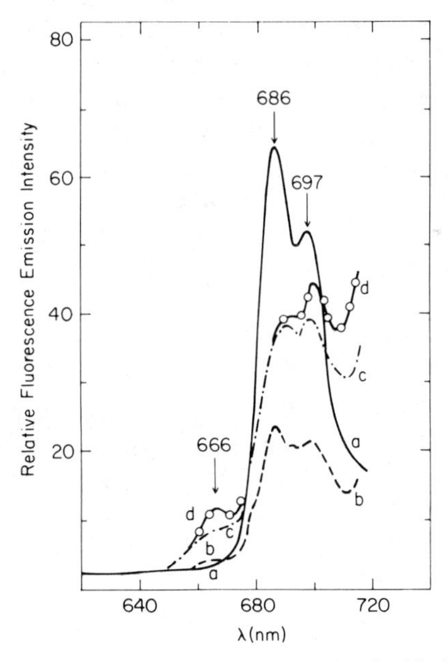

Fig. 6. Tracings of the fluorescence emission spectra of unfortified chloroplasts recorded at 77 K after various periods of incubation at room temperature. Chloroplasts were incubated as described in Fig. 2. Spectra were elicited by a 480-nm excitation in order to maximize the Chl *b* fluorescence; all other conditions were as in Fig. 3(a). Scale: 0.3 × ordinate. Arrows point to wavelengths of interest. Bcfore incubation (a); after 2 hr of incubation (b); after 3 hr of incubation (c); after 5 hr of incubation (d).

excitation energy of Chl *b* was no longer transferred to Chl *a* but was reemitted as Chl *b* fluorescence at 666 nm, which, in turn, indicated an alteration in the molecular architecture of the incubated chloroplasts.

B. Biochemical Detection of Early Degradative Events in Isolated Chloroplasts

In addition to early changes in the spectroscopic properties of Chl it was conjectured that the onset of the molecular disorganization of chloroplast membranes may be accompanied by subtle alterations of the Chl chromophore. As was mentioned earlier, it is also very likely that these biochemical alterations of the Chl chromophore may precede the disappearance of Chl. Although the disappearance of Chl is detectable by conventional spectrophotometry, the formation of small amounts of Chl degradation products has not been readily detectable by conventional spectrophotometry. Indeed Chl and some of its degradation products exhibit nearly identical red absorption maxima and consequently cannot be distinguished in organic extracts of the aging chloroplasts. On the other hand, extensive chromatographic manipulations prior to spectroscopic determinations are not advisable since chromatographic yields of porphyrins and phorbins are usually low and small amounts of segregated tetrapyrroles cannot be detected by conventional spectrophotometry [51, 52]. Moreover, because of the lability of the Chl extracted in organic solvents, small amounts of metabolic Chl by-products may be easily obscured by degradation products formed from the Chl during chromatography. It was therefore essential to develop sensitive analytical techniques for the early detection of metabolic degradation products of the Chl, with minimum processing of the chloroplast extract in order to avoid the formation of chemical artifacts. Some of these techniques are described below.

1. Early Detection and Quantitative Determination of Chlorophyllides and Pheophorbides during the Onset of Chloroplast Degradation

Chlorophyllides (Chlides), that is, chlorophylls that have lost their phytol and pheophorbides (Chlides that have lost their central Mg atom) are known metabolic degradation products of Chl metabolism [53]. For reasons discussed above the early detection of these compounds during the onset of chloroplast degradation has not been possible by conventional spectrophotometry.

In the past two years we have succeeded in developing sensitive spectrofluorometric techniques for the early detection and quantitative determination of picomole amounts of some known and hitherto unknown degradation products of Chl metabolism. It was demonstrated earlier that spectrofluorometry was much more sensitive than conventional spectrophotometry and exhibited a far superior resolution in the analysis of tetrapyrroles [51, 52]. These techniques were based on the observation that although Chlide *a* exhibited similar red emission and absorption maxima as pheophorbide *a* (pheo *a*) and Chlide *b* exhibited similar red emission and absorption maxima as pheo *b*, these four tetrapyrroles exhibited distinct soret excitation maxima (Fig. 7). This observation was used for the derivation of four simultaneous equations for the determination of picomole amounts of Chlide *a*, Chlide *b*, pheo *a*, and pheo *b* in mixtures of these tetrapyrroles without prior chromatographic segregation (M. B. Bazzaz and C. A. Rebeiz, in preparation).

The incubated chloroplasts were extracted in acetone: $0.1N$ NH_4OH (9:1 v/v) and the intact Chl was removed by extracting the acetone extract with hexane as previously described [51, 52]. The chlorophyllides and pheophorbides remained in the more polar, hexane-extracted acetone fraction. Two soret excitation spectra were recorded, one at F674 nm and the other at F660 nm (i.e., at the emission maxima of Chlide *a* + pheo *a* and Chlide *b* + pheo *b*, respectively). The fluorescence excitation amplitudes contributed solely by Chlide *a*, pheo *a*, Chlide *b*, and pheo *b* were then calculated from

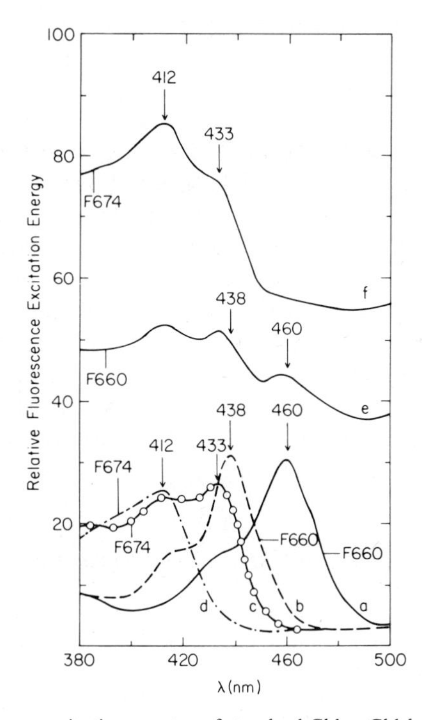

Fig. 7. Fluorescence soret excitation spectra of standard Chl *a*, Chl *b*, pheo *a*, and pheo *b* and of the mixture of these four tetrapyrroles in hexane-extracted acetone. Spectra were recorded, either at F660 or F674 nm as indicated on each spectrum, in cylindrical microcells 4 mm in diameter at room temperature. Scale 0.33 × ordinate. Arrows point to wavelengths of interest. Chl *b* (a); pheo *b* (b); Chl *a* (c); pheo *a* (d); 195 pmol Chl *b* + 170 pmol pheo *b* + 198 pmol Chl *a* + 198 pmol pheo *a* ml solution (e, f). For the convenience of presentation the vertical arrangement of the spectra was arbitrary.

the recorded soret excitation spectra with the newly derived simultaneous equations. The calculated fluorescence excitation amplitudes contributed solely by the individual tetrapyrroles were then converted to concentrations by reference to standard curves calibrated in pmol standard Chlide *a*, pheo *a*, Chlide *b*, or pheo *b* vs fluorescence excitation amplitudes. The simultaneous equations used in the calculation of the fluorescence excitation amplitudes are given below:

$$
\begin{aligned}
\text{Chlide } a \text{ (F674 E433)} &= 1.248\text{(F674 E433)} - 0.145\text{(F674 E412)} \\
&\quad - 0.068\text{(F660 E460)} \\
&\quad - 0.313\text{(F660 E438)} \\
\text{pheo } a\text{(F674 E412)} &= 1.198\text{(F674 E412)} - 1.100\text{(F674 E433)} \\
&\quad + 0.057\text{(F660 E460)} \\
&\quad + 0.110\text{(F660 E438)} \\
\text{Chlide } b\text{(F660 E460)} &= 0.067\text{(F674 E433)} + 1.028\text{(F660 E460)} \\
&\quad - 0.086\text{(F660 E438)} \\
&\quad - 0.061\text{(F674 E412)} \\
\text{pheo } b \text{ (F660 E438)} &= 0.001\text{(F674 E412)} + 1.152\text{(F660 E438)} \\
&\quad - 0.423\text{(F674 E433)} \\
&\quad - 0.420\text{(F600 E460)}
\end{aligned}
$$

In these equations, Chlide *a* (F674 E433) equals the soret excitation amplitude of Chlide *a* at 433 nm (E433), which is recorded at an emission wavelength of 674 nm (F674) in hexane-extracted acetone. Similarly (F674 E412) equals the soret excitation amplitude at 412 nm (E412) of the hexane-extracted acetone fraction containing the mixture of tetrapyrroles and recorded at an emission wavelength of 674 nm (F674), etc.

With the help of these equations, it was readily demonstrated that the early changes in the spectroscopic properties of Chl that were described in Figures 2 and 4–6 were indeed accompanied by the accumulation of small amounts of Chlides *a* and *b* and pheo *a*. These tetrapyrroles started to accumulate from the onset of incubation.

2. *Detection of Other Chlorophyll Degradation Products*

We have recently reported [54] that a hitherto unreported product of Chl catabolism accumulated during chloroplast degradation *in vitro*. This compound was also detected by its fluorescence emission and excitation properties in the hexane-extracted acetone fraction of the chloroplasts. It exhibited a red emission maximum at about 633 nm and a soret excitation maximum at about 435 nm. It was observed, however, that the accumulation of this compound in aging chloroplasts occurred after the initial accumulation of chlides and pheophorbides. Evidently the accumulation of this product denoted a more advanced state of early chloroplast degradation.

V. DETECTION OF REPAIR IN MATURE UNFORTIFIED CHLOROPLASTS

In addition to the retardation of degradative processes in cultured chloroplasts, the successful prolongation of chloroplast survival *in vitro* will very likely depend on the effectiveness of Chl repair in the isolated plastids.

In vivo repair of functional chloroplasts involves the replacement of the chlorophyll and of the photosynthetic membranes that become degraded during the lifetime of a green leaf. Perhaps the most obvious manifestation of repair in green leaves is the accumulation of protochlorophyllide (a precursor of Chl) in the dark and its subsequent conversion to Chl in the light [55, 56]. A less obvious, but equally important manifestation of chloroplast repair is the continuous turnover of the protein components of the photosynthetic membranes in green leaves [57]. As a marker of Chl repair in isolated chloroplasts we have used the conversion of exogenous δ-aminolevulinic acid (ALA) to protochlorophyllide (Pchlide) in the dark and its subsequent conversion to Chl in the light. This choice was motivated by the availability, expediency, and sensitivity of spectrofluorometric and radiochemical techniques that were developed in our laboratory for monitoring several single step and multistep reactions of the Chl biosynthetic pathway *in vitro* [51, 52, 58–69]. The same techniques that were developed to monitor the above biosynthetic events in developing chloroplasts were used here to describe the repair potential of unfortified mature chloroplasts. Unfortified chloroplasts are plastids to which none of the exogenous cofactors which were previously shown to be required for Chl biosynthesis *in vitro*, were added [61, 62, 64, 65].

A. Operation of the Chlorophyll Biosynthetic Pathway between δ-Aminolevulinic Acid and Protochlorophyllide

The operation of the reactions between ALA and Pchlide was monitored by determining the dark-incorporation of [$^{14}C_4$] ALA into [^{14}C] Pchlide. As shown in Table I, the incorporation of [$^{14}C_4$] ALA into [^{14}C] Pchlide was detectable. It was completely inhibited by 50 μmol of levulinic acid. The latter is a well-documented inhibitor of the conversion of ALA into porphobilinogen [70, 71]. These results indicated that in isolated, unfortified, mature chloroplasts, the reactions between ALA and Pchlide were still functional.

B. Operation of the Chlorophyll Biosynthetic Pathway between Protochlorophyllide and Chlorophyll

A complete Chl repair capacity would also involve the ability to convert Pchlide into Chl. The operation of the reactions between Pchlide and Chl were therefore monitored by determining the conversion of exogenous [^{14}C]

TABLE I

Dark-Incorporation of [$^{14}C_4$] ALA into [^{14}C] Pchlide in the Absence and Presence of Levulinic Acid*

Components	Net ^{14}C-incorporation into Pchlide after 1 h of incubation[a]
	10^{-3} X dpm/100 mg protein[b]
Complete	16.6 ± 2.6
+ 50 μmol Levulinic acid	-3.6 ± 2.8
+ 100 μmul Levulinic acid	-0.5 ± 2.9

* Mature chloroplasts were isolated from fully expanded green cotyledons and were incubated in the dark at 28°C for 1 hr. Complete reaction mixture contained in a total volume of 3 ml: 2.5 μc (100 nmol) [$^{14}C_4$] ALA, 400 μmol Tris-HCl, pH 7.7, 1 mmol sucrose, and 2 ml plastids (3.4 mg protein). Levulinic acid was adjusted to pH 6.5 with KOH before addition to the incubation mixture. This table was excerpted from Ref. 54.

[a]Net ^{14}C-incorporation represents the difference between the dpm recovered in the Pchlide band of the unheated and heat inactivated plastids after 1 hr of dark incubation. Heat inactivation was achieved by heating the incubation mixture containing the plastids for 5 min in a 100°C water bath just prior to adding the ^{14}C-labeled substrate. Protochlorophyllide was segregated on thin layers of silica gel H [21].

[b] Uncertainties were calculated as the square root of the sum of squares of the standard errors of the heat-inactivated control and unheated plastids [69]. The standard errors were calculated as the square root of the sum of the squares of the standard deviations of the background and sample counts per minute [69].

Pchlide to [^{14}C] Chl as recently described by Mattheis and Bebeiz [69, 72]. Essentially this assay involves the following sequential steps: (a) the dark incubation of exogenous [^{14}C] Pchlide with isolated plastids for 30 to 60 min to allow the binding of the substrate to the appropriate apoprotein; (b) a short light treatment (15 sec) to phototransform the [^{14}C] Pchlide; and (c) an additional dark incubation of 30 min to allow the phytylation of the newly formed [^{14}C] Chlide to form [^{14}C] Chl *a* [69, 72].

The results of such an experiment are reported in Table II. It is evident that the unfortified chloroplasts did not possess the ability to convert exogenous [^{14}C] Pchlide into [^{14}C] Chl *a*, which indicated that mature chloroplasts isolated in an unfortified Tris-sucrose medium were not capable of complete Chl *a* repair *in vitro*.

C. Net Tetrapyrrole Biosynthetic Competence of the Isolated Chloroplasts

To assess the ful measure of the Pchlide biosynthetic competence of the isolated plastids, the tetrapyrrole net biosynthetic activity of the mature chloroplasts was evaluated *in vitro*. Mature unfortified chloroplasts were isolated from fully expanded green cotyledons and were incubated with or without 100 nmol of ALA in 0.5*M* sucrose, 0.2*M* Tris-HCl, pH 7.7, for various periods of time, in darkness. A net synthesis of Zn-uroporphyrin

TABLE II

Conversion of Added [^{14}C] Pchlide into [^{14}C] Chl by Isolated Chloroplasts*

Experiment	^{14}C-labeled Pchlide added	^{14}C-recovered in Chl		
	10^{-3} x dpm/100 mg plastid protein			
		After 90 min of dark incubation	After 60 min of dark incubation, 15 sec of illumination and 30 min of dark incubation	Net ^{14}C recovered in Chl[a]
A	5 400	7.2 ± 5.1[b]	7.7 ± 5.5	0.5 ± 7.5[c]
B	3 400	14.6 ± 18.4	14.1 ± 4.7	-0.5 ± 19

* 5 ml chloroplasts suspension, in a total reaction mixture of 7.5 ml were incubated in a 50 ml Erlenmeyer flask in the dark at 28°C with [^{14}C] Pchlide having a specific radioactivity of 169 dmp/pmol. The added [^{14}C] Pchlide was dissolved in 0.075 ml of methanol. After 1 hr of incubation, the sample was divided into two equal portions. One portion was incubated for an additional 1/2 hr in the dark while the other portion was illuminated with 320 μW cm^{-2} of white fluorescent light for 15 sec at 0°C followed by an additional 1/2 hr incubation in the dark. At the end of the second dark incubation, the samples were extracted with 75% acetone. All pigments were transferred to ether as described elsewhere [69]. ^{14}C-Labeled Pchlide and Chl were purified on thin layers of silica gel H as described in [69]. This table was excerpted from Ref. 54.

[a] Refers to the difference between the illuminated and nonilluminated samples.

[b] Standard error was calculated as in Table I.

[c] Uncertainties were calculated as the square root of the sum of squares of the standard errors of the illuminated and nonilluminated samples.

and coproporphyrin was readily observed when the chloroplasts were incubated with ALA. However, no net synthesis of tetrapyrrole intermediates beyond coproporphyrin was detectable.

Altogether the above results indicated that mature chloroplasts to which no exogenous cofactors were added had a limited capacity for repair, which was manifested by their ability to catalyze the incorporation of [^{14}C] ALA into [^{14}C] Pchlide. It is hoped that this repair potential may be rendered more efficient and more complete in the future by biochemical manipulations of the incubation medium.

VI. CONCLUSIONS

Future work on the implementation of cell-free agriculture will emphasize, among other things, the biochemical manipulation of the culture medium in order to retard the chloroplast degradative processes and to enhance the Chl repair potential. The techniques described in this work should prove convenient and useful in evaluating the progress achieved in this field. Investigations of the output and photosynthetic efficiency of the partially stabilized chloroplasts cultured in the reactor will be equally important. Hopefully, we may one day know enough about the biosynthesis of photosynthetic membranes to bioengineer a more efficient and longer-lived photosynthetic membrane with which to populate a long-lived photosynthetic reactor.

Nomenclature

ALA	α-aminolevulinic acid
Chl	chlorophyll
PSI	photosystem I
PSII	photosystem II
Pchlide	protochlorophyllide

References

[1] D. Pimental, E. C. Terhune, R. Dyson-Hudson, S. Rocherau, R. Samis, A. E. Smith, D. Denman, D. Reifschneider, and N. Shepard, *Science 194*, 149 (1976).

[2] M. Calvin, *Photochem. Photobiol 23*, 425–444 (1976).

[3] C. A. Rebeiz, *Ill. Res. 16*, 3–4 (1974).

[4] M. Calvin, in *Progress in Photobiology*, G. O. Schenk (Ed.) (Deut. Gesell. Lichtforsch., Hamburg, German, 1974).

[5] G. Heichel, *Technology Review 76*, 1–9 (1974).

[6] S. Lien and A. San Pietro, *NSF/RANN Report*, Indiana University, Bloomington, 1975, pp. 50.

[7] J. A. Bassham, M. Kirk, and R. G. Jensen, *Biochem. Biophys. Acta. 153*, 211–218 (1968).

[8] E. S. Bamberger and R. B. Park, *Plant Physiol. 41*, 1591–1600 (1966).

[9] G. Constantopoulos and C. N. Kenyon, *Plant Physiol. 43*, 531–536 (1968).

[10] R. L. Heath and L. Packer, *Arch. Biochem. Biophys. 125*, 1189–198 (1968).

[11] R. L. Heath and L. Packer, *Arch. Biochem. Biophys. 125*, 850–857 (1968).

[12] P. A. Siegenthaler and P. Vaucher-Bonjour, *Planta. 100*, 106–123 (1971).

[13] P. A. Siegenthaler, *Act. Soc. Hel. Sci. Nat.*, 159–163 (1972).

[14] H. T. Choe and K. V. Thimann, *Planta. 121*, 201–203 (1974).

[15] T. Takaoki, J. Torres-Pereira, and L. Packer, *Biochim. Biophys. Acta. 352*, 260–267 (1974).

[16] G. Kulandaivelu and D. O. Hall, *Z. Naturforsch. 31c*, 452–455 (1976).

[17] L. Packer, *FEBS Lett. 64*, 17–19 (1976).

[18] H. T. Choe and K. V. Thimann, *Planta. 135*, 101–107 (1977).

[19] N. K. Boardman and S. W. Thorne, *Biochim. Biophys. Acta. 253*, 222–231 (1971).

[20] J. T. O. Kirk, and R. A. E. Tilney-Basset, *The Plastids* (Freeman, San Francisco, 1967).

[21] G. P. Gurinovich, A. N. Sevchenko, and K. N. Soloniev, *Spectroscopy of Chlorophyll and Related Compounds* (Izdatel'stvo Nauka i Tekknita), 520 pp; Transl. No. AEC tr-7199. Nat. Techn. Inform. Serv. U.S. Dept. Commerce, Springfield, Virginia (1968).

[22] M. E. Deroche and C. Costes, *Ann. Physiol. Veg. 8*, 223–254 (1966).

[23] C. S. French, J. S. Brown, and M. C. Lawrence, *Plant Physiol. 49*, 421–429 (1972).

[24] J. S. Brown, *Ann. Rev. Plant. Physiol. 23*, 73–86 (1972).

[25] J. C. Goedheer, *Biochem. Biophys. Acta. 88*, 304–317 (1964).

[26] G. I. Garab, G. Horvath, and A. Faludi-Daniel, *Biochem. Biophys. Res. Comm. 56*, 1004–1009 (1974).

[27] K. Sauer and M. Calvin, *J. Mol. Biol. 4*, 451–466 (1962).

[28] J. F. Becker, N. E. Geacintov, F. Van Nostrand, and R. Van Metter, *Biochem. Biophys. Res. Comm. 51*, 597–602 (1973).

[29] J. Breton, J. T. Becker, and N. E. Geacintov, *Biochem. Biophys. Res. Comm. 54*, 1403–1409 (1973).

[30] M. Lutz and J. Breton, *Biochem. Biophys. Res. Comm. 53*, 513–418 (1972).
[31] N. K. Boardman, S. W. Thorne, and J. M. Anderson, *Proc. Natl. Acad. Sci. USA 56*, 586–593 (1966).
[32] J. S. Brown, *Photochem. Photobiol. 26*, 319–326 (1977).
[33] J. B. Thomas and C. Van Hardeveed, *Acta. Bot. Neerl. 17*, 199–202 (1968).
[34] J. B. Thomas and H. H. Nijhuis, *Biochem. Biophys. Acta. 153*, 868–877 (1968).
[35] B. Nathanson and M. Brody, *Photochem. Photobiol. 12*, 469–479 (1970).
[36] S. W. Thorne and N. K. Boardman, *Plant Physiol. 47*, 252–261 (1971).
[37] E. S. Bamberger and R. B. Park, *Plant Physiol. 41*, 1591–1600 (1966).
[38] J. B. Thomas and J. W. Bielen, *Progr. Photosynth. Res. II*, 646–654 (1969).
[39] W. L. Butler and M. Kitajima, *Biochim. Biochim. Biophys. Acta 396*, 72–85 (1975).
[40] N. Murata, *Biochim. Biophys. Acta. 189*, 171–181 (1969).
[41] P. Homann, *Plant Physiol. 44*, 932–936 (1969).
[42] C. Bonaventura and J. Myers, *Biochim. Biophys. Acta 189*, 366–383 (1969).
[43] S. S. Brody, M. Brody, and J. H. Levine, *Biochim. Biophys. Acta 94* 310–312 (1965).
[44] M. Bazzaz and Govindjee, *Plant Sci. Lett. 16*, 201–206 (1973).
[45] B. M. Nathanson and M. Brody, *Photochem. Photobiol. 12*, 469–479 (1970).
[46] W. L. Butler, *Arch. Biochem. Biophys. 92*, 287–295 (1961).
[47] J. C. Goedheer, *Biochem. Biophys. Acta 51*, 494–504 (1961).
[48] J. C. Goedheer, *Biochem. Biophys. Acta 102*, 73–89 (1965).
[49] N. K. Boardman and S. W. Thorne, *Biochem. Biophys. Acta 153*, 448–458 (1971).
[50] L. N. M. Duysens, Ph.D. Thesis, (Univ. of Utrecht, The Netherlands, 1952) (unpublished).
[51] C. A. Rebeiz, J. R. Mattheis, B. B. Smith, C. C. Rebeiz, and D. F. Dayton, *Arch. Biochem. Biophys., 166*, 446–465 (1975).
[52] C. A. Rebeiz, J. R. Mattheis, B. B. Smith, C. C. Rebeiz, and D. F. Dayton, *Arch. Biochem. Biophys. 171*, 549–567 (1975).
[53] C. O. Chichester and T. O. Nakayama, in *Chemistry and Biochemistry of Plant Pigments*, T. W. Goodwin (Ed.) (Academic, New York, 1965), pp. 439–457.
[54] M. B. Bazzaz and C. A. Rebeiz, *Biochim. Biophys. Acta. 504*, 310–323 (1978).
[55] F. F. Litvin, A. A. Krasnovsky, and G. T. Rikhireva, *Doklady Akad. Nauk. SSR. 127*, 203–205 (1959).
[56] H. I. Virgin, *Physiol. Plant. 13*, 155–164 (1960).
[57] F. Henriques and R. Park, *Plant Physiol. 54*, 386–391 (1974).
[58] C. A. Rebeiz, B. B. Smith, J. R. Mattheis, C. C. Rebeiz, and D. F. Dayton, *Arch. Biochem. Biophys. 167*, 351–365 (1975).
[59] B. B. Smith and C. A. Rebeiz, *Photochem. Photobiol. 26*, 527–532 (1977).
[60] C. A., Rebeiz, M. Abou Haidar, M. Yaghi, and P. A. Castelfranco, *Plant Physiol. 46*, 543–549 (1970).
[61] C. A. Rebeiz and P. A. Castelfranco, *Plant Physiol. 47*, 24–32 (1971).
[62] C. A. Rebeiz and P. A. Castelfranco, *Plant Physiol. 47*, 33–37 (1971).
[63] C. A. Rebeiz, J. C. Crane, and C. Nishijima, *Plant Physiol. 50*, 185–186 (1972).
[64] C. A. Rebeiz, J. C. Crane, C. Nishijima, and C. C. Rebeiz, *Plant Physiol. 51*, 660–666 (1973).
[65] C. A. Rebeiz, S. Larson, T. E. Weier, and P. A. Castelfranco, *Plant Physiol. 51*, 651–659 (1973).
[66] B. B. Smith and C. A. Rebeiz, *Arch. Biochem. Biophys. 18*, 178–185 (1977).
[67] J. R. Mattheis and C. A. Rebeiz, *J. Biol. Chem. 252*, 4022–4024 (1977).
[68] J. R. Mattheis and C. A. Rebeiz, *J. Biol. Chem. 252*, 8347–8349 (1977).
[69] J. R. Mattheis and C. A. Rebeiz, *Arch. Biochem. Biophys. 184*, 189–196 (1977).
[70] D. L. Nandi and D. J. Shemin, *Biol. Chem. 243*, 1236–1242 (1968).
[71] S. I. Beale, *Plant Physiol. 48*, 316–319 (1971).
[72] J. R. Mattheis and C. A. Rebeiz, *Photochem. Photobiol. 28*, 55–60 (1978).
[73] C. E. Cohen, and C. A. Rebeiz, *Plant Physiol. 61*, 824–829.

Chlorophyll on Plasticized Particles—A New Model System

G. R. SEELY
Charles F. Kettering Research Laboratory, Yellow Springs, Ohio 45387

INTRODUCTION

The investigation of model systems for photosynthesis is undertaken for two reasons. The first is to further understanding of how the natural system operates by excluding from the model those features considered irrelevant. The second, and less often avowed, is to learn how to develop totally synthetic systems for efficient capture and conversion of solar energy. The efficiency with which natural photosynthetic units (PSU) of green plants convert sunlight into electrochemical energy in the primary photoreaction affords a strong incentive for developing model systems containing as many as possible of those features which allow the PSU to operate so well.

The PSU's in their natural state are integrally associated with a lipoprotein lamellar membrane. The membranes form closed, flattened sacs called thylakoids, which are packed in various ways within the chloroplast. The action of light probably drives electrons and ions (e.g., H^+) across the membrane, effecting separation of oxidation and reduction products and phosphorylation of ADP.

The roles of chlorophyll in photosynthesis are three: to absorb sunlight effectively, to transfer its energy efficiently to a reaction center, and there to react by electron transfer with a primary acceptor and a donor. It is a not trivial coincidence that one molecular species fulfills all three roles in higher plants. Most or all of the chlorophyll molecules are associated in complexes with proteins, three or four kinds of which are known or strongly suspected [1]. The chlorophyll–protein complexes can in turn be assigned to one of the two photosystems that make up the PSU, though perhaps not uniquely. Most of the chlorophyll molecules (and their protein complexes) are engaged in the roles of light absorption and energy transfer only (the "antenna" chlorophyll), and just a few (four per PSU) are photochemically active in reaction centers.

The absorption spectrum of chlorophyll *in vivo*, which is mainly that of antenna chlorophyll, shows a complexity of structure that becomes more evident at low temperature (77K). The spectrum of chlorophyll *a* then can be resolved into bands, separated by intervals of about 8 nm, which belong

Biotechnology and Bioengineering Symp. No. 8, 473–481 (1978)

0572-6565/78/0008-0473$01.00

to at least six distinguishable spectral forms of the pigment. [2]. The forms appear to be universal, though their relative proportions vary with the species and the age of the plant. Chlorophyll–protein complexes show a similar spectral resolution [3]; it seems likely that the different forms arise through electronic interaction of somewhat separated chlorophyll molecules.

The fluorescence of chloroplasts at low temperatures emanates in three distinct bands near 685, 695, and 730 nm, which are probably characteristic of different chlorophyll–protein assemblages [4].

These observations, and others not detailed here, encourage the inference that chlorophyll molecules are quite precisely located and oriented in the PSU, to the end that they might (1) transfer energy rapidly, (2) suppress fluorescence quenching (as explained below), and (3) control distribution of energy between the two photosystems. But the form of the arrangement is not known, nor what structural features of the PSU are needed to assure it. It is quite possible that interaction with the other lipid constituents of the membrane is important, and this provides the rationale for introduction of the present model system.

MODELS FOR PHOTOSYNTHESIS

In a model for chlorophyll–protein complexes put forth recently by Anderson [5], the porphyrin part of the pigment is embedded in the protein, while the lipophilic phytyl group lies outside the protein and communicates with the lipid bilayer. Chlorophyll thereby serves as a boundary lipid. In the chloroplast lamellae, the most abundant lipid species is galactolipid [6]. Structural features of both chlorophyll and galactolipids have led to suggestions that there are specific complexes *in vivo* between them [7, 8]. Whether or not specific complexes exist, and indeed whether or not Anderson's model is valid [it may not represent the manner in which most of the chlorophyll is bound (personal communication, J. M. Anderson)], the integrity and activity of the lamellar membrane is clearly dependent on the presence of galactolipid [7, 9], and there is reason to learn a great deal more than we now know about the interaction of chlorophyll with it, and with other lipids as well.

A great many model systems have been proposed and studied for the role of chlorophyll in photosynthesis. Before introducing a new one, it is necessary to point out some of the problems with the existing ones. In all models, it is considered desirable to retain as many features of the PSU as possible without sacrificing simplicity. Unfortunately, this is not always easy.

On the basis of a review of the role of model systems in the investigation of photosynthesis [8], it is possible to classify most of them into two groups, which we shall call the "essentially liquid" and "essentially solid" systems, according to the amount of freedom to move or to diffuse that the chlorophyll has during its excited state lifetime.

Systems of the former group, which includes chlorophyll in solution [10–12], micelles [13], liposomes [14], bilayers [15], and certain mixed monolayers [16], and bound to dissolved polymers [17], have the following properties in common: (1) absorption spectra are relatively simple, suggesting little interaction between chlorophylls; (2) when the chlorophyll concentration is low, the fluorescence and photochemical activity have their normal high values for dilute homogeneous solution; (3) at high concentration, both fluorescence and (triplet state) photochemistry are quenched. This quenching is most probably a bimolecular self-quenching, the mechanism of which is speculative, but which is a consequence of the freedom of motion of chlorophyll within the system [10, 18]. It appears that a primary function of the lipoprotein membrane is to keep the chlorophyll molecules a short distance apart, to prevent loss of energy by quenching during transfer.

The "essentially solid" systems include chlorophyll in dimeric, oligomeric, and "crystalline" aggregates, [19], in suspension [20], in dense or two-phase monolayers [21], and films [22, 23]. In these systems, the chlorophyll molecules are typically locked into place when the system is formed, usually into configurations with strong interaction. Absorption spectra are often complex, but fluorescence and photochemical activity are usually weak or absent.

One noteworthy class of exceptions to this generalization exists. Chlorophyll or a related pigment adsorbed to the surface of particles of plastic (polystyrene [24], polycaprolactam [25], and polyacrylonitrile [26]) has a complex absorption spectrum, fluorescence when dilute, and photochemical activity even at concentrations where fluorescence is quenched. By these signs, chlorophyll on plastic particles might seem a propitious model, but it fails to imitate the PSU in one important respect: the phytyl chains are unable to penetrate the solid plastic matrix as they would a lipid bilayer. Our present model system attempts to remedy this defect by plasticizing particles of polymer with a low molecular weight diluent, which softens the surface enough to permit anchoring of chlorophyll and other lipids by their hydrocarbon chains, in the same way as they are believed to be anchored in bilayer lipid membranes. Thus, we hope to provide enough mobility to permit equilibration of chlorophyll with other polar lipids at the surface, while retaining enough rigidity to conserve fluorescence and photochemical activity at high pigment density. Such particles, suspended in water, could be a fairly realistic model of the interface between the lipid bilayer and the surrounding aqueous phase, a model which might eventually be amenable to development for solar energy conversion.

PREPARATION OF PARTICLES

In the present system the polymeric support consists of polyethylene particles plasticized with undecane. These are made simply by dissolving commercial polyethylene powder in undecane at an elevated temperature, and cooling through the cloud point (~77°). After filtration, a methanol rinse to remove adhering undecane, and drying in air, the increase in weight of the recovered particles corresponded to a content of 78% undecane. The particles were angular or prismatic in appearance, with dimensions in the range 2–5 μm.

The normal procedure for coating the particles with chlorophyll in the presence of other lipids is to suspend them in a solvent such as 90% methanol and 10% water, then to increase the water content by adding 60% methanol until the chlorophyll has deposited on them. The particles are wetted by 90% methanol but do not suspend well unless a surfactant or polar lipid is present. Lipids not soluble in 90% methanol can be added to the particles in chloroform, and after evaporation of most of the chloroform, the particles can be suspended in aqueous methanol.

After the particles are coated with chlorophyll, they are recovered by filtration and resuspended in water. To prevent the particles from clumping while spectra are taken, they are better suspended in an aqueous medium made viscous by addition of a polyelectrolyte such as alginate, xanthan, or chitosan. Absorption spectra of optically dense mulls were recorded in a specially constructed cell which fits into the exit window of a Cary 14 spectrophotometer, to increase the amount of multiply scattered light received by the detector. The reference beam was attenuated by screens. Fluorescence spectra of optically thin mulls are recorded on an apparatus assembled in this laboratory by Dr. B. C. Mayne.

Finally, suspensions of the particles in alginate can be gelled by contact with solutions of Ca salts or of other divalent metal ions, for further investigation under heterogeneous conditions.

The chlorophyll used in these preparations was isolated from brown algae and purified chromatographically [27]. Other chemicals, including polyethylene (USI Microthene) were good quality commercial materials, used as received.

RESULTS AND DISCUSSION

When water is added to a suspension of particles in 90% methanol with no other lipid or surfactant than chlorophyll present, the absorption spectrum of the green particles is dominated by two equal bands of 679 and 714 nm, which probably can be identified as belonging to chlorophyll hydrate dimers and oligomers of types noted in other wet hydrocarbon systems [23]. The fluorescence is weak, but at 77 K the spectrum has bands (682 and 715.5 nm) corresponding to the two species of chlorophyll hydrate seen in absorption.

If surfactants are present which appear to have little interaction with chlorophyll, such as dipalmitoyl phosphatidyl ethanolamine and dodecyl trimethyl ammonium chloride (DTMA), similar spectra are observed. With DTMA, the 715-nm band is dominant, but shoulders are evident near 740 and 680 nm (Fig. 1). The fluorescence is weak, but about three times as strong at 77 K as at 295 K. Its spectrum at either temperature evidently reflects a monomeric chlorophyll component which contributes to the absorption band shoulder near 670 nm. The bulk of the chlorophyll in this case is nonfluorescent.

In contrast, when a polar lipid or surfactant is present which might be expected to interact strongly with the porphyrin part of chlorophyll, characteristic and unique spectra are often obtained. *N*,*N*-Dimethylmyristamide (DMMA) has an amide group which might interact with chlorophyll in much the same way as the peptide groups of proteins. In its presence, the spectrum of chlorophyll on particles is dominated by the bands of a probably monomeric form (660.5 nm) (Fig. 2). The fluorescence is strong at both 295 K and 77 K, but at the higher temperature, emission of the monomeric form (669 nm) dominates, whereas at the lower temperature, the emission of an aggregated species (725 nm) appears equally strong. In other systems, appearance of an emission band near 725 nm has been associated with a presumably dimeric chlorophyll species absorbing near 700 nm [11, 28]. Such a species might be present in the long-wave extension of the absorption band in Figure 2.

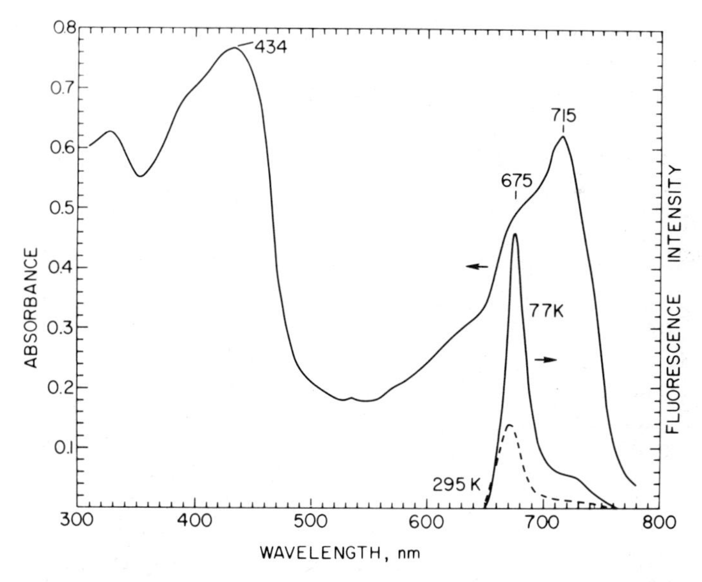

Fig. 1. Absorption and fluorescence spectra of chlorophyll with dodecyl trimethyl ammonium chloride (DTMA) on polyethylene particles plasticized with undecane. Absorption spectrum is of an optically dense mull in chitosan solution at 295 K; fluorescence spectra are of a dilute mull of the same sample at 295 K and 77 K, at the same detector sensitivity setting, and are not corrected for variation of photomultiplier sensitivity with wavelength. Sample was prepared in a mixture containing a weight ratio of particles to DTMA of 20 and mole ratio of DTMA to chlorophyll of 147.

Spectra of chlorophyll on particles in the presence of the glycolipids digitonin, phosphatidyl inositol, and monogalactosyl diglyceride contain complex features which have been reported elsewhere [29]. Usually, the presence of polar lipids has a disaggregating effect on the absorption spectrum and promotes forms of chlorophyll that fluoresce strongly, even at room temperature.

A desirable property of a system for studying interaction of chlorophyll with other lipids is reversibility of binding of chlorophyll. By this is meant that during the process of coating the particles with chlorophyll, there is an equilibrium between pigment adsorbed and pigment in solution. If this were not so, then the amount and spectrum of adsorbed chlorophyll would depend on the manner in which the composition of the solvent was altered. Normally, the solvent is made continually more aqueous in order to drive the chlorophyll onto the particles. However, Figure 3 contains evidence that this procedure may be varied without changing the result.

In Figure 3, the amount of chlorophyll in solution is plotted against the water content of the solvent in the presence and in the absence of particles. The relative absorbance of chlorophyll at 665 nm is used as an indicator of chlorophyll in solution. It will first be noted that in the absence of particles (upper trace), there is no change in dissolved chlorophyll until the solubility limit is exceeded at about 17% water. Then the absorbance drops as a

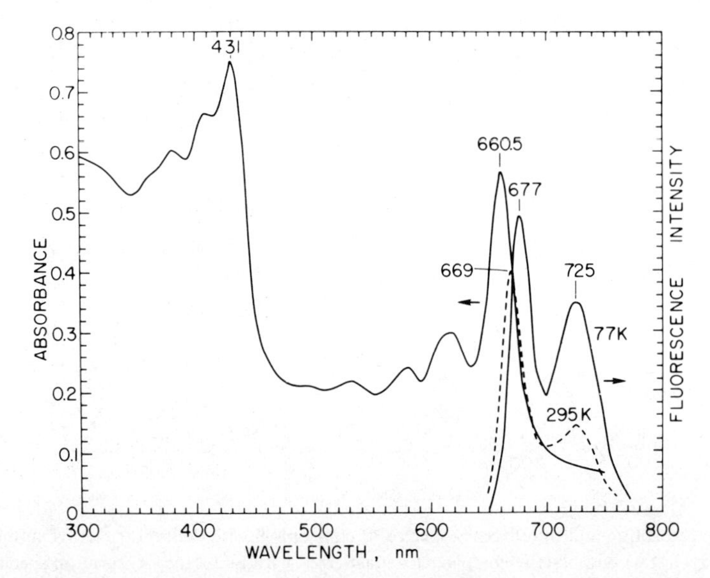

Fig. 2. Absorption and fluorescence spectra of chlorophyll with *N,N*-dimethylmyristamide (DMMA) on plasticized polyethylene, prepared in a mixture containing a weight ratio of particles to DMMA of 10, and a mole ratio of DMMA to chlorophyll of 185. Absorption and fluorescence spectra are of Na alginate mulls. Fluorescence spectra are corrected for variation of photomultiplier response with wavelength.

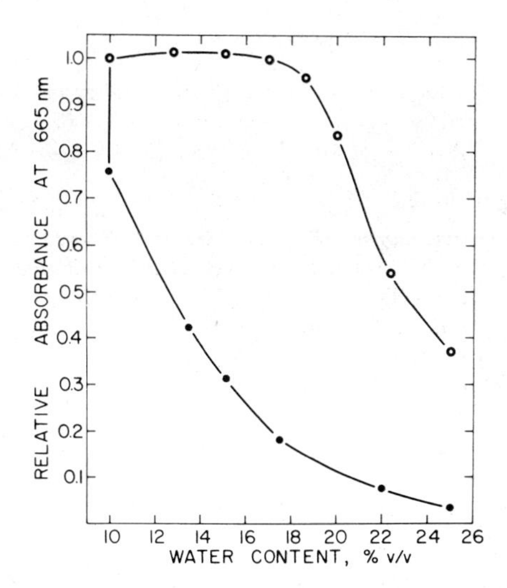

Fig. 3. Illustration of the adsorption of chlorophyll to plasticized polyethylene particles in the presence of *N,N*-dimethylmyristamide, and a test of reversibility of the binding. Relative absorbance at 665 nm of chlorophyll remaining in the supernatant (reduced to constant volume) is plotted against the percent of water in the aqueous methanol medium, in the presence (●) and in the absence (○) of particles. See the text for procedural details concerning the lower trace.

colloidal chlorophyll suspension separates. However, when particles are added to a chlorophyll solution in 90% methanol, there is an immediate drop of about 25% in chlorophyll concentration due to adsorption. At increasing water content, the chlorophyll in solution falls smoothly toward zero (lower trace). But the points in this figure were not all obtained in order of increased water content. First, water was added to make the solution about 13.5%, then 17.5%, aqueous. Then methanol was added until the solution was 15% aqueous. Finally, addition of water brought it to 22% and to 25% aqueous. The fact that all points lie on a fairly smooth curve indicates that the binding of chlorophyll is reversible in this critical range of aggregate formation.

The ability of chlorophyll-coated particles to sensitize photochemistry efficiently has not yet been established. It has become obvious that the chlorophyll is bleached rather rapidly by light in the presence of air, which is certainly photochemistry of a sort. Technical difficulties in following photochemical reactions in an anaerobic heterogeneous system must be overcome before quantum yields can be determined. However, since concentration quenching of chlorophyll fluorescence seems not to be a limitation of this system, as it is not in the PSU, it is possible that significant photochemistry will follow.

The assistance of Dr. B. C. Mayne in the fluorescence measurements is much appreciated.

References

[1] J. P. Thornber, R. S. Alberte, F. A. Hunter, J. A. Shiozawa, and K.-S. Kan, *Brookhaven Symp. Biol.*, *28*, 132 (1976).

[2] C. S. French, *Proc. Natl. Acad. Sci. USA 68*, 2893 (1971); B. A. Gulyaev and F. F. Litvin, *Biofizika*, *15*, 670 (1970).

[3] J. A. Shiozawa, R. S. Alberte, and J. P. Thornber, *Arch. Biochem. Biophys.*, *165*, 388 (1974); R. L. Van Metter, *Biochim. Biophys. Acta*, *462*, 642 (1977).

[4] W. L. Butler and M. Kitajima, in *Proceedings of the Third International Congress of Photosynthesis*, M. Avron (Ed.) (Elsevier, Amsterdam, 1974), p. 13; R. J. Strasser and W. L. Butler, *Biochim. Biophys. Acta*, *462*, 307 (1977).

[5] J. M. Anderson, *Nature* (*Lond.*), *253*, 536 (1975).

[6] H. K. Lichtenthaler and R. B. Park, *Nature* (*Lond.*), *198*, 1070 (1963).

[7] A. Rosenberg, *Science*, *157*, 1191 (1967).

[8] G. R. Seely in *Primary Processes of Photosynthesis*, J. Barber (Ed.) (Elsevier, Amsterdam, 1977), Vol. 2, p. 1.

[9] O. Hirayama and T. Matsui, *Biochim. Biophys. Acta*, *423*, 540 (1976); P. G. Roughan and N. K. Boardman, *Plant Physiol. 50*, 31 (1972); Z. Krupa and T. Baszynski, *Biochim. Biophys. Acta*, *408*, 26 (1975); A. B. Shaw, M. M. Anderson, and R. E. McCarty, *Plant Physiol.*, *57*, 724 (1976).

[10] G. S. Beddard and G. Porter, *Nature* (*Lond.*), *260*, 366 (1976).

[11] S. S. Brody, *Biophys. J.*, *8*, 210 (1968).

[12] R. Livingston, *Quart. Rev.* (*Lond.*), *14*, 174 (1960); A. R. Kelly and G. Porter, *Proc. R. Soc.* (*Lond.*), *A 315*, 149 (1970).

[13] P. Massini and G. Voorn, *Biochim. Biophys. Acta*, *153*, 589 (1968); E. I. Zen'kevich, A. P. Losev, and G. P. Gurinovich, *Mol. Biol.* (*Moscow*), *6*, 824 (1972); E. Lehoczki, K. Csatorday, L. Szalay, and J. Szabad, *Biofizika*, *20*, 44 (1975).

[14] D. Chapman and P. G. Fast, *Science*, *160*, 188 (1968); A. G. Lee, *Biochemistry*, *14*, 4397 (1975); K. Colbow, *Biochim. Biophys. Acta*, *318*, 4 (1973); P. Nicholls, J. West, and A. D. Bangham, *Biochim. Biophys. Acta*, *363*, 190 (1974).

[15] C.-H Chen and D. S. Berns, *Photochem. Photobiol.*, *24*, 255 (1976); A. Steinemann, N. Alamuti, W. Brodmann, O. Marschall, and P. Läuger, *J. Membrane Biol.*, *4*, 284 (1971); H. P. Ting, W. A. Huemoeller, S. Lalitha, A. L. Diana, and H. T. Tien, *Biochim. Biophys. Acta*, *163*, 439 (1968).

[16] A. G. Tweet, G. L. Gaines, Jr., and W. D. Bellamy, *J. Chem. Phys.*, *40*, 2596 (1964); T. Trosper, R. B. Park and K. Sauer, *Photochem. Photobiol.*, *7*, 451 (1968); S. M. de B. Costa, J. R. Froines, J. M. Harris, R. M. LeBlanc, B. H. Orger, and G. Porter, *Proc. R. Soc.* (*Lond.*) *A*, *326*, 530 (1972).

[17] G. R. Seely, *J. Phys. Chem.*, *71*, 2091 (1967); *75*, 1667 (1971).

[18] G. R. Seely, in *Current Topics in Bioenergetics*, D. R. Sanadi and L. P. Vernon (Eds.), (Academic, New York, 1978), Vol. 7, p. 3; G. Porter, *Naturwissenschaften*, *63*, 207 (1976).

[19] J. R. Norris, H. Scheer, and J. J. Katz, *Ann. N.Y. Acad. Sci.*, *244*, 260 (1975); M. R. Wasielewski, M. H. Studier, and J. J. Katz, *Proc. Natl. Acad. Sci. USA*, *73*, 4282 (1976); E. E. Jacobs, A. S. Holt, R. Kromhaut, and E. Rabinowitch, *Arch. Biochem. Biophys. 72*, 495 (1957); A. P. Losev, E. I. Zen'kevich, and G. P. Gurinovich, *Zh. Prikl. Spektrosk.*, *19*, 262 (1973).

[20] L. P. Vernon, *Acta Chem. Scand.*, *15*, 1639 (1961); V. B. Evstigneev and V. A. Gavrilova, *Biofizika*, *4*, 641 (1959); A. A. Krasnovskii and E. V. Pakshina, *Dokl. Akad. Nauk, SSSR*, *127*, 913 (1959).

[21] B. Ke, in *The Chlorophylls*, L. P. Vernon and G. R. Seely (Eds.) (Academic, New York, 1966), Chap. 8.

[22] G. Sherman and S.-F. Wang, *J. Org. Chem.*, *31*, 1465 (1966); M. I. Bystrova and A. A. Krasnovskii, *Mol. Biol.* (*Moscow*), *2*, 847 (1968).

[23] K. Ballschmiter and J. J. Katz, *Nature* (*Lond.*), *220*, 1231 (1968).
[24] R. A. Cellarius and D. Mauzerall, *Biochim. Biophys. Acta*, *112*, 235 (1966).
[25] G. G. Komissarov, V. A. Gavrilova, L. I. Nekrasov, N. I. Kobozev, and V. B. Evstigneev, *Dokl. Akad. Nauk USSR*, *150*, 174 (1963).
[26] R. Kapler and L. I. Nekrasov, *Biofizika*, *11*, 420 (1966).
[27] G. R. Seely, M. J. Duncan, and W. E. Vidaver, *Marine Biol.*, *12*, 184 (1972).
[28] F. K. Fong and V. J. Koester, *Biochim. Biophys. Acta*, *423* 52 (1976); S. G. Boxer and G. L. Closs, *J. Am. Chem. Soc.*, *98*, 5406 (1976).
[29] G. R. Seely, in *Chlorophyll Organization and Energy Transfer in Photosynthesis*, Ciba Symposium No. 61 (in press).

Direct Electron Transfer at an Immobilized Cofactor Electrode: Approaches and Progress

LEMUEL B. WINGARD, JR. and JOSEPH L. GURECKA, JR.
Department of Pharmacology, School of Medicine, University of Pittsburgh, Pittsburgh, Pennsylvania 15261

INTRODUCTION

Since 1968 we have been studying the use of enzymes to catalyze the *in vitro* conversion of chemical energy to electrical energy [1–3] . Although our studies have shown such a fuel cell type of process to be possible, the efficiency of conversion for the glucose–glucose oxidase system has been extremely poor[3]. The present study was undertaken to try to define the cause of the poor efficiencies and to suggest approaches for obtaining improved transfer of chemical to electrical energy with enzyme catalysts. Such an improvement could prove useful in medical, analytical, and industrial applications.

Biochemical fuel cells can be divided into at least two classes: direct and indirect. In the direct type the enzyme catalyst acts as an integral part of the anode or cathode to oxidize or reduce the appropriate substrate. The enzyme provides reactant specificity as well as allowing the reaction to proceed under relatively mild temperatures. In the indirect type of cell, the enzyme catalyzes the conversion of a primary reactant into a compound that can be utilized by a conventional fuel cell having nonenzymatic or nonbiological electrodes. The work discussed here pertains only to the direct type of biochemical fuel cell.

For an enzyme-catalyzed anodic reaction, the half-cell reaction can be represented by

$$S \rightarrow P + e^- \qquad (1)$$

where S is the substrate and P is the product. The objective is to get the electron onto an external circuit and extract work before undergoing a reduction half-cell reaction at the cathode. Since oxidation–reduction enzymes require cofactors, Eq. (1) can also be written

$$S + C_{ox} \rightarrow P + C_{red} \rightarrow C_{ox} + e^- + H^+ \qquad (2)$$

C represents the oxidized or reduced cofactor, which is normally the acceptor of the electrons and protons released by oxidation of the substrate.

Biotechnology and Bioengineering Symp. No. 8, 483–487 (1978)
 0572-6565/78/0008-0483$01.00

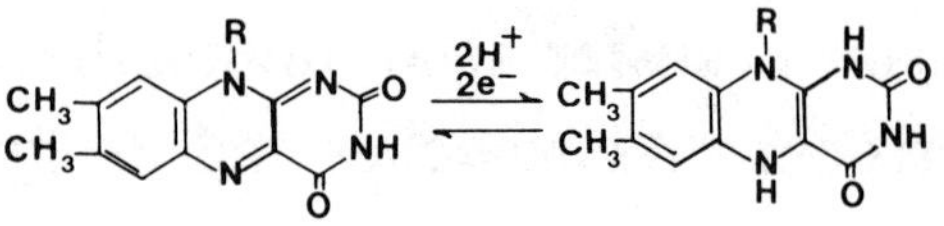

Fig. 1. Oxidized and reduced forms of riboflavin. R stands for a ribose group.

For a functioning enzyme-catalyzed anode the cofactor must revert rapidly to the oxidized state, thus releasing the electrons to the surrounding medium or to an external wire, or the reduced cofactor must migrate to the electrode support and there give up the electrons to the external circuit. At least two major possibilities exist for low efficiency. First, the electroactivity, or ease by which the cofactor transfers electrons to the solid electrode surface, may be low; and second, the rate of transport of the reduced cofactor from the enzyme molecule to the electrode surface may be slow [4]. This paper deals only with the electroactivity aspect.

Several approaches for improving the cofactor electroactivity include immobilizing the enzyme in a semiconductor gel [5], adding a redox couple or charge-transfer mediator of high electroactivity [6–8], or binding the cofactor directly to the electrode surface [4]. A novel approach for the covalent attachment of flavin cofactors to electron-conducting supports is described in the present work.

COFACTOR IMMOBILIZATION

The flavin cofactors are exemplified by riboflavin, of which the oxidation–reduction chemistry is summarized in Figure 1. Particular attention is drawn to the aromatic ring on which the methyl groups are attached at positions 7 and 8. The π orbitals of this ring overlap with those of the sites of electron and proton addition during reduction of riboflavin. Therefore, if the methyl group at position 8 could be coupled through a double bond to an electron-conducting support, then a pathway of overlapping π orbitals should be formed between the site of electron addition and the external circuit. Since overlapping π orbitals are thought to provide a pathway for easier electron conduction, the resulting immobilized flavin might serve as a highly electroactive cofactor with an appropriate enzyme and substrate. The postulated structure for riboflavin covalently attached to carbon is shown in Figure 2.

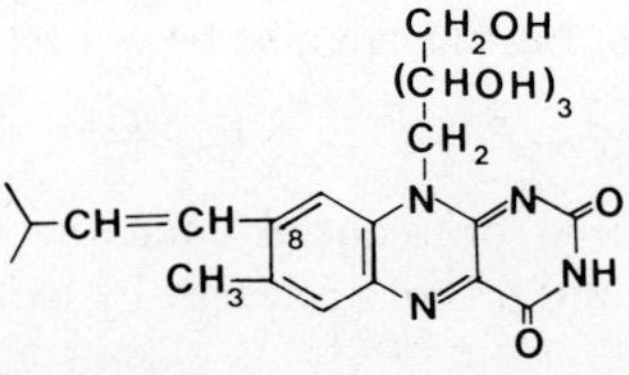

Fig. 2. Postulated structure of riboflavin attached to carbon in a manner to give overlapping π orbitals and, therefore, possible enhanced electron conduction.

There is considerable evidence that flavin cofactors are covalently coupled to certain apoenzymes in the natural enzymes and that the coupling is through a number 8 position methyl group [9]. Therefore, there is a good likelihood that the Figure 2 immobilized cofactor, or a similar structure with a spacer between the flavin and the carbon support, will show activity when incubated with an appropriate apoenzyme and substrate.

RESULTS AND DISCUSSION

The general approach for chemical synthesis of the immobilized flavin analog involves activation and coupling through the number 8 methyl group of the riboflavin molecule. This general scheme was described very briefly in late 1977 [4, 10], before most of the structural verification studies had been started. An outline of the verification work is given here; however, the detailed reaction conditions and characterization data will be presented elsewhere.

Selective activation of the number 8 position methyl group of riboflavin can be accomplished by bromination [11, 12]. However, it is first necessary to block the hydroxyl groups on the ribosyl appendage, for example by acetylation, in order that the hydroxyls will not be brominated. The reaction sequence for activation of the riboflavin is shown in Figure 3. The parent riboflavin, the acetylated analog, and the brominated analog were each isolated and the structures verified using nuclear magnetic resonance (NMR) spectroscopy at 250 MHz. The experimental spectra agreed with those in the literature [11, 12]. Subsequent reaction of the bromo–acetyl–riboflavin with triphenyl phosphine produces the phosphonium salt of the modified riboflavin. The NMR spectra of this salt were obtained and they agreed with the expected chemistry; however, no reference spectra could be found in the literature for this material. Phosphonium salts can be converted to ylides by the addition of a base. The advantage of this modification is that certain ylides will couple to an aldehyde in a Wittig-type reaction to give the desired double bond of Figure 2. We are carrying out these steps by reacting the flavin– phosphonium salt with a base and then with a known aldehyde. The product has been isolated

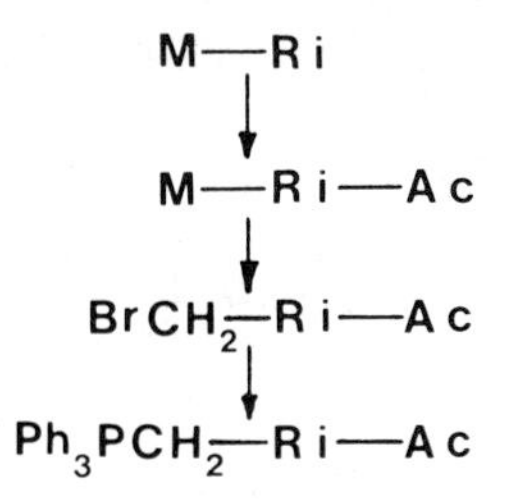

Fig. 3. Activation of riboflavin for subsequent coupling to an aldehyde-functionalized solid support. Symbols: M is 8-position methyl, Ri is remainder of riboflavin molecule, Ac is acetylated hydropyl groups, and Ph is phenyl.

and characterized so far using a 60-MHz NMR unit. These preliminary results indicate that the coupling reaction was successfully carried out; however, the present higher resolution studies with the 250-MHz NMR should provide a more definitive answer.

Carbon was selected for the support because of its good electron conductivity, availability of surface functional groups, and relatively low cost. A reaction sequence for generation of aldehyde groups on the surface of carbon is shown in Figure 4. Our initial work was done with activated carbon powder type GW, kindly provided as a gift from the Pittsburgh Activated Carbon Division of the Calgon Corporation. The carbon powder was incubated with concentrated nitric acid for 15 min at room temperature followed by concentrated sulfuric acid for 60–90 min at 170°C. The resulting particles were washed; titration with sodium hydroxide showed a carboxylic acid density of 66 μequiv/g dry carbon. These groups were converted to aldehyde functions by treatment with thionyl chloride and tri-tertiary–butoxy aluminum hydride. A 2,4-dinitrophenylhydrazine test for aldehydes gave positive results. The aldehyde–carbon powder and the activated riboflavin were incubated under the conditions for coupling; however, it is a very difficult task to show chemically the nature of compounds attached in low concentrations to solid supports. At present we are exploring electron spectroscopy, primarily electron spectroscopy for chemical analysis (ESCA), for characterizing the presence of specific compounds on the carbon surface. In the case of the flavin-powdered carbon, it was difficult to conclude whether or not the flavin was coupled to the carbon surface because of the difficulty in washing away all of the adsorbed reactants from the large surface area of the activated carbon powder.

Our present work is with solid glassy carbon, wherein the problem of reactant adsorption due to large surface areas should be minimal since glassy carbon is relatively nonporous. However, the production of suitably

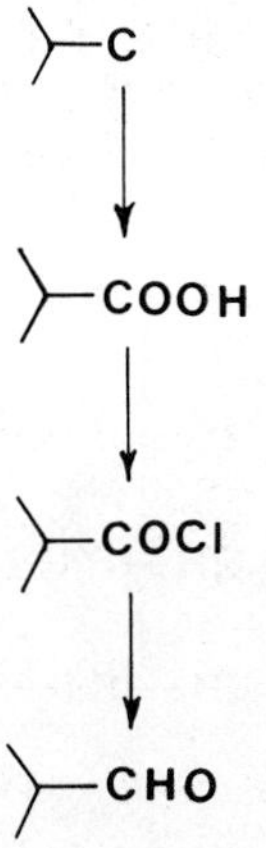

$$\succ\!-C \rightarrow \succ\!-COOH \rightarrow \succ\!-COCl \rightarrow \succ\!-CHO$$

Fig. 4. Activation of solid carbon support for subsequent coupling of activated riboflavin.

large concentrations of surface aldehyde groups may be more difficult with the glassy carbon. This part of the work is in progress.

In summary, preliminary results indicate that the proposed chemistry for attaching riboflavin to a solid support to give overlapping π orbitals across the point of attachment is workable. A major question at present is how to prepare the surface of solid glassy carbon to obtain sufficient levels of aldehyde groups for coupling of the activated flavin.

This work is supported by grant No. ENG 7516403 from the National Science Foundation. The authors are indebted to S. Danishefsky of the Chemistry Department for helpful discussions.

References

[1] L. B. Wingard, Jr. and C. C. Liu, *Proc. 8th Intl. Conf. Med. Biological Eng.*, *26*, 10 (1969).

[2] L. B. Wingard, Jr., C. C. Liu, and N. L. Nagda, *Biotechnol. Bioeng.*, *13*, 629 (1971).

[3] E. J. Lahoda, C. C. Liu, and L. B. Wingard, Jr., *Biotechnol. Bioeng.*, *17*, 413 (1975).

[4] L. B. Wingard, Jr., presented at the Workshop on Biotechnology; Electron Transfer Proceedings, Philadelphia, November 5–9, 1977; sponsored by Intl. Fed. Inst. Adv. Study, Stockholm, Sweden.

[5] I. V. Berezin (personal communication, 1977).

[6] S. D. Varfolomeev, A. I. Yaropolov, I. V. Berezin, M. R. Tarasevich, and V. A. Bogdanovskaya, *Bioelectrochem. Bioenerg.*, *4*, 314 (1977).

[7] J. Mizuguchi, S. Suzuki, and F. Takahashi, *Bull. Tokyo Inst. Technol.*, *78*, 27 (1966).

[8] G. Johansson, First Ann. Proj. Review, Report 7, Intl. Fed. Inst. Adv. Study, Stockholm, Sweden, 1975, p. 41.

[9] D. E. Edmondson and T. P. Singer, *FEBS Lett.*, *64*, 255 (1976).

[10] L. B. Wingard, Jr., presented at 3rd US–USSR Enzyme Conference, November 28–December 2, 1977, Tallinn, Estonian SSR.

[11] W. H. Walker, T. P. Singer, S. Ghisla, and P. Hemmerich, *Eur. J. Biochem.*, *26*, 279 (1972).

[12] D. B. McCormick, *J. Heterocyclic Chem.*, *7*, 447 (1970).

Use of Cell-Free Biological Systems for Hydrogen Production*

B. ZANE EGAN and C. D. SCOTT
Oak Ridge National Laboratory, Oak Ridge, Tennessee 37830

INTRODUCTION

Many algae and bacteria that produce hydrogen have been isolated and characterized [1, 2]. Several investigators have discussed the continuous production of hydrogen by intact biological systems [3–7], including immobilized whole cells [8, 9]. Hydrogen evolution has been demonstrated in cell-free systems using ferredoxin and hydrogenase in combination with chemical reductants such as dithionite and pyruvate [10–16]. Hydrogen was formed as the reduction product from water, and the substrate was oxidized. Benemann et al. [17–19] demonstrated the photoproduction of hydrogen from water using an illuminated suspension of spinach chloroplasts, hydrogenase, and ferredoxin. Each of these processes, one utilizing chemical reductants and the other utilizing the photochemical reducing potential of illuminated chloroplasts, involves the reduction of ferredoxin followed by the hydrogenase-catalyzed reaction between the reduced ferredoxin and protons to produce hydrogen gas. Related studies have been reviewed by Lien and San Pietro [20].

The ability of a cell-free system, such as one composed of chloroplasts, ferredoxin, and hydrogenase, to produce hydrogen provides an approach which avoids the requirements and inefficiencies of metabolic interactions in intact cell systems. The evolution of hydrogen in an *in vitro* illuminated chloroplast-hydrogenase-ferredoxin system has been shown to continue for 6.5 hr when oxygen was chemically removed [21] and up to 20 hr at a reduced rate at lower temperatures [22]. Various conditions have been used to obtain higher hydrogen production rates [23]. Advances have also been made in the immobilization of intact cells and various active fractions including chloroplasts [4, 24] and hydrogenase [25]. Such developments allow these systems to be investigated as potential processes for hydrogen production. A conceptual process scheme would include bioreactors with chloroplasts immobilized in one reactor and hydrogenase immobilized in the second. Reduced ferredoxin would flow through the hydrogenase bed, where

* Research sponsored by the Office of Basic Energy Sciences, U.S. Department of Energy, under contract No. W-7405-eng-26 with the Union Carbide Corporation.

Biotechnology and Bioengineering Symp. No. 8, 489–500 (1978)

0572-6565/78/0008-0489$01.00

hydrogen would be generated. In the chloroplast reactor, oxidized ferredoxin circulating from the hydrogenase bed would be reduced by photosynthetic interactions, and oxygen would be formed. Figure 1 shows a simplified schematic illustrating this concept.

We have considered several aspects of developing such a system into a feasible process. These include thermal stability of the enzyme system, the rates of oxidation and reduction of ferredoxin, and immobilization of the enzyme system. Because higher hydrogen production rates can be obtained at higher temperatures, it would be desirable to operate a process at the maximum temperature allowable without inactivating the enzyme system. The relative rates of reduction and oxidation of ferredoxin help to define the requirements for flow rates and methods for separating the oxygen and reduced ferredoxin produced in the chloroplast bioreactor. Rapid and efficient removal of the oxygen formed would be required to prevent reoxida-

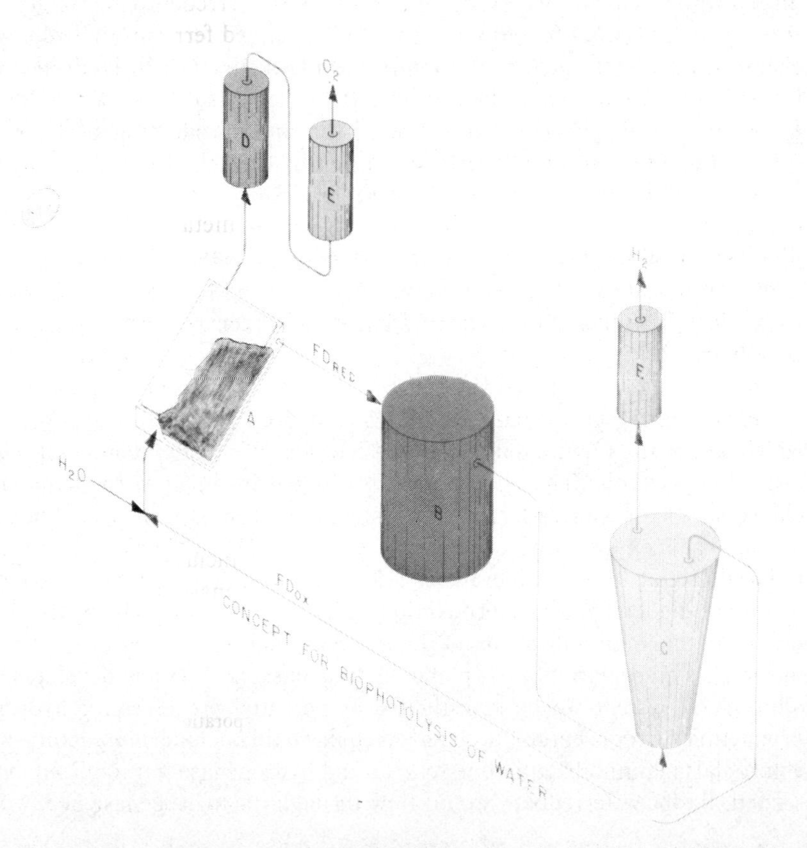

Fig. 1. Simplified schematic for photosynthetic hydrogen production. A, immobilized chloroplast bioreactor; B, surge tank; C, immobilized hydrogenase bioreactor; D, catalytic converter; E, dehumidifiers.

tion of the reduced ferredoxin and inhibition of the hydrogenase. Immobilization of the hydrogenase would allow its isolation from the oxygen-rich environment and alleviate the problem of deactivation of the hydrogenase by the oxygen produced.

MATERIALS AND METHODS

Reactions and preparations were carried out in a glove box or in closed vessels under an argon atmosphere.

Samples of ferredoxin from *Clostridium pasteurianum* and from spinach were obtained from Sigma Chemical Co. (St. Louis, Mo.). Cell extracts containing hydrogenase and ferredoxin were prepared from *C. pasteurianum* either by extraction of ground, lyophilized cells [13] with phosphate buffer or by ultrasonic disruption of the cells [26] at 0°C and subsequent centrifugation. Enzyme extracts remained active for several weeks when stored frozen (−20°C) in 0.05*M* phosphate buffer under argon or vacuum in septum-equipped vials.

Hydrogen gas was analyzed by gas chromatography using a molecular sieve 5 A column, argon carrier gas, and a thermal conductivity detector.

Sodium dithionite was obtained from Matheson, Coleman and Bell (Norwood, Ohio) and stored in a glove box containing argon. Sodium pyruvate was purchased from Aldrich Chemical Co. (Milwaukee, Wis.).

RESULTS AND DISCUSSION

Thermal Stability of the Ferredoxin—Hydrogenase System

The optimum temperature was determined for the hydrogenase–ferredoxin enzyme system by measuring the rate of hydrogen production from pyruvate or dithionite substrate. Reaction vessels were 25 ml flasks that contained a Teflon-coated stirring bar and were sealed with a rubber septum. Typically, 10 ml of freshly prepared, buffered reductant solution were added to the reaction vessel in the glove box, and the vessel was sealed. The vessel was then removed from the glove box and placed in a constant-temperature bath. The enzyme extract, maintained under an argon atmosphere, was thawed just prior to use and injected through the septum into the stirred solution. At timed intervals, 10 to 30 min, the gas phase was analyzed for hydrogen.

Generally, higher temperatures favor higher hydrogen production rates; of course, the maximum temperature is determined by thermal inactivation of the enzyme system. When sodium pyruvate was used as a reductant for ferredoxin, the hydrogen production rate increased with temperature in the range of 25–50°C (Fig. 2); however, at 50°C the hydrogen production rate decreased considerably after approximately 1 hr. Similar behavior was observed when sodium dithionite was used as the reductant. Also, the rate

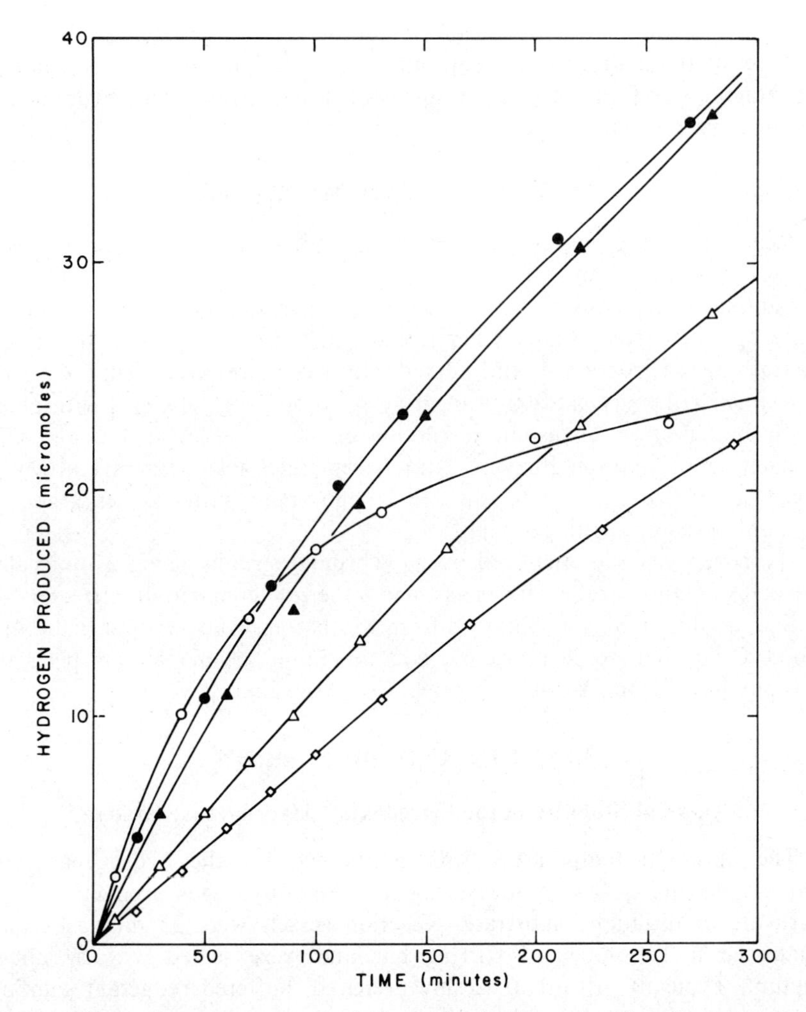

Fig. 2. Effect of temperature on the hydrogen production rate with sodium pyruvate as reductant for ferredoxin. In addition to ferredoxin and hydrogenase, the solution contained 1.0 g sodium pyruvate/liter 0.05*M* phosphate buffer, pH 6.6. (●) 42°C; (▲) 35°C; (△) 30°C; (○) 50°C; (◇) 25°C.

of hydrogen production using cell extract that had been incubated at 50°C for 2.5 hr was about 25–30% lower than with fresh enzyme.

Data obtained from initial rates can be correlated in a plot of hydrogen production rate versus $1/T$ (Fig. 3). The apparent activation energy, 11.5–12.5 kcal/mol H_2, obtained from the slope of the curves is similar regardless of the reductant used.

These results indicate that the optimum temperature for the ferredoxin–hydrogenase system is 40–45°C.

Immobilization of Hydrogenase

The hydrogen-producing reaction between reduced ferredoxin and protons, catalyzed by hydrogenase, can be separated from the oxygen-producing reaction by immobilizing the hydrogenase in a separate bioreactor. Immobilization of hydrogenase by covalent binding to glass beads has been reported [25]. Although the activity was low (2 to 5%), a reduction in oxygen sensitivity was observed. We investigated several techniques for immobilizing the hydrogenase–ferredoxin system, including covalent binding to other solid supports, entrapment in polyacrylamide gels, containment in hollow fibers, and adsorption. A cell extract containing both ferredoxin and hydrogenase was used in each experiment. The activity of the immobilized enzymes was measured by the hydrogen produced from reaction with sodium dithionite substrate.

Although hydrogenase was immobilized to some extent by each of the procedures, an enzyme activity of only 1% or less was obtained in most cases. In the polyacrylamide gels, best results were obtained with 10/30 gels (10% monomer plus *N*,*N'*-methylenebisacrylamide, of which 30% is *N*,*N'*-methylenebisacrylamide [27]). Although the activity was low, the gel-entrapped enzyme system remained active for several days.

A hollow fiber cartridge (HlD P-10, Amicon Corporation, Lexington, Mass.) was filled with *C. pasteurianum* extract containing ferredoxin and hydrogenase. A solution of sodium dithionite was circulated through the outer jacket. Hydrogen evolution was observed for two days with the system at room temperature.

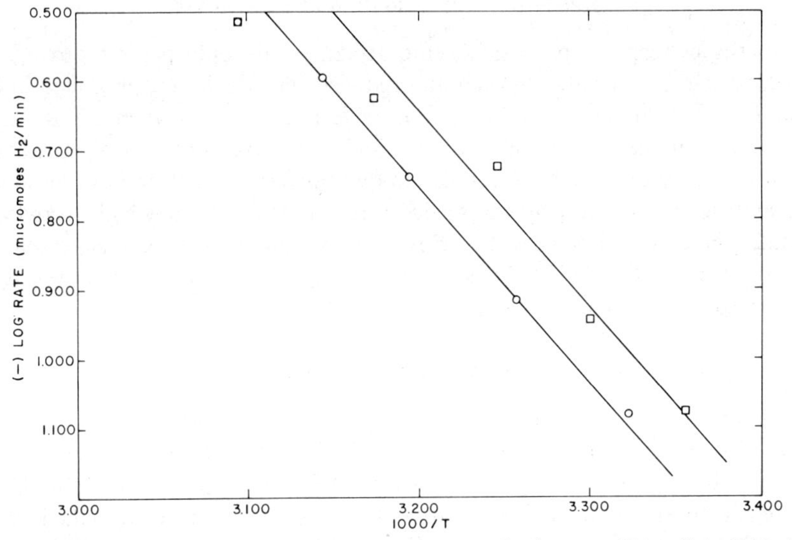

Fig. 3. Effect of temperature on the initial hydrogen production rate: (○) sodium dithionite; (□) sodium pyruvate.

The activity of the enzyme system adsorbed on alumina and covalently bound [28] to alumina, Chromosorb W, and Kieselguhr G was also no greater than 1%, compared with that of the free enzyme system.

The most active immobilized preparation was obtained by adsorption of the hydrogenase-ferredoxin on RPC-5, which consists of polychlorotrifluoroethylene powder coated with methyltricaprylylammonium chloride [29]. The RPC-5 was washed successively with buffer solution, buffered sodium dithionite solution, and buffer solution. After being mixed with cell extract containing ferredoxin and hydrogenase for 2 hr at room temperature, the solid was recovered by centrifugation. Less than 5% of the enzyme activity remained in solution, and 10% of the retained enzyme remained active on the support. Figure 4 shows typical results in which the hydrogen evolution was measured for a given batch of hydrogenase-ferredoxin in successive contacts with fresh dithionite solution over a period of 7 days. Curve 1 shows the initial contact with dithionite. Curves 2, 3, 4, and 5 were constructed by using data obtained from successive contacts of the adsorbed enzymes with fresh dithionite solutions after 2, 3, 5, and 7 days, respectively. Oxygen sensitivity of the adsorbed enzymes was not determined since this would not be a major consideration in a process where immobilized hydrogenase is physically separated from the oxygen-producing reaction. We expect that the capacity of the RPC-5, and therefore the specific activity of adsorbed enzyme, can be increased considerably by using purified hydrogenase. Development of a more active immobilized hydrogenase is continuing.

Reduction and Oxidation of Ferredoxin

In the conceptual process scheme utilizing chloroplasts, ferredoxin, and hydrogenase, molecular oxygen and reduced ferredoxin are produced in the chloroplast bioreactor. The oxidation of ferredoxin by oxygen is postulated to be one of the causes for low rates and rapid decay of hydrogen production in the system. It has been stated qualitatively that "the limiting factor in hydrogen evolution by the basic system is the rate at which ferredoxin reduction counterbalances ferredoxin oxidation by the oxygen produced during the reaction" [17]. Thus, the relative rates of reduction and oxidation of ferredoxin become very important.

Reduction of Ferredoxin

Oxidized ferredoxin has a significantly higher absorbance near 415 nm than the reduced form [26]. Both the oxidized and the reduced species showed linear behavior in a Beer's law plot oı absorbance at 415 nm. These spectral characteristics were utilized to measure the rate of reduction of oxidized ferredoxin by dithionite. Evidence has been reported [30] that the reduction of ferredoxin by dithionite involves SO_2^-. Oxidized ferredoxin was assured by exposing the solution to air or bubbling air through it. Solu-

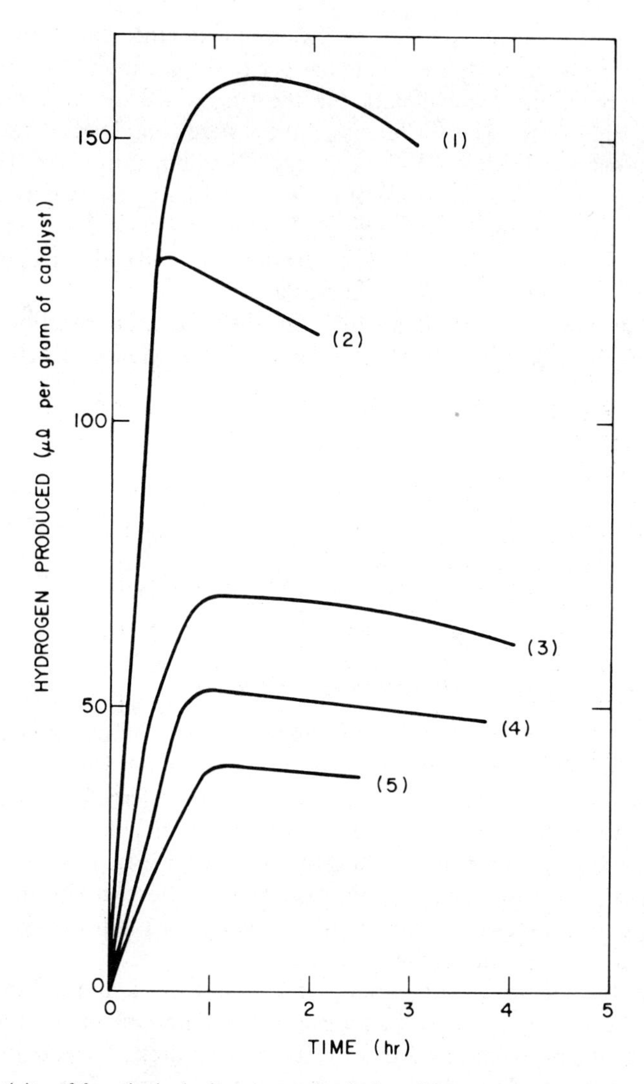

Fig. 4. Activity of ferredoxin–hydrogenase adsorbed on RPC-5. Reaction mixture contained 2.0 g RPC-5-enzymes and 35 mg sodium dithionite in 10 ml 0.05*M* phosphate buffer, pH 6.6, at 40°C.

tions of sodium dithionite were prepared in oxygen-free, argon-saturated phosphate buffer, pH 6.6, in a glove box just prior to use.

A Centrifugal Fast Analyzer [31] was used to mix the reagents and measure the adsorbance changes. Rate measurements were made in 0.05*M* phosphate buffer, pH 6.6, at 30°C. The total volume in each case was 135 μl. Each reaction mixture contained 55 μl of 0.0115*M* $Na_2S_2O_4$ (final concentration, $4.68 \times 10^{-3}M$) and varying amounts of ferredoxin. Rotors were purged with argon and loaded in a glove box under an argon atmosphere, and taped to prevent oxidation of the dithionite by air. Absorbance measurements were made at 0.1 to 0.5 sec intervals after initiation of the reaction. Typical data for the first 7 sec of the reaction are shown in Figure 5. The different curves were obtained with various concentrations of ferredoxin and an excess of sodium dithionite. The curve that is shown near an absorbance of 0.600 and remains essentially constant with time was obtained for a reference solution containing ferredoxin without dithionite. The initial concentration of oxidized ferredoxin was obtained from the intercept for the curves at zero time.

Kinetic parameters were evaluated for the reduction reaction from a Lineweaver–Burk plot (Fig. 5). Assuming Michaelis–Menten kinetics,

$$\frac{1}{v} = \frac{K_m}{V_m} \frac{1}{c} + \frac{1}{V_m}$$

where v is the reaction velocity or initial rate, c is the ferredoxin concentration, K_m is the Michaelis constant, and V_m is maximum velocity. From Figure 6, under the conditions described, $V_m = 7.0 \times 10^{-5}$ mol/liter sec, and $K_m = 2.2 \times 10^{-5} M$, based on a molar absorptivity of 27,600 for oxidized ferredoxin [26].

Oxidation of Ferredoxin

Molecular oxygen exerts various inhibitory effects on enzyme-catalyzed, hydrogen-producing reactions, affecting both the ferredoxin and the hydrogenase. Oxidation, oxygenation, and autooxidation of ferredoxin may occur. Consequently, it has generally been necessary to include oxygen traps in order to sustain hydrogen production. In order to more quantitatively define the problem, we investigated the oxidation of ferredoxin by gaseous molecular oxygen. Results of previous studies [32, 33] indicate that the oxidation of ferredoxin probably involves O_2^-.

Reduced ferredoxin was prepared by reaction with sodium dithionite. The amount of dithionite required to reduce a given amount of ferredoxin was determined by spectrophotometric titration of oxidized ferredoxin and use of the Centrifugal Fast Analyzer as described above. The reduced ferredoxin solution was placed in a cuvet which was sealed with a Teflon cap containing an outlet port and an inlet tube. The absorbance at 415 nm was measured continuously while air or oxygen was bubbled through the solution. The gas supply was equipped with a flow controller, a rotameter, a sparger for saturating the gas with water, and a three-way valve.

Oxidation rates were measured under conditions that included different flow rates of air and oxygen, different concentrations of ferredoxin, and samples of ferredoxin from both spinach and *C. pasteurianum*. Figure 7 shows representative curves obtained for the oxidation of ferredoxin from spinach (Curve 1) and *C. pasteurianum* (Curve 2). The dashed lines indicate the initial absorbance of the oxidized ferredoxin. The spinach ferredoxin

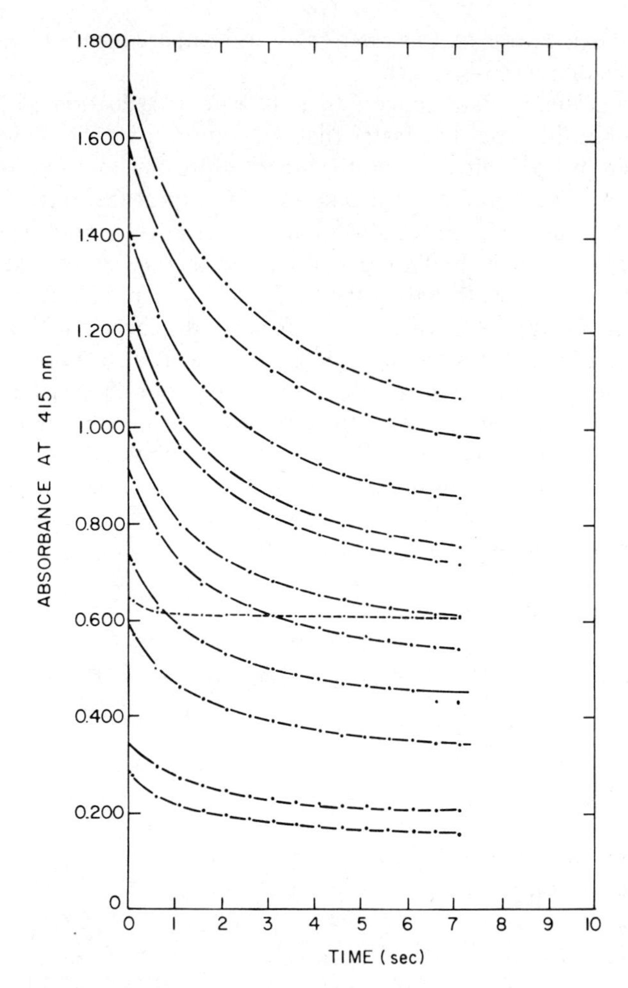

Fig. 5. Initial decrease in absorbance of oxidized ferredoxin on reduction with sodium dithionite. Different curves represent different concentrations of ferredoxin in the range 10^{-5}–$10^{-4}M$.

solution contained 1.20 mg of ferredoxin in 3.6 ml of $0.05M$ phosphate buffer, pH 6.6. The *C. pasteurianum* solution contained 0.60 mg of ferredoxin in 3.0 ml of buffer. The oxygen flow rate was about 12 cm^3 (STP)/min. The two samples behaved similarly; maximum oxidation was achieved in about 25 sec, and the ferredoxin was not completely reoxidized in either case. Summarizing several experiments, the time required to achieve maximum oxidation ranged from 30 to 40 sec in air and 8 to 25 sec in oxygen. These variations probably result from problems associated with mixing and diffusion of the gases into the solution.

We also observed that reduced samples of ferredoxin sealed under an argon atmosphere showed a gradual increase in absorbance over a period of several minutes. However, this rate is slow as compared with the rate of oxidation in the presence of oxygen.

These preliminary data suggest that the rate of reduction of ferredoxin with dithionite is somewhat faster than the rate of oxidation with gaseous oxygen. We are presently attempting to measure the rate of oxidation of ferredoxin by dissolved oxygen in solution. We expect this rate to be faster. In any case, a successful process will require a relatively rapid separation of the reduced ferredoxin and oxygen produced in the chloroplast reaction. Potential methods for alleviating the oxidation problem include chemical or biochemical removal of oxygen as it is formed, displacement of the oxygen with an inert gas, diffusion through a membrane, adsorption, and rapid transit of the reduced ferredoxin through the oxygen-rich environment. Displacement with an inert gas or diffusion through a membrane may be too slow to prevent reoxidation of the ferredoxin. Adsorbed oxygen remains highly active; thus, adsorption of the oxygen from solution would probably

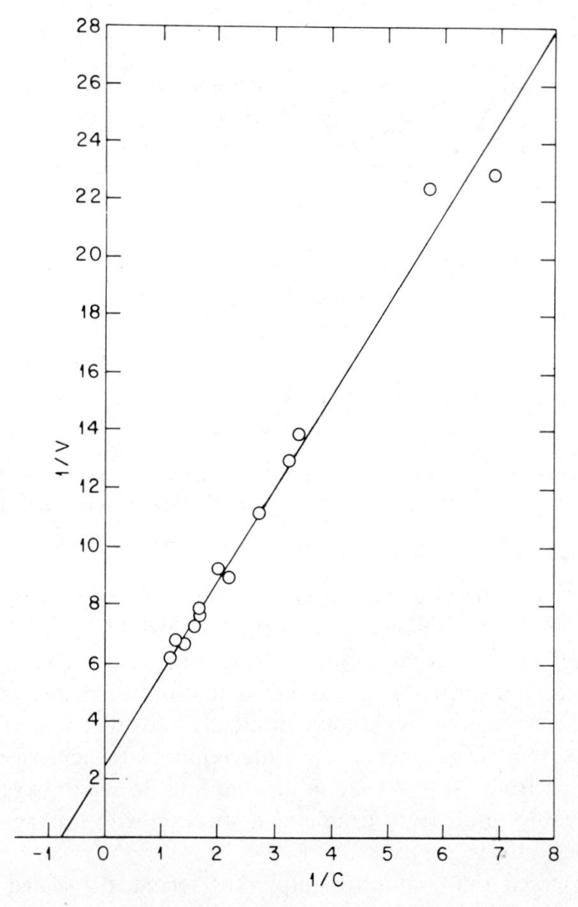

Fig. 6. Lineweaver–Burk plot correlating the rate of reduction of ferredoxin with ferredoxin concentration.

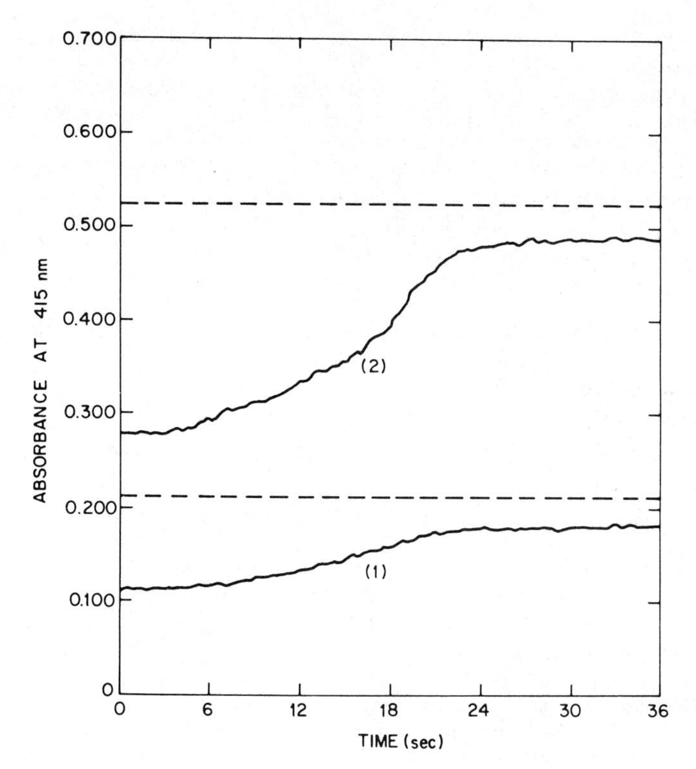

Fig. 7. Oxidation of ferredoxin from spinach (curve 1) and from *Clostridium pasteurianum* (curve 2) with molecular oxygen. Dashed lines indicate the initial absorbance of the oxidized ferredoxin samples before reduction.

not prevent reaction if the adsorber were left in contact with the ferredoxin solution. If the relative rates of reduction and oxidation of ferredoxin are such that these physical methods of oxygen removal are insufficient, then chemical removal will be required. An interesting possibility would be compounds which reversibly bind oxygen [34–38]. Some of these compounds exhibit oxygen-binding properties similar to those of hemoglobin and myoglobin. Some of them also show temperature- and pressure-dependent decomposition. The suitability of applying this technique to a process has yet to be demonstrated.

J. P. Eubanks provided excellent technical assistance in this study. G. D. Novelli and E. F. Phares graciously supplied *Clostridium pasteurianum*. We thank J. E. Mrochek, R. K. Genung, and J. B. Overton for help with the Centrifugal Fast Analyzer.

References

[1] C. T. Gray and H. Gest, *Science, 148*, 186 (1965).
[2] W. D. McElroy and B. Glass, Eds., *A Symposium on Inorganic Nitrogen Metabolism* (Johns Hopkins Press, Baltimore, 1956).

[3] A. Mitsui, S. Miyachi, A. San Pietro, and S. Tamura, Eds., *Biological Solar Energy Conversion* (Academic, New York, 1977).
[4] A. Hollaender, K. J. Monty, R. M. Pearlstein, F. Schmidt-Bleek, W. T. Snyder, and E. Volkin, *An Inquiry into Biological Energy Conversion* (The University of Tennessee, Knoxville, 1972).
[5] M. Gibbs, A. Hollaender, B. Kok, L. O. Krampitz, and A. San Pietro, *Proceedings of the Workshop on Bio-Solar Conversion*, Bethesda, Md., 1973.
[6] G. W. Strandberg, *Dev. Ind. Microbiol.*, *18*, 649 (1977).
[7] J. D. Keenan, *Energy Convers.*, *16*, 95 (1977).
[8] H. H. Weetall and M. A. Bennett, Fifth International Fermentation Symposium, Berlin, 1976.
[9] I. Karube, T. Matsunaga, S. Tsuru, and S. Suzuki, *Biochim. Biophys. Acta*, *444*, 338 (1976).
[10] K. Tagawa and D. I. Arnon, *Nature*, *195*, 537 (1962).
[11] L. E. Mortenson, R. C. Valentine, and J. E. Carnahan, *Biochem. Biophys. Res. Commun.*, *7*, 448 (1962).
[12] G. Nakos and L. Mortenson, *Biochim. Biophys. Acta*, *227*, 576 (1971).
[13] H. D. Peck, Jr., and H. Gest, *J. Bacteriol.*, *71*, 70 (1956).
[14] E. Knight, Jr., A. J. D'Eustachio, and R. W. F. Hardy, *Biochim. Biophys. Acta*, *113*, 626 (1966).
[15] R. C. Valentine, L. E. Mortenson, and J. E. Carnahan, *J. Biol. Chem.*, *238*, 1141 (1963).
[16] B. Z. Egan and C. D. Scott, *Chemical Technology*, *8*, 304 (1978).
[17] J. R. Benemann, J. A. Berenson, N. O. Kaplan, and M. D. Kamen *Proc. Natl. Acad. Sci. (USA)*, *70*, 2317 (1973).
[18] J. R. Benemann, *Fed. Proc.*, *32*, 632 (1973).
[19] J. R. Benemann and N. M. Weare, *Science*, *184*, 174 (1974).
[20] S. Lien and A. San Pietro, *An Inquiry into Biophotolysis of Water to Produce Hydrogen*, NSF/RANN Report, 1975.
[21] K. K. Rao, L. Rosa, and D. O. Hall, *Biochem. Biophys. Res. Commun.*, *68*, 21 (1976).
[22] I. Fry, G. Papageorgiou, E. Tel-Or, and L. Packer, *Z. Naturforsch.*, *32c*, 110 (1977).
[23] D. Hoffman, R. Thauer, and A. Trebst, *Z. Naturforsch.*, *32c*, 257 (1977).
[24] H. Ochiai, H. Shibata, T. Matsuo, K. Hashinokuchi, and M. Yakawa, *Agric. Biol. Chem. (Jpn)*, *41*, 721 (1977).
[25] D. A. Lappi, F. E. Stolzenbach, N. O. Kaplan, and M. D. Kamen, *Biochem. Biophys. Res. Commun.*, *69*, *878* (*1976*).
[26] B. E. Sobel and W. Lovenberg, *Biochem.*, *5*, 6 (1966).
[27] O. R. Zaborsky, *Immobilized Enzymes* (CRC Press, Cleveland, Ohio, 1973), pp. 83–92.
[28] H. H. Weetal and A. M. Filbert, *Methods Enzymol.*, *34*, 59 (1974).
[29] R. L. Pearson, J. F. Weiss, and A. D. Kelmers, *Biochim. Biophys. Acta*, *228*, 770 (1971).
[30] D. O. Lambeth and G. Palmer, *J. Biol. Chem.*, *248*, 6095 (1973).
[31] J. E. Mrochek, C. A. Burtis, W. F. Johnson, M. L. Bauer, D. G. Lakomy, R. K. Genung, and C. D. Scott, *Clin. Chem.*, *23*, 1416 (1977).
[32] W. H. Orme-Johnson and H. Beinert, *Biochem. Biophys. Res. Commun.*, *36*, 905 (1969).
[33] H. P. Misra and I. Fridovich, *J. Biol. Chem.*, *246*, 6886 (1971).
[34] L. Vaska, *Acc. Chem. Res.*, *9*, 175 (1976).
[35] J. P. Collman, *Acc. Chem. Res.*, *10*, 265 (1977).
[36] F. Basolo, B. M. Hoffman, and J. A. Ibers, *Acc. Chem. Res.*, *8*, 384 (1975).
[37] N. Farrell, D. H. Dolphin, and B. R. James, *J. Am. Chem. Soc.*, *100*, 324 (1978).
[38] E. Bayer and G. Holzbach, *Angew. Chem. Int. Ed. Engl.*, *16*, 117 (1977).

Application of a Biochemical Fuel Cell to Wastewaters

SHUICHI SUZUKI, ISAO KARUBE, and
TADASHI MATSUNAGA

Research Laboratory of Resources Utilization, Tokyo Institute of Technology, Nagatsuta-cho, Midori-ku, Yokohama, 227 Japan

INTRODUCTION

The high-energy, electron-rich substances such as carbohydrates, lipids, and proteins are not usually electroactive in a fuel cell, but the intermediates produced during biological oxidation may often be active at an electrode. The following are some of the suggested bio-anode reactions and bacteria involved [1, 2]: carbohydrate → ethyl alcohol (*Saccharomyces* sp., *Pseudomonas lindneri*); carbohydrate → hydrogen (*Clostridium butyricum*); sulfate → hydrogen sulfide (*Desulfovibrio desulfuricans*). Hydrogen exhibits excellent reactivity in these electroactive materials. Furthermore, hydrogen is now attracting attention as one of the clean fuel resources. Various bacteria and algae produce hydrogen under anaerobic conditions [1]. One of the first biochemical fuel cell systems designed to use hydrogen-producing bacteria was developed by Rohrback et al. [3], who reported on the production of hydrogen from glucose by *C. butyricum*. The production of hydrogen from the fermentation of glucose by *C. butyricum* and the utilization of hydrogen as a reactant at the anode in a hydrogen-oxygen (air) fuel cell were also mentioned by DelDuca et al. [4]. However, because the hydrogen evolution system, especially hydrogenase, in microorganisms is unstable, it is difficult to use whole cells for continuous hydrogen production [5].

Many enzymes and bacteria have been immobilized by various methods and then applied to industrial production [6–8]. The immobilization method offers the considerable advantages of stability and continuous use of enzymes and bacteria. Reports on the immobilization of most enzymes and bacteria have dealt with a single-step substrate transformation. However, the conversion of carbohydrates to hydrogen is achieved by a multienzyme system. Therefore, the immobilization of hydrogen-producing bacteria has a great value. Multienzyme systems, cofactors such as NAD, and ATP in bacteria can be used for production of hydrogen. Recently, the authors' laboratory immobilized *C. butyricum* in polyacrylamide gel [9], and the immobilized whole cells continuously evolved hydrogen from glucose under anaerobic conditions. Production of hydrogen using immobilized *Rhodospirillum rubrum* has also been reported by Bennett and Weetall [10]. Immobi-

Biotechnology and Bioengineering Symp. No. 8, 501–511 (1978)

0572-6565/78/0008-0501$01.00

lized cells of *C. butyricum* have been applied to a biochemical fuel cell [11] which was left on for 15 days and from which a current of 1.2 to 1.1 mA was obtained continuously during that period. Glucose was used as a nutrient for immobilized cells of *C. butyricum* in the biochemical fuel cell. However, because glucose is expensive as a nutrient for the bacteria, wastewaters were applied as nutrients for the immobilized whole cells in the present study. *C. butyricum* was immobilized in agar gel, and the immobilized whole cells were employed for the production of hydrogen. The system employed in this study was composed of a reactor for hydrogen production by the immobilized cells of *C. butyricum*, a fuel cell, and a reactor for wastewater treatment by the immobilized microorganisms. Industrial wastewaters were applied to the system.

MATERIALS AND METHODS

Materials

Yeast extract was purchased from Difco Laboratories. Peptone (from casein) and agar were purchased from Kyokuto Pharmaceutical Company. Acrylamide and *N,N'*-methylenebisacrylamide were obtained from Wako Pure Chemicals Ind. Other reagents were commercially available reagents or laboratory-grade materials. Deionized water was used in all procedures. Untreated wastewaters were obtained from a slaughterhouse (Shibaura, Tokyo), a food factory (Kawasaki), and an alcohol factory (Inage, Chiba). The wastewaters were diluted with deionized water, and the pH of the wastewaters was adjusted to 7 with 0.1*M* NaOH or 0.1*M* HCl.

Culture of Microorganisms

An authentic culture of *C. butyricum* IFO 3847 was maintained in a cooked meat medium, and a medium (pH 7.0) reported previously [9] was employed for cultivation. After inoculation, the culture flask containing 1 liter medium and 0.5 g $FeSO_4$ was exhausted by a vacuum pump, and the medium was degassed. Bacteria were cultivated anaerobically at 37°C for 8 hr. The cells were isolated by centrifugation at 5°C and 5000 *g* for 10 min, after which the cells were washed twice with physiological saline (pH 7.0, 5°C, oxygen free).

The bacteria isolated from activated sludge (Niko River, Nagoya) were cultured under aerobic conditions at 30°C for 12 hr in 80 ml medium (pH 7.0) containing 1% glucose, 1% peptone, 1% beef extract, and tap water. The cells were then centrifuged at 5°C and 8000 *g* and washed twice with physiological saline (pH 7.0).

Immobilization of Microorganisms

For the entrapment of *C. butyricum* and of bacteria isolated from activated sludge in agar gel, 0.8 g agar was dissolved in 36 ml physiological saline in a flask at 100°C and cooled to 50°C. Then, 4 ml physiological saline containing 4 g wet, intact cells was added to the flask, mixed, and cooled to 37°C. The gelled, immobilized whole cells were then cut into small blocks (8 mm^3) and washed thoroughly with saline.

The immobilization of *C. butyricum* in polyacrylamide gel was described previously [9]. These gelled, immobilized whole cells were also cut into small blocks.

Determination of Hydrogen and Glucose

Hydrogen was determined by a gas chromatograph (Shimadzu Seisakujo, Model GC 6 AM) [9]. Glucose was determined by the method of Bergmeyer and Bernt [12].

Biochemical Fuel Cell System

A schematic diagram of the system is presented in Figure 1. A reactor of acrylic plastic with a 90-ml capacity (diameter 2.2 cm × 24 cm) was used for the immobilized cells of *C. butyricum*. About 40 g of the immobilized whole cells were packed into the reactor. The fuel cell consisted of an anode chamber (10 cm × 10 cm × 3 cm) and a cathode chamber (10 cm × 10 cm × 0.5 cm), separated by an anion exchange membrane (Selemion type AMV, Asahi Glass Company). The anode was a platinum black electrode (10 cm × 20 cm); the anolyte was 150 ml wastewater from the alcohol factory. The cathode was a carbon electrode (7.5 cm × 8.0 cm × 3.0 cm); the catholyte was 50 ml 0.1*M* phosphate buffer (pH 7.0). The reactor for wastewater treatment was a glass vessel (diameter 9.2 cm × 45 cm). About 300 g immobilized whole cells were added to this reactor, and the wastewater was saturated with dissolved oxygen and stirred magnetically. The temperature of this biochemical fuel cell system was maintained at 37 ± 1°C. The current, the anode potential, and the cell voltage were measured by a millivolt–milliammeter (Kikusui Electronics, model 114) and displayed on a recorder (TOA, model EPR-100A). The anode potential was determined with reference to a saturated calomel electrode (SCE).

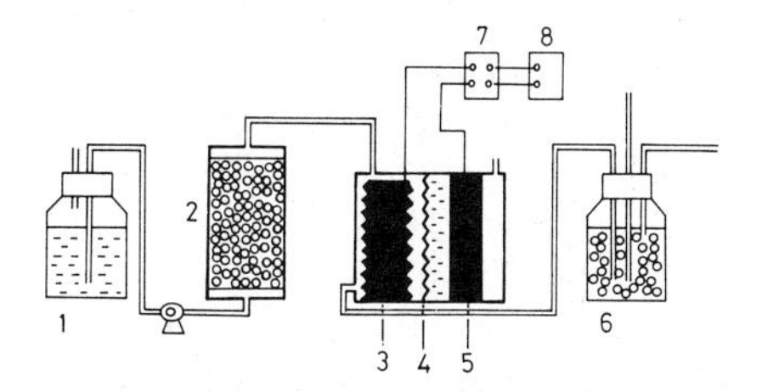

Fig. 1. Schematic diagram of biochemical fuel cell system. (1) Reservoir for the wastewater. (2) Packed-bed reactor for immobilized *C. butyricum*. (3) Platinum black anode. (4) Ion-exchange membrane. (5) Carbon cathode. (6) Continuously stirred tank reactor for immobilized aerobic microorganisms. (7) Ammeter. (8) Recorder.

Microbial Electrode for BOD Estimation

The microbial electrode for BOD estimation was prepared by a method described previously [13]. A solution (pH 7.0) containing glucose (150 mg/liter) and glutamic acid (150 mg/liter) was employed as a standard wastewater. The microbial electrode was inserted into a sample solution (60 ml), and the sample solution was saturated with dissolved oxygen and stirred magnetically while measurements were taken. The five-day BOD (BOD_5) of the sample solution was measured by the "Standard Method" according to the Japanese Industrial Standard Committee [14].

RESULTS

Hydrogen Production by Immobilized Whole Cells

Figure 2 shows hydrogen produced by the agar gel–entrapped *C. butyricum* and by the polyacrylamide gel–entrapped ones. Condensed wastewater from an alcohol factory was used for experiments because the consumption of organic compounds did not affect the hydrogen production by the immobilized whole cells; polyacrylamide gel [9] and agar gel were employed for immobilization carriers of bacteria. As shown in Figure 2, the amount of hydrogen produced by the agar gel-entrapped whole cells was three times larger than that by the polyacrylamide gel-entrapped ones. Therefore, agar gel was selected for subsequent studies.

Hydrogen Production from Wastewaters

Hydrogen produced from wastewaters and from glucose by the immobilized *C. butyricum* are shown in Table I. The experiments were performed as follows: 10 ml of wastewater (BOD 660 ppm, pH 7.0) or of 0.1% glucose (pH 7.0) were introduced into vessels containing 0.4 g of the immobilized whole cells. The vessel was exhausted by a vacuum pump and sealed with a stopcock. The immobilized whole cells were then incubated at 37°C for 3 hr. Hydrogen production was determined by gas chromatography. As

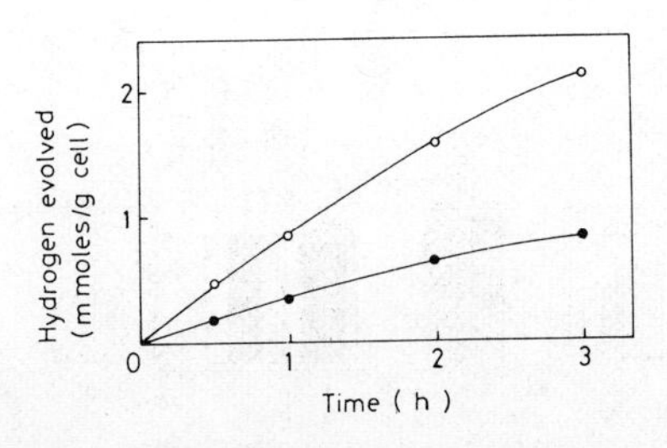

Fig. 2. Hydrogen production by immobilized whole cells. 4 g gel (0.4 g wet bacteria) were incubated with 10 ml wastewater (pH 7.0, BOD 16,400 ppm) from an alcohol factory at 37°C. (O——O) Agar gel–entrapped whole cells, (●——●) polyacrylamide gel–entrapped whole cells.

TABLE I
Hydrogen Evolution From Various Waste Waters and Glucose

	Hydrogen evolved (μ mol/g cell·h)
Glucose	55
Alcohol factory	72
Food factory	18
Slaughterhouse	8

shown in Table I, the rate of hydrogen production by immobilized whole cells in the alcohol factory's wastewater was higher than that in the other wastewaters and in the glucose solution. The wastewater of the alcohol factory was, therefore, employed for further studies.

Reusability of Immobilized Whole Cells

Reusability of immobilized *C. butyricum* for hydrogen production was tested as follows. The immobilized whole cells were placed in a vessel containing the wastewater and incubated for 6 hr at 37°C, and the operation was repeated for 20 consecutive days. The results of this repetition are shown in Figure 3. The immobilized whole cells produced hydrogen continuously over a 20-day period. Therefore, the immobilized whole cells were applied to a packed-bed reactor.

Effect of Flow Rate on Hydrogen Production

The effect of the flow rate of the wastewater on hydrogen production was examined with a packed-bed reactor. Figure 4 shows the relationship between the flow rate and the amount of hydrogen produced as flow rates varied from 1.5 to 12 ml/min. The optimum hydrogen evolution was observed at flow rates between 5 and 10 ml/min. Further decrease or increase of the flow rate decreased the amount of hydrogen produced.

The utilization ratio of organic compounds by immobilized *C. butyricum* was examined. Glucose in the wastewater was chosen for an indicator of

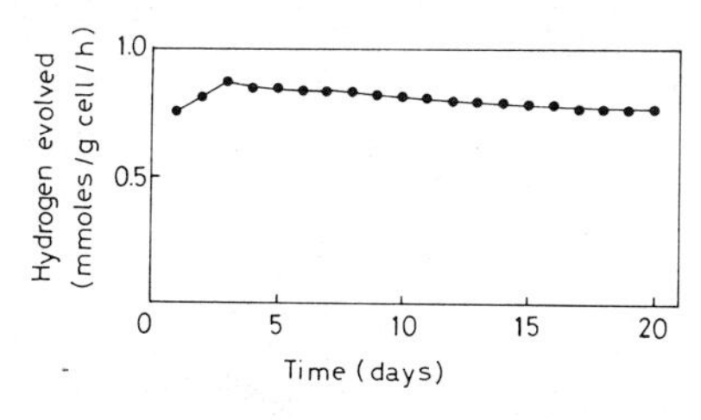

Fig. 3. Reusability of immobilized whole cells. Experimental conditions are as described in Figure 2 except for the incubation time employed.

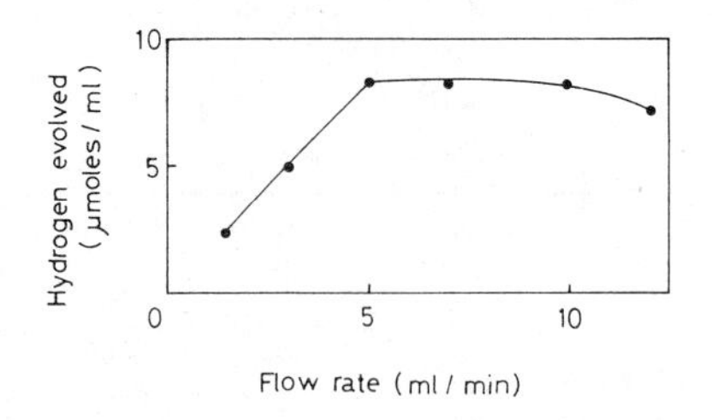

Fig. 4. Relationship between flow rate and the amount of hydrogen produced. Wastewater (pH 7.0, BOD 660 ppm) from an alcohol factory was transferred to the packed-bed reactor containing 40 g immobilized whole cells (4 g wet *C. butyricum*) at various flow rates.

organic compounds and determined by the method of Bergmeyer and Bernt [12]. Table II shows the utilization ratio of glucose in the wastewater; the utilization ratio decreased as the flow rate of the wastewater increased.

Effect of BOD on Hydrogen Production

Figure 5 shows the effect of the BOD of the wastewater on the hydrogen production by the immobilized whole cells when the condensed wastewaters were employed for experiments. The amount of hydrogen produced increased with the increase of the BOD. However, as shown in Table II, the utilization ratio of organic compounds in the wastewater also decreased with increasing BOD. Therefore, an average wastewater concentration of about 660 ppm and a flow rate of 5 ml/min were employed for further work.

Anode Potential and Cell Voltage of Fuel Cell

The wastewater from the packed-bed reactor was transferred to a fuel cell at a flow rate of 5 ml/min. The anode potential became more negative because the anode was saturated with the hydrogen produced by the immo-

TABLE II

Utilization Ratio of Glucose in Wastewater[a]

BOD (ppm)	Flow rate (ml/min)	Glucose utilized (%)
660	5	60
660	7	43
660	10	32
660	12	29
660	5	60
1300	5	28
2600	5	13
3300	5	10

[a] pH 7.0, 37°C.

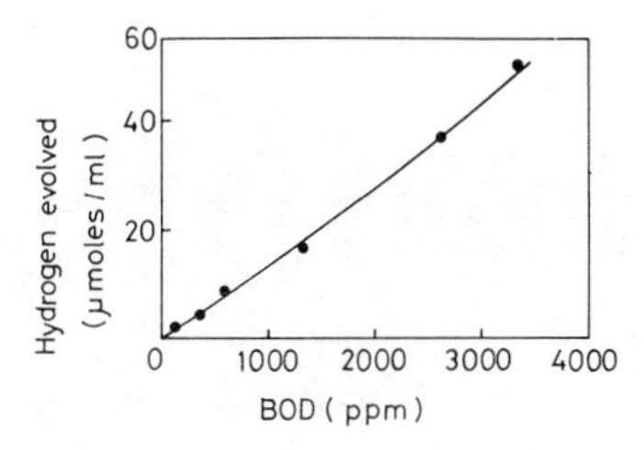

Fig. 5. Effect of BOD on hydrogen production. Wastewater (pH 7.0) from an alcohol factory was transferred to the packed-bed reactor at a flow rate of 5 ml/min.

bilized whole cells. The anode potential was -0.5 V (vs. SCE), and the cell voltage of the fuel cell was 0.63 V at pH 7.0. The relationship between the current density and the anode potential is shown in Figure 6. The limiting current density changed from 10^{-3} to 10^{-1} mA/cm^2 as the resistance between the electrodes changed from 5 to 20,000 Ω. The maximum current density of 1 mA/cm^2 and anode potential of -0.63 V were obtained using the condensed wastewater (BOD 3300 ppm) and a platinum black electrode (2 cm × 4 cm).

Effect of Flow Rate and BOD on Current

The effect of the flow rate on the current of the fuel cell is shown in Figure 7. As flow rates were varied from 1.5 to 12 ml/min, the current increased with increasing flow rate, until the maximum current was obtained at a flow rate of 10 ml/min. On the other hand, the current increased with the increasing BOD of the wastewater, as is shown in Figure 8. The maximum current was about 40 mA when the BOD of the concentrated wastewater was 3300 ppm. Hydrogen transferred to the anode chamber might increase with increasing flow rate and BOD of the wastewater. However, as shown in Table II, the utilization ratio of organic compounds in the wastewater decreased with increasing flow rate and BOD. Therefore, a flow rate of 5 ml/min and a BOD of 660 ppm were employed for the fuel cell.

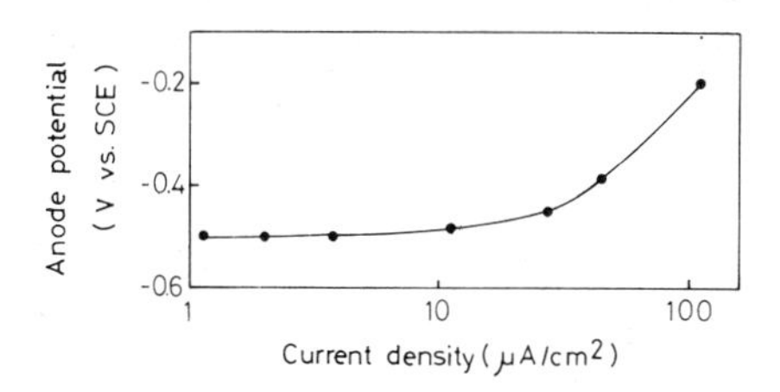

Fig. 6. Relationship between current density and anode potential. Wastewater (pH 7.0, BOD 660 ppm) from an alcohol factory was employed and transferred at a flow rate of 5 ml/min. Measurements were carried out after the anode potential became constant.

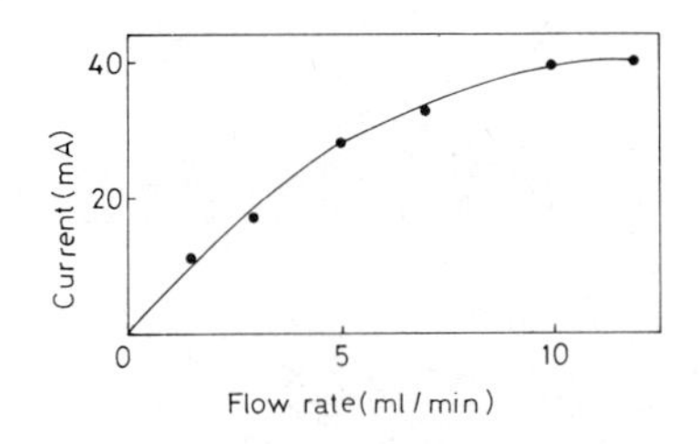

Fig. 7. Effect of flow rate on current. Experimental conditions are as described in Figure 4.

Wastewater Treatment by Immobilized Microorganisms

The BOD of the wastewater employed for the system, as determined at the output of the reactor and fuel cell, was 660 ppm. No significant decrease of the BOD of the wastewater was observed in the reactor and the fuel cell. The immobilized *C. butyricum* produced organic acids as metabolites [9]. Most of these organic acids are not electroactive at the anode; therefore, they may contribute to the BOD of the wastewater. Microorganisms isolated from activated sludge were immobilized in agar gel, and the immobilized microorganisms were applied to the wastewater treatment. About 300 g immobilized microorganisms were placed in the reactor. Figure 9 shows the effect of flow rate on the BOD of the wastewater, as flow rates varied from 1.5 to 10 ml/min.

The BOD of the wastewater, which decreased with decreasing flow rate, was about 45 ppm at a flow rate of 5 ml/min. In practice, the BOD of industrial wastewaters of the Tokyo area is restricted to below 50 ppm. Therefore, about 300 g of immobilized microorganisms were used in the system.

Continuous Operation of System

The biochemical fuel cell system was operated at the optimum conditions described above. The fuel cell was left on for 20 days, and a current from 15 to 13 mA was obtained continuously over a 20-day period. This continuous operation indicates that the *C. butyricum* in the agar gel was living and had maintained hydrogen evolution activity for a long time. At the same time, the BOD of the wastewater could be maintained below 50 ppm (Fig. 10).

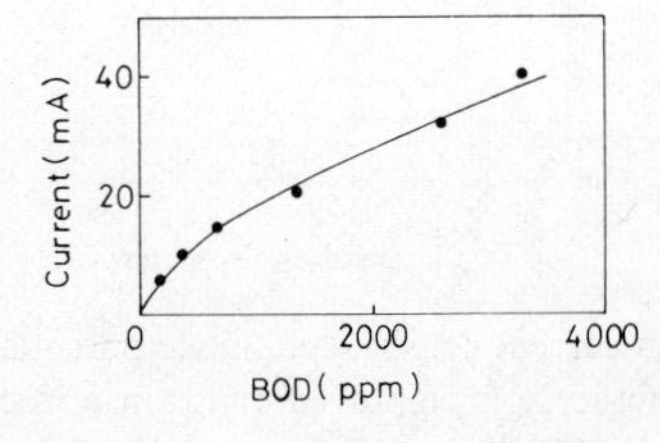

Fig. 8. Effect of BOD on current. Experimental conditions are as described in Figure 5.

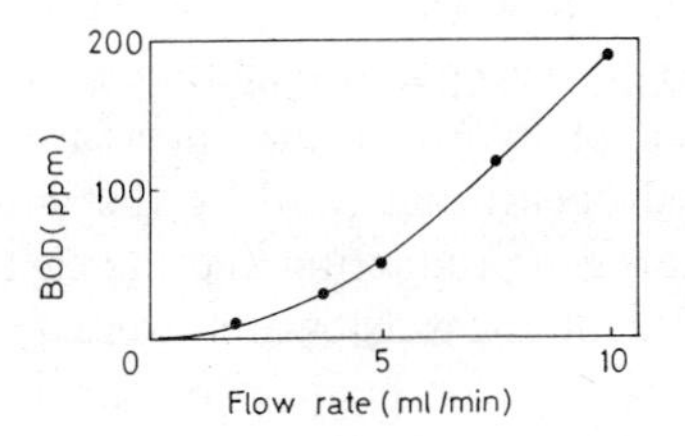

Fig. 9. Effect of flow rate on BOD of wastewater. Wastewater (pH 7.0, BOD 660 ppm) from an alcohol factory was transferred to the continuously stirred tank reactor at a flow rate of 5 ml/min. Reactor was aerated at 4 liter/min.

DISCUSSION

Hydrogen-producing bacteria, *C. butyricum*, were immobilized in agar gel and polyacrylamide gel. However, the rate of hydrogen production by the agar gel–entrapped whole cells was higher than that by the polyacrylamide gel–entrapped ones. The bacteria might be inactivated with reagents used for polymerization of the acrylamide.

Under ideal conditions nearly 2 mol hydrogen/mol glucose can be obtained using *Clostridia*. Of the total glucose consumed, 63% was converted to hydrogen in this immobilized whole cell system at optimum conditions. This conversion ratio is twice that reported previously [9]. Hydrogen was also produced from wastewaters by the immobilized whole cells. The largest amount of hydrogen was obtained from the wastewater of an alcohol factory, and the conversion rate was higher than that obtained from glucose. Because molasses was used as a raw material for alcohol fermentation, the wastwater contained carbohydrates and other nutrients. In our observation, bacteria could grow in the agar gel. Therefore, the high rate of hydrogen production in the wastewater may be caused from the growth of the bacteria in agar gel. No significant decrease of hydrogen production was observed over a 20-day period. This result suggested that no inhibitor for *C. butyricum* existed in the wastewater.

The optimum flow rate for hydrogen production was observed with a packed-bed reactor. As the bacteria produced organic acids, the wastewater

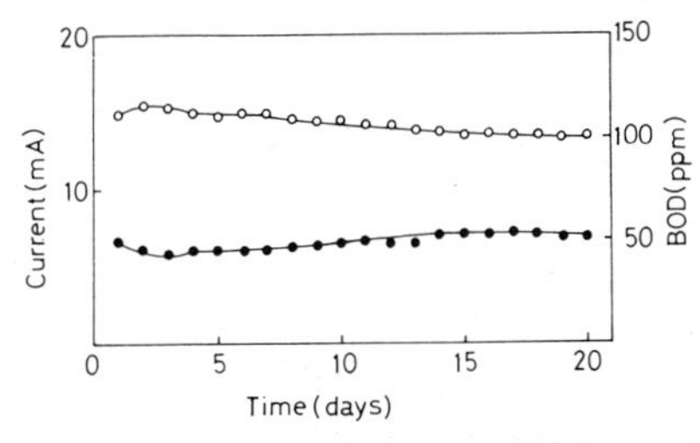

Fig. 10. Continuous operation of system. Wastewater (pH 7.0, BOD 660 ppm) from an alcohol factory was employed and transferred to the system at a flow rate of 5 ml/min.

became acidic at the low flow rate. This increased acid level may be the reason for the decrease of hydrogen production at the low flow rate. On the other hand, the amount of hydrogen also decreased at the high flow rate because the rate of hydrogen production by the immobilized bacteria was limited. The rate of hydrogen production and the current increased when the flow rate and the BOD of the wastewater were increased. However, the utilization ratio of organic compounds by immobilized *C. butyricum* decreased at those conditions. Therefore, a BOD of 660 ppm, which was an average BOD of the untreated wastewater in the alcohol factory, was chosen for the continuous operation of the fuel cell system. As previously reported, the current obtained from the biochemical fuel cell was low (1 mA) [11]. In this case, a platinum black electrode covered with gel-entrapped microorganisms was employed for the fuel cell. Therefore, the diffusion of hydrogen to the electrode surface and the amount of hydrogen produced were the rate-determining steps of the current generation.

The system was composed of three devices, the reactor for the hydrogen production, the fuel cell, and the reactor for wastewater treatment. The chemical reactions involved in three devices are as follows:

Reactor (1):

$$\text{wastewater} \xrightarrow{C.\ butyricum} H_2 + \text{formic acid} + \text{other compounds}$$

Fuel cell:

$$H_2,\ \text{formic acid} \longrightarrow \text{current}$$

Reactor (2):

$$\text{other compounds} \xrightarrow{\text{microorganisms}} CO_2 + H_2O + \text{unused compounds}$$

Formic acid produced by immobilized *C. butyricum* also contributed to the current generation [11]. The organic acids and other organic compounds produced by the immobilized *C. butyricum* were utilized by the immobilized microorganisms in the continuously stirred tank reactor, and the BOD of the wastewater finally decreased below 50 ppm. The continuous operation of the system was attempted over a 20-day period. A current of 15 mA was obtained continuously from the system. However, no attempt has been made to optimize the biochemical fuel cell system.

This study provides an application of immobilized whole cells to energy conversion and wastewater treatment. Further developmental studies in this laboratory are being directed toward optimizing and scaling up the system.

The authors would like to thank Dr. T. Iijima for supplying the carbon electrode.

References

[1] K. Lewis, *Bacteriol. Rev. 30*, 101 (1966).
[2] W. C. Potter, *Proc. R. Soc. (Lond.) Ser. B, 84*, 260 (1912).

[3] G. H. Rohrback, W. R. Scott, and J. H. Canfield, *Proceedings of the 16th Annual Power Sources Conference*, 1962, p. 18.
[4] M. G. DelDuca, J. M. Friscoe, and R. W. Zurilla, *Developments in Industrial Microbiology* (American Institute of Biological Sciences, Washington, D.C., 1963), Vol. 4, p. 81.
[5] C. T. Gray and H. Gest, *Science 148*, 186 (1965).
[6] H. H. Weetall and S. Suzuki (Eds.), *Immobilized Enzyme Technology* (Plenum, New York, 1975).
[7] I. Chibata, T. Tosa, T. Sato, T. Mori, and Y. Matsuo, *Proc. IV IFS: Ferment. Technol. Today*, 1972, p. 383.
[8] I. Chibata, T. Tosa, and T. Sato, *Appl. Microbiol. 27*, 878 (1974).
[9] I. Karube, T. Matsunaga, S. Tsuru, and S. Suzuki, *Biochim. Biophys. Acta 444*, 338 (1976).
[10] M. A. Bennet and H. H. Weetall, *J. Solid-Phase Biochem. 1*, 137 (1976).
[11] I. Karube, T. Matsunaga, S. Tsuru, and S. Suzuki, *Biotechnol. Bioeng. 19*, 1727 (1977).
[12] H. U. Bergmeyer and E. Bernt, *Methods of Enzymatic Analysis*, H. U. Bergmeyer (Ed.) (Academic, New York, 1974), p. 1205.
[13] I. Karube, T. Matsunaga, S. Mitsuda, and S. Suzuki, *Biotechnol. Bioeng. 19*, 1535 (1977).
[14] Japanese Industrial Standard Committee, *Testing Methods for Industrial Waste Water* (JIS K 0102, Tokyo, Japan, 1974), p. 33.

Author Index

Published Symposia of Biotechnology & Bioengineering

1969 No. 1 Global Impacts of Applied Microbiology II
Edited by Elmer L. Gaden, Jr.

1971 No. 2 Biological Waste Treatment
Edited by Raymond P. Canale

1972 No. 3 Enzyme-Engineering
Edited by Lemuel B. Wingard, Jr.

1973 No. 4 Advances in Microbial Engineering (Part 1)
Edited by B. Sikyta, A. Prokop, and M. Novák

1974 No. 4 Advances in Microbial Engineering (Part 2)
Edited by B. Sikyta, A. Prokop, and M. Novák

1975 No. 5 Cellulose as a Chemical and Energy Resource
Edited by C. R. Wilke

1976 No. 6 Enzymatic Conversion of Cellulosic Materials: Technology and Applications
Edited by Elmer L. Gaden, Jr., Mary H. Mandels, Elwyn T. Reese, and Leo A. Spano

1977 No. 7 Single Cell Protein from Renewable and Nonrenewable Resources
Edited by Arthur E. Humphrey and Elmer L. Gaden, Jr.

1978 No. 8 Biotechnology in Energy Production and Conservation
Edited by Charles D. Scott